# Rich Quasiparticle Properties in Layered Graphene-related Systems

# Rich Quasiparticle Properties in Layered Graphene-related Systems

Chiun-Yan Lin, Ching-Hong Ho,
Hsien-Ching Chung, Lu-Yao Wang,
Chih-Wei Chiu & Ming-Fa Lin

*National Cheng Kung University, Taiwan*

World Scientific

EW JERSEY • LONDON • SINGAPORE • BEIJING • SHANGHAI • HONG KONG • TAIPEI • CHENNAI • TOKYO

*Published by*

World Scientific Publishing Co. Pte. Ltd.
5 Toh Tuck Link, Singapore 596224
*USA office:* 27 Warren Street, Suite 401-402, Hackensack, NJ 07601
*UK office:* 57 Shelton Street, Covent Garden, London WC2H 9HE

**British Library Cataloguing-in-Publication Data**
A catalogue record for this book is available from the British Library.

**RICH QUASIPARTICLE PROPERTIES IN LAYERED GRAPHENE-RELATED SYSTEMS**



ISBN 978-981-12-7778-8 (hardcover)
ISBN 978-981-12-7779-5 (ebook for institutions)
ISBN 978-981-12-7780-1 (ebook for individuals)

For any available supplementary material, please visit
https://www.worldscientific.com/worldscibooks/10.1142/13453#t=suppl

Desk Editor: Rhaimie Wahap

Typeset by Stallion Press
Email: enquiries@stallionpress.com

# PREFACE

This book is devoted to comprehensively understanding quasiparticle properties under intrinsic and extrinsic interactions in graphene-related materials. A theoretical framework is developed to unify the properties across single-particle to many-particle scales in the presence of various mechanisms, such as the intrinsic interactions of charges and spins, the crystal symmetries of chemical bonds, the quantum confinement of edge structures, external charge doping and external temperature, electric, magnetic and mechanical fields. The current quasiparticle framework could be gradually extended to include the concept of hybridized and/or composite quasiparticles and explored as a whole. The research topics include the many-body phenomena affected by long-range Coulomb interactions, namely the Coulomb excitations and scattering rates, dynamic charge density waves and Friedel oscillations, as well as the optical reflection and transmission spectra of thin films. In addition, a phenomenological model is used to explore the origin of quasiparticle topology and its evolution in quantum systems. The calculated results are consistent with the latest experimental measurements.

# CONTENTS

Chapter 1

# INTRODUCTION

Chiun-Yan Lin, Ching-Hong Ho, Hsien-Ching Chung, Lu-Yao Wang, Chih-Wei Chiu, and Ming-Fa Lin

In the field of modern condensed matter physics, the behavior of quasiparticles has been a major area of research with the advancement of technology and many-body theory. Optical spectroscopy, using ultrafast laser pulses, has become a popular tool for generating and detecting dynamic oscillations of electric charges [1–13]. In 2001, femtosecond spectroscopy demonstrated that dynamic screening takes a finite time (around 10 fs) to establish in GaAs, causing the collective behavior of electrons [13]. Furthermore, a new microscopy technique integrating ultrashort laser pulses and scanning tunneling microscopy (STM) has enabled real-space imaging of transient carrier dynamics in nanostructures [11]. Recently, coherent spectroscopy with nanometer spatial and femtosecond temporal resolution has observed ultrafast coherent dynamics and related nanoscale phenomena [11, 14–16]. As near-field optical microscopes achieve nanoscale imaging, there has been increased interest in plasmon propagation in real space and time [14–16]. The confinement and enhancement of plasma waves in large optical fields can be used to design active plasmonic devices [17–29].

The concept of quasiparticles is essential for understanding many-body properties in condensed matter systems [30–32]. The key difference between quasiparticles and real particles arises from the many-body interactions that lead to the limited lifetime and energy range of quasiparticles during elastic and inelastic scattering in the crystal potential [30, 33–36].

In contrast, real particles exhibit a one-to-one correspondence between state energy and wavevector. The behavior of well-behaved quasiparticles near the Fermi surface is of great interest in condensed matter physics due to phenomena such as Coulomb excitations and decays [35–48], dynamic charge density waves [3, 7, 11, 49, 50], and Friedel oscillations [51–60]. This book provides detailed discussions and explanations of the effects of long-range Coulomb interactions on these phenomena.

The self-energy of quasiparticles, which defines their state energy and lifetime, can be calculated using the self-consistent Dyson equation [30]. The exchange and correlation energies of electron–electron Coulomb scatterings play a significant role in self-energy. Research on electron gas systems in three-dimensional (3D) [30, 61, 62], two-dimensional (2D) [30, 33, 62–66], one-dimensional (1D) [67–70], and zero-dimensional (0D) [71–73] systems has illustrated the charge response phenomenon of different dimensions, such as the effective screening length related to the free carrier density [30, 33, 63], conservation of the total induced charge, and dimension-related Friedel oscillations that reflect the long-distance charge oscillation in association with the Fermi surface [61]. In this book, we have developed a modified random-phase approximation and modified self-energy to understand the Coulomb scattering rate of quasiparticles in layered graphene systems. The theoretical model and framework are presented in Chapter 2, which is suitable for studying other fundamental phenomena that are strongly affected by both exchange and correlation interactions, such as the dynamic propagation of charge density waves and Friedel oscillations. The effective Coulomb interactions, both short-range and long

Graphene has been one of the most widely studied 2D materials since monolayer graphene sheets were first fabricated and investigated in 2004 by Andre Geim and Konstantin Novoselov [77]. It is a flexible 2D film composed of atom-thick hexagonal lattices of carbon atoms, which can be rolled or folded without losing its unique electronic and atomic structural characteristics [77, 78]. As the building blocks of other 0D–3D related materials, such as fullerenes [79], graphene bubbles [80], carbon nanotubes [81], carbon nanoscrolls [82], multilayer graphenes [83–94] and graphites [95–97], they are very suitable for exploring and verifying essential physical phenomena related to various dimensionalities. The attractive force between interplanes is derived from the van der Waals interactions of the $2p_z$ orbitals. In general, the stacking orders of graphene sheets in the layered graphites and graphite compounds include the typical sequences, namely AA [97], AB and ABC stackings [95, 96]. The first stacking configuration has been artificially made either from intercalated graphite compounds [97] or from few-layer graphene grown on diamond (111) or on the SiC (000$\bar{1}$) plane [83, 84]. However, the AB and ABC configurations are common in natural graphite, with estimated volume fractions of 80% and 14%, respectively [95, 96]. The remaining 6% of natural graphite is composed of randomly stacked graphene sheets, called turbostratic configuration. The honeycomb lattice and stacking configuration are responsible for the unique properties of graphite, e.g., the semi-metallic behavior due to the hexagonal symmetry and the interlayer atomic interactions [77, 87, 88, 90, 92]. It has been verified that the massless Dirac quasiparticles in monolayer graphene lead to peculiar physical properties, such as the novel half-integer quantum Hall effect (QHE) [98, 99], Berry's phase [99] and the anomalous Landau level (LL) spectrum [98, 100, 101]. Following the isolation of monolayer graphene by mechanical exfoliation for the first time in 2004, various experimental methods have successfully produced high-quality multilayer graphene with different stacking configurations [83–94] and graphites [95–97]. They have hexagonal symmetry and nanometer-scale size, leading to many remarkable properties of 2D materials, e.g., the feature-rich energy bands [103, 244], special optical selection rules [100, 104–110], unique magnetic quantization [111–120], anomalous QHEs [121–127] and multi-mode plasmons [128, 129]. The essential properties related to the graphene quasiparticles are dramatically changed by the intercalation of various atoms or molecules [77, 130–136].

In Chapter 3, the Coulomb decay rate is investigated in detail through de-excitations of an electron–hole pair or a plasmon. The band-structure effects play an important role in the excitation and de-excitation processes. In monolayer graphene, the nonzero decay rate at the Dirac point under electron doping indicates its robustness. Moreover, plasmons provide efficient channels for excited holes to decay, especially in the region of $|E^{c,v}| < 0.5\, E_F$, where the quasiparticles show fast Coulomb decay [34, 35]. On the other hand, the decay rate spectra of AA-, AB- and ABC-stacked graphenes are in sharp contrast with each other due to their unique quasiparticle excitation behavior. The isotropic and symmetric properties of the Coulomb decay rate are studied in terms of the quasiparticle energy and the wavefunction for any stacking configuration. As known in the conventional 2D systems, the prominent acoustic plasmon mode triggers a rapid decay rate of quasiparticles and is therefore expected to make graphene systems unique under different electron concentrations

[33]. The influence of plasmon resonance played by the optical plasma mode on Coulomb scattering can affect the excited states over a large energy range; however, the effect of Landau damping may overshadow its importance. In AA-stacked graphene, the hybrid nature of the multi-Dirac cones can make the distinct features of Coulomb scattering from the massless Dirac quasiparticles [128]. The AB stack is the most common stacking structure, and the bilayer graphene exhibits two pairs of parabolic bands [95, 96]. As the number of staked layers increases, the electronic properties can be regarded as a linear combination of monolayer and modified bilayer graphenes. This deduction cannot be applied to fundamental physical phenomena, such as carrier-density-dependent chemical potential [137, 138], the fermion distribution function near the Fermi level [139, 140], the QHE [141, 142], the renormalization factor near the Fermi momentum [143, 144], the dynamic propagation of charge density wave [7, 11, 49, 50], the spin-dependent pair distribution function [7, 69, 145] and Friedel oscillation [52–61]. Moreover, the additional de-excitation channels due to band hybridization can enrich and diversify the Coulomb decay rate in multilayer graphene. The physical pictures of electron–hole and electron–plasma couplings can be used to explore the dynamics of Dirac quasiparticles in related materials, such as layered Group-IV materials [146, 147], semimetals ZrSiSe [148], $Bi_2Te_3$ [149] and 3D HgTe [150, 151].

Friedel oscillation is also one of the most typical phenomena of the static screening effect caused by defects or impurities in metallic systems [51–60]. When the charges are scattered by an impurity, the screened potential rapidly decays away from the impurity in the Yukawa potential form; moreover, the induced charge density oscillates at a distance far greater than the Thomas–Fermi screening length [30]. In metals, the Friedel oscillation is the primary mechanism for the long-range electron–electron Coulomb interactions, dominating carrier transport and residual resistivity at extremely low temperatures [152–155]. This quantum phenomenon reveals that the properties of quasiparticles are highly dependent on the Fermi surface structure. STM results have shown standing wave patterns due to Friedel oscillations with a 2D electron-like Fermi surface on the surfaces of Cu(111) [156], Au(111) [157] and Be(0001) [156]. However, the observation of anisotropic screening charge density oscillations on the Be($10\bar{1}0$) surface using STM clearly shows a consequence of the anisotropic Fermi surface [158]. The unique information of the oscillation of the charge density (Friedel oscillation) in graphene provides an insight into the chirality [159] and Berry phase [57, 160, 161] of quasiparticles. In Chapter 4, the charge screening phenomena are thoroughly discussed in the static limit $\omega \to 0$ for monolayer and multilayer graphenes. A comparison with 1D carbon nanotubes is also made. All the critical effects are unified in the theoretical framework, such as the intralayer and interlayer orbital hybridizations in cylindrical or planar honeycomb lattices and the Coulomb interactions between charged particles. The electric polarization is closely related to the momentum and frequency of the external electric field and energy bands of the system [75]. The effective Coulomb potential and induced charge density are calculated using the self-consistent Dyson formula. The real-space Friedel oscillation, which is the subject of this chapter, is straightforwardly derived by using the Fourier transform. The calculated effective screening length and the long-range oscillation of the charge density are discussed in detail. The intravalley and intervalley Coulomb scatterings give rise to multi-mode oscillations in graphene-related systems. Previous studies have shown that the selection rule for transitions between electronic states is highly dependent on the number of layers and stack structure in the graphene system [74, 75]. These will cause changes in the polarization structure, such as the number and location of divergence points, and reflect on the screening phenomenon. It is worthwhile to systematically study the dominant mechanism of electron scatterings among the layered graphene systems, 1D carbon nanotubes and electron gas, especially for effective screening length and long-range charge oscillation. Direct examination of various phenomena caused by the Coulomb field through high-resolution STM/STS tools has become an urgent issue.

In Chapter 5, the dynamic screening of quasiparticles from single-particle and collective excitations is investigated in detail in terms of the wave propagation in time and position domains. Up to now, apart from the electron-gas systems, there have been few theoretical studies on the dynamic charge oscillation phenomenon that focus on the frequency response of external perturbation of a step potential [63–65]. The screening of electrons in a medium is based on the concept of a quasiparticle in quantum mechanics. Graphene systems are outstanding candidates for clearly illustrating the diverse carrier dynamics due to the diversity of stacked configurations [74, 75]. Previous studies have shown that electron concentration, temperature, gate voltage, tube radius, chirality and stacking configuration are critical mechanisms for the screening effect in 1D carbon nanotubes and 2D few-layer graphene systems

[44–48, 74, 75, 128]. It should be noted that all the aforementioned effects on the screened response functions are unified in the developed theoretical framework (details in Chapter 2). Moreover, the clear classification of collective and single-particle excitations allows for the charge response to be decomposed into two parts: one caused by the plasma mode and the other related to the electron–hole pair. The approach is one of the most important aspects of quasiparticles. At a sufficiently high electron density, plasmons dominate the elementary excitations, and the wave propagation corresponds to the resonance with prominent plasmons. Also, the duration and strength of the external perturbation play an important role in the excitation spectrum and plasma waves, which are expected to be greatly diversified in the position and time domains according to the various stacking configurations of multilayer graphene. This causes interesting changes in the energy loss spectrum, which can be verified through high-resolution pump-probe femtosecond optical spectroscopies [8, 9, 11].

Graphene-related systems are also among the mainstream materials for studying the optical scatterings in layered systems because of their various crystal symmetries [162–180]. The number of layers, stacking configurations, and external fields can cause different phenomena [164, 181], such as the simple modulation of vertical optical transitions. There are many theoretical [181–200] and [201–218] experimental studies on low-layer graphene systems. Major concerns include low-frequency absorption structures with/without gate voltages and magnetic fields. In particular, magneto-optic theory has been developed through a generalized tightening model based on the two sublattices of each honeycomb crystal. Diverse magneto-quantization systems of 2D materials [181, 182], such as anomalous LLs and magneto-optical selection rules, have been studied. This theoretical framework helps to fully understand the electronic [162, 163], optical [181–184, 201–205], transport [177], magnetic [182, 201] and Coulomb-excitation [178] characteristics of various atomic interactions and field disturbances. However, the Kubo formula is not sufficient to explore the reflection and transmission spectra of thin films with prominent boundary effects. In Chapter 6, an optical scattering model is developed within the unified framework. The calculation results can be examined by high-resolution optical measurements [201–218].

The most intriguing physics in condensed matter is topology-based energy dispersions [239, 240], which consist of topologically protected localized states (LSs) on either the boundaries of 2D materials [221–223] or the surfaces of 3D materials [224–226]. Such topological LSs within the specific materials which provide strong spin–orbit coupling (SOC) are called topological insulators (TIs) or quantum spin Hall effects due to the insulating bulk carrying robust propagating modes against disorder [239, 240]. The topological LSs occurring near the edges of TI materials have recently attracted a lot of attention, not only for their fundamental physical importance but also because of their potential application in spintronics [227]. TIs are made from semiconductor- or alloy-based materials and have been realized experimentally [228, 229]. The edge states of these TIs reveal helical characteristics arising from a spin-momentum locking mechanism where the edge states with opposite spins propagate in opposite directions [239, 240]. The topological nature of such edge states existing in gapped 2D TI materials can be conventionally characterized by the topological index, Chern number. Fortunately, one of the best candidates for 2D materials is graphene due to its unique electron properties and wide applications [77, 244]. The novel topological edge states in a 2D graphene incorporating the intrinsic SOC have been theoretically studied by Kane and Mele [230, 231]. The SOC mixing of electron spins on different valleys (K, K′-valleys) induces Dirac cones to open, forming a small global gap. The topological edge states launch from an upper K (K′) valley to another lower K′ (K) valley [230, 231] within the gap. Ideally, the topological edge states in 2D TI materials are immune to backscattering by local nonmagnetic disorders due to the spin-momentum locking feature [232]. However, the topological edge states in graphene with intrinsic SOC are too insignificant to be observed because of the very small gap opening. Thus, it is an interesting issue to generate topologically protected LSs in a graphene sheet by purely electric means without SOC [233].

Chapter 7 discusses how applying a gate potential based on $N_c$, the number of chains, along the armchair chain (directed along the $y$-axis) in 2D graphene can produce topological LSs. The wave functions of the LSs show strongly localized behavior around the gate and decay away from the gate. The number of LSs is shown to depend on the number of chains, $N_c$. It is remarkable that there are one branch and two branches of LSs launching from the Dirac cone for single-chain ($N_c = 1$) and multi-chain gates ($N_c \geq 2$), respectively. Such topological LSs are formed as long as the gate-induced potential $V_0$ is nonzero. The topological origin of the formation of LSs can be interpreted through the mechanism of pseudospin rotations, i.e., the rotation of one valley-dependent pseudospin is equivalent to that of another valley-dependent pseudospin. Notably, both pseudospins are on the same side of the gate potential. A picture of pseudospin rotations can be extracted from the LS secular equation, which is built up due to the scattering process induced by the gate potential in a graphene sheet. As the energy $E$ varies across the entire energy gap for a given $k_y$, the rotation angle of the pseudospin-rotation operator fits within the variation range of the relative angle $\Delta\theta_p$ between two pseudospins. These LS branches reveal a Dirac-point signature, with dispersion relations emerging from the Dirac point (at $k_y = 0$). For general multiple-chain ($N_c > 1$) potential cases, the number of LS branches, $N_{LS}$, increases with the potential strength $V_0$ up to a maximum of $N_{LS,max} = N_c$. It is important to note that there are only two LS branches coupling to a Dirac-point signature in the case of $N_c > 1$.

A description of the topological properties of quasiparticles is also included. As is well known, quasiparticles appear as Dirac fermions or their variants in electronic matter if they can be effectively described by the Dirac equation. Such Dirac matter can be endowed with a topological character [234–237]. A highlight of the recent research on condensed matter is that abundant topological phases have been analytically classified and characterized in a number of models and materials, even in the single-particle regime, ranging from Chern insulators displaying QHEs in the presence or absence of a magnetic field, time reversal symmetry (TRS) protected TIs/superconductors with gapped bulk bands, to topological semimetals/nodal superconductors that are gapless with nodes in the bulk [238–243]. Here, graphene plays a central role [239–244], where 2D Dirac fermions are hosted as the quasiparticles about the ground states. In this regard, some topological aspects of 2D Dirac fermions existing in 2D or 3D systems are introduced in Chapters 8 and 9 of this book.

In Chapter 8, a discussion on 2D Dirac fermions characterized by binary $Z_2$ topological invariants is set forth. It starts with a description of Dirac fermions, particularly being concerned with the disparity between their spatial dimensions, i.e., 2D and 3D. Another preliminary is given for the origin of topology in the Bloch band structures, where the so-called Berry curvature integrated over specific gapped manifolds (Berry flux) makes sense [245]. The topological invariant can be an integer, called the Chern number, and nonzero values are acquired by Chern insulators without any symmetry constraints. Besides the nonzero integer, $Z$ (Chern number), acquired by Chern insulators without any symmetry constraints, the $Z_2$ invariant is described, where the zero-valued Berry flux is further differentiated into distinguishable sectors in the eigenspace of electron spin or some kinds of pseudospin, with an intuitive interpretation instead of the conventional formidable derivation. It is known that certain topological semimetals belong to the $Z_2$ class, as do TRS protected TIs. After the preliminaries, examples of $Z_2$ topological matter are introduced, including the 2D TRS protected TIs that show the quantum spin Hall effect, and spinless graphene, bilayer graphene, and rhombohedral graphite having Dirac nodal lines. Finally, some aspects of how chiral symmetry (CS), if present, may act are addressed.

A special topic discussing the zero-mode LL of massless 2D Dirac fermions is given in Chapter 9. Since the research breakthrough of graphene, it has been believed that the CS present in graphene can lead to the zero-mode LL, which in turn is a prerequisite for the half-integer QHE [244]. The first section addresses the following two features of graphene: the zero-mode LL can be stably present even in the absence of CS; and the zero-mode LL is not the unique premise for the half-integer QHE. Next, other possibilities of topological protection for the zero-mode LL are introduced, attesting to the release from the necessity for CS. Finally, tilted massless 2D Dirac fermions in 2D and 3D systems are given as a specialized case.

In Chapter 10, the generalized tight-binding model is used to explore the magneto-electronic properties of monolayer buckled GaAs, where the buckled structure, multi-orbital chemical bondings, SOC, electric field, and magnetic field are considered simultaneously. The rich magnetic quantization covers three groups of spin-polarized LLs near the Fermi level, with unique initial energies, degeneracy of LLs, energy spacings, dependence of the magnetic field, and splitting of spin. Moreover, the probabilities of Landau states show specific oscillation patterns, being composed of the localization centers, node regularities and energy-dependent variations of the dominating orbitals. The density of states directly reflects the major features of the LL energy spectra in the number, form, height, and frequency of the spin-split delta-function-like peaks. The electric field results in monotonous/nonmonotonous LL energy dispersions, LL crossing behavior, gap modulation, phase transition and enhancement of spin splitting. The complex gap modulations and semiconductor–semimetal transitions are attributed to the strong competition between the intrinsic spin–orbit interactions and external fields (magnetic and electric fields).

The current book is focused on the quasiparticle properties under intrinsic and many-particle interactions in low-dimensional nanosystems. The first chapter introduces the overall progress of graphene-related materials with different lattice symmetries, stacking configurations and dimensionalities. Also, the various experimental spectroscopies and phenomenological models used to describe the origin of the topological formation of quasiparticles in quantum systems are briefly introduced. The development of the unified theoretical framework is addressed in Chapter 2 from the perspective of single-particle to many-particle in the presence of various mechanisms, such as the intrinsic interactions of charges and spins, the crystal symmetries of chemical bonds, the quantum confinement of edge structure, and the external charge doping, temperature and the electric, magnetic and mechanical fields. In the tight-binding model, the Group-IV materials are excellent candidates for exhibiting diverse Hamiltonians provided by such mechanisms, which greatly enrich the quasiparticle modes and topological properties in condensed matter systems. The significant couplings between multi-orbital hybridizations and spin interactions produce unusual quasiparticle behaviors under the many-body effects, which require further study of physical phenomena due to optical scattering and static and dynamic Coulomb scattering. For quasi-1D armchair-edge graphene nanoribbons, the anisotropic hopping integrals give rise to edge states, and the topological nature is proved by the Zak phase [239, 240]. These topological edge states can be transformed into a bulk Hamiltonian in armchair-edge graphene without local tunable parameters. The binary $Z_2$ topological invariant is described by the 2D Dirac fermion, which has a Berry flux in the distinguishable sector in the eigenspace of the electron spin or some pseudospin. The latest experimental measurement results are compared with theoretical predictions. Furthermore, it is clear from the presented ideas in this book that the direct combination of phenomenological models and experimental observations can promote the progress of the basic and applied sciences.

The main focus of Chapters 3–5 is the in-depth discussions and interpretations of interesting many-body phenomena affected by long-range Coulomb interactions, i.e., Coulomb excitations and scattering decays, dynamic charge density waves and Friedel oscillations. The optical reflection and transmission spectra of thin films are studied for layered graphenes with prominent boundary effects in Chapter 6, revealing the composite quasiparticles due to the significant couplings between electrons and photons under electromagnetic-field perturbations. In the generalized tight-binding model based on layers and sublattices, all important mechanisms of the Hamiltonian are considered to be consistent with the modified RPA and self-energy methods, as described in Section 2.1.2, in the sense up to entire order of two-particle scatterings. Various dominant mechanisms of charge scatterings among the layered graphene system, 1D carbon nanotubes and electron gas are discussed in detail under the long-range Coulomb field. It should be noted that different characteristics of multi-plasmon modes play an important role in the quasiparticle properties, especially in the particle phase diagram where the scattering resonates with the plasmon frequency. Meanwhile, the physical pictures of single and many particles can be used to explain the dynamics of Dirac quasiparticles in related materials. From the aspects of plasmon intensity, dispersion relation, quantity, Landau damping and characteristics of electron–hole excitation, the roles played by the elementary excitations are thoroughly studied from the perspective of the geometric symmetry of the studied system.

The topological properties of the massless 2D Dirac fermions are studied under the influence of gate potential, SOC, edge structure and geometric symmetry and dimensionality. In most condensed matter lattices, near the single-particle ground state, fermionic quasiparticles have continuous effective Hamiltonians mimicking the Dirac Hamiltonian of Dirac fermions in quantum vacuum. The disparity between them lies in the fact that quantum vacuum is continuous, while the naturally existing condensed matter lattices are discrete. In the language of quantum field theory for both cases, a single continuous Dirac field has some kinds of quantum anomalies, which are the origin of topology. However, both Dirac fields suffer from ultraviolet divergence, so they must be regularized by some means. The condensed matter lattices provide a natural means; however, topology would now diminish with the diminished quantum anomaly. Fortunately, there are various symmetries or other ways to recover topology in condensed matter lattices. That is why abundant topological phases are found in condensed matter. In Chapter 7, the gate potential applied to a graphene sheet can produce topological LSs. Dirac-point characteristics of the topological LS branches are manifestations of both the absence of a finite threshold potential for their formation and the zero gap in the bulk graphene. In Chapters 8 and 9, 2D Dirac fermions and the origin of the binary $Z_2$ topological invariants are discussed. The zero-mode LL of massless 2D Dirac fermions is addressed under CS, implying additional possibilities for its topological protection. Meanwhile, the applicability of the generalized tight-binding model is used in Chapter 10 to discuss in depth the magnetoelectronic spectrum of monolayer buckled GaAs, which shows six subgroups of spin-polarized peaks with rich van Hove singularities. Under the influence of an external electric field, there are drastic changes in the magnetoelectronic energy dispersions, LL crossings, enhancement of spin splitting, gap modulation, and phase transition. In particular, the complex gap modulation and phase transition based on the fairly strong competition/cooperation between the magnetic field and the electric field are studied in detail. Moreover, the $E_z$–$B_z$ phase diagram clearly shows the critical electric field under a specific magnetic field. Chapter 11 contains concluding remarks on all topics in Chapters 3–10. Finally, Chapter 12 presents near-term perspectives and unresolved issues.

## References

[1] Cocker T.L., Peller D., Yu P., Repp J. and Huber R. 2016. Tracking the ultrafast motion of a single molecule by femtosecond orbital imaging. *Nature* **539** 263.
[2] Eisele M., Cocker T.L., Huber M.A., Plankl M., Viti L. *et al.* 2014. Ultrafast multi-terahertz nano-spectroscopy with sub-cycle temporal resolution. *Nat. Photonics* **8** 841.
[3] Sola I.J., Mével E., Elouga L., Constant E., Strelkov V. *et al.* 2006. Controlling attosecond electron dynamics by phase-stabilized polarization gating. *Nat. Phys.* **2** 319.
[4] Cavalieri A.L., Müller N., Uphues T., Yakovlev V.S., Baltuška A. *et al.* 2003. Attosecond control of electronic processes by intense light fields. *Nature* **421** 611.
[5] Baltuška A., Udem T., Uiberacker M., Hentschel M., Goulielmakis E. *et al.* 2007. Attosecond spectroscopy in condensed matter. *Nature* **449** 1029.
[6] Zahn D., Hildebrandt P.-N., Vasileiadis T., Windsor Y.W., Qi Y. *et al.* 2007 Anisotropic nonequilibrium lattice dynamics of black phosphorus. *Nano Lett.* **20** 3728.
[7] Bao J.M., Pfeiffer L.N., West K.W. and Merlin R. 2007. Ultrafast dynamic control of spin and charge density oscillations in a GaAs quantum well. *Phys. Rev. Lett.* **92** 236601.
[8] Roskos H.G., Thomson M.D., Kreß M. and Loffler T. 2007. Broadband THz emission from gas plasmas induced by femtosecond optical pulses: From fundamentals to applications. *Laser Photonics Rev.* **4** 349.
[9] Margetic V., Pakulev A., Stockhaus A., Bolshov M., Niemax K. *et al.* 2000. A comparison of nanosecond and femtosecond laser-induced plasma spectroscopy of brass samples. *Spectrochim. Acta B* **55** 31771.
[10] Cocker T.L., Jelic V., Gupta, M., Molesky S.J. and Jacob A.J. 2013. An ultrafast terahertz scanning tunnelling microscope. *Nat. Photonics* **7** 620.
[11] Terada Y., Yoshida S., Takeuchi O. and Shigekawa H. 2010. Real-space imaging of transient carrier dynamics by nanoscale pump-probe microscopy. *Nat. Photonics* **4** 869.

[12] Jiang T., Kravtsov V., Tokman M., Belyanin A. and Raschke M.B. 2019. Ultrafast coherent nonlinear nanooptics and nanoimaging of graphene. *Nat. Nanotechnol.* **14** 838.
[13] Huber R., Tauser F., Brodschelm A., Bichler M., Abstreiter G. *et al.* 2001. How many-particle interactions develop after ultrafast excitation of an electron-hole plasma. *Nature* **414** 286.
[14] Fei Z., Rodin A.S., Andreev G.O., Bao W., McLeod A.S. *et al.* 2012. Gate-tuning of graphene plasmons revealed by infrared nano-imaging. *Nature* **487** 82.
[15] Ma W., Alonso-González P., Li S., Nikitin A.Y., Yuan J. *et al.* 2018. In-plane anisotropic and ultra-low-loss polaritons in a natural van der Waals crystal. *Nature* **562** 557.
[16] Ni G.X., McLeod A.S., Sun Z., Wang L., Xiong L. *et al.* 2018. Fundamental limits to graphene plasmonics. *Nature* **557** 530.
[17] Wang L., Kafshgari M.H. and Meunier M. 2020. Optical properties and applications of plasmonic-metal nanoparticles. *Adv. Funct. Mater.* **30** 2005400.
[18] Park N.R., Kim H.N., Jin Y.H., Kim M., Lee K.S. *et al.* 2020. Extreme field confinement in zigzag plasmonic crystals. *Nanotechnology* **31** 495206.
[19] Liu J.J., Jalali M., Mahshid S. and Wachsmann-Hogiu S. 2020. Are plasmonic optical biosensors ready for use in point-of-need applications? *Analyst* **145** 364.
[20] Zhang Y., Min C., Dou X., Wang X. and Urbach H.P. 2021. Plasmonic tweezers: For nanoscale optical trapping and beyond. *Light Sci. Appl.* **10** 59.
[21] Davis T.J., Gomez D.E. and Roberts A. 2017. 2021. Plasmonic circuits for manipulating optical information. *Nanophotonics* **6** 543.
[22] Choi D., Shin C.K., Yoon D., Chung D.S., Jin Y.W. *et al.* 2014. Plasmonic optical interference. *Nano Lett.* **14** 3374.
[23] Liu Z.Q., Fu G.L., Huang Z.P. and Chen J. 2016. Plasmonic crystals with sharp optical transmission behaviors. *Mater. Lett.* **185** 519.
[24] Sun Y.H., Jiang L., Zhong L.B., Jiang Y.Y. and Chen X.D. 2015. Towards active plasmonic response devices. *Nano Res.* **8** 406.
[25] Mehrnegar M.M., Darbari S., Ramezani H. and Moravvej-Farshi M.K. 2019. Designing graphene-based multi-mode acousto-plasmonic devices. *J. Light. Technol.* **8** 406.
[26] Liang Z.Q., Sun J., Jiang Y.Y., Jiang L. and Chen X.D. 2014. Plasmonic enhanced optoelectronic devices. *Plasmonics* **9** 859.
[27] Kumar U., Viarbitskaya S., Cuche A., Girard C., Bolisetty S. *et al.* 2018. Designing plasmonic eigenstates for optical signal transmission in planar channel devices. *ACS Photonics* **5** 2328.
[28] Seren H.R., Zhang J.D., Keiser G.R., Maddox S.J., Zhao X.G. *et al.* 2016. Nonlinear terahertz devices utilizing semiconducting plasmonic metamaterials. *Plasmonics* **4** 16078.
[29] Iqbal T., Khalil S., Ijaz M., Riaz K.N., Khan M.I. *et al.* 2019. Optimization of 1D plasmonic grating of nanostructured devices for the investigation of plasmonic bandgap. *Plasmonics* **14** 775.

[30] Mahan G.D. 1990. *Many-Particle Physics* (New York: Plenum).
[31] Nozières P. and Luttinger J.M. 1962. Derivation of the Landau theory of Fermi liquids. I. Formal preliminaries. *Phys. Rev.* **127** 1423.
[32] Schrieffer J.R. 1970. What is a quasi-particle? *J. Res. Natl. Bur. Stand. A* **74** 537.
[33] Giuliani G.F. and Quinn J.J. 1982. Lifetime of a quasiparticle in a two-dimensional electron gas. *Phys. Rev. B* **26** 4421.
[34] Chiu C.W., Chung Y.L., Yang C.H., Liu C.T. and Lin C.Y. 2020. Coulomb decay rates in monolayer doped graphene. *RSC Adv.* **10** 2337.
[35] Shung K.W.-K. 1986. Lifetime effects in low-stage intercalated graphite systems. *Phys. Rev. B* **34** 2.
[36] Bostwick A., Ohta T., Seyller T., Horn K. and Rotenberg E. 2007. Quasiparticle dynamics in graphene. *Nat. Phys.* **3** 36.
[37] Hong T., Qiu Y., Matsumoto M., Tennant D.A., Coester K. *et al.* 2017. Field induced spontaneous quasiparticle decay and renormalization of quasiparticle dispersion in a quantum antiferromagnet. *Nat. Commun.* **8** 15148.
[38] Poterba J.M. and Summers L.H. 1987. Finite lifetimes and the effects of budget deficits on national saving. *J. Monet. Econ.* **20** 369.
[39] Tokuda S., Souma S., Segawa K., Takahashi T., Ando Y. *et al.* 2021. Unveiling quasiparticle dynamics of topological insulators through Bayesian modelling. *J . Phys. Commun.* **4** 170.
[40] Sessi P., Fan F.R., Kuster F., Manna K., Schroter, N.B.M. *et al.* 2020. Handedness-dependent quasiparticle interference in the two enantiomers of the topological chiral semimetal PdGa. *Nat. Commun.* **11** 3507.
[41] Zhensheng T., Cong C., Tibor S., Mark K., Manos M. *et al.* 2016. Direct time-domain observation of attosecond final-state lifetimes in photoemission from solids. *Science* **353** 62.
[42] Xu S., Cao J., Miller C.C., Mantell D.A., Miller R.J.D. *et al.* 1996. Energy dependence of electron lifetime in graphite observed with femtosecond photoemission spectroscopy. *Phys. Rev. Lett.* **76** 483.
[43] Moos G., Gahl C., Fasel R., Wolf M. and Hertel T. 2001. Anisotropy of quasiparticle lifetimes and the role of disorder in graphite from ultrafast time-resolved photoemission spectroscopy. *Phys. Rev. Lett.* **87** 267402.
[44] Chiu C.W., Shyu F.L., Chang C.P., Chen R.B., Chuu D.S. and Lin M.F. 2004. Electron lifetime in armchair carbon nanotubes. *J. Phys. Soc. Jpn.* **73** 2936.
[45] Chiu C.W., Chiu Y.H., Shyu F.L., Chang C.P., Chuu D.S. and Lin M.F. 2005. Temperature-dependent carrier dynamics in metallic carbon nanotubes. *Phys. Lett. A* **346** 347.
[46] Ho J.H., Chiu C.W. and Lin M.F. 2006. Temperature-dependent self-energy of single-walled carbon nantubes. *Physica E* **35** 212.
[47] Zamkov M., Woody N., Shan B., Chang Z. and Richard P. 2005. Lifetime of charge carriers in multiwalled nanotubes. *Phys. Rev. Lett.* **94** 056803.
[48] Lin M.F. and Chuu D.S. 1997. Elementary excitations in a carbon nanotube. *J. Phys. Soc. Jpn.* **66** 757.
[49] Rahnejat K.C., Howard C.A., Shuttleworth N.E., Schofield S.R., Iwaya K. *et al.* 2011. Charge density waves in the graphene sheets of the superconductor $CaC_6$. *Nat. Commun.* **2** 558.

[50] Bürgi L., Jeandupeux O., Brune H., Kern K. *et al.* 1999. Probing hot-electron dynamics at surfaces with a cold scanning tunneling microscope. *Phys. Rev. Lett.* **82** 22.
[51] Liu W.H., Luo J.W., Li S.S. and Wang L.W. 2019. Impurity diffusion induced dynamic electron donors in semiconductors. *Phys. Rev. B* **100**, 165203.
[52] Simion G.E. and Giuliani G.F. 2005. Friedel oscillations in a Fermi liquid. *Phys. Rev. B* **72** 045127.
[53] Sprunger P.T., Petersen L., Plummer E.W., Lægsgaard E. and Besenbacher F. 1997. Giant Friedel oscillations on the beryllium(0001) surface. *Science* **275** 1764.
[54] Riechers K., Hueck K., Luick N., Lompe T., Moritz H. *et al.* 2017. Detecting Friedel oscillations in ultracold Fermi gases. *Eur. Phys. J. D* **71** 232.
[55] Kanisawa K., Butcher M.J., Yamaguchi H. and Hirayama Y. 2001. Imaging of Friedel oscillation patterns of two-dimensionally accumulated electrons at epitaxially grown InAs(111)A surfaces. *Phys. Rev. Lett.* **86** 3384.
[56] Zhang R.F., Argon A.S. and Veprek S. 2009. Friedel oscillations are limiting the strength of superhard nanocomposites and heterostructures. *Phys. Rev. Lett.* **102** 015503.
[57] Zhang S.H., Yang J., Shao D.F., Wu Z. and Yang W. 2021. Robust wave-front dislocations of Friedel oscillations in gapped graphene. *Phys. Rev. B* **103** L161407.
[58] Wunsch B., Stauber T., Sols F. and Guinea F. 2006. Dynamical polarization of graphene at finite doping. *New J. Phys.* **100** 165203.
[59] Bácsi Á. and Virosztek A. 2010. Local density of states and Friedel oscillations in graphene. *Phys. Rev. B* **82** 193405.
[60] Hwang E.H. and Sarma S.D. 2008. Screening, Kohn Anomaly, Friedel oscillation, and RKKY interaction in bilayer graphene. *Phys. Rev. Lett.* **101** 156802.
[61] Friedel J. 1958. Metallic alloys. *Il Nuovo Cimento* **7** 287.
[62] Halperin, B.I. 1987. Possible states for a three-dimensional electron gas in a strong magnetic field. *Jpn. J. Appl. Phys.* **26** S3.
[63] Stern F. 1967. Polarizability of a two-dimensional electron gas. *Phys. Rev. Lett.* **18** 546.
[64] Bagchi A. 1977. Transverse dielectric response of a semi-infinite metal: Surface effect. *Phys. Rev. B* **15** 3060.
[65] Canright G.S. 1988. Time-dependent screening in the electron gas. *Phys. Rev. B* **38** 1647.
[66] Maldague P.F. 1978. Many-body corrections to the polarizability of the two-dimensional electron gas. *Surf. Sci.* **73** 296.
[67] Sato O., Tanaka Y., Kobayashi M. and Hasegawa A. 1993. Exotic behavior of the dielectric function and the plasmons of an electron gas on a tubule. *Phys. Rev. B* **48** 1947.
[68] Lin M.F. and Shung K.W.-K. 1997. Quasiparticle properties of an electron gas in a cylindrical tubule. *J. Phys. Soc. Jpn.* **66** 3543.
[69] Thomas K.J., Nicholls J.T., Simmons M.Y., Pepper M., Mace D.R. *et al.* 1996. Possible spin polarization in a one-dimensional electron gas. *Phys. Rev. Lett.* **77** 135.

[70] Ferrari G., Bertoni A., Goldoni G. and Molinari E. 2008. Cylindrical two-dimensional electron gas in a transverse magnetic field. *Phys. Rev. B* **78** 115326.

[71] Hallam D., Weis J. and Maksym P.A. 1996. Screening of the electron-electron interaction by gate electrodes in semiconductor quantum dots. *Phys. Rev. B* **53** 1452.

[72] March N.H. 1952. Thomas-Fermi fields for molecules with tetrahedral and octahedral symmetry. *Math. Proc. Cambridge Philos. Soc.* **48** 665.

[73] Yannouleas C., Bogachek E.N. and Landman U. 1952. Collective excitations of multishell carbon microstructures: Multishell fullerenes and coaxial nanotubes. *Phys. Rev. B* **53** 10225.

[74] Lin C.Y., Chen R.B., Ho Y.H. and Lin M.F. 2017. *Electronic and Optical Properties of Graphite-Related Systems* (CRC Press).

[75] Lin C.Y., Do T.N., Huang Y.K. and Lin M.F. 2017. Electronic and optical properties of graphene in magnetic and electric fields. *IOP Concise Physics.* San Raefel, CA, USA: Morgan & Claypool Publishers.

[76] Lin C.Y., Lee M.H. and Lin M.F. 2018. Coulomb excitations in ABC-stacked trilayer graphene. *Phys. Rev. B* **98** 041408(R).

[77] Novoselov K.S., Geim A.K., Morozov S.V., Jiang D., Zhang Y. *et al.* 2004. Electric field effect in atomically thin carbon films. *Science* **306** 666.

[78] Lee C., Wei X., Kysar J.W. and Hone J. 2008. Measurement of the elastic properties and intrinsic strength of monolayer graphene. *Science* **321** 385.

[79] Kroto H.W., Heath J.R., O'Brien S.C., Curl R.F. and Smalley R.E. 1985. $C_{60}$: Buckminsterfullerene. *Nature* **318** 162.

[80] Lim C.H.Y.X., Sorkin A., Bao Q.L., Li A., Zhang K., Nesladek M., *et al.* 2013. A hydrothermal anvil made of graphene nanobubbles on diamond. *Nat. Commun.* **4** 1556.

[81] Iijima S. 1991. Helical microtubules of graphitic carbon. *Nat. Nanotechnol.* **354** 56.

[82] Cresti A., Fogler M.M., Guinea F., Castro Neto A.H., Roche S. 2012. Quenching of the quantum Hall effect in graphene with scrolled edges. *Phys. Rev. Lett.* **108** 166602.

[83] Lee J.K., Lee S.C., Ahn J.P., Kim S.C., Wilson J.I.B. *et al.* 2008. The growth of AA graphite on (1 1 1) diamond. *J. Chem. Phys.* **129** 234709.

[84] Borysiuk J., Sołtys J. and Piechota J. 2011. Stacking sequence dependence of graphene layers on SiC ($000\bar{1}$)-experimental and theoretical investigation. *J. Appl. Phys.* **109** 093523.

[85] Sutter P.W., Flege J.I. and Sutter E.A. 2008. Epitaxial graphene on ruthenium. *Nat. Mater.* **7** 406.

[86] Kim K.S., Zhao Y., Jang H., Lee S.Y., Kim J.M. *et al.* 2009. Large-scale pattern growth of graphene films for stretchable transparent electrodes. *Nature* **457** 706.

[87] Yan K., Peng H.L., Zhou Y., Li H., Liu Z.F. *et al.* 2011. Formation of bilayer bernal graphene: Layer-by-layer epitaxy via chemical vapor deposition. *Nano Lett.* **11** 1106.

[88] Liu L., Zhou H., Cheng R., Yu W.J., Liu Y. *et al.* 2012. High-yield chemical vapor deposition growth of high-quality large-area AB-stacked bilayer graphene. *ACS Nano* **6** 8241.
[89] Zhang X., Li H. and Ding F. 2014. Self-assembly of carbon atoms on transition metal surfaces-chemical vapor deposition growth mechanism of graphene. *Adv. Mater.* **26** 5488.
[90] Zhou Z., Habenicht B.F. and Guo Q. 2013. Graphene moire structure grown on a pseudomorphic metal overlayer supported on Ru(0001). *Surf. Sci.* **611** 67.
[91] Li X., Cai W., An J., Kim S., Nah J. *et al.* 2009. Large-area synthesis of high-quality and uniform graphene films on copper foils. *Science* **324** 1312.
[92] Zhou H., Yu W.J., Liu L., Cheng R., Chen Y. *et al.* 2013. Chemical vapour deposition growth of large single crystals of monolayer and bilayer graphene. *Nat. Commun.* **4** 2096.
[93] Su C.Y., Lu A.Y., Wu C.Y., Li Y.T., Liu K.K. *et al.* 2011. Direct formation of wafer scale graphene thin layers on insulating substrates by chemical vapor deposition. *Nano Lett.* **11** 3612.
[94] Bae S., Kim H., Lee Y., Xu X., Park J.S. *et al.* 2010. Roll-to-roll production of 30-inch graphene films for transparent electrodes. *Nat. Nanotechnol.* **5** 574.
[95] Bernal J.D. 1924. The structure of graphite. *Proc. R. Soc. Lond. Ser. A* **106** 749.
[96] Lipson H. and Stokes A.R. 1942. The structure of graphite. *Proc. R. Sot. Lond. Ser. A* **181** 101.
[97] Kambe N., Dresselhaus M.S., Dresselhaus G., Basu S., McGhie A.R. *et al.* 1979. Intercalate ordering in first stage graphite-lithium. *Mater. Sei. Eng.* **40** 1.
[98] Novoselov K.S., Jiang Z., Zhang Y., Morozov S.V., Stormer H.L. *et al.* 2007. Room-temperature quantum hall effect in graphene. *Science* **315** 1379.
[99] Zhang Y.B., Tan Y.W., Stormer H.L. and Kim P. 2005. Experimental observation of the quantum Hall effect and Berry's phase in graphene. *Nature* **438** 201.
[100] Jiang Z., Henriksen E.A., Tung L.C., Wang Y.-J., Schwartz M.E. *et al.* 2007. Infrared spectroscopy of Landau levels of graphene. *Phys. Rev. Lett.* **98** 197403.
[101] Ho J.H., Lai Y.H., Chiu Y.H. and Lin M.F. 2008. Landau levels in graphene. *Physica E* **40** 1722.
[102] Castro Neto A.H., Guinea F., Peres N.M.R., Novoselov K.S. and Geim A.K. 2009. The electronic properties of graphene. *Rev. Mod. Phys.* **81** 109.
[103] McCann E. and Koshino M. 2013. The electronic properties of bilayer graphene. *Rep. Prog. Phys.* **76** 056503.
[104] Do T.N., Shih P.H., Chang C.P., Lin C.Y. and Lin M.F. 2016. Rich magneto-absorption spectra of AAB-stacked trilayer graphene. *Phys. Chem. Chem. Phys.* **18** 17597.
[105] Abergel D.S.L. and Fal'ko V.I. 2007. Optical and magneto-optical far-infrared properties of bilayer graphene. *Phys. Rev. B* **75** 155430.
[106] Ho Y.H., Chiu Y.H., Lin D.H., Chang C.P. and Lin M.F. 2010. Magneto-optical selection rules in bilayer Bernal graphene. *ACS Nano* **4** 1465.

[107] Orlita M., Faugeras C., Borysiuk J., Baranowski J.M., Strupiński W. *et al.* 2011. Magneto-optics of bilayer inclusions in multilayered epitaxial graphene on the carbon face of SiC. *Phys. Rev. B* **83** 125302.

[108] Kuhne P., Darakchieva V., Yakimova R., Tedesco J.D., Myers-Ward R.L. *et al.* 2013. Polarization selection rules for inter-Landau-level transitions in epitaxial graphene revealed by the infrared optical Hall effect. *Phys. Rev. Lett.* **111** 077402.

[109] Orlita M., Faugeras C., Plochocka P., Neugebauer P., Martinez G. *et al.* 2008. Approaching the Dirac point in high-mobility multilayer epitaxial graphene. *Phys. Rev. Lett.* **101** 267601.

[110] Plochocka P., Faugeras C., Orlita M., Sadowski M.L., Martinez G. *et al.* 2008. High-energy limit of massless Dirac fermions in multilayer graphene using magneto-optical transmission spectroscopy. *Phys. Rev. Lett.* **100** 087401.

[111] Huang Y.K., Chen S.C., Ho Y.H., Lin C.Y. and Lin M.F. 2014. Feature-rich magnetic quantization in sliding bilayer graphenes. *Sci. Rep.* **4** 7509.

[112] Lin C.Y., Wu J.Y., Ou Y.J., Chiu Y.H. and Lin M.F. 2015. Magneto-electronic properties of multilayer graphenes. *Phys. Chem. Chem. Phys.* **17** 26008.

[113] Lai Y.H., Ho J.H., Chang C.P. and Lin M.F. 2008. Magnetoelectronic properties of bilayer Bernal graphene. *Phys. Rev. B* **77** 085426.

[114] McCann E. and Fal'ko V.I. 2006. Landau-level degeneracy and quantum hall effect in a graphite bilayer. *Phys. Rev. Lett.* **96** 086805.

[115] Min H. and MacDonald A.H. 2008. Chiral decomposition in the electronic structure of graphene multilayers. *Phys. Rev. B* **77** 155416.

[116] Sena S.H.R., Pereira J.M., Peeters F.M. and Faria G.A. 2011. Landau levels in asymmetric graphene trilayers. *Phys. Rev. B* **84** 205448.

[117] Lin C.Y., Wu J.Y., Chiu Y.H., Chang C.P. and Lin M.F. 2014. Stacking-dependent magnetoelectronic properties in multilayer graphene. *Phy. Rev. B* **90** 205434.

[118] Lin Y.P., Wang J., Lu J.M., Lin C.Y. and Lin M.F. 2014. Energy spectra of ABC-stacked trilayer graphene in magnetic and electric fields. *RSC Adv.* **4** 56552.

[119] Tran N.T.T., Lin S.Y., Glukhova O.E. and Lin M.F. 2015. Configuration-induced rich electronic properties of bilayer graphene. *J Phys. Chem. C* **119** 10623.

[120] Do T.N., Lin C.Y., Lin Y.P., Shih P.H. and Lin M.F. 2015. Configuration-enriched magnetoelectronic spectra of AAB-stacked trilayer graphene. *Carbon* **94** 619.

[121] Novoselov K.S., McCann E., Morozov S.V., Fal'ko V.I., Katsnelson M.I. *et al.* 2006. Unconventional quantum Hall effect and Berry's phase of $2\pi$ in bilayer graphene. *Nat. Phys.* **2** 177.

[122] Taychatanapat T., Watanabe K., Taniguchi T. and Herrero P.J. 2008. Quantum Hall effect and Landau-level crossing of Dirac fermions in trilayer graphene. *Nat. Phys.* **7** 621.

[123] Zhang L., Zhang Y., Camacho J., Khodas M. and Zaliznyak I. 2011. The experimental observation of quantum Hall effect of $l = 3$ chiral quasiparticles in trilayer graphene. *Nat. Phys.* **7** 953.

[124] Nakamura M., Hirasawa L. and Imura K.I. 2008. Quantum Hall effect in bilayer and multilayer graphene with finite Fermi energy. *Phys. Rev. B* **78** 033403.
[125] Liu J., Ma Z., Gao J. and Dai X. 2019. Quantum valley Hall effect, orbital magnetism, and anomalous Hall effect in twisted multilayer graphene systems. *Phys. Rev. X* **9** 031021.
[126] Koshino M. and McCann E. 2011. Landau level spectra and the quantum Hall effect of multilayer graphene. *Phys. Rev. B* **83** 165443.
[127] Lin C.Y., Do T.N., Wu J.Y., Shih P.H., Lin S.Y. *et al.* 2019. *Quantization Phenomena in Emergent Layered Materials* (CRC Press).
[128] Ho J.H., Lu C.L., Hwang C.C., Chang C.P. and Lin M.F. 2006. Coulomb excitations in AA- and AB-stacked bilayer graphites. *Phys. Rev. B* **74** 085406 (R).
[129] Lin C.Y., Wu J.Y., Chiu C.W. and Lin M.F. 2019. *Coulomb Excitations and Decays in Graphene-Related Systems* (CRC Press).
[130] Otpmebs D.M. and Rase H.F. 1966. Potassium grahites prepared by mixed-reaction technique. *Carbon* **4** 125.
[131] Emery N., Herold C., Mareche J.F. and Lagrange P. 2008. Synthesis and superconducting properties of $CaC_6$. *Sci. Tech. Adv. Mater.* **9** 044102.
[132] Dresselhaus M.S. and Dresselhaus G. 2002. Intercalation compounds of graphite. *Adv. Phys.* **51** 1.
[133] Dresselhaus M.S. and Dresselhaus G. 1980. Intercalation compounds of graphite. *Adv. Phys.* **30** 139.
[134] Geim A.K. and Novoselov K.S. 2007. The rise of graphene. *Nat. Mater.* **6** 183.
[135] Gierz I., Riedl C., Starke U., Ast C.R. and Kern K. 2008. Atomic hole doping of graphene. *Nano Lett.* **8** 4603.
[136] Shin S.Y., Kim N.D., Kim J.G., Kim K.S., Noh D.Y. *et al.* 2011. Control of the pi plasmon in a single layer graphene by charge doping. *App. Phys. Lett.* **99** 082110.
[137] Rossi E. and Sarma S.D. 2008. Ground state of graphene in the presence of random charged impurities. *Phys. Rev. Lett.* **101** 166803.
[138] Kim D., Li Q., Syers P., Butch N.P., Paglione J., Sarma S.D. and Fuhrer M.S. 2012. Intrinsic electron-phonon resistivity of $Bi_2Se_3$ in the topological regime. *Phys. Rev. Lett.* **109** 166801.
[139] Ericson T. 1958. Level density of degenerate Fermi systems. *Nucl. Phys. News* **8** 265.
[140] Davenport J.L., Ge Z., Liu J., Nuñez-Lobato C. and Moon S. 2019. Probing the electronic structure of graphene near and far from the Fermi level via planar tunneling spectroscopy featured. *Appl. Phys. Lett.* **115** 163504.
[141] Zibrov A.A., Rao P., Kometter C., Spanton E.M., Li J.I.A. *et al.* 2018. Emergent Dirac gullies and gully-symmetry-breaking quantum Hall states in ABA trilayer graphene. *Phys. Rev. Lett.* **121** 167601.
[142] Bao W., Jing L., Velasco Jr. J., Lee Y., Liu G. *et al.* 2011. Stacking-dependent band gap and quantum transport in trilayer graphene. *Nat. Phys.* **7** 948.
[143] Chubukov A.V., Galitski V.M. and Yakovenko V.M. 2005. Quantum critical behavior near a density-wave instability in an isotropic Fermi liquid. *Phys. Rev. Lett.* **94** 046404.

[144] González J. 2012. Electron self-energy effects on chiral symmetry breaking in graphene. *Phys. Rev. B* **85** 085420.

[145] Bulutay C. and Tanatar B. 2002. Spin-dependent analysis of two-dimensional electron liquids. *Phys. Rev. B* **65** 195116.

[146] Li Z.Q., Henriksen E.A., Jiang Z., Hao Z. and Martin M.C. 2008. Dirac charge dynamics in graphene by infrared spectroscopy. *Nat. Phys.* **4** 532.

[147] Matthes L., Pulci O. and Bechstedt F. 2013. Massive Dirac quasiparticles in the optical absorbance of graphene, silicene, germanene, and tinene. *J. Condens. Matter Phys.* **25** 395305.

[148] Gatti G., Crepaldi A., Puppin M., Tancogne-Dejean N. and Xian L. 2020. Light-induced renormalization of the Dirac quasiparticles in the nodal-line semimetal ZrSiSe. *Phys. Rev. Lett.* **125** 076401.

[149] Miao L., Wang Z.F., Ming W.M., Yao M.Y. and Wang M.X. 2013. Quasiparticle dynamics in reshaped helical Dirac cone of topological insulators. *PNAS* **110** 2758.

[150] Liu C., Bian G., Chang T.R., Wang K. and Xu S.Y. 2020. Tunable spin helical Dirac quasiparticles on the surface of three-dimensional HgTe. *Phys. Rev. B* **92** 115436.

[151] Jost A., Bendias M., Bottcher J., Hankiewicz E. and Brune C. 2017. Electron-hole asymmetry of the topological surface states in strained HgTe. *PNAS* **114** 3381.

[152] DiVincenzo D.P. and Mele E.J. 1984. Self-consistent effective-mass theory for intralayer screening in graphite intercalation compounds. *Phys. Rev. B* **29** 1685.

[153] Cheianov V.V. and Fal'ko V.I. 2006. Friedel oscillations, impurity scattering, and temperature dependence of resistivity in graphene. *Phys. Rev. Lett.* **97** 226801.

[154] Babić E., Krsnik R., Leontić B., Očko M., Vučić Z. *et al.* 1972. Temperature dependent impurity resistivity in A1-based 3-d transition metal alloys. *Solid State Commun.* **10** 691.

[155] Bhatia A.B. and Thornton D.E. 1970. Structural aspects of the electrical resistivity of binary alloys. *Phys. Rev. B* **2** 3004.

[156] Crommie M.F., Lutz C.P. and Eigler D.M. 1993. Imaging standing waves in a two-dimensional electron gas. *Nature* **363** 524.

[157] Hasegawa Y. and Avouris P. 1993. Direct observation of standing wave formation at surface steps using scanning tunneling spectroscopy. *Phys. Rev. Lett.* **71** 1071.

[158] Hofmann P., Briner B.G., Doering M., Rust H.-P., Plummer E.W. *et al.* 1997. Anisotropic two-dimensional Friedel oscillations. *Phys. Rev. Lett.* **79** 265.

[159] Brihuega I., Mallet P., Bena C., Bose S., Michaelis C. *et al.* 2021. Quasiparticle chirality in epitaxial graphene probed at the nanometer scale. *Phys. Rev. Lett.* **101** 206802.

[160] Dutreix C., Gonzalez-Herrero H., Brihuega I., Katsnelson M.I., Chapelier C. *et al.* 2019. Measuring the Berry phase of graphene from wavefront dislocations in Friedel oscillations. *Nature* **574** 219.

[161] Zhang Y., Su Y. and He L. 2020. Local berry phase signatures of bilayer graphene in intervalley quantum interference. *Phys. Rev. Lett.* **125** 116804.
[162] Geim A.K. and Novoselov K.S. 2007. The rise of graphene. *Nat. Mater.* **6** 183.
[163] Castro Neto A.H., Guinea F., Peres N.M.R., Novoselov K.S. and Geim A.K. 2009. The electronic properties of graphene. *Rev. Mod. Phys.* **81** 109.
[164] Bae S. *et al.* 2010. Roll-to-roll production of 30-inch graphene films for transparent electrodes. *Nat. Nanotechnol.* **5** 574.
[165] Guinea F. 2007. Charge distribution and screening in layered graphene systems. *Phys. Rev. B* **75** 235433.
[166] Gutierrez F.A., Fernandes F.C.B., Rivas G.A. and Bueno P.R. 2017. Mesoscopic behaviour of multi-layered graphene: The meaning of supercapacitance revisited. *Phys. Chem. Chem. Phys.* **19** 6792–6806.
[167] Chen H.A., Hsin C.L., Huang Y.T., Tang M.L., Dhuey S., Cabrini S., Wu W.W. and Leone S.R. 2013. Measurement of interlayer screening length of layered graphene by plasmonic nanostructure resonances. *J. Phys. Chem. C* **117** 22211–22217.
[168] Englman R. and Vértesi T. 2008. Large Berry phases in layered graphene. *Phys. Rev. B* **78** 205311.
[169] Ansón-Casaos A., Puértolas J.A., Pascual F.J., Hernández-Ferrer J., Castell P., Benito A.M., Maser W.K. and Martínez M.T. 2014. The effect of gamma-irradiation on few-layered graphene materials. *Appl. Surf. Sci.* **301** 264–272.
[170] Masyukov M., Vozianova A., Grebenchukov A., Gubaidullina K., Zaitsev A. and Khodzitsky M. 2020. Optically tunable terahertz chiral metasurface based on multi-layered graphene. *Sci. Rep.* **10** 3157.
[171] Yuan G., Lin D., Wang Y., Huang X., Chen W., Xie X., Zong J., Yuan Q.Q., Zheng H., Wang D., Xu J., Li S.C., Zhang Y., Sun J., Xi X. and Gao L. 2020. Proton-assisted growth of ultra-flat graphene films. *Nature* **577** 204–208.
[172] Wang R., Shi M., Brewer B., Yang L., Zhang Y., Webb D.J., Li D. and Xu Y.Q. 2018. Ultrasensitive graphene optoelectronic probes for recording electrical activities of individual synapses. *Nano Lett.* **18** 5702–5708.
[173] Park H., Lim C., Lee C.J., Kang J., Kim J., Choi M. and Park H. 2018. Optimized poly(methyl methacrylate)-mediated graphene-transfer process for fabrication of high-quality graphene layer. *Nanotechnology* **29** 415303.
[174] Mendez J.P., Ramos F.A., Ortiz M. and Ariza M.P. 2018. Charge carrier transport across grain boundaries in graphene. *Acta. Mater.* **154** 199–206.
[175] Kang M., Kim J., Jang B., Chae Y., Kim J.H. and Ahn J.H. 2017. Graphene-based three-dimensional capacitive touch sensor for wearable electronics. *ACS Nano* **11** 7950–7957.
[176] Matruglio A., Nappini S., Naumenko D., Magnano E., Bondino F., Lazzarino M. and Zilio S.D. 2016. Contamination-free suspended graphene structures by a Ti-based transfer method. *Carbon* **103** 305–310.
[177] Jimenez-Villacorta F., Climent-Pascual E., Ramirez-Jimenez R., Sanchez-Marcos J., Prieto C. and de Andrés A. 2016. Graphene-ultrasmall silver nanoparticle interactions and their effect on electronic transport and Raman enhancement. *Carbon* **101** 305–314.

[178] Grushina A.L., Ki D.K., Koshino M., Nicolet A.A.L., Faugeras C., McCann E., Potemski M. and Morpurgo A.F. 2015. Insulating state in tetralayers reveals an even-odd interaction effect in multilayer graphene. *Nat. Commun.* **6** 6419.
[179] Tuan D.V., Ortmann F., Soriano D., Valenzuela S.O. and Roche S. 2014. Pseudospin-driven spin relaxation mechanism in graphene. *Nat. Phys.* **10** 857.
[180] Nandkishore R., Levitov L.S. and Chubukov A.V. 2012. Chiral superconductivity from repulsive interactions in doped graphene. *Nat. Phys.* **8** 158–163.
[181] Lin C.Y., Chen R.B., Ho Y.H. and Lin M.F. 2018. *Electronic and Optical Properties of Graphite-Related Systems* (CRC Press).
[182] Ho Y.-H., Chiu Y.-H., Lin D.-H., Chang C.-P. and Lin M.-F. 2010. Magneto-optical selection rules in bilayer Bernal graphene. *ACS Nano* **4** 1465–1472.
[183] Gusynin V.P., Sharapov S.G. and Carbotte J.P. 2007. Anomalous absorption line in the magneto-optical response of graphene. *Phys. Rev. Lett.* **98** 157402.
[184] Abergel D.S.L. and Fal'ko V. 2007. Optical and magneto-optical far-infrared properties of bilayer graphene. *Phys. Rev. B* **75** 155430.
[185] Shin J., Park Y., Chittari B.L., Sun J.H. and Jung J. 2021. Electron-hole asymmetry and band gaps of commensurate double moire patterns in twisted bilayer graphene on hexagonal boron nitride. *Phys. Rev. B* **103** 075423.
[186] Pack J., Russell B.J., Kapoor Y., Balgley J., Ahlers J., Taniguchi T., Watanabe K. and Henriksen E.A. 2020. Broken symmetries and Kohn's theorem in graphene cyclotron resonance. *Phys. Rev. X* **10** 041006.
[187] Ando T. 2019. Theory of weak-field magnetoresistance in bilayer graphene. *J. Phys. Soc. Jpn.* **88** 044707.
[188] Avetissian H.K. and Mkrtchian G.F. 2016. Coherent nonlinear optical response of graphene in the quantum Hall regime. *Phys. Rev. B* **94** 045419.
[189] Savostianova N.A. and Mikhailov S.A. 2015. Giant enhancement of the third harmonic in graphene integrated in a layered structure. *Appl. Phys. Lett.* **107** 181104.
[190] Carbotte J.P., LeBlanc J.P.F. and Ashby P.E.C. 2013. Impact of electron-phonon coupling on near-field optical spectra in graphene. *Phys. Rev. B* **87** 045405.
[191] Koshino M. 2013. Stacking-dependent optical absorption in multilayer graphene. *New J. Phys.* **15** 015010.
[192] Zhu J., Badalyan S.M. and Peeters F.M. 2012. Electron-phonon bound states in graphene in a perpendicular magnetic field. *Phys. Rev. Lett.* **109** 256602.
[193] Stauber T., Peres N.M.R. and Castro Neto A.H. 2008. Conductivity of suspended and non-suspended graphene at finite gate voltage. *Phys. Rev. B* **78** 085418.
[194] Mikhailov S.A. 2021. Nonperturbative quasiclassical theory of graphene photoconductivity. *Phys. Rev. B* **103** 245406.
[195] Fatima, Inerbaev T., Xia W. and Kilin D.S. 2021. Magnetic-field-driven electron dynamics in graphene. *J. Phys. Chem. Lett.* **12** 4749.
[196] Noro M. and Ando T. 2016. Weak-field Hall effect in graphene with long-range scatterers. *J. Phys. Soc. Jpn.* **85** 014708.

[197] Klimchitskaya G.L. and Mostepanenko V.M. 2016. Quantum electrodynamic approach to the conductivity of gapped graphene. *Phys. Rev. B* **94** 195405.

[198] Klimchitskaya G.L. and Mostepanenko V.M. 2016. Conductivity of pure graphene: Theoretical approach using the polarization tensor. *Phys. Rev. B* **93** 245419.

[199] Tahir M., Vasilopoulos P. and Peeters F.M. 2015. Magneto-optical transport properties of monolayer phosphorene. *Phys. Rev. B* **92** 045420.

[200] Orlita M., Crassee I., Faugeras C., Kuzmenko A.B., Fromm F., Ostler M., Seyller T., Martinez G., Polini M. and Potemski M. 2012. Classical to quantum crossover of the cyclotron resonance in graphene: A study of the strength of intraband absorption. *New J. Phys.* **14** 095008.

[201] Orlita M., Faugeras C., Plochocka P., Neugebauer P., Martinez G., Maude D.K., Barra A.-L., Sprinkle M., Berger C., de Heer W.A. and Potemski M. 2008. Approaching the Dirac point in high-mobility multilayer epitaxial graphene. *Phys. Rev. Lett.* **101** 267601.

[202] Mak K.F., Shan J. and Heinz T.F. 2010. Electronic structure of few-layer graphene: Experimental demonstration of strong dependence on stacking sequence. *Phys. Rev. Lett.* **104** 176404.

[203] Wang F., Zhang Y., Tian C., Girit C., Zettl A., Crommie M. and Shen Y.R. 2008. Gate-variable optical transitions in graphene. *Science* **320** 206.

[204] Sensale-Rodriguez B., Yan R., Rafique S., Zhu M., Li W., Liang X., Gundlach D., Protasenko V., Kelly M.M., Jena D., Liu L. and Xing H.G. 2012. Extraordinary control of terahertz beam reflectance in graphene electro-absorption modulators. *Nano Lett.* **12** 4518–4522.

[205] Skulason H.S., Gaskell P.E. and Szkopek T. 2010. Optical reflection and transmission properties of exfoliated graphite from a graphene monolayer to several hundred graphene layers. *Nanotechnology* **21** 295709.

[206] Ju L., Wang L., Li X., Moon S., Ozerov M., Lu Z., Taniguchi T., Watanabe K., Mueller E., Zhang F., Smirnov D., Rana F. and McEuen P.L. 2020. Unconventional valley-dependent optical selection rules and landau level mixing in bilayer graphene. *Nat. Commun.* **11** 2941.

[207] Russell B.J., Zhou B., Taniguchi T., Watanabe K. and Henriksen E.A. 2018. Many-particle effects in the cyclotron resonance of encapsulated monolayer graphene. *Phys. Rev. Lett.* **120** 047401.

[208] Nakamura D., Saito H., Hibino H., Asano K. and Takeyama S. 2020. Quantum limit cyclotron resonance in monolayer epitaxial graphene in magnetic fields up to 560 T: The relativistic electron and hole asymmetry. *Phys. Rev. B* **101** 115420.

[209] Nedoliuk I.O., Hu S., Geim A.K. and Kuzmenko A.B. 2019. Colossal infrared and terahertz magneto-optical activity in a two-dimensional Dirac material. *Nat. Nanotechnol.* **14** 756.

[210] Massabeau S., Baillergeau M., Phuphachong T., Berger C., de Heer W.A., Dhillon S., Tignon J., de Vaulchier L.A., Ferreira R. and Mangeney J. 2017. Evidence of Fermi level pinning at the Dirac point in epitaxial multilayer graphene. *Phys. Rev. B* **95** 085311.

[211] Wendler F., Mittendorff M., König-Otto J.C., Brem S., Berger C., de Heer W.A., Böttger R., Schneider H., Helm M., Winnerl S. and Malic E. 2017. Symmetry-breaking supercollisions in Landau-quantized graphene. *Phys. Rev. Lett.* **119** 067405.
[212] König-Otto J.C., Mittendorff M., Winzer T., Kadi F., Malic E., Knorr A., Berger C., de Heer W.A., Pashkin A., Schneider H., Helm M. and Winnerl S. 2016. Slow noncollinear Coulomb scattering in the vicinity of the Dirac point in graphene. *Phys. Rev. Lett.* **117** 087401.
[213] Razado-Colambo I., Avila J., Chen C., Nys J.P., Wallart X., Asensio M.C. and Vignaud D. 2015. Probing the electronic properties of graphene on C-face SiC down to single domains by nanoresolved photoelectron spectroscopies. *Phys. Rev. B* **92** 035105.
[214] Mittendorff M., Wendler F., Malic E., Knorr A., Orlita M., Potemski M., Berger C., de Heer W.A., Schneider H., Helm M. and Winnerl S. 2015. Carrier dynamics in Landau-quantized graphene featuring strong Auger scattering. *Nat. Phys.* **11** 75.
[215] Malko D., Neiss C., Viñes F. and Görling A. 2012. Competition for graphene: Graphynes with direction-dependent Dirac cones. *Phys. Rev. Lett.* **108** 086804.
[216] Orlita M., Faugeras C., Grill R., Wysmolek A., Strupinski W., Berger C., de Heer W.A., Martinez G. and Potemski M. 2011. Carrier scattering from dynamical magnetoconductivity in quasineutral epitaxial graphene. *Phys. Rev. Lett.* **107** 216603.
[217] Bisti V.E. and Kirova N.N. 2011. Coulomb interaction and electron-hole asymmetry in cyclotron resonance of bilayer graphene in a high magnetic field. *Phys. Rev. B* **84** 155434.
[218] Crassee I., Levallois J., Walter A.L., Ostler M., Bostwick A., Rotenberg E., Seyller T., van der Marel D. and Kuzmenko A.B. 2011. Giant Faraday rotation in single- and multilayer graphene. *Nat. Phys.* **7** 48.
[219] Hasan M.Z. and Kane C.L. 2010. *Rev. Mod. Phys.* **82** 3045.
[220] Qi X.L. and Zhang S.C. 2011. *Rev. Mod. Phys.* **83** 1057.
[221] Haldane F.D.M. 1988. *Phys. Rev. Lett.* **61** 2015.
[222] Bernevig B.A., Hughes T.L. and Zhang S.C. 2006. *Science* **314** 1757.
[223] Bernevig B.A. and Zhang S.C. 2006. *Phys. Rev. Lett.* **96** 106802.
[224] Fu L., Kane C.L. and Mele E.J. 2007. *Phys. Rev. Lett.* **98** 106803.
[225] Hsieh D., Qian D., Wray L., Xia Y., Hor Y.S. *et al.* 2008. *Nature* **452** 970.
[226] Hsieh D., Xia Y., Qian D., Wray L., Dil J.H. *et al.* 2009. *Nature* **460** 1101.
[227] Fan Y. and Wang K.L. 2016. *Spin* 6 1640001.
[228] König M., Wiedmann S., Brüne C., Roth A., Buhmann H. *et al.* 2007. Quantum spin Hall insulator state in HgTe quantum wells science. *Science* **318** 766.
[229] Chen Y.L., Analytis J.G., Chu J.-H., Liu Z.K., Mo S.-K. *et al.* 2009. *Science* **325** 178.
[230] Kane C.L. and Mele E.J. 2005. *Phys. Rev. Lett.* **95** 226801.
[231] Kane C.L. and Mele E.J. 2005. *Phys. Rev. Lett.* **95** 146802.
[232] Wu Q., Du L. and Sacksteder IV V.E. 2013. *Phys. Rev. B* **88** 045429.

[233] Wang L.Y., Chang C.Y. and Chu C.S. 2018. *Phys. Rev. B* **98** 235427.
[234] Volovik G.E. 2003. *The Universe in a Helium Droplet* (Oxford: Oxford University Press).
[235] Cayssol J. 2013. *C. R. Phys.* **14** 760.
[236] Vafek O. and Vishwanath A. 2014. *Annu. Rev. Condens. Matter Phys.* **5** 83.
[237] Wehling T.O., Black-Schaffer A.M. and Balatsky A.V. 2014. *Adv. Phys.* **63** 1.
[238] Hatsugai Y. 1997. *J. Phys. Condens. Matter* **9** 2507.
[239] Hasan M.Z. and Kane C.L. 2010. *Rev. Mod. Phys.* **82** 3045.
[240] Qi X.L. and Chang S.C. 2011. *Rev. Mod. Phys.* **83** 1057.
[241] Chiu C.K., Teo J.C.Y., Schnyder A.P. and Ryu S. 2016. *Rev. Mod. Phys.* **88** 035005.
[242] Armitage N.P., Mele E.J. and Vishwanath A. 2018. *Rev. Mod. Phys.* **90** 015001.
[243] Schnyder A.P. and Brydon P.M.R. 2015. *J. Phys. Condens. Matter* **27** 243201.
[244] Castro Neto A.H., Guinea F., Peres N.M.R., Novoselov K.S. and Geim A.K. 2009. *Rev. Mod. Phys.* **81** 109.
[245] Xiao D., Chang M.C. and Niu Q. 2009. *Rev. Mod. Phys.* **82** 1959.

Chapter 2

# A QUASIPARTICLE FRAMEWORK OF SCIENCE RESEARCHES

Chih-Wei Chiu, Hsien-Ching Chung, Chiun-Yan Lin,
Ching-Hong Ho, Lu-Yao Wang, and Ming-Fa Lin

Combining phenomenological models [1], first-principles simulations [2] and experimental observations [3] can significantly advance progress in both basic [4] and applied sciences [5]. The relationships between them may vary in clarity, depending on charge- [6] and spin-based [7] interactions that determine essential properties [8, 9]. A comprehensive understanding of physical, chemical and material mechanisms can be achieved through these three approaches only when the intrinsic interactions are well characterized under normal orbital hybridizations [10] and spin configurations [11]. Current research indicates that graphene-related $sp^2$-bonding systems are ideal for exploring unique quasiparticle phenomena, such as bulk Bernal [12] and rhombohedral [13] graphites, layered graphenes with AAA/ABA/ABC/AAB stackings (Figs. 2.1(a)–(d)) [14–17], coaxial carbon nanotubes/carbon nanotube bundles [18, 19], planar/curved/rolled/folded graphene nanoribbons [20–23] and carbon tori [24]. Their diverse quasiparticle properties, which include geometric [25], electronic [26], magnetic [27], optical [28], transport [29], Coulomb-excitation [30] and mechanical properties [31, 32], can be examined through theoretical predictions [33, 34] and experimental examinations [4, 35], with consistent results.

The success of this strategy is attributed to the well-behaved $\pi$ and $\sigma$ bonding of active C–C chemical bonds on various surface morphologies in the absence of magnetic configurations. However, after chemical

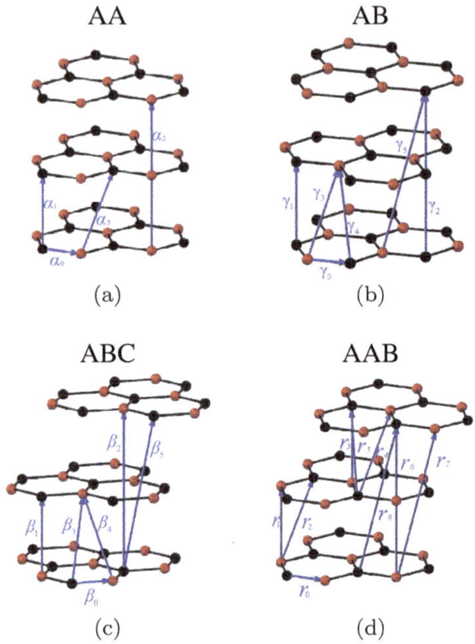


Fig. 2.1 Trilayer graphene systems with (a) AAA, (b) ABA, (c) ABC and (d) AAB stackings.

absorptions [36], adatom/molecule intercalations [37], boundary decorations and guest-atom substitutions [38], the single-/multi-orbital mixings become more complicated. Moreover, Hubbard-like on-site spin Coulomb interactions may exist [39], leading to super-high barriers for the reliable development of qualitative models and experimental verifications. For example, it is challenging to obtain suitable hopping integrals for hydrogenated graphene compounds [40] due to $ssp^3$ in C–H bonds (five-orbital hopping integrals) [40]. In addition, the calculated band structure cannot be tested by angle-resolved photoemission spectroscopy (ARPES) [41] due to non-negligible zone-folding effects [42, 43]. Non-consistent cases among the three approaches frequently occur in most condensed matter systems. Therefore, linking them more efficiently should be a primary focus of present and future studies.

## 2.1 Quasiparticle

Quasiparticles, which are generalized classical particles with one-to-one correspondence between momentum and energy, frequently experience elastic and inelastic scatterings within physical and chemical environments. Their energy spectra and widths, as clearly illustrated in previously published books [44], are very sensitive to the various single- and bi-particle interactions associated with charge and spin density distributions. Such quantized particles consist of the same and/or different quasiparticles. Electrons, phonons and photons, respectively, correspond to the propagating wave packets [45], the coherent atom vibrations [46] and the transverse electric and magnetic fields [47]. Many composite quasiparticles, with comparable properties among/between the same and/or distinct ones, are able to combine under external field perturbations. For example, a 2D plasmon mode in an alkali-doped graphene is the oscillation superposition of all conduction electron screenings and is purely induced by the low-energy $\pi^*$ electron. More interestingly, the theoretical calculations and EELS measurements also present the $\pi$ and $\sigma$ plasmon modes at the middle and high frequencies, respectively [48, 49]. The second and third modes mainly come from the whole $\pi$- and $\sigma$-electronic valence states [50]. The orbital-mediated viewpoints need to be taken into account for the diversified plasmon modes within the active frequency ranges, where their coherent oscillation frequencies strongly depend on the orbital-hybridization strengths [51]. This strategy is very successful in exploring optimal geometric symmetries [52], featured electronic properties [53], diverse magnetic quantization phenomena [54], rich Coulomb/magneto-Coulomb excitations [30, 55], unusual quantum transports [56] and optical/magneto-optical selection rules [57, 58]. Most importantly, certain modified quasiparticles have been proposed to account for the dynamic scattering processes. For example, the stable and intermediate orbital hybridizations of hydrogenated Group-IV materials [40], which are delicately identified from each prominent optical absorption channel through VASP simulations [40], might arise simultaneously, where the orbital–photon composite quasiparticles play a critical role in the vertical transitions [59]. The current quasiparticle framework could be gradually enlarged by generalizing the concepts of hybridized and/or composite quasiparticles, as thoroughly explored in Chapters 3 and 10.

### 2.1.1 *The tight-binding model/the generalized tight-binding model*

The critical mechanisms, the intrinsic interactions of charges and spins [6, 7], the crystal symmetries of chemical bonds, the quantum confinement of edge structure [60] and the external electric [60], magnetic and [60]

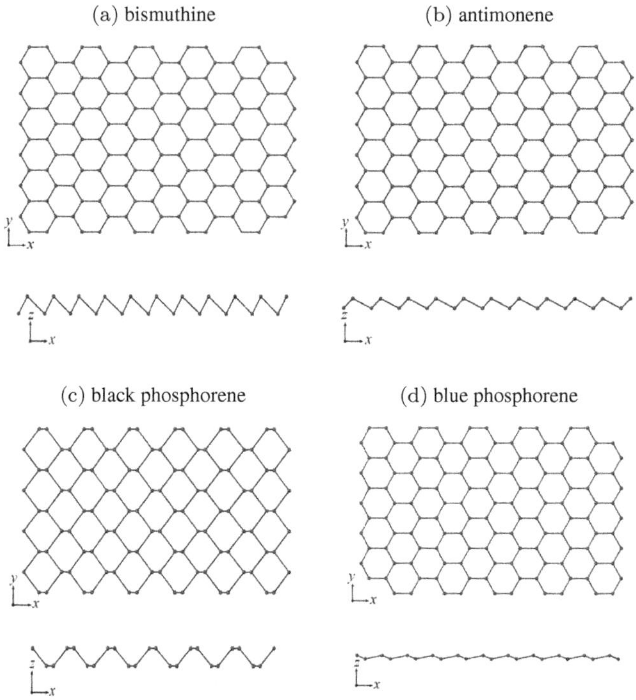


Fig. 2.2 Monolayer (a) top-view graphene, as well as, side-view (b) graphene, (c) silicene, (d) germanene, (e) tinene and (f) plumbene.

mechanical fields [60] can greatly diversify the quasiparticle modes. Such factors might frequently arise in the tight-binding models of distinct condensed matter systems [61]. Group-IV materials (Figs. 2.2(a)–(f)) [62], which possess rich dimensionalities [63], layer numbers [64], stacking configurations [65] and boundary conditions, are outstanding candidates for providing diverse Hamiltonians. The up-to-date theoretical predictions have shown that monolayer graphene/silicene/germanene [66–68] and tinene/plumbene [69, 70], respectively, exhibit the featured low-energy physics through the dominating $\pi$- and $sp^3$-orbital bondings, i.e., their essential properties are clearly revealed through the single- and four-orbital tight-binding models, respectively. Furthermore, single-particle spin–orbital couplings play non-negligible roles in non-graphene buckled systems, especially for heavy tin and lead atoms [71]. The significant couplings between the multi-orbital hybridizations and the spin-dominated interactions have created unusual quasiparticle behaviors, e.g., the unique band structures of tinene and plumbene in terms of bandgap [72], stable valleys [73], energy dispersions [74], band-edge states [75], split energy subbands [71] and crossings/anticrossings near the Fermi level [71]. This strategy could be relatively easily generalized to the layered graphene systems with the $sp^2$ chemical bondings, as shown by the band structures of trilayer graphene with AAA, ABA, ABC and AAB configurations in Fig. 2.3 [76]. However, the opposite is true for the few-/multi-layer

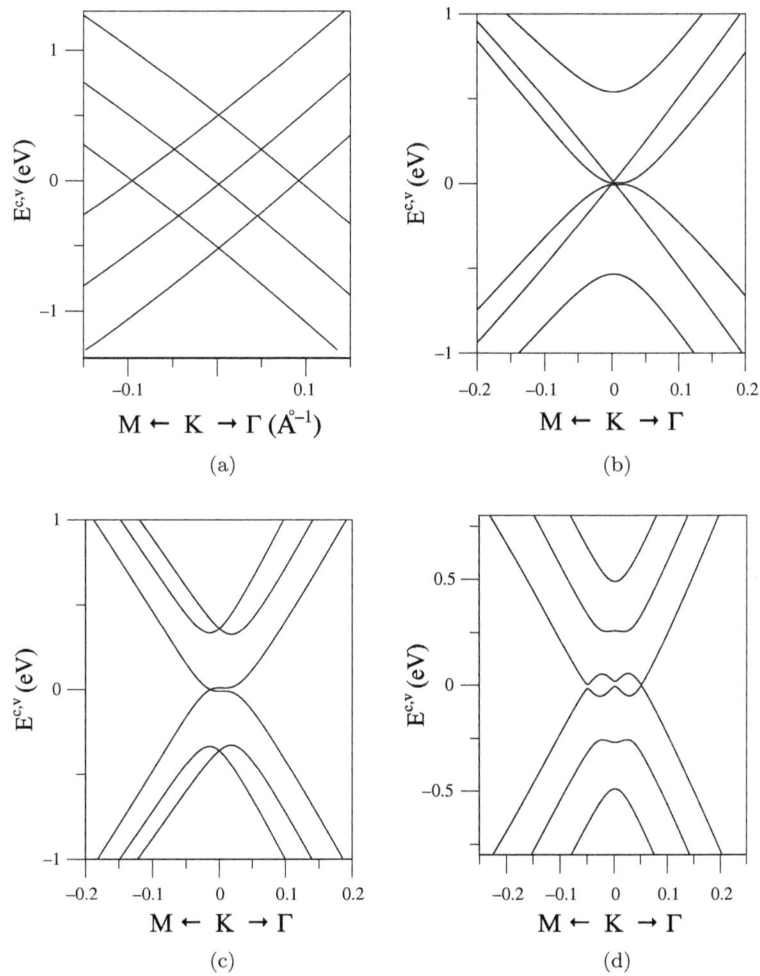


Fig. 2.3 Unusual band structures of (a) AAA-, (b) AAB-, (c) ABC- and (d) AAB-stacked trilayer graphenes.

silicene/germanene/tinene/plumbene. The strong couplings of buckling and stacking, which indicate the intralayer and interlayer $sp^3$ mixings [77], the non-perpendicular $\pi$- and $\sigma$-orbital hybridizations [10], and the complicated spin configurations, lead to very complicated band structures. The previous VASP simulations, which were conducted on sliding bilayer silicene systems, display well-behaved, perturbed, strongly modified and totally destroyed $\pi$- and $\sigma$-electronic states [78]. In addition, the up-to-date tight-binding model cannot characterize the low-lying valence and conduction energy subbands even under high-symmetry stackings, e.g., band structures of AB-bb and AB-bt [78, 79]. Systematic investigations are urgently needed for these emergent materials. In addition, similar studies could be generalized to mainstream cathodes [80], electrolytes [81], separators [82] and anodes [83] of ion-based batteries [84].

It is well known that a 1D hollow carbon nanotube [85]/a 1D graphene nanoribbon [86] could be regarded as a rolled-up/longitudinally cut monolayer graphene in the cylindrical/planar carbon honeycomb structure. The former and the latter, respectively, have periodic and open boundary conditions. Both of them exhibit a number of 1D valence and conduction energy subbands due to the 1D quantum confinement. However, only a cylindrical system displays the translation symmetry along the transverse direction (the azimuthal one [87]) and thus the angular-momentum-dominated band structures [88] and wave functions [89]. These features are directly reflected in the dynamic Coulomb- and electromagnetic-field excitations [90], such as the angular-momentum-decoupled electron–hole pair excitations [91] and plasmon modes [91], as well as the optical selection rules [57]. Such transverse translation symmetry is absent in all graphene nanoribbons with/without hydrogen edge passivations. The nonconservation of momentum transfer leads to super-high barriers to the model formulas, numerical evaluations and delicate analyses. The theoretical predictions on the dielectric responses of graphene-nanoribbon-related materials have been almost absent up to now. How to solve them is one of the focuses of current studies. In general, the pristine $sp^2$-bonding carbon systems do not show the spin-created magnetic configurations. Specifically, the zigzag open edges might create the ferromagnetic and/or anitferromagnetic spin configurations through the VASP simulations/tight-binding models [61, 92], e.g., zigzag 1D graphene nanoribbons [93] and 0D finite-length carbon nanotubes [94]. The Hubbard-like on-site electron–electron interactions are fully identified as a critical mechanism, i.e., the orbital- and spin-dependent intrinsic interactions cooperate/compete to create the diverse quasiparticle spectra and wave functions (the extended, localized and entangled probability distributions) [95]. Similar studies could be generalized to other Group-IV materials, e.g., silicene/germanene/tinene/plumbene [96–99] and nanoribbons under the significant buckling effects. Diversified behaviors are expected to appear among them [96–99].

The external electric and magnetic fields are able to modulate the quasiparticle properties, especially for the rich quantization phenomena [100]. The vector potential of a uniform perpendicular magnetic field leads to a super-large Moire superlattice [101] since the Peierls phases between the position-dependent hopping integrals create an extra period (as detailed in the previous books [100]). The simultaneous considerations of all intrinsic interactions and the perturbed fields have been successfully developed through the generalized tight-binding model, with the multi-orbital hybridizations, the greatly enlarged interlayer atomic interactions due to stackings [65] and layer numbers [64], the spin–orbital couplings [102], gate voltages [103] and magnetic fields [104]. Such a non-perturbed method, which is band on the intralayer and interlayer sublattices, is operated under the crystal symmetries. More interestingly, this delicate model cannot be replaced by the effective-mass approximations [105]. The latter is directly built from the perturbation viewpoint of low-lying simple band structures about the high-symmetry point, so it is unable to deal with the magnetic quantization for the unusual energy dispersions, e.g., the unique Landau levels (LLs) in the partially flat, oscillatory, sombrero-shaped and degenerate energy dispersions in ABC- [16] and AAB-stacked [17] graphene systems. The previous delicate analyses have covered the following unique magneto-electronic properties: (a) a localized probability oscillation of an LL state within the sufficiently wide magnetic length [106], (b) a good quantum number characterized from its zero point, (c) three kinds of LLs with the well-defined, perturbed or undefined zero-point numbers [107], (d) the non-crossing/crossing/anticrossing magnetic-field dependences [107], (e) the specific relations between the quantum number/magnetic-field strength and LL energies (consistent with the pristine energy dispersion relations) [108], several LL subgroups with different dominating sublattices and spin configurations [11], their significant entanglements [109], gate-induced valley–LL splittings [110], rich valley- and spintronics [111, 112] and crystal-symmetry-enriched LLs [113]. The above-mentioned strategy could be generalized to a number of mainstream materials, e.g., the ternary/quaternary lithium oxides in batteries [114] and the lithium-, phosphorus- and sulfur-related allotrope crystals [115].

### 2.1.2 *The modified random-phase approximation/self-energy*

The Coulomb-field perturbations play critical roles in the basic sciences [116] since they can create many-body effects due to the electron–electron interactions in all condensed matter systems [117]. Furthermore, the charge screenings provide complete information about the dramatic transformation between classical and quantum electrodynamics [118, 119]. An electric field generates the charge fluctuations at any time and position and thus the related one. Within the linear superposition, the total electric field/Coulomb potential mainly comes from the multi-pole contribution, in which the electric dipole moment is the dominant one [120]. In general, the electric polarization, i.e., the induced dipole moment per volume, is important in characterizing the dielectric constant of an insulating material [121]. Such charge screening behaviors could be replaced by electron–electron interactions under the particle scattering mechanics. At zero temperature, the occupied electronic states are excited to the unoccupied ones by getting the momentum and frequency from the time- and position-dependent charge densities (the Coulomb-field perturbations along the longitudinal direction) [122]; therefore, the Pauli exclusion principle and the conservation laws of inelastic scatterings need to be included in the model simulations [123]. The dielectric function ($\epsilon(\mathbf{q},\omega)$) [124], being defined in the momentum and frequency spaces [124], is suitable for describing the bare charge response ability. After covering all charge screening effects, the screened response function (the energy loss function) [125] is defined as $Im[-1/\epsilon]$. This is consistent with the high-precision measurements through energy loss spectroscopy (discussed later in detail) [126]. More interestingly, the bare and screened response functions, respectively, are available for thoroughly exploring the electron–hole pair excitations [127] and the coherent plasmon modes [128]. A number of successful Coulomb excitation phenomena have been presented in previously published books/works, such as the diverse electronic excitations in 3D graphite [129], 2D layered graphenes with different stacking configurations and layer numbers [130], 1D coaxial carbon nanotubes [131] and monolayer silicene/germanene [132, 133]. In addition, further studies are urgently needed in the near future for other single-layer emergent materials.

Interestingly, multi-layer/multi-wall graphene-related systems can exhibit the unusual excitation behaviors of simultaneous interlayer and interlayer charge responses. That is, the excited electrons and holes of the intermediate states would emerge in the same and distinct layers. Within the linear Coulomb-field perturbations, the 2D random-phase approximation has been successfully modified to include the intralayer and interlayer atomic interactions [134], as well as electron–electron ones [117]. The band-structure effects, which are due to energy spectra and wave functions, are directly reflected in bare response functions [122] through the sublattice-based tight-binding functions [30]. Furthermore, there exist linear relations among the screened charges and induced Coulomb potentials at any layer [135]. This strategy is also suitable for magneto-electronic excitations (details in Ref. [136]). Through delicate calculations and analyses, the self-consistent 2D Dyson equations are obtained within the linear perturbation viewpoint, leading to well-characterized responsible functions. The layer-dependent tensor of dielectric function (polarization) [137] clearly illustrates the whole charge screenings on all layered structures/coaxial nanotubes [138, 139] and thus indicates the main features of electron–hole pair excitations. Moreover, the Born approximation is available for characterizing the dimensionless energy loss function, which is consistent with the high-resolution inelastic scattering probability of EELS [126]. The calculated results could be verified from the latter.

The modified RPA is very successful in fully exploring the graphene-related systems with the dominating $sp^2$ bondings. The diverse (momentum/angular momentum, frequency)-excitation phase diagrams are clearly revealed in layered graphene systems [140] and coaxial carbon nanotubes [141]. The carbon-$2p_z$ orbitals are sufficient for determining the low- and middle-frequency electronic structures, such as 2D acoustic plasmons [142] and optical $\pi$ plasmons [143]. It is well known that the essential properties are very sensitive to the geometric symmetries and the external fields (gate voltages and magnetic fields) [103, 104]. Stacking configurations of AAA/ABA/ABC/AAB [14–17] and layer numbers are identified to play a critical role in electronic structures and Coulomb excitations, as thoroughly investigated in the previously published books [136]. The featured quasiparticle excitation behaviors cover the whole excitation phase diagrams of single-particle and collective excitations using the EELS intensities [126], the diversified screening responses easily modulated by the intrinsic [144] and external mechanisms [144], the specific (momentum/angular, frequency) ranges of electron–hole pair excitations associated with the band-edge and Fermi-momentum states [145], the emergence/serious Landau dampings/absence of significant plasmon modes [146], the acoustic/optical/irregular charge oscillations through the longitudinal Coulomb-field interactions (the plasma waves with various group velocities; [100]) and the low-/middle-energy coherent charge vibrations due to the Dirac valence and conduction carriers/all $\pi$ valence electrons (the orbital-dependent coherent charge oscillations) [133]. As to the static impurity screenings, the long-range Friedel oscillations, which are created by the Fermi surfaces, are revealed in the real-space total/induced Coulomb potential and screening charge densities [147]. Such unusual charge screenings only appear in metallic/semimetallic graphene systems and carbon nanotubes, while the opposite is true for semiconducting ones. The interesting results will be shown in Chapter 4, where the composite Friedel oscillations are clearly illustrated by the interlayer Coulomb interactions and orbital hybridizations [10]. In addition, these theoretical predictions could be further utilized to understand the residual resistivity at zero temperature [148].

All quasiparticles, electrons, phonons and photons possess finite lifetimes through the frequent elastic and inelastic scatterings in any physical environment. The quasiparticle decay rates can be well characterized by the imaginary part of self-energies (the energy widths in Ref. [149]), in which the whole interaction due to the intrinsic and external mechanisms is included. It is well known that it is impossible to derive exact formulas for the many-particle interactions [150], e.g., the electron–electron Coulomb [151] and electron–phonon interactions [152, 153]. A number of approximate models have been successfully developed for the reliable self-energies, as thoroughly done for 3D/2D/1D electron gases [154–156], metals [157] and doped semiconductor compounds [158]. The screened exchange self-energy, the RPA one [159], is frequently chosen for a standard method about the Coulomb decay rates of electronic valence and conduction states [160]. The excited quasiparticles correspond to the excited conduction electrons/holes above/below the Fermi level for EGSs, and there also exist excited valence holes for others, such as the graphene-related materials with the $\pi$ valence bands. For example, the $sp^2$-dominated systems, without/with the *n*- or *p*-type dopings, exhibit diversified wavevector dependences, covering monolayer graphene/silicene/germanene [66–68] and metallic/narrow-gap/middle-gap single-walled carbon nanotubes [161–163]. More interestingly, the various de-excitation pictures, which are associated with the Pauli exclusion principle [136], the sufficient decay energy [164], the momentum/angular momentum transfer consistent with the translation/rotation symmetry, are thoroughly identified to achieve the intermediate excitation configurations and the unusual decay rate [163, 164]. 2D graphene, which is totally different from 2D EGS [76], has shown unique behaviors at the Fermi level, the Dirac point and large wavevectors. There are certain important differences among graphene, silicene and germanene in Coulomb excitations and decay rates [130, 132, 133]. Moreover, these calculated results agree with the recent ARPES measurements of spectral widths [165–167].

More interestingly, the quasiparticle self-energies in few-layer and coaxial systems have been successfully solved by using the layer- and sublattice-projected tight-binding functions [168]. The screened exchange self-energy, which is developed through layer decomposition, is consistent with the generalized tight-binding model [169]. Its unusual behaviors directly reflected the geometric symmetries, the featured band structures and the unique electronic excitations. Within this reliable strategy, all available Coulomb deexcitation channels cover the intralayer and interlayer electron–hole pairs and plasmon modes. The previous book [136], which thoroughly explored the double-walled armchair carbon nanotubes of $C_{5h}$, $D_{5h}$ and $S_{5h}$ (Fig. 2.4), clearly shows the rich Coulomb decay phenomena. The main features cover the stacking-enriched quasiparticle lifetimes, the excited conduction electrons, valence electrons/conduction holes and valence holes in one/another pair of modified linear energy subbands [170], the excited conduction electrons and valence holes in a linearly intersecting band structure of a single-walled armchair nanotube [171], the intra- and inter-valley de-excitation channels, the highly anisotropic behaviors, the sensitive wavevector dependences near and away from the Fermi level/the Dirac point and the totally different from 1D cylinder electron gases [172]. Specifically, how to overcome the difficulties of numerical calculations and achieve a concise decay picture needs to be strongly investigated in future research. Such a suitable strategy will be utilized to thoroughly explore the diversified inelastic Coulomb decays in few-layer graphene systems, as clearly illustrated in Chapters 3, 4 and 5. More interestingly, this method could be generalized to the quasiparticle energy spectra of bilayer silicene/germanene/tinene/plumbene emergent materials [67–70].

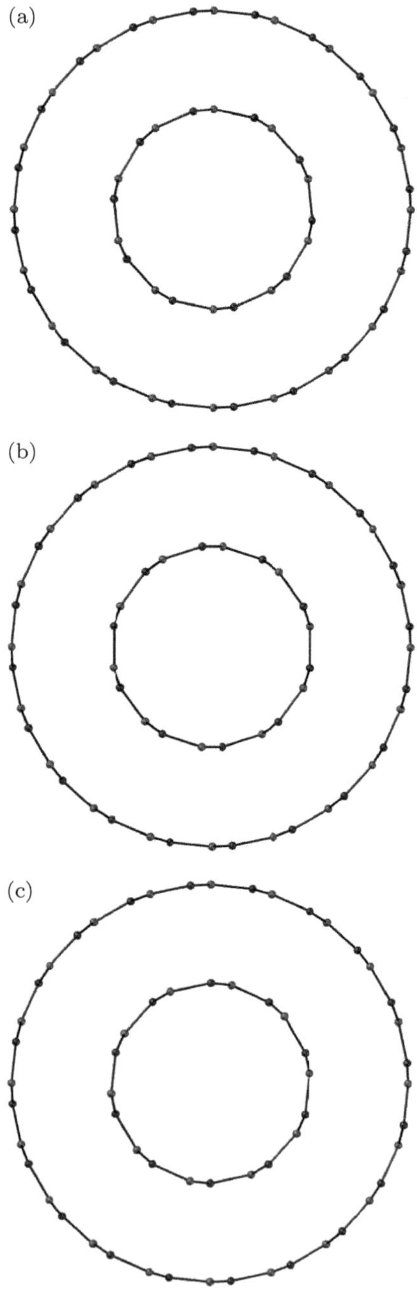


Fig. 2.4 The double-walled armchair carbon nanotubes of (a) $C_5h$, (b) $D_5h$ and (c) $S_{5h}$.

### 2.1.3 *The dynamic/static Kubo formulas*

The composite quasiparticles of significant photon–electron couplings [173–175], as clearly illustrated in the previously published books [176–178], frequently survive through the electromagnetic-field perturbations. Since charge carriers naturally experience time- and position-dependent external electric fields, there exist strong and rapid fluctuations in charge density distributions [206, 207]. The induced current density and the total electric field have a linear relation with the Kubo formula [180], which has been verified by several optical examinations [181]. The dynamic charge screening behaviors could be regarded as the scattering processes of classical particles. That is, the occupied valence states are vertically excited to the unoccupied conduction ones after getting almost zero momentum and the finite energies of photons. When the Pauli exclusion principle of the identical electrons is considered in the scattering probability [182, 183], the imaginary part of the dielectric function is obtained from the Fermi golden rule [205] or the linear Kubo formula [180]. Next, its real part is obtained through principal-value integration because of a physical event (the Kramer–Kroning relations) [184, 185]. It is well known that the special structures in the imaginary- and real-part dielectric functions only reflect the zero joint density of states (the vanishing group velocity/van Hove singularities) [186, 187]. The single-particle optical excitations would be transferred into the many-particle ones with all the whole charge screening effects only if the external electric fields were replaced by the total ones. The energy loss functions are very suitable for characterizing the coherent charge oscillations and their coexistence with electron–hole pair excitations (the Landau damping phenomenon) [188]. Such optical quasiparticle frameworks have been successfully identified by the systematic investigation of graphene-related systems, covering AA- [208–210], AB- [211, 212] and ABC-stacked [213] bulk graphite materials; the layer-number- [214–216], stacking- [217, 218], doping- [219–221], gate- [222, 223] and magnetic-field-enriched [224, 225] graphenes; radius-, chirality- and multi-wall-related carbon nanotubes [226, 232]; edge- and width-dominated graphene nanoribbons [233–239]; and carbon tori [240, 241]. The rich and unique optical properties are obviously shown by the numbers, frequencies, forms and intensities of the featured excitation peaks, for which the concise physical pictures are responsible. The featured magneto-optical properties are fully revealed in bilayer sliding graphene systems with the various stacking configurations [195, 217, 242–244] and trilayer

AAA/ABA/ABC/AAB stackings [245]/[246, 247]/[248–250]/[251]. The current studies show that only the direct linking of the dynamic Kubo formula (Fermi's golden rule) and the generalized tight-binding model [189] is able to deal with the vertical transitions of the discrete LLs because of the extensive numerical calculations pertaining to the mageto-electronic states [190]. The development of the exact diagonalization method and the gradient approximation for the electric dipole moment are responsible for the reliable evaluations. The well-defined, perturbed and undefined LLs, surviving in sliding bilayer structures, respectively, display the normal, extra and vanishing magneto-optical selection rules [191, 192] through the delta-function-like symmetric peaks. All magneto-absorption peak structures in trilayer systems could be classified into nine categories, in which there are three intra-subgroup and six inter-subgroup absorption spectra. Furthermore, the unusual structures might take the double-peak and/or pair forms. Such results mainly arise from the magnetic quantization phenomena of the unique energy dispersions. As a result, the close relations among band structures, magnetic states and magneto-optical properties/magneto-Coulomb excitations are clearly established within a unified quasiparticle framework. The various composite quasiparticles are created by the changes in the physical/chemical environments [260–262].

The quantum Hall transports present another composite quasiparticle phenomenon through the various phases of the magnetic hopping integrals (the LLs) [193, 194] and their scattering selection rules [195]. The direct combination of the generalized tight-binding model (details discussed earlier in Section 2.1.1 and the static Kubo formula) is able to verify the high-resolution measurements of the bias-voltage- and magnetic-field-dependent Hall conductivities of the graphene-related systems, such as those measurements of monolayer [252, 253], AB bilayer [254, 255] and ABA/ABC trilayer [256, 257]/[258, 259]. In addition, the effective-mass approximation cannot delicately evaluate the Hall effects since the rough magnetic quantization is unable to get the reliable LL wave function. A suitable strategy is used to successfully explore the inter-LL scattering processes during the static carrier transport processes. The initial and final magnetic states are thoroughly examined to satisfy the magnetic selection rules, as clearly illustrated for the magneto-optical excitations [196, 197]. The symmetric or antisymmetric probability distributions of the initial and final LLs dominate the scattering amplitudes and thus dominate their existence. The LL degeneracy is directly reflected in the values of electrical conductivities [198]. Furthermore, the complicated LL energy spectra have strong effects on the sensitive dependencies of external perturbations [199]. The current methods could be easily generalized to layered graphene materials; however, they would face serious challenges in other complex systems, such as chemically modified systems and other Group-IV elements with significant couplings of stackings and bucklings [200–202]. The urgent requirements are to establish the complicated intrinsic interactions due to the diversified hopping integrals [203, 204].

### 2.1.4 *Topological Dirac fermions*

The concept of quasiparticles [263, 264], first proposed by Lev Landau in the early days and cast later in the theory of interacting Fermi liquids [265, 266], is one of the most fundamental concepts in condensed matter physics. After Landau's fermionic quasiparticles, other kinds of quasiparticles in condensed matter, such as bosonic and anyonic quasiparticles, and various composites governed by different statistics, were predicted and realized successively. Recently, physicists have found conclusive evidence for Abelian anyons, which are central to the fractional quantum Hall effect, a topological phenomenon in the many-body regime [267].

In the fermionic single-particle regime, an essential assumption is noted [263, 264]: About the Fermi surface, strongly interacting elementary fermion excitations can show one-to-one correspondence with weakly interacting fermionic quasiparticle excitations, with the Coulomb interactions between elementary fermions incorporated into certain parameters, e.g., effective mass. The parameters appear to be phenomenological and often deviate from the true values dramatically [263, 264]. However, they are specified in terms of true microscopic scattering amplitudes. A desirable aspect of this is the fact that momentum and spin are kept unchanged because translation symmetry is respected by interactions and spin is intrinsic. Hence, quasiparticles are labeled by the wavevectors $\mathbf{k}$ and spin $\mathbf{s}$ ($s = \pm 1/2$ for orientations of electrons or holes). In metals and other lattice systems, momentum reduces to crystal momentum and the wavevectors $\mathbf{k}$ are discrete rather than continuous. Moreover, the single-particle states are additionally labeled by the band indices $n$. Presumably, the excited quasiparticles occupy those single-particle states, which are not strictly the eigenstates of the weakly interacting system, but closely approximate them for a long period if they are in proximity to the Fermi surface. This is reasonable since, at low temperatures, only a few quasiparticles are excited with low energy, so they interact one another rarely. In a similar sense, the independent-particle approximation is known to work well, say, for normal metals at a temperature of $T \ll T_F \sim 10^5$ K, with $T_F$ being the Fermi temperature, where only low-lying excited states of noninteracting particles are involved [268]. Thereby, excited quasiparticles behave as if they were independent particles to some extent, while Coulomb interactions, though weak, should be treated in a self-consistent way because, in this theory, only the average interaction is considered [264]. The density parameter $r_s = 3/4\pi n(\mathbf{r})$ is a common means of measuring the ratio of Coulomb potential to kinetic energy [269]. It is exciting to note that the single-particle approximation can be well used even for $r_s > 1$ [263, 264]. For example, the mean-field theory can be properly used in the single approximation for metals [264]. That is why and how the single-particle approximation works for fermionic quasiparticles.

A category of fermionic quasiparticles is frequently encountered in narrow-gap semiconductors or semimetals with the Fermi surfaces sitting in the bandgap or at the band touching points [270–273]. Such quasiparticles are described by the Dirac equation and are termed Dirac fermions, although there is some distinction from elementary Dirac fermions in a quantum vacuum, such as the presence of fermion doubling and the absence of Lorentz invariance [271–274]. Consider quasiparticles described by the Schrödinger equation in a lattice with the Hamiltonian $H = H^0 + H^{\text{int}}$, where $H^0$ and $H^{\text{int}}$ are the noninteracting and weakly interacting Hamiltonians, respectively. That is, near the Fermi surface, an effective Hamiltonian $H^0_{\text{eff}}$ can be derived in association with the Dirac equation. This condensed matter analog of Dirac fermions possesses topology in a more feasible way since the Bloch wave functions distributing in momentum can induce a gauge and lead to a curvature field [275]. Along with $H^{\text{int}}$ weakly interacting Hamiltonian, spontaneous symmetry breaking could occur and peculiar topological phases could show up [276, 277]. For a comparison, it is noted that the topology of elementary Dirac fermions in quantum could only take place when certain gauge fields come into coupling with symmetry breaking due to a quantum anomaly [278]. In addition to Dirac fermions, there are also Weyl fermions and Majorana fermions, known as the variants, appearing in relevant systems [271, 274, 279, 280].

## 2.2 First-Principles Simulations

The first-principles method can fully explore most condensed matter systems, except for giant Moire superlattices due to drawbacks [281], deformations [282], chemical modifications [283] and modulated electric and magnetic fields [284]. It needs to solve the Hamiltonian diagonalization with the single- and bi-particle interactions, respectively, arising from the crystal potential and the electron–electron Coulomb fields [284]. Specifically, the latter covers the many-body exchange and correlation energies [150], which are very sensitive to the position-dependent charge density distributions, as indicated by the flowchart of the VASP simulations shown in Fig. 2.5. This term is closely related to the single-/multi-orbital hybridizations of chemical bonds. That their concise physical/chemical pictures are achieved from the consistent relevant quantities

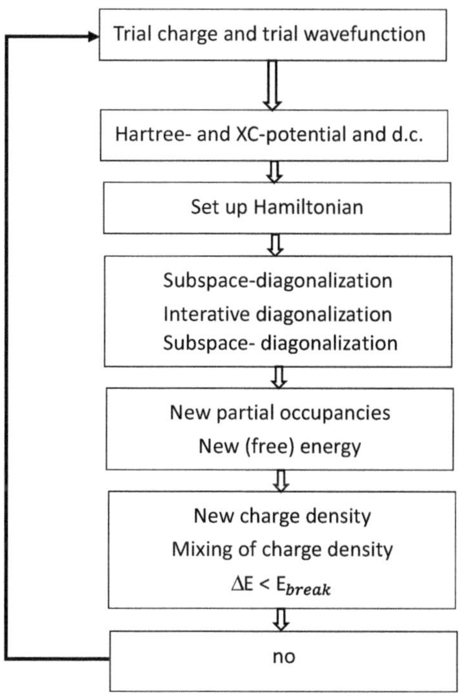


Fig. 2.5 A flowchart depicting the VASP simulations.

is one of the focuses of the study. Also, note that the spin–orbital couplings (the single-particle forms) [284] and the on-site spin-related Coulomb interactions (the Hubbard-like ones) [150] are included in the VASP calculations for the magnetic configurations, e.g., the significant mechanisms in monolayer silicene/germanene/tinene/plumbene and zigzag graphene nanoribbons, respectively [285]. After the reliable determination of a ground state energy, an optimal crystal structure presents high translation and rotation symmetries [177], such as AB- and ABC-stacked graphites, AAA-, ABA-, ABC- and AAB-stacked trilayer graphenes [286], $C_5/D_{5h}/S_{5h}$ double-walled armchair carbon nanotubes (Fig. 2.4) [287] and planar/curved/folded/scrolled graphene nanoribbons [286]. More interestingly, layered Group-IV systems and Group-IV–IV compounds, with various stacking configurations and buckling structures simultaneously, come under the current systematic investigations.

The charge- and spin-dominated interactions, as clearly illustrated in previously published books [286], are greatly diversified by the various atom and orbital dependencies. The significant quasiparticle viewpoints are delicately developed through thorough examinations of physical and chemical environments, covering the position-dependent chemical bonds in a crystal structure, the atom-, orbital- and spin-determined energy spectra and wave functions, the spatial charge/spin density distributions, the atom- and orbital-decomposed magnetic moments and the atom-, orbital- and spin-projected density of states [286]. The successful simulations have indicated that the merged/split van Hove singularities are responsible for the single-/multi-orbital hybridizations/the specific spin configurations. There are more active orbital hybridizations and spin configurations, which are initiated by the following different cases: (I) the geometric symmetries of layer numbers/stacking configurations/bucklings/deformations/finite-size quantum confinements/foldings/scrollings/dimensionalities, (II) the multi-element syntheses in creating various compounds, (III) an unusual heterojunction between sample and substrate, (IV) gate voltages and (V) the chemical modifications about the significant adatom adsorptions, guest-atom substitutions, atom/molecule intercalations and open edge decorations. The reliable VASP simulations have successfully illustrated the diverse orbital mixings and spin arrangements, e.g., the non-perpendicular $\pi$ and $\sigma$ bondings in pristine non-graphene Group-IV systems and the $sp^3$ bondings in hydrogenated ones. More interestingly, the atom- and orbital-determined ferromagnetic, antiferromagnetic and nonmagnetic configurations [Refs] frequently arise in transition-/rare-earth-metal compound materials and silicene/germanene-metal substrate heterojunctions through delicate calculations and analyses.

Importantly, the optical excitations of the VASP simulations, with/without excitonic effects [286], are capable of providing the new composite quasiparticles and the significant couplings of orbitals and photons. This strategy is currently under investigation. When a condensed matter system is perturbed by an electromagnetic field, the strong absorption structures mainly arise from the joint van Hove singularities with finite dipole matrix elements [150]. The photon-created excited valence holes and conduction electrons might be bound together by the attracted Coulomb potentials, in which all charge screening effects would determine the existence of stable or quasi-stable bound states. In general, the excitonic effects can create new absorption structures below or near the threshold frequency and even change the numbers, frequencies, forms and intensities of the single-particle excitation peaks. These many-body phenomena frequently appear in semiconducting/insulating materials with sufficiently large bandgaps, e.g., in fully hydrogenated Group-IV systems, but not in narrow-gap or semi-metallic pristine materials. Whether the excitonic effects are easily observed in the experimental measurements strongly depends on the carrier effective mass and static screening ability, according to the hydrogen-like picture. The featured optical excitations are achieved from the VASP calculations on the frequency- and polarization-dependent dielectric functions, screened response functions (energy loss functions in Ref. [286]), reflectance spectra and absorption coefficients. Both stable and intermediate orbital hybridizations, which survive during the dynamic vertical transitions between the initial occupied and final unoccupied states, are successfully proposed under an expanded quasiparticle framework, e.g., the diverse optical excitations in double- and single-side full hydrogenations of Group-IV materials (the close relations among the geometric, electronic and optical properties illustrated in Figs. 2.6–2.9). These viewpoints will be very useful in fully exploring the single- and many-particle optical properties and thus largely enhancing their potential applications.

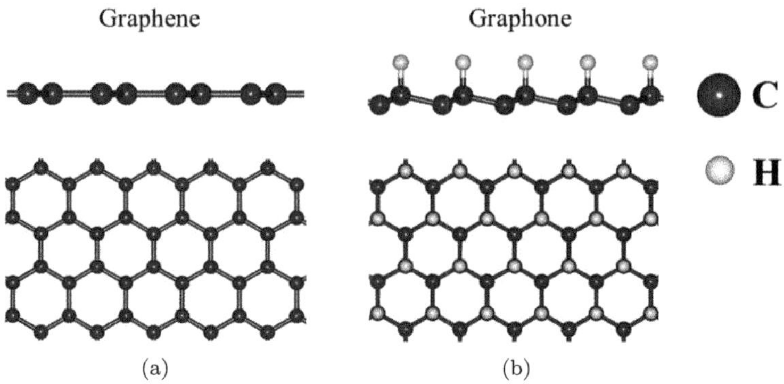


Fig. 2.6 The optimal geometries for (a) graphene and (b) graphone.

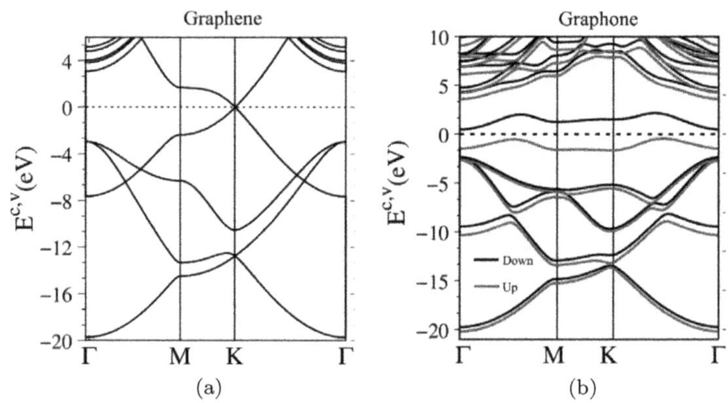


Fig. 2.7 Electronic energy spectra of (a) graphene and (b) graphone.

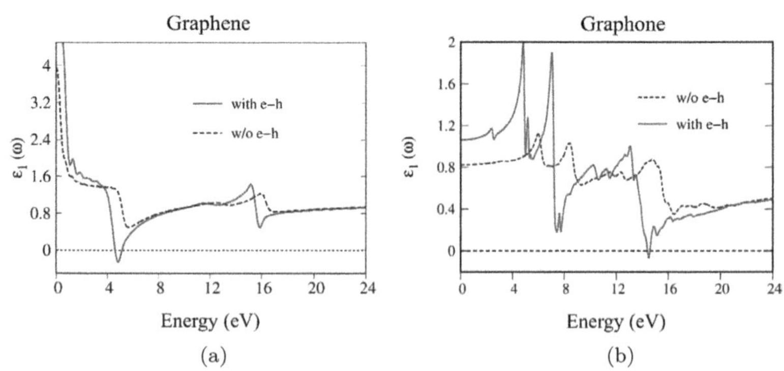


Fig. 2.8 The real-part dielectric functions for (a) graphene and (b) graphone in the presence/absence of excitonic effects (denoted by the solid/dashed curves).

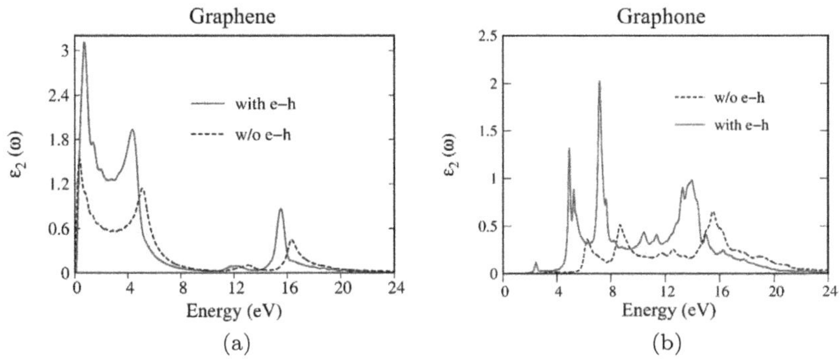


Fig. 2.9 Similar plots as in Fig. 2.8 but shown for the imaginary-part dielectric functions.

## 2.3 Experimental Measurements

The close relationships among the high-precision experiments, the first-principles simulations and the phenomenological models are responsible for much progress in the basic and applied sciences. Generally, scanning tunneling spectroscopy/reflection low-energy electron diffraction/tunneling electron microscopy/X-ray (STM, Fig. 2.10 [288]), ARPES (Fig. 2.11 [290]), RLEED [289], electron energy loss spectroscopy (EELS, Fig. 2.12 [289]) and reflectance/absorption/transmission/photoluminescence spectroscopies (Fig. 2.13 [289]) are, respectively, available for thoroughly exploring the main features of crystal symmetries, occupied energy spectra and quasiparticle lifetimes [150], van Hove singularities with zero group velocities above/below the Fermi levels and the single- and many-particle vertical transitions through the excitonic effects. More interestingly, the composite quasiparticles due to the significant couplings of the perturbed and intrinsic fields are very suitable for characterizing the diversified response phenomena, as clearly demonstrated for the emergent materials in the previously published books [177].

The X-ray patterns, which correspond to the earliest experimental observations [291], are frequently utilized to identify the bulk crystal structures, even with many atoms/molecules in a primitive unit cell. Each incident photon beam experiences elastic scattering events with the periodic charge density distribution associated with the significant chemical bonds. The linear superposition of neighboring ones will determine the diffraction intensity through the conservation of momentum transfer: the change

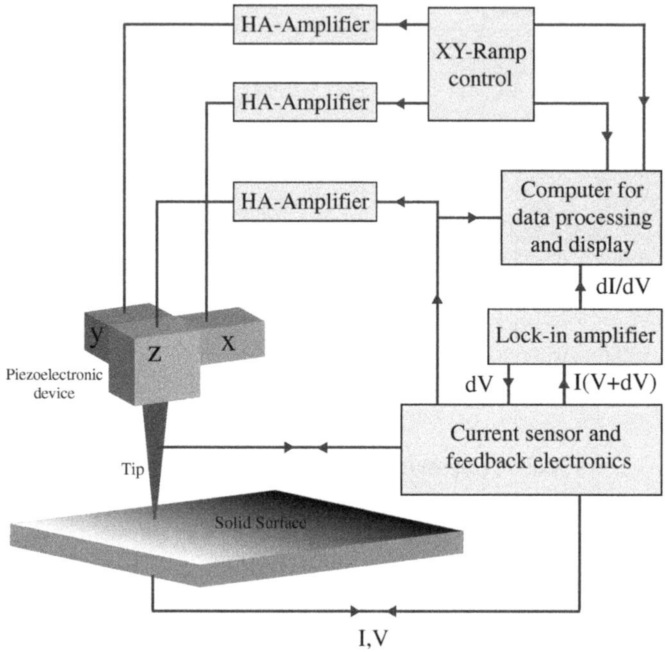


Fig. 2.10 Scanning tunneling microscopy.

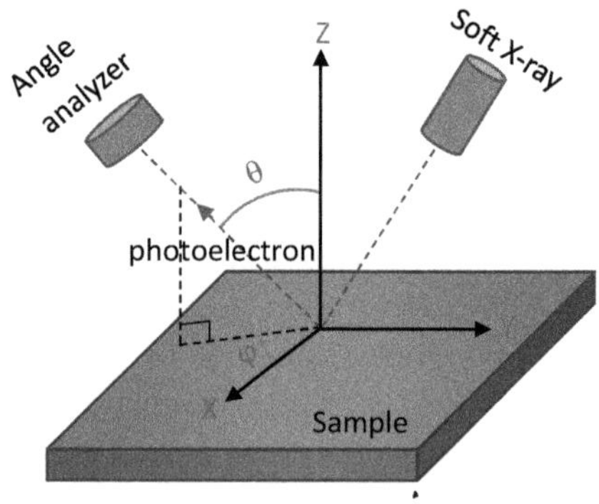


Fig. 2.11 Angle-resolved photoemission spectroscopy.

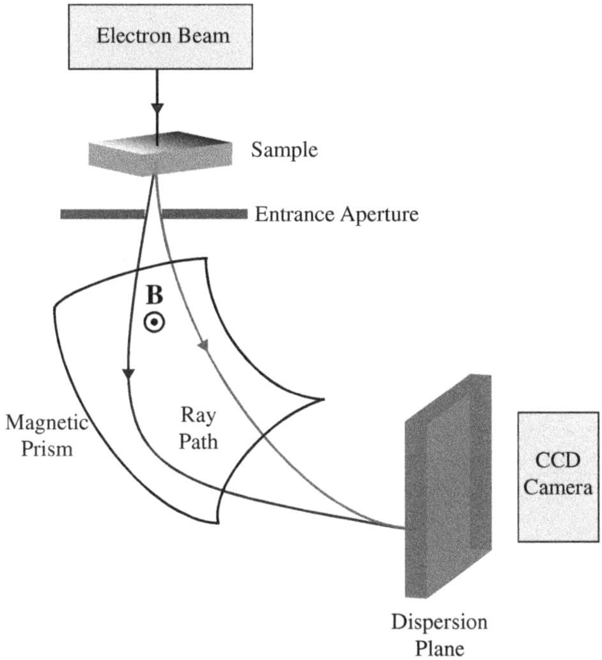


Fig. 2.12 Reflection electron energy loss spectroscopy.

in the initial and final photon wavevectors equals the reciprocal lattice vectors. Notably, the composite quasiparticles of photons and electrons account for the maximum diffraction peaks. Recently, X-ray techniques have been frequently utilized to identify the initial and final saturated crystal structures of cathode/electrolyte/anode materials in ion-based batteries, which are very useful in understanding their performances and commercial applications. For example, X-ray measurements have been successfully done for LiFeO/LiCoO/LiNiO, LiSiO/LiGeO/LiSnO and LiTiO in lithium-ion-related batteries, respectively. However, it should present a number of continuous intermediate structures mainly arising from the lithium-ion transports (the intercalations or de-intercalations during the rapidly charging/discharging processes). How to exactly simulate the time- and position-dependent intermediate configurations among the

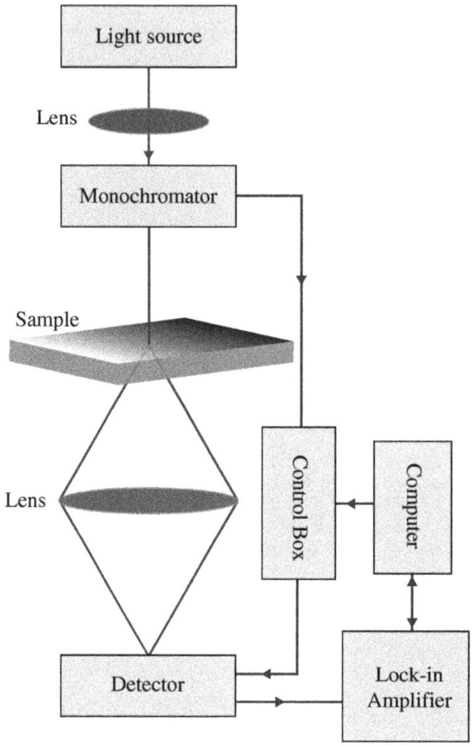


Fig. 2.13 Photoluminescence optical spectroscopy.

three core components is a very interesting research topic. This requires full cooperation between theoretical and experimental studies.

## References


[1] Foulkes W.M.C. and Haydock R. 1989. Tight-binding models and density-functional theory. *Phys. Rev. B* **39** 12520–12526.

[2] Ahuja R., Auluck S., Wills J.M., Alouani M., Johansson B. and Eriksson O. 1997. Optical properties of graphite from frst-principles calculations. *Phys. Rev. B* **55** 4999.

[3] Cheng C.M., Hsu C.J., Peng J.L., Chen C.H., Yuh J.Y. and Tsuei K.D. 2015. Tightbinding parameters of graphite determined with angle-resolved photoemission spectra. *Appl. Surf. Sci.* **354** 229–234.

[4] Zhang Y., Tan Y.W., Stormer H.L. and Kim P. 2005. Experimental observation of the quantum Hall effect and Berry's phase in graphene. *Nature* **438** 201.

[5] Zhang X., Rajaraman B.R.S., Liu H. and Ramakrishna S. 2014. Graphene's potential in materials science and engineering. *RSC Adv.* **4** 28987.
[6] Lee C.H., Yang C.K., Lin M.F., Chang C.P. and Su W.S. 2011. Structural and electronic properties of graphene nanotube-nanoribbon hybrids. *Phys. Chem. Chem. Phys.* **13** 3925–3931.
[7] Konschuh S., Gmitra M. and Fabian J. 2010. Tight-binding theory of the spin-orbit coupling in graphene. *Phys. Rev. B* **82** 245412.
[8] Mele E.J. 2010. Commensuration and interlayer coherence in twisted bilayer graphene. *Phys. Rev. B* **81** 161405.
[9] Son Y.W., Choi S.M., Hong Y.P., Woo S. and Jhi S.H. 2011. Electronic topological transition in sliding bilayer graphene. *Phys. Rev. B* **84** 155410.
[10] Sreeprasad T.S. and Berry V. 2013. How do the electrical properties of graphene change with its functionalization? *Small* **9** 341–350.
[11] Chang S.L., Wu B.R., Wong J.H. and Lin M.F. 2014. Configuration-dependent geometric and electronic properties of bilayer graphene nanoribbons. *Carbon* **77** 1031–1039.
[12] Lai Y.H., Ho J.H., Chang C.P. and Lin M.F. 2008. Magnetoelectronic properties of bilayer Bernal graphene. *Phys. Rev. B* **77** 085426.
[13] Ho C.H., Ho Y.H., Liao Y.Y., Chiu Y.H., Chang C.P., Lin M.F. 2012. Diagonalization of Landau level spectra in rhombohedral graphite. *J. Phys. Soc. Jpn.* **81** 024701.
[14] Lin C.Y., Chen R.B., Ho Y.H. and Lin M.F. 2018. *Electronic and Optical Properties of Graphite-Related Systems* (Boca Raton, FL: CRC Press).
[15] Lin C.Y., Lin M.C., Ho C.H., Wu J.Y. and Lin M.F. 2020. Unusual electronic excitations in ABA trilayer graphene. *Sci. Rep.* 11106.
[16] Lu C.L., Chang C.P., Huang Y.C., Ho J.H., Hwang C.C. and Lin M.F. 2007. Electronic properties of AA- and ABC-stacked few-layer graphites. *J. Phys. Soc. Jpn.* **76** 024701.
[17] Do T.N., Lin C.Y., Lin Y.P., Shih P.H. and Lin M.F. 2015. Configuration-enriched magnetoelectronic spectra of AAB-stacked trilayer graphene. *Carbon* **94** 619–632.
[18] Yannouleas C., Bogachek E.N. and Landman U. 1996. Collective excitations of multishell carbon microstructures: Multishell fullerenes and coaxial nanotubes. *Phys. Rev. B* **53** 10225.
[19] Lin M.F., Shyu F.L. and Chen R.B. 2000. Optical properties of well-aligned multiwalled carbon nanotube bundles. *Phys. Rev. B* **61** 14114–14118.
[20] Gomez C.V., Pisarra M., Gravina M., Pitarke J.M. and Sindona A. 2016. Plasmon modes of graphene nanoribbons with periodic planar arrangements. *Phys. Rev. Lett.* **117** 116801.
[21] Li T.S., Lin M.F., Lin C.Y., Chang S.C. and Yang S.P. 2013. Electronic properties of curved graphene nanoribbons. *Synth Met* **171** 7–14.
[22] Viculis L.M., Mack J.J., Kaner R.B. 2003. A chemical route to carbon nanoscrolls. *Science* **299** 1361.
[23] Chang S.L., Wu B.R., Yang P.H. and Lin M.F. 2016. Geometric, magnetic and electronic properties of folded graphene nanoribbons. *RSC Adv.* **6** 64852–64860.

[24] Lin M.F. 1998. Low-frequency optical excitations in zigzag carbon tori. *Phys. Rev. B* **58** 3629–3632.

[25] Ho Y.H. and Lin M.F. 2008. Effects of geometric structures and temperature on electronic properties of double-walled armchair carbon nanotubes. *Diam. Relat. Mat.* **17** 1550–1553.

[26] Nguyen D.K., Nguyen T.T., Tran N.T.T., Lin M.F. 2018. Diverse electronic and magnetic properties of chlorination-related graphene nanoribbons. *Sci. Rep.* **8** 17859.

[27] Wu J.Y., Ho Y.H. and Lin M.F. 2010. Magnetic-modulation effects in bilayer Bernal graphene. *J. Phys. Chem. C* **114** 11940–11945.

[28] Shyu F.L. and Lin M.F. 1999. Pi-electronic excitations in multiwalled carbon nanotubes. *J. Phys. Soc. Jpn.* **68** 3806–3809.

[29] Li T.S., Lee C.H. and Lin M.F. 2007. Electron transport in double-walled carbon nanotubes. *Eur. Phys. J. B* **60** 45–50.

[30] Lin C.Y., Lee M.H. and Lin M.F. 2018. Coulomb excitations in trilayer ABC-stacked graphene. *Phys. Rev. B* **98** 041408(R).

[31] Yan W., He W.Y., Chu Z.D., Liu M., Meng L., Dou R.F. *et al.* 2013. Strain and curvature induced evolution of electronic band structures in twisted graphene bilayer. *Nat. Commun.* **4** 2159.

[32] Wong J.H., Wu B.R. and Lin M.F. 2012. Strain effect on the electronic properties ofsingle layer and bilayer graphene. *J. Phys. Chem. C* **116** 8271.

[33] Gusynin V.P., Sharapov S.G. and Carbotte J.P. 2007. Anomalous absorption line in the magneto-optical response of graphene. *Phys. Rev. Lett.* **98** 157402.

[34] Polini M., Asgari R., Borghi G., Barlas Y., Pereg-Barnea T. and MacDonald A.H. 2008. Plasmons and the spectral function of graphene. *Phys. Rev. B* **77** 081411.

[35] Ahn S., Hwang E.H. and Min H. 2014. Inelastic carrier lifetime in a coupled graphene/electron-phonon system: Role of plasmon-phonon coupling. *Phys. Rev. B* **90** 245436.

[36] Tran N.T.T., Dahal D., Gumbs G., Lin M.F. 2017. Adatom doping-enriched geometric and electronic properties of pristine graphene: A method to modify the band gap. *Struct. Chem.* **28** 5 1311–1318.

[37] Lin M.F., Huang C.S. and Chuu D.S. 1997. Plasmons in graphite and stage-1 graphite intercalation compounds. *Phys. Rev. B* **55** 13961–13971.

[38] Lv R., Li Q., Botello-Méndez A.R., Hayashi T., Wang B., Berkdemir A., Hao Q., Elías A.L., Cruz-Silva R. and Gutiérrez H.R. 2012. Nitrogen-doped graphene: Beyond single substitution and enhanced molecular sensing. *Sci. Rep.* **2** 586.

[39] Hubbard J. 1958. The description of collective motions in terms of many-body perturbation theory. II. The correlation energy of a free-electron gas. *Proc. R. Soc. Lond.* **14** 336–352.

[40] Huang H.C., Lin S.Y., Wu C.L., Lin M.F. 2016. Configuration- and concentration-dependent electronic properties of hydrogenated graphene. *Carbon* **103** 84–93.

[41] Kordyuk A.A. 2014. Arpes experiment in fermiology of quasi-2d metals. *Low Temp. Phys.* **40** 286.

[42] Barker A.S. *et al.* 1978. Study of zone-folding effects on phonons in alternating monolayers of GaAs-AlAs. *Phys. Rev. B* **17** 3181.
[43] Sato K., Saito R., Cong C., Yu T. and Dresselhaus M.S. 2012. Zone folding effect in Raman G-band intensity of twisted bilayer graphene. *Phys. Rev. B* **86** 125414.
[44] Pham H.D., Dien V.K., Hanh T.T.T., Tran N.T.T., Ho C.H. and Lin M.F. 2021. *Diverse Quasiparticle Properties of Emergent Materials* (IOP Publishing).
[45] Dolcini F., Iotti R.C., Montorsi A. and Rossi F. 2016. Photoexcitation of electron wave packets in quantum spin Hall edge states: Effects of chiral anomaly from a localized electric pulse. *Phys. Rev. B* **94** 165412.
[46] Kuznetsov A.V. and Stanton C.J. 1994. Theory of coherent phonon oscillations in semiconductors. *Phys. Rev. Lett.* **73** 3243.
[47] Zeilinger A., Weihs G., Jennewein T. and Markus Aspelmeyer M. 2005. Happy centenary, photon. *Nature* **433** 230–238.
[48] Lin M.F. and Chuu D.S. 1998. The pi plasmons in carbon nanotube bundles. *Phys. Rev. B* **57** 10183–10187.
[49] Chiu C.W., Lee S.H., Chen S.C. and Lin M.F. 2009. Electronic excitations in doped monolayer graphenes. *J. Appl. Phys.* **106** 113711(6).
[50] Lu C.L., Chang C.P., Huang Y.C., Chen R.B. and Lin M.F. 2006. Influence of an electric field on the optical properties of few-layer graphene with AB stacking. *Phys. Rev. B* **73** 144427.
[51] Diaz H.C., Avila J., Chen C., Addou R., Asensio M.C., Batzill M. *et al.* 2015. Direct observation of interlayer hybridization and Dirac relativistic carriers in graphene/MoS2 van der Waals heterostructures. *Nano Lett.* **15** 1135–1140.
[52] Huang B.L., Chuu C.P. and Lin M.F. 2019. Asymmetry-enriched electronic and optical properties of bilayergraphene. *Sci. Rep.* **9** 859.
[53] Lin S.Y., Lin Y.T., Tran N.T.T., Su W.P. and Lin M.F. 2017. Feature-rich electronic properties of aluminum-absorbed graphenes. *Carbon* **120** 209–218.
[54] Huang Y.K., Chen S.C., Ho Y.H., Lin C.Y. and Lin M.F. 2014. Feature-rich magnetic quantization in sliding bilayer graphenes. *Sci. Rep.* **4** 7509.
[55] Wu J.Y., Chen S.C., Roslyak O., Gumbs G. and Lin M.F. 2011. Plasma excitations in graphene: Their spectral intensity and temperature dependence in magnetic field. *ACS Nano* **5** 1026.
[56] Fallahazad B., Hao Y., Lee K., Kim S., Ruoff R.S. and Tutuc E. 2012. Quantum Hall effect in Bernal stacked and twisted bilayer graphene grown on Cu by chemical vapor deposition. *Phys. Rev. B* **85** 201408.
[57] Chung H.C., Lee M.H., Chang C.P. and Lin M.F. 2011. Exploration of edge-dependent optical selection rules for graphene nanoribbons. *Opt. Express* **19** 23350.
[58] Ho Y.H., Chiu Y.H., Lin D.H., Chang C.P. and Lin M.F. 2010. Magneto-optical selection rules in bilayer Bernal graphene. *ACS Nano* **4** 1465–1472.
[59] Ferrer A.F.J., Cerezo J., Stendardo E., Improta R. and Santoro F. 2013. Insights for an accurate comparison of computational data to experimental absorption and emission spectra: Beyond the vertical transition approximation. *J. Chem. Theory Comput.* **9** 2072–2082.

[60] Huang Y.C., Chang C.P. and Lin M.F. 2007. Magnetic and quantum confinement effects on electronic and optical properties of graphene ribbons. *Nanotechnology* **18** 495401(8).
[61] Saito R., Fujita M., Dresselhaus G. and Dresselhaus M.S. 1998. Electronic structure of chiral graphene tubules. *Appl. Phys. Lett.* **60** 2204.
[62] Matusalem F., Marques M., Teles L.K. and Bechstedt F. 2015. Stability and electronic structure of two-dimensional allotropes of group-IV materials. *Phys. Rev. B* **92** 045436.
[63] Friedlein R., Pichler T., Knupfer M., Golden M.S., Mukhopadhyay K., Sugai T., Shinohara H. and Fink J. 1999. The effects of dimensionality on the pi-plasmon-dispersion in multi-wall carbon nanotubes. *AIP Conf. Proc.* **486** 351.
[64] Do T.N., Chang C.P., Shih P.H., Lin M.F. 2017. Stacking-enriched magneto-transport properties of few-layer graphenes. *Phys. Chem. Chem. Phys.* **19** 29525–29533.
[65] Chuang Y.C., Wu J.Y. and Lin M.F. 2012. Analytical calculations on low-frequency excitations in aa-stacked bilayer graphene. *J. Phys. Soc. Jpn.* **81** 124713.
[66] Geim A.K. and Novoselov K.S. 2007. The rise of graphene. *Nature Materials* **6** 183–191.
[67] Shu H., Cao D., Liang P., Wang X., Chen X. and Lu W. 2014. Two-dimensional silicene nucleation on a Ag(111) surface: Structural evolution and the role of surface diffusion. *Phys. Chem. Chem. Phys.* **16** 304.
[68] Dávila M.E. and Lay G.L. 2016. Few layer epitaxial germanene: A novel two-dimensional Dirac material. *Sci. Rep.* **6** 20714.
[69] Zhu F.F. *et al.* 2015. Epitaxial growth of two-dimensional stanene. *Nat. Mater.* **14** 1020.
[70] Yuhara J. *et al.* 2019. Graphene's latest cousin: Plumbene epitaxial growth on a "ano WaterCube". *Adv. Mater.* **31** 1901017.
[71] Chen S.C., Wu C.L., Wu J.Y. and Lin M.F. 2016. Magnetic quantization of $sp^3$ bonding in monolayer gray tin. *Phys. Rev. B* **94** 045410.
[72] Cai B. *et al.* 2015. Tinene: A two-dimensional Dirac material with a 72 meV band gap. *Phys. Chem. Chem. Phys.* **17** 12634–12638.
[73] Matthes L., Pulci O. and Bechstedt F. 2013. Massive Dirac quasiparticles in the optical absorbance of graphene, silicene, germanene, and tinene. *J. Phys. Condens. Matter* **25** 395305.
[74] Lu J., Loh K.P., Huang H., Chen W. and Wee A.T.S. 2009. Plasmon dispersion on epitaxial graphene studied using high-resolution electron energy-loss spectroscopy. *Phys. Rev. B* **80** 113410.
[75] Zhang L. *et al.* 2018. Discovery of a new quantum spin Hall phase in bilayer plumbene. *Chem. Phys. Lett.* **712** 78–82.
[76] Castro Neto A.H. *et al.* 2009. The electronic properties of graphene. *Rev. Mod. Phys.* **81** 109.
[77] Do T.N., Shih P.H., Gumbs G., Huang D., Chiu C.W. and Lin M.F. 2018. Diverse magnetic quantization in bilayer silicene. *Phys. Rev. B* **97** 125416.

[78] Fu H.X., Zhang J., Ding Z.J., Li H. and Menga S. 2014. Stacking-dependent electronic structure of bilayer silicene. *Appl. Phys. Lett.* **104** 131904.

[79] Yaokawa R., Ohsuna T., Morishita T., Hayasaka Y., Spencer M.J.S. and Nakano H. 2016. Monolayer-to-bilayer transformation of silicenes and their structural analysis. *Nat. Comm.* **7** 10657.

[80] Mukherjee S., Schuppert N., Bates A., Lee S.C. and Park S. 2017. Novel mesoporous microspheres of Al and Ni doped LMO spinels and their performance as cathodes in secondary lithium ion batteries. *Int. J. Green Energy* **14** 656–664.

[81] Song J.Y., Wang Y.Y. and Wan C.C. 1999. Review of gel-type polymer electrolytes for lithium-ion batteries. *J. Power Sources* **77** 183–197.

[82] Inaguma Y. and Nakashima M. 2013. A rechargeable lithium-air battery using a lithium ion-conducting lanthanum lithium titanate ceramics as an electrolyte separator. *J. Power Sources* **228** 250–255.

[83] Liu K., Kong B., Liu W., Sun Y., Song M.-S., Chen J., Liu Y., Lin D., Pei A. and Cui Y. 2018. Stretchable lithium metal anode with improved mechanical and electrochemical cycling stability. *Joule* **2** 1857–1865.

[84] Saroja A., Samantaray S.S. and Sundara R. 2019. A room temperature multivalent rechargeable iron ion battery with an ether based electrolyte: A new type of post-lithium ion battery. *Chem. Commun.* **55** 10416–10419.

[85] Lai P.L., Chen S.C. and Lin M.F. 2008. Electronic properties of single-walled carbon nanotubes under electric and magnetic fields. *Physica E* **40** 2056–2058.

[86] Barone V., Hod O. and Scuseria G.E. 2006. Electronic structure and stability of semiconducting graphene nanoribbons. *Nano Lett.* **6** 2748–2754.

[87] Barros E.B. *et al.* 2006. Review on the symmetry-related properties of carbon nanotubes. *Phys. Rep.* **431** 261–302.

[88] Lin M.F., Chuu D.S., Huang C.S., Lin Y.K. and Shung K.W.-K. 1996. Collective excitations in a single-layer carbon nanotube. *Phys. Rev. B* **53** 15493.

[89] Hernando D.H., Guinea F. and Brataas A. 2006. Spin-orbit coupling in curved graphene, fullerenes, nanotubes, and nanotube caps. *Phys. Rev. B* **74** 155426.

[90] Lee C.H., Su W.S., Chen R.B. and Lin M.F. 2009. Low-energy electronic properties of finite double-walled carbon nanotubes under external fields. *Physica E* **41** 1226–1231.

[91] Yang C.H., Lin C.Y., Chiu C.W., Liu C.T., Lin M.F. 2020. Dimension-and lattice-diversified Coulomb excitations in 3D, 2D, 1D-nanotube electron gases, graphene and carbon nanotube. arXiv preprint arXiv:2004.08209.

[92] Lin S.Y., Tran N.T.T., Chang S.L., Su W.P. and Lin M.F. 2018. *Structure- and Adatom- Enriched Essential Properties of Graphene Nanoribbons* (CRC Press).

[93] Chang C.P., Chiu C.W., Shyu F.L., Chen R.B. and Lin M.F. 2002. Magnetoband structures of AB-stacked zigzag nanographite ribbons. *Phys. Lett. A.* **306** 137–143.

[94] Chen R.B., Chang C.P., Shyu F.L., Huang J.S. and Lin M.F. 2004. Optical excitations of finite carbon nanotubes. *Carbon* **42** 531–535.

[95] Huang Y.K., Chen S.C., Ho Y.H., Lin C.Y. and Lin M.F. 2014. Feature-rich magnetic quantization in sliding bilayer graphenes. *Sci. Rep.* **4** 7509.

[96] Zberecki K. *et al.* 2013. Thermoelectric effects in silicene nanoribbons. *Phys. Rev. B* **88** 115404.
[97] Wang Y. *et al.* 2012. Half-metallic silicene and germanene nanoribbons: Towards high-performance spintronics device. *Nano* **7** 1250037.
[98] Xiong W. *et al.* 2016. Spin–orbit coupling effects on electronicstructures in stanene nanoribbons. *Phys. Chem. Chem. Phys.* **18** 6534.
[99] Li Y. *et al.* 2019. Constructive coupling effect of topological states and topological phase transitions in plumbene. *Phys. Rev. B* **99** 195402.
[100] Lin C.Y., Ho C.H., Wu J.Y., Do T.N., Shih P.H., Lin S.Y. and Lin M.F. 2020. *Diverse Quantization Phenomena in Layered Materials* (CRC Press).
[101] Chiu Y.H., Ho J.H., Chang C.P., Chuu D.S. and Lin M.F. 2008. Low-frequency magneto-optical excitations of a graphene monolayer: Peierls tight-binding model and gradient approximation calculation. *Phys. Rev. B* **78** 245411(10).
[102] Krasovskii E.E. 2011. Spin-orbit coupling at surfaces and 2D materials. *Phys. Chem. Chem. Phys.* **13** 11929.
[103] Tsai S.J., Chiu Y.H., Ho Y.H. and Lin M.F. 2012. Gate-voltage-dependent Landau levels in AA-stacked bilayer graphene. *Chem. Phy. Lett.* **550** 104–110.
[104] Ho J.H., Lai Y.H., Chiu Y.H. and Lin M.F. 2008. Modulation effects on Landau levels in a monolayer graphene. *Nanotechnology* **19** 035712(6).
[105] Divincenzo D.P. and Mele E.J. 1984. Self-consistent effective-mass theory for intralayer screening in graphite intercalation compounds. *Phys. Rev. B* **29** 1685–1694.
[106] Laughlin R.B. 1983. Quantized motion of three two-dimensional electrons in a strong magnetic field. *Phys. Rev. B* **27** 3383.
[107] Ho C.H., Tsai S.J., Chen R.B., Chiu Y.H., Lin M.F. 2011. Low-energy Landau level spectrum in ABC-stacked trilayer graphene. *J. Nanosci. Nanotechnol.* **11** 4938–4947.
[108] Ho Y.H., Wang J., Chiu Y.H., Lin M.F. and Su W.P. 2011. Characterization of Landau subbands in graphite: A tight-binding study. *Phys. Rev. B* **83** 121201.
[109] Jurcevic P., Lanyon B.P., Hauke P., Hempel C., Zoller P., Blatt R. and Roos C.F. 2014. Quasiparticle engineering and entanglement propagation in a quantum many-body system. *Nature* **511** 202.
[110] Chung H.C., Su W.P. and Lin M.F. 2012. Electric-field-induced destruction of quasi-Landau levels in bilayergraphene nanoribbons. *Phys. Chem. Chem. Phys.* **15** 868–875.
[111] Chu R.L., Li X., Wu S., Niu Q., Yao W., Xu X. and Zhang C. 2014. Valley-splitting and valley-dependent inter-landau-level optical transitions in monolayer mos 2 quantum hall systems. *Phys. Rev. B* **90** 045427.
[112] Katayama-Yoshida H. and Sato K. 2003. Materials design for semiconductor spintronics by ab initio electronic-structure calculation. *Physica B Condens. Matter* **327** 337–343.
[113] Ho C.H., Chang C.P. and Lin M.F. 2014. Landau subband wave functions and chirality manifestation in rhombohedral graphite. *Solid State Commun.* **197** 11–15.
[114] Lin M.F. and Hsu W.D. 2019. *Green Energy Materials Handbook* (Boca Raton: CRC Press).

[115] Li C., He L., Luo J., Tang J., Cheng N. and Chen Z. 2020. Theoretical anchoring effect of new phosphorus allotropes for lithium-sulfur batteries. *Nanoscale* **12** 11095–11111.
[116] Novoselov K.S., Geim A.K., Morozov S.V., Jiang D., Zhang Y., Dubonos S.V., Grigorieva I.V. and Firsov A.A. 2004. Electric field effect in atomically thin carbon film. *Science* **306** 666.
[117] Zhao H.B. and Mazumdar S. 2004. Electron-electron interaction effects on the optical excitations of semiconducting single-walled carbon nanotubes. *Phys. Rev. Lett.* **93** 157402.
[118] Jackson J.D. 1975. *Classical Electrodynamics* (New York: Wiley).
[119] Schwinger J. 1958. *Selected Papers on Quantum Electrodynamics* (Dover Publications).
[120] Lin M.F. and Shung K.W.K. 1994. Plasmons and optical properties of carbon nanotubes. *Phys. Rev. B* **50** 17744.
[121] Adler S.L. 1962. Quantum theory of the dielectric constant in real solids. *Phys. Rev.* **126** 413.
[122] Lin M.F. and Shung K.W.-K. 1993. Elementary excitations in cylindrical tubules. *Phys. Rev. B* **47** 6617.
[123] Nechaev I.A., Echenique P.M. and Chulkov E.V. 2010. Inelastic decay rate of quasiparticles in a two-dimensional spin-orbit coupled electron system. *Phys. Rev. B* **81** 195112.
[124] Penn D.R. 1962. Wave-number-dependent dielectric function of semiconductors. *Phys. Rev.* **128** 2093.
[125] Chung H.C., Lee M.H., Chang C.P., Huang Y.C. and Lin M.F. 2011. Effects of transverse electric fields on quasi-Landau levels in zigzag graphene nanoribbons. *J. Phys. Soc. Jpn.* **80** 044602.
[126] Lu J., Loh K.P., Huang H., Chen W. and Wee A.T.S. 2009. Plasmon dispersion on epitaxial graphene studied using high-resolution electron energy-loss spectroscopy. *Phys. Rev. B* **80** 113410.
[127] Kravets V.G., Grigorenko A.N., Nair R.R., Blake P., Anissimova S., Novoselov K.S. *et al.* 2010. Spectroscopic ellipsometry of graphene and an exciton-shifted van Hove peak in absorption. *Phys. Rev. B* **81** 155413.
[128] Scholz A., Stauber T. and Schliemann J. 2012. Dielectric function, screening, and plasmons of graphene in the presence of spin-orbit interactions. *Phys. Rev. B* **86** 195424.
[129] Ho J.H., Lu C.L., Hwang C.C., Chang C.P. and Lin M.F. 2006. Coulomb excitations in AA- and AB-stacked bilayer graphites. *Phys. Rev. B* **74** 085406(8).
[130] Lin C.Y., Huang B.L., Ho C.H., Gumbs G. and Lin M.F. 2018. Geometry-diversified Coulomb excitations in trilayer AAB stacking graphene. *Phys. Rev. B* **98** 195442.
[131] Chiu C.W., Shyu F.L., Chang C.P., Chuu D.S. and Lin M.F. 2006. Coulomb scattering rates of excited carriers in moderate-gap carbon nanotubes. *Phys. Rev. B* **73** 235407(5).
[132] Wu J.Y., Chen S.C., Gumbs G. and Lin M.F. 2016. Feature-rich electronic excitations of silicene in external fields. *Phys. Rev. B* **94** 205427.

[133] Shih P.H., Chiu Y.H., Wu J.Y., Shyu F.L., Lin M.F. 2017. Coulomb excitations of monolayer germanene *Sci. Rep.* **7** 40600.

[134] Dang N.D. and Zelevinsky V. 2001. Improved treatment of ground-state correlations: Modified random phase approximation. *Phys. Rev. C* **64** 064319.

[135] Pines D. and Bohm D. 1952. A collective description of electron interactions: II. Collective vs individual particle aspects of the interactions. *Phys. Rev.* **85** 338.

[136] Lin C.Y., Wu J.Y., Chiu C.W. and Lin M.F. 2019. *Coulomb Excitations and Decays in Graphene-Related Systems* (CRC).

[137] Chuang Y.C., Wu J.Y. and Lin M.F. 2013. Electric field dependence of excitation spectra in AB-stacked bilayer graphene. *Sci. Rep.* **3** 1368.

[138] Hwang E. and Sarma S.D. 2007. Dielectric function, screening, and plasmons in two-dimensional graphene. *Phys. Rev. B* **75** 205418.

[139] Lin M.F., Chuu D.S., Huang C.S., Lin Y.K. and Shung K.W.-K. 1996. Collective excitations in a single-layer carbon nanotube. *Phys. Rev. B* **53** 15493.

[140] Lin C.Y., Do T.N., Huang Y.K. and Lin M.F. 2017. *Electronic and Optical Properties of Graphene in Magnetic and Electric Fields*. IOP Concise Physics (San Raefel, CA: Morgan & Claypool Publishers)

[141] Lin C.Y., Yang C.H., Chiu C.W., Chung H.C., Lin S.Y., LinM F. 2020. *Many-Particle Interactions in Carbon Nanotubes* (IOP Publishing)

[142] Politano A. *et al.* 2011. Evidence for acoustic-like plasmons on epitaxial graphene on Pt(111). *Phys. Rev. B* **84** 033401.

[143] Kramberger C., Hambach R., Giorgetti C., Rummeli M.H., Knupfer M., Fink J., Buchner B., Reining L., Einarsson E., Maruyama S., Sottile F., Hannewald K., Olevano V., Marinopoulos A.G. and Pichler T. 2008. Linear plasmon dispersion in single-wall carbon nanotubes and the collective excitation spectrum of graphene. *Phys. Rev. Lett.* **100** 196803.

[144] Tsai S.J., Sheu J.K., Ho C.H., Chiu Y.H. and Lin M.F. 2011. Modulation effects of periodic potentials on the electronic properties of bilayer Bernal graphene: Tight-binding model. *J. Phys. Soc. Jpn.* **81** 014705 (11).

[145] Hagen A., Steiner M., Raschke M.B., Lienau C., Hertel T., Qian H., Meixner A.J. and Hartschuh A. 2005. Exponential decay lifetimes of excitons in individual single-walled carbon nanotubes. *Phys. Rev. Lett.* **95** 197401.

[146] Shung K.W.-K. 1986. Dielectric function and plasmon structure of stage-1 intercalated graphite. *Phys. Rev. B* **34** 979.

[147] Cheianov V.V. and Fal'Ko V.I. 2006. Friedel oscillations, impurity scattering, and temperature dependenceof resistivity in graphene. *Phys. Rev. Lett.* **97** 226801.

[148] Wang Z., Luo M., Yan D., Ying H. and Li W. 1995. Electronic structure and transport properties of fullerene nanotubules. *Phys. Rev. B* **51** 13833.

[149] Lin M.F. and Shung K.W.-K. 1996. Self-energy of electrons in graphite intercalation compounds. *Phys. Rev. B* **53** 1109.

[150] Mahan G.D. 1990. *Many-Particle Physics* (New York: Plenum)

[151] Kane C., Balents L. and Fisher M.P. 1997. Coulomb interactions and mesoscopic effects in carbon nanotubes. *Phys. Rev. Lett.* **79** 5086.

[152] Calandra M. and Mauri F. 2007. Electron-phonon coupling and electron self-energy in electron-doped graphene: Calculation of angular-resolved photoemission spectra. *Phys. Rev. B* **76** 205411.
[153] Park C.-H., Giustino F., Cohen M.L. and Louie S.G. 2007. Velocity renormalization and carrier lifetime in graphenefrom the electron-phonon interaction. *Phys. Rev. Lett.* **99** 086804.
[154] Caruso F., Novko D. and Draxl C. 2018. Phonon-assisted damping of plasmons in three- and two-dimensional metals. *Phys. Rev. B* **97** 205118.
[155] Stern F. 1967. Polarizability of a two-dimensional electron gas. *Phys. Rev. Lett.* **18** 546.
[156] Thomas K.J., Nicholls J.T., Simmons M.Y., Pepper M., Mace D.R. and Ritchie D.A. 1996. Possible spin polarization in a one- dimensional electron gas. *Phys. Rev. Lett.* **77** 135.
[157] Caruso F., Novko D. and Draxl C. 2018. Phonon-assisted damping of plasmons in three- and two-dimensional metals. *Phys. Rev. B* **97** 205118.
[158] Cox P.A., Egdell R.G., Goodenough J.B., Hamnett A. and Naish C.C. 1983. The metal-to-semiconductor transition in ternary ruthenium (IV) oxides: A study by electron spectroscopy. *J. Phys. C: Solid State Phys.* **16** 6221.
[159] Ehrenreich H. and Cohen M.H. 1959. Self-consistent field approach to the many-electron problem. *Phys. Rev.* **115** 786–790.
[160] Winzer T., Knorr A. and Malic E. 2010. Carrier multiplication in graphene. *Nano Lett.* **10** 4839–4843.
[161] Lin M.F., Chuu D.S. and Shung K.W.-K. 1997. Low-frequency plasmons in metallic carbon nanotubes. *Phys. Rev. B* **56** 1430–1439.
[162] Chiu C.W. and Lin M.F. 2007. Electronic excitations and deexcitations in narrow-gap carbon nanotubes. *Nanotechnology* **18** 435401(7).
[163] Chiu C.W., Ho Y.H., Chen S.C., Lee C.H., Lue C.S. and Lin M.F. 2006. Electronic decay rates in semiconducting carbon nanotubes. *Physica E* **34** 658–661.
[164] Ho J.H., Chang C.P., Chen R.B. and Lin M.F. 2006. Electron decay rates in a zero-gap graphite layer. *Phys. Lett. A* **357** 401–406.
[165] Sprinkle M. *et al.* 2009. First direct observation of a nearly ideal graphene band structure. *Phys. Rev. Lett.* **103** 226803.
[166] Vogt P. *et al.* 2012. Silicene: Compelling experimental evidence for graphene-like two-dimensional silicon. *Phys. Rev. Lett.* **108** 155501.
[167] Dávila M.E. and Lay G.L. 2016. Few layer epitaxial germanene: A novel two-dimensional Dirac material. *Sci. Rep.* **6** 20714.
[168] Petersen L. and Hedegard P. 2000. A simple tight-binding model of spin–orbit splitting of sp-derived surface states. *Surf. Sci.* **459** 49–56.
[169] Lin C.Y., Wu J.Y., Ou Y.J., Chiu Y.H. and Lin M.F. 2015. Magnetoelectronic properties of multilayer graphenes. *Phys. Chem. Chem. Phys.* **17** 26008.
[170] Tran N.T.T., Lin S.Y., Lin C.Y. and Lin M.F. 2017. *Geometric and Electronic Properties of Graphene-Related Systems: Chemical Bonding Schemes* (CRC Press).
[171] Chiu C.W., Lee S.H. and Lin M.F. 2010. Inelastic Coulomb scatterings of doped armchair carbon nanotubes. *J. Nanosci. Nanotechnol.* **10** 2401–2408.

[172] Lin M.F., Huang C.S. and Chuu D.S. 1998. Electronic excitations of an electron-gas cylinder bundle. *Solid. State. Communs.* **105** 161–164.
[173] Xia F., Mueller T., Lin Y.M., Valdes-Garcia A. and Avouris P. 2009. Ultrafast graphene photodetector. *Nat. Nanotech.* **4** 839–843.
[174] Mueller T., Xia F. and Avouris P. 2010. Graphene photodetectors for high-speed optical communications. *Nat. Photonics* **4** 297–301.
[175] Nguyen B.H., Nguyen V.H., Bui D.H. and Le T.T.P. 2015. Theory of photon-electron interaction in single-layer graphene sheet. *Adv. Nat. Sci. Nanosci. Nanotechnol.* **6** 045009.
[176] Lin C.Y., Do T.N., Wu J.Y., Shih P.H., Lin S.Y., Ho C.H. and Lin M.F. 2019. *Diverse Quantization Phenomena in Emergent Layered Materials* (CRC Press).
[177] Lin C.Y., Wu J.Y., Chiu C.W. and Lin M.F. 2019. *Coulomb Excitations and Decays in Graphene-Related Systems* (CRC Press).
[178] Lin C.Y., Chen R.B., Ho Y.H. and Lin M.F. 2018. *Electronic and Optical Properties of Graphite-Related Systems* (CRC Press).
[179] Griesmaier R. 2011. A general perturbation formula for electromagnetic fields in presence of low volume scatterers. *ESAIM: M2AN* **45** 1193–1218.
[180] Kubo R. 1957. Statistical-mechanical theory of irreversible processes. I. General theory and simple applications to magnetic and conduction problems. *J. Phys. Soc. Jpn.* **12** 570–586.
[181] De Filippis G., Cataudella V., de Candia A., Mishchenko A.S. and Nagaosa N. 2014. Alternative representation of the Kubo formula for the optical conductivity: A shortcut to transport properties. *Phys. Rev. B* **90** 014310.
[182] Imamoğlu A. 1998. Phase-space filling and stimulated scattering of composite bosons. *Phys. Rev. B* **57** R4195(R).
[183] Combescot M. and Betbeder-Matibet O. 2004. Scattering rates and lifetime of exact and boson excitons. *Phys. Rev. Lett.* **93** 016403.
[184] Kronig R.D.L. 1926. On the theory of dispersion of X-rays. *J. Opt. Soc. Am.* **12** 547–557.
[185] Kramers H.A. 1927. La diffusion de la lumière par les atomes. *Atti Congr. Int. Fis. Como* **2** 545–557.
[186] Fan X.D., Peng J.L. and Bursill L.A. 1998. Joint density of states of wide-band-gap materials by electron energy loss spectroscopy. *Mod. Phys. Lett. B* **12** 541–554.
[187] García de Abajo F.J. 2010. Optical excitations in electron microscopy. *Rev. Mod. Phys.* **82** 209.
[188] Mouhot C. and Villani C. 2011. On Landau damping. *Acta Math.* **207** 29–201.
[189] Hancock Y., Uppstu A., Saloriutta K., Harju A. and Puska M.J. 2010. Generalized tight-binding transport model for graphene nanoribbon-based systems. *Phys. Rev. B* **81** 245402.
[190] Chen S.C., Wu J.Y. and Lin M.F. 2018. Feature-rich magneto-electronic properties of bismuthine. *New J. Phys.* **20** 062001.
[191] Wu J.Y., Chen S.C., Do T.N., Su W.P., Gumbs G. and Lin M.F. 2018. The diverse magneto-optical selection rules in bilayer black phosphorus. *Sci. Rep.* **8** 13303.

[192] Ho Y.H., Chiu Y.H., Lin D.H., Chang C.P. and Lin M.F. 2010. Magneto-optical selection rules in bilayer Bernal graphene. *ACS Nano* **4** 1465–1472.
[193] Barreteau C., Spanjaard D., Desjonquères M.C. 2016. An efficient magnetic tight-binding method for transition metals and alloys. *C. R. Physique* **17** 406–429.
[194] Higuchi K., Hamal D.B. and Higuchi M. 2015. Relativistic tight-binding approximation method for materials immersed in a uniform magnetic field: Application to crystalline silicon. *Phys. Rev. B* **91** 075122.
[195] Ju L., Wang L., Li X., Moon S., Ozerov M., Lu Z., Taniguchi T., Watanabe K., Mueller E., Zhang F., Smirnov D., Rana F. and McEuen P.L. 2020. Unconventional valley-dependent optical selection rules and landau level mixing in bilayer graphene. *Nat. Commun.* **11** 2941.
[196] Chiu Y.H., Ho J.H., Chang C.P., Chuu D.S. and Lin M.F. 2008. Low-frequency magneto-optical excitations of a graphene monolayer: Peierls tight-binding model and gradient approximation calculation. *Phys. Rev. B* **78** 245411.
[197] Kuan W.H., Tang C.S. and Chang C.H. 2007. Spectral properties and magneto-optical excitations in semiconductor double rings under Rashba spin-orbit interaction. *Phys. Rev. B* **75** 155326.
[198] Noky J., Gooth J., Sun Y. and Felser C. 2021. (Quasi-)Quantization of the electrical, thermal, and thermoelectrical conductivities in two and three dimensions. *J. Phys. Commun.* **5** 045007.
[199] Edery A. and Audin Y. 2019. New degeneracies and modification of Landau levels in the presence of a parallel linear electric field. *J. Phys. Commun.* **3** 025013.
[200] Takeda K. and Shiraishi K. 1994. Theoretical possibility of stage corrugation in Si and Ge analogs of graphite. *Phys. Rev. B* **50** 14916.
[201] Zhu F.F., Chen W.J., Xu Y., Gao C.L., Guan D.D., Liu C.H., Qian D., Zhang S.C. and Jia J.F. 2015. Epitaxial growth of two-dimensional stanine. *Nat. Mater.* **14** 1020–1025.
[202] Liu B. and Zhou K. 2019. Recent progress on graphene-analogous 2D nanomaterials: Properties, modeling and applications. *Prog. Mater. Sci.* **100** 99.
[203] Liu C.C., Jiang H. and Yao Y. 2011. Low-energy effective Hamiltonian involving spin-orbit coupling in silicene and two-dimensional germanium and tin. *Phys. Rev. B* **84** 195430.
[204] Ezawa M. 2012. A topological insulator and helical zero mode in silicene under an inhomogeneous electric field. *New J. Phys.* **14** 033003.
[205] Dirac P.A.M. 1927. The quantum theory of emission and absorption of radiation. *Proc. R. Soc. A* **114** 243–265.
[206] Avetisyan A.A., Partoens B. and Peeters F.M. 2009. Electric-field control of the band gap and Fermi energy in graphene multilayers by top and back gates. *Phys. Rev. B* **80** 195401.
[207] Krstajić P.M., Van Duppen B. and Peeters F.M. 2013. Plasmons and their interaction with electrons in trilayer graphene. *Phys. Rev. B* **88** 195423.
[208] Chiu C.W., Lee S.H., Chen S.C., Shyu F.L. and Lin M.F. 2010. Absorption spectra of AA-stacked graphite. *New J. Phys.* **12** 083060.

[209] Chen R.B., Chiu Y.H. and Lin M.F. 2013. A theoretical evaluation of the magneto-optical properties of AA-stacked graphite. *Carbon* **54** 268–276.

[210] Chen R.B., Chiu Y.H. and Lin M.F. 2015. Beating oscillations of magneto-optical spectra in simple hexagonal graphite. *Comput. Phys. Commun.* **189** 60–65.

[211] Shyu F.L. and Lin M.F. 2000. Plasmons and optical properties of semimetal graphite. *J. Phys. Soc. Jpn.* **69** 3781–3784.

[212] Ho Y.H., Chiu Y.H., Su W.P. and Lin M.F. 2011. Magneto-absorption spectra of Bernal graphite. *Appl. Phys. Lett.* **99** 011914.

[213] Chiu C.W., Huang Y.C., Chen S.C., Lin M.F. and Shyu F.L. 2011. Low-frequency electronic and optical properties of rhombohedral graphite. *Phys. Chem. Chem. Phys.* **13** 6036–6042.

[214] Skulason H.S., Gaskell P.E. and Szkopek T. 2010. Optical reflection and transmission properties of exfoliated graphite from a graphene monolayer to several hundred graphene layers. *Nanotechnology* **21** 295709.

[215] Holovský J., Nicolay S., De Wolf S., Ballif C. 2015. Effect of the thin-film limit on the measurable optical properties of graphene. *Sci. Rep.* **28** 15684.

[216] García-Ruiz A., Slizovskiy S., Mucha-Kruczyński M. and Fal'ko V.I. 2019. Spectroscopic signatures of electronic excitations in Raman scattering in thin films of rhombohedral graphite. *Nano Lett.* **19** 6152–6156.

[217] Abergel D.S.L. and Fal'ko V. 2007. Optical and magneto-optical far-infrared properties of bilayer graphene. *Phys. Rev. B* **75** 155430.

[218] Wang Y., Ni Z., Liu L., Liu Y., Cong C., Yu T., Wang X., Shen D. and Shen Z. 2010. Stacking-dependent optical conductivity of bilayer graphene. *ACS Nano* **4** 4074–4080.

[219] Yang L. 2011. Excitonic effects on optical absorption spectra of doped graphene. *Nano Lett.* **11** 3844–3847.

[220] Liang Y. and Yang L. 2012. Quasiparticle energy and optical excitations of gated bilayer graphene. *Phys. Rev. B* **86** 205423.

[221] Mak K.F., da Jornada F.H., He K., Deslippe J., Petrone N., Hone J., Shan J., Louie S.G. and Heinz T.F. 2014. Tuning many-body interactions in graphene: The effects of doping on excitons and carrier lifetimes. *Phys. Rev. Lett.* **112** 207401.

[222] Wang F., Zhang Y., Tian C., Girit C., Zettl A., Crommie M. and Shen Y.R. 2008. Gate-variable optical transitions in graphene. *Science* **320** 206.

[223] Sensale-Rodriguez B., Yan R., Rafique S., Zhu M., Li W., Liang X., Gundlach D., Protasenko V., Kelly M.M., Jena D., Liu L. and Xing H.G. 2012. Extraordinary control of terahertz beam reflectance in graphene electro-absorption modulators. *Nano Lett.* **12** 4518–4522.

[224] Gusynin V.P., Sharapov S.G. and Carbotte J.P. 2007. Anomalous absorption line in the magneto-optical response of graphene. *Phys. Rev. Lett.* **98** 157402.

[225] Chizhova L.A., Burgdörfer J. and Libisch F. 2015. Magneto-optical response of graphene: Probing substrate interactions. *Phys. Rev. B* **92** 125411.

[226] Kataura H., Kumazawa Y., Maniwa Y., Umezu I., Suzuki S., Ohtsuka Y. and Achiba Y. 1999. Optical properties of single-wall carbon nanotubes. *Synth. Met.* **103** 2555–2558.

[227] Murakami Y., Einarsson E., Edamura T. and Maruyama S. 2005. Polarization dependence of the optical absorption of single-walled carbon nanotubes. *Phys. Rev. Lett.* **94** 087402.
[228] Jeong S.H., Kim K.K., Jeong S.J., An K.H., Lee S.H. and Lee Y.H. 2007. Optical absorption spectroscopy for determining carbon nanotube concentration in solution. *Synth. Met.* **157** 570–574.
[229] Blancon J.-C., Paillet M., Tran H.N., Than X.T., Guebrou S.A., Ayari A., San Miguel A., Phan N.-M., Zahab A.-A., Sauvajol J.-L., Del Fatti N. and Vallée F. 2013. Direct measurement of the absolute absorption spectrum of individual semiconducting single-wall carbon nanotubes. *Nat. Commun.* **4** 2542.
[230] Zhang D., Yang J., Hasdeo E.H., Liu C., Liu K., Saito R. and Li Y. 2016. Multiple electronic Raman scatterings in a single metallic carbon nanotube. *Phys. Rev. B* **93** 245428.
[231] Wei X., Tanaka T., Akizuki N., Miyauchi Y., Matsuda K., Ohfuchi M. and Kataura H. 2016. Single-chirality separation and optical properties of (5, 4) single-wall carbon nanotubes. *J. Phys. Chem. C* **120** 10705–10710.
[232] van Bezouw S., Arias D.H., Ihly R., Cambré S., Ferguson A.J., Campo J., Johnson J.C., Defillet J., Wenseleers W. and Blackburn J.L. 2018. Diameter-dependent optical absorption and excitation energy transfer from encapsulated dye molecules toward single-walled carbon nanotubes. *ACS Nano* **12** 6881–6894.
[233] Hsu H. and Reichl L.E. 2007. Selection rule for the optical absorption of graphene nanoribbons. Phys. Rev. B **76** 045418.
[234] Gundra K. and Shukla A. 2011. Theory of the electro-optical properties of graphene nanoribbons. *Phys. Rev. B* **83** 075413.
[235] Chernov A.I., Fedotov P.V., Talyzin A.V., Lopez I.S., Anoshkin I.V., Nasibulin A.G., Kauppinen E.I. and Obraztsova E.D. 2013. Optical properties of graphene nanoribbons encapsulated in single-walled carbon nanotubes. *ACS Nano* **7** 6346–6353.
[236] Berahman M., Asad M., Sanaee M. and Sheikhi M.H. 2015. Optical properties of chiral graphene nanoribbons: A first principle study. *Opt. Quant. Electron.* **47** 3289–3300.
[237] Chung H.-C., Chang C.-P., Lin C.-Y. and Lin M.-F. 2016. Electronic and optical properties of graphene nanoribbons in external fields. *Phys. Chem. Chem. Phys.* **18** 7573–7616.
[238] Saroka V.A., Shuba M.V. and Portnoi M.E. 2017. Optical selection rules of zigzag graphene nanoribbons. *Phys. Rev. B* **95** 155438.
[239] Uryu S. 2020. Excitons and plasmons of graphene nanoribbons in infrared frequencies in an effective-mass approximation. *Phys. Rev. B* **102** 155409.
[240] Shyu F.L. 2009. Effect of external fields on low-frequency optical properties of achiral carbon tori. *J. Phys. Condens. Matter* **22** 025302.
[241] Lin M.F. 1998. Low-frequency optical excitations in zigzag carbon tori. *Phys. Rev. B* **58** 3629.
[242] Koshino M. and Ando T. 2008. Magneto-optical properties of multilayer graphene. *Phys. Rev. B* **77** 115313.

[243] Ho Y.-H., Wu J.-Y., Chen R.-B., Chiu Y.-H. and Lin M.-F. 2010. Optical transitions between Landau levels: AA-stacked bilayer graphene. *Appl. Phys. Lett.* **97** 101905.

[244] Liu J. and Dai X. 2020. Anomalous Hall effect, magneto-optical properties, and nonlinear optical properties of twisted graphene systems. *njp Comput. Mater.* **6** 57.

[245] Kumar A., Sachdeva G., Pandey R. and Karna S.P. 2020. Optical absorbance in multilayer two-dimensional materials: Graphene and antimonene. *Appl. Phys. Lett.* **116** 263102.

[246] Rashidian Z., Bludov Y.V., Ribeiro R.M., Peres N.M.R. and Vasilevskiy M.I. 2014. Optical conductivity of ABA stacked graphene trilayer: Mid-IR resonance due to band nesting. *J. Phys. Condens. Matter* **26** 395301.

[247] Mei H., Xu W., Wang C., Yuan H., Zhang C., Ding L., Zhang J., Deng C., Wang Y. and Peeters F.M. 2018. Terahertz magneto-optical properties of bi- and tri-layer graphene. *J. Phys. Condens. Matter* **30** 175701.

[248] Xiao Y.M., Xu W., Zhang Y.Y. and Peeters F.M. 2013. Optoelectronic properties of ABC-stacked trilayer graphene. *Phys. Status Solidi B* **250** 86–94.

[249] Sobol O.O., Gorbar E.V. and Gusynin V.P. 2014. Dynamical polarization in ABC-stacked multilayer graphene in a magnetic field. *Phys. Rev. B* **90** 085420.

[250] Lin Y.P., Lin C.Y., Ho Y.H., Do T.N. and Lin M.F. 2015. Magneto-optical properties of ABC-stacked trilayer graphene. *Phys. Chem. Chem. Phys.* **17** 15921–15927.

[251] Do T.N., Shih P.H., Chang C.P., Lin C.Y. and Lin M.F. 2016. Rich magneto-absorption spectra of AAB-stacked trilayer graphene. *Phys. Chem. Chem. Phys.* **18** 17597–17605.

[252] Bolotin K.I., Ghahari F., Shulman M.D., Stormer H.L. and Kim P. 2009. Observation of the fractional quantum Hall effect in graphene. *Nature* **462** 196–199.

[253] Novoselov K.S., Jiang Z., Zhang Y., Morozov S.V., Stormer H.L., Zeitler U., Maan J.C., Boebinger G.S., Kim P. and Geim A.K. 2007. Room-temperature quantum Hall effect in graphene. *Science* **315** 1379.

[254] Novoselov K.S., McCann E., Morozov S.V., Fal'ko V.I., Katsnelson M.I., Zeitler U., Jiang D., Schedin F. and Geim A.K. 2006. Unconventional quantum Hall effect and Berry's phase of $2\pi$ in bilayer graphene. *Nat. Phys.* **2** 177–180.

[255] Ki D.-K., Fal'ko V.I., Abanin D.A. and Morpurgo A.F. 2014. Observation of even denominator fractional quantum Hall effect in suspended bilayer graphene. *Nano Lett.* **14** 2135–2139.

[256] Kumar A., Escoffier W., Poumirol J.M., Faugeras C., Arovas D.P., Fogler M.M., Guinea F., Roche S., Goiran M. and Raquet B. 2011. Integer quantum Hall effect in trilayer graphene. *Phys. Rev. Lett.* **107** 126806.

[257] Henriksen E.A., Nandi D. and Eisenstein J.P. 2012. Quantum Hall effect and semimetallic behavior of dual-gated ABA-stacked trilayer graphene. *Phys. Rev. X* **2** 011004.

[258] Bao W., Jing L., Velasco Jr. J., Lee Y., Liu G., Tran D., Standley B., Aykol M., Cronin S.B., Smirnov D., Koshino M., McCann E., Bockrath M. and Lau C.N. 2011. Stacking-dependent band gap and quantum transport in trilayer graphene. *Nat. Phys.* **7** 948–952.
[259] Lee Y., Tran D., Myhro K., Velasco Jr. J., Gillgren N., Poumirol J.M., Smirnov D., Barlas Y. and Lau C.N. 2016. Multicomponent quantum Hall ferromagnetism and Landau level crossing in rhombohedral trilayer graphene. *Nano Lett.* **16** 227–231.
[260] Crassee I., Levallois J., van der Marel D., Walter A.L., Seyller T. and Kuzmenko A.B. 2011. Multicomponent magneto-optical conductivity of multilayer graphene on SiC. *Phys. Rev. B* **84** 035103.
[261] Chizhova L.A., Burgdörfer J. and Libisch F. 2015. Magneto-optical response of graphene: Probing substrate interactions. *Phys. Rev. B* **92** 125411.
[262] Solodov A., Neklyudov V.V., Shayimova J., Amirov R.R. and Dimiev A.M. 2018. Magneto-optical properties of the magnetite-graphene oxide composites in organic solvents. *ACS Appl. Mater. Interfaces* **10** 40024–40031.
[263] Nozières P. and Luttinger J.M. 1962. *Phys. Rev.* **127** 1423.
[264] Schrieffer J.R. 1970. *J. Res. Natl. Bur. Stand. A* **74** 537.
[265] Landau L. 1957. *Sov. Phys. JETP* **3** 920.
[266] Landau L. 1957. *Sov. Phys. JETP* **5** 101.
[267] Nakamura J., Liang S., Gardner G.C. and Manfra M.J. 2020. *Nat. Phys.* **16** 931.
[268] Ashcroft N.W. and Mermin N.D. 1976. *Solid State Physics* (Philadelphia: Harcourt College Publishers).
[269] Hohenberg P. and Kohn W. 1964. *Phys. Rev.* **136** B864.
[270] Volovik G.E. 2003. *The Universe in a Helium Droplet* (Oxford: Oxford University Press).
[271] Cayssol J. 2013. *C. R. Phys.* **14** 760.
[272] Vafek O. and Vishwanath A. 2014. *Annu. Rev. Condens. Matter Phys.* **5** 83.
[273] Wehling T.O., Black-Schaffer A.M. and Balatsky A.V. 2014. *Adv. Phys.* **63** 1.
[274] Armitage N.P., Mele E.J. and Vishwanath A. 2018. *Rev. Mod. Phys.* **90** 015001.
[275] Xiao D., Chang M.C. and Niu Q. 2009. *Rev. Mod. Phys.* **82** 1959.
[276] Hou C.Y., Chamon C. and Mudry C. 2007. *Phys. Rev. Lett.* **98** 186809.
[277] Chamon C., Hou C.Y., Jackiw R., Mudry C., Pi S.Y. and Semenoff G. 2008. *Phys. Rev. B* **77** 235431.
[278] Fujikawa K. and Suzuki H. 2004. *Path Integrals and Quantum Anomalies* (Oxford: Oxford University Press).
[279] Chamon C., Hou C.Y., Mudry C., Ryu S. and Santos L. 2012. *Phys. Scr.* **T146** 014013.
[280] Pientka F., Peng Y., Glazman L. and von Oppen F. 2015. *Phys. Scr.* **T164** 014008.
[281] Taoa S.X., Kalisvaartb W.P., Danaieb M., Mitlinb D., Nottena P.H.L. and van R.A. 2011. First principle study of hydrogen diffusion in equilibrium rutile, rutile with deformation twins and fluorite polymorph of Mg hydride. *Int. J. Hydrog. Energy* **36** 11802.

[282] Gülseren O., Yildirim T., Ciraci S. and Kılıç Ç. 2002. Reversible band-gap engineering in carbon nanotubes by radial deformation. *Phys. Rev. B* **65** 155410.

[283] Jo J.Y. and Kim B.G. 2012. Carbon allotropes with triple bond predicted by first-principle calculation: Triple bond modified diamond and T-carbon. *Phys. Rev. B* **86** 075151.

[284] Nguyen C.V. and Hieu N.N. 2016. Effect of biaxial strain and external electric field on electronic properties of MoS2 monolayer: A first-principle study. *Chem. Phys.* **468** 9.

[285] Ye X.S., Shao Z.G., Zhao H., Yang L. and Wang C.L. 2014. Intrinsic carrier mobility of germanene is larger than graphene's: First-principle calculations. *RSC Adv.* **4** 21216.

[286] Tran N.T.T., Lin S.Y., Lin C.Y. and Lin M.F. 2018. *Geometric and Electronic Properties of Graphene-Related Systems: Chemical Bonding Schemes* (CRC Press).

[287] Ho Y.H., Ho G.W., Chen S.C., Ho J.H. and Lin M.F. 2007. Low-frequency excitation spectra in double-walled armchair carbon nanotubes. *Phys. Rev. B* **76** 115422.

[288] Allison T., Puce A. and McCarthy G. 2000. Social perception from visual cues: Role of the STS region. *Trends Cogn. Sci.* **4** 267–278.

[289] Mccash E.M., Parker S.F., Pritchard J. and Chesters M.A. 1989. The adsorption of atomoc-hydrogen on Cu(111) investigated by reflection-absorption infrared-spectroscopy, electron-energy loss spectroscopy and low-energy electron-diffraction. *Surf. Sci.* **215** 363–377.

[290] Lv B.Q., Qian T. and Ding H. 2019. Angle-resolved photoemission spectroscopy and its application to topological materials. *Nat. Rev. Phys.* **1** 609–626.

[291] Sun T.X. and Ding X.L. 2015. Confocal X-ray technology based on capillary X-ray optics. *Rev. Analyt. Chem.* **34** 45–59.

Chapter 3

# COULOMB SCATTERING OF QUASIPARTICLES

Chiun-Yan Lin, Ching-Hong Ho, Hsien-Ching Chung,
Lu-Yao Wang, Chih-Wei Chiu, and Ming-Fa Lin

Quasiparticles play a crucial role in describing the properties of many-particle systems in condensed matter physics. As quasiparticles move through a crystal, they experience frequent elastic and inelastic scatterings due to many-particle interactions, resulting in a finite lifetime within a certain energy range [1–4]. At zero temperature, electron–electron (e–e) Coulomb scatterings are the primary source of many-particle interactions. The quasiparticle concept differs from that of real particles in that it only demonstrates a one-to-one correspondence between state energy and wavevector in the dispersion relation [5]. The properties of quasiparticles are determined by their self-energies, which have real and imaginary components that are precisely defined by the self-consistent Dyson equation [6].

Although the exchange energy, which corresponds to first-order Coulomb interactions, has an analytic formula and has been derived for three-dimensional (3D), two-dimensional (2D) and one-dimensional (1D) electron gases [7–10], it cannot be solely used to evaluate many-body effects, such as quasiparticle energy spectra, as the correlation energy covering the entire two-particle interaction orders, except for the leading term, does not have an exact solution for any condensed matter system [11]. Additionally, higher-order Coulomb interactions cannot be calculated even through numerical simulations. As a result, several approximate models have been developed to understand the energy and decay rate of quasiparticles, such as those proposed by Singwi *et al.* [13], Nozieres and Pines [14], and Hubbard [12]. These models help to qualitatively and semi-quantitatively explain the quasiparticle energies and decay rates.

Moreover, exchange and correlation interactions strongly affect other fundamental properties, such as the carrier-density-dependent chemical potential [15, 16], the fermion distribution function near the Fermi level [17], the renormalization factor near the Fermi momentum [18], the dynamic propagation of charge density wave [19], the spin-dependent pair distribution function [20] and Friedel oscillation [21]. These interactions are highly sensitive to local charge density and have been formulated as a function of $\rho(\mathbf{r})$ for a homogeneous gas through Monte Carlo quantum calculations [22]. The density functional formula has become the standard method for first-principles numerical calculations, such as those used in the VASP program [23].

In condensed matter, the electronic properties can be changed by varying the type or concentration of doping. It has been shown that the tunability of the electronic properties can be realized in layered graphenes by means of both electrical [24, 25] and chemical [26–28] doping without deterioration of the unique nature of graphene. Either n-type or p-type electrical doping is conveniently controlled by electrical and static electric gating [24, 25], and the carrier concentration can be varied over a wide range simply by changing the bias voltage. Chemical doping is also an effective method to change these characteristics, in addition to electrical doping. Electrons can be transferred from intercalants or dopants to graphene and vice versa due to their different electronegativities. The carrier concentration can be manipulated for both n-type and p-type graphenes by adsorbing metal atoms or gas molecules on the surface [26–28]. The aforementioned experiments prove the 2D nature of the charge carriers in graphene, with the Fermi level determined according to the rigid shift model. The plasmon dispersions have been widely studied within the framework of the random-phase approximation (RPA) [29]. Many relevant theoretical papers on plasmons in monolayer graphenes [30, 31] and few-layer graphenes [108–110] are based on these considerations and adopt the rigid shift model to adjust the Fermi level in the doping case. The theoretical predictions on the low-energy plasmons [30, 31] have been confirmed through electron energy loss spectroscopy (EELS) in n-type monolayer graphene doped with potassium atoms [28] and then epitaxially grown on SiC(0001) [36]. The agreement between experiment and theory shows important advances in many-body e–e interactions in graphenes.

Quasiparticle properties are a mainstream research topic in condensed matter systems, such as topological insulators [2, 37–39], doped 3D semiconductor compounds [40, 41], 2D quantum wells [42–44], transition metal dichalcogenides [91], 1D quantum wires [46, 47], and 3D–1D carbon-$sp^2$ bonding systems, e.g., carbon nanotubes [48], graphene [49, 50] and graphite intercalation compounds [51–53]. The main research focus is to thoroughly clarify the quasiparticle energy and lifetime near the Fermi level in relation to the e–e interaction, Bloch wave functions, energy dispersions and Fermi surfaces. For example, the 3D stage-1 graphite intercalation compound is an outstanding candidate, which can clearly illustrate the many-body effect due to the coupling of intraband and interband excitations [54]. The quasiparticles near the Fermi surface have an energy width proportional to $(k - k_F)^2 ln|k - k_F|$ (similar to 2D electron gas) [7–9], but the excited conduction electron in the p-type compound appears as an abnormal characteristic closely affected by the inter-zone Coulomb excitation [7]. The many-body effects that have been neglected so far in graphite systems may reduce or increase the effective mass of the Fermi-momentum state, depending on the specific Fermi energy. The study of charge carriers in graphene can be considered as an attempt to understand the dynamics of quasiparticles in graphene and its related systems under the many-body interactions of e–e, electron–plasma and electron–phonon couplings [50]. Unveiling the quasiparticle dynamics of topological insulators reveals the origin of the gapped Dirac states in topological insulators [37, 38]. That is, such materials can provide the complete information required to clarify the dynamics of quasiparticles of quantum materials. In this chapter, all available Coulomb decay channels are included in the modified RPA and the modified self-energy. Also, the Coulomb interaction between different layers or within the same layer is considered. The linear Dirac cone structure with a red- or blue-shift with respect to the Fermi level is suitable for describing the electronic structure of the ground state under electron or hole doping. The dependence of the quasiparticle energy spectrum and lifetime on wavevector and doped carrier density is discussed in detail, which should manifest itself as many-body properties in the optical absorption spectra [55, 56], transport conductivities [57], electron energy loss spectra [58, 59] and angle-resolved photoemission spectra [60].

The doped monolayer graphene, silicene and germanene have been chosen for modeling using the tight-binding model [61] and the effective-mass approximation [62] to understand the quasiparticle behaviors near the Dirac point. The former is more suitable and reliable for realizing the low-energy Dirac-like Hamiltonian under many-particle interaction. The significant difference between the three systems is whether the gap is opened at the Dirac point due to the spin–orbital couplings [63]. Germanene owns the smallest hopping integral of the $\pi$ bondings, so the essential electronic properties and the intrinsic interaction, such as spin–orbit interaction, are relatively easily modulated by external factors, such as carrier doping as well as electric and magnetic fields. On the other hand, due to the strongest hopping integral ($\sim$2.6 eV), graphene has a robust gapless Dirac cone, consisting of a pair of linearly intersecting valence and conduction bands, which has been regarded as a good material to explore the rich and unique physical properties of Dirac quasiparticles [64, 65]. Graphene and germanene exhibit different intrinsic effects in terms of electronic excitation and Coulomb decay rate. In germanene, the acoustic plasmon mode, which could survive in both systems, presents fully undamped responses at small momentum transfers but experiences very strong interband Landau dampings and even gradually disappears under large $\mathbf{q}$. The acoustic plasmon mode that can survive in the 2D systems presents an undamped response at small momentum transfers; however, under a large momentum, it experiences very strong interband Landau damping and gradually disappears. In addition, it is predicted that the doped germanene exhibits anisotropic excitation spectra, electron–hole (e–h) excitation boundaries and extra second and third plasmas due to spin–orbital interactions. These features do not exist in doped graphene [66, 67].

The Coulomb decay rates are investigated for both excited conduction electrons or holes and valence holes. Their behavior is abnormal for doped graphene and germanene systems that contrast sharply with each other [66, 67, 116]. The low-energy quasiparticles just near the Fermi level decay associated with intraband single-particle excitations are similar to those in a 2D electron gas [7–9]. Only the occupied valence holes near the Fermi level can decay through the undamped acoustic plasmon, so that they present very fast Coulomb de-excitations, nonmonotonous energy dependence and anisotropic behavior. That is, such a decay process can occur for plasmons outside the e–h boundary, where plasmons are well-defined quasiparticles. The higher-energy conduction state and the deeper-energy valence one with sufficient momentum transfer to the de-excitation channel of e–h pairs behave similarly in terms of the dependence of the wavevector on the Coulomb decay rate. By measuring the energy dispersion and linewidths of the quasiparticles along the *KM* and

$K\Gamma$ directions, high-resolution angle-resolved photoelectron spectroscopy (ARPES) measurements are performed in monolayer graphene with potassium adsorption, thereby verifying the scattering decay rate in graphene [50, 69, 70]. The Lorentzian peak in the measured momentum distribution curve is centered on the energy of the quasiparticle, and the Lorentzian linewidth is $-2Im\Sigma^{(\text{res})}$, which is identified as the scattering decay rate, where $\Sigma^{(\text{res})}$ is self-energy. In addition, the Coulomb decay rate of the Fermi-momentum state is described by its relationship with the temperature $T$, $T^2 ln[T]$, as observed in a 2D electron gas [71]. The dynamics of the quasiparticle are suggested to be explained by the single particle, plasmon and phonon scatterings at finite temperatures, while the ARPES measurements performed at extremely low temperatures may provide Coulomb scattering pictures to support the theoretical predictions [72]. Regarding monolayer doped germanene, experimental studies are still needed to clarify the main characteristics of the Coulomb decay rate and the de-excitation mechanism. In the previously published works [109, 110], the calculations consider the electronic excitations coming only from the first pair of valence and conduction bands nearest to the Fermi level [109] or coupling only by the interlayer Coulomb interactions [110]. On the other hand, in this work, we incorporate the full charge screening due to all the pairs of energy bands through the development of the modified RPA. The layer-based Coulomb potential, the polarization function, the dielectric function and the energy loss function are characterized by the analytic equations, in which all the interlayer atomic interactions and Coulomb interactions are taken into account. The reliable Coulomb excitation spectra are very useful in understanding the many-body effects in layered graphenes.

It is worthwhile to systematically investigate monolayer graphene [35] as well as bilayer and trilayer graphenes with AA, AB or ABC stacking configurations in terms of the Coulomb decay rates. The main characteristics of the electronic state and elementary excitations are briefly introduced in Section 4.1. The de-excitation channels for carriers include the single-particle and collective excitations of different angular-momentum modes, with the latter especially being a very effective decay channel. Generally, conduction or valence carriers far from the Fermi level have more de-excitation channels, or shorter quasiparticle lifetimes. Certain states of each conduction subband have very long lifetimes, which are attributed to intraband single-particle channels under small transferred momentum, and there is no simple relationship between the inverse of lifetime and the state energy. Temperature and free carrier doping can affect single-particle and collective excitations, leading to a more complex de-excitation process [74]. As for monolayer graphene [35], the temperature-induced free carriers affect the excitation near the Fermi level and cause the low-frequency intraband and interband e–h excitations to become effective decay channels for the excited states of the low-lying subbands. Specifically, temperature also triggers the Coulomb de-excitation channels and decay rates specified in type-I, type-II and type-III carbon nanotubes [75]; in addition to temperature, these characteristics also depend on chirality and radius. The electronic states of type-II carbon nanotubes present the highest decay rate near the Fermi level at room temperature among the three types of carbon nanotubes [73–78] . In addition, temperature produces few free carriers [75] for the Coulomb scattering rate of excited carriers in type-III carbon nanotubes. However, due to the medium gap, the intraband e–h excitations are the only de-excitation channels and are very weak. The Coulomb decay rates of double-walled carbon nanotubes [79, 80] with different stacking configurations are fully explored under Coulomb interactions and the intrinsic atomic interactions in the tube. In general, two plasmon-assisted peaks occur at the momentum distribution curve of the decay rate spectrum at different Fermi levels. The results show that the symmetry of the interlayer stacking configuration plays different roles between the cylindrical and planar carbon honeycomb lattices, as explained in the following [81]. The above statement should consider the e–e inelastic scattering of the multi-band structure in the layered system, which is reflected in the main characteristics of the single-particle excitations and the multi-mode plasmons in the elementary excitation spectrum. The inelastic lifetime of doped excited carriers above the Fermi level can be attributed to single-particle inelastic scattering, and other processes by emission of plasmons, in general, may be of importance via charge collective excitations. As for N-layer graphene [82, 83], the induced free carriers in the low-energy bands cause several low-frequency plasmon modes in addition to the intraband and interband e–h excitations, which become effective decay channels for the excited states of the lowest conduction and valence subbands. As expected, the Coulomb decay rate of the excited state must be sensitive to changes in its energy and wave function, de-excitation process, temperature, number of layers and stacking configuration. The de-excitation process is worthy of detailed study on the conditions under which the excited carriers decay through e–h pairs or collective charges.

The quasiparticle band structure of low-dimensional materials has been carefully examined through various experimental methods, such as ARPES [84, 86, 101] and EELS [87–90]. The remarkable work by Ohta *et al.* [101] involved using ARPES to examine the Coulomb screening effect and interlayer atomic interactions in multilayer graphene by comparison with tight-binding calculations. Jin *et al.* [86] also used the same technique to determine the thickness dependence of the $MoS_2$ electronic band structure. Meanwhile, Ulstrup *et al.* also investigated the variations in the lateral electronic structure in a lateral heterostructure of semiconducting $WS_2$ placed on monolayer graphene or bilayer graphene by using ARPES [91]. EELS has played an equally impressive role in verifying the electronic properties. For example, EELS is used to study the effect of strain on the optical conductivity of graphene [87]. A plasmon mode with a negative dispersion of 1.1 eV energy is revealed on the Cu(111) surface by high-resolution EELS [88]. This collective mode is suggested to come from the electronic oscillation of the 2D Shockley surface state confined to the Cu(111) surface. In a previous work, the plasmon dispersion of epitaxial graphene was measured [90]. The most important thing is that the $\pi$ bonding between the two $p_z$ orbitals in graphene dominates the low-energy band, as shown by first-principles calculations [46]. These results show that the plasmons for monolayer, bilayer and trilayer graphenes have significant differences in the in-phase and out-of-phase modes as a result of the different band structures between graphenes.

ARPES is the most powerful and efficient technique to measure the energy dispersions and scattering rate of quasiparticles within the first Brillouin zone. The former measurement can be used to verify the calculated band structure, whereas the latter can be used to identify the screened exchange self-energy and therefore directly explore many-body interactions. Experimental measurements have verified the band structures in a condensed matter system related to graphene. For example, the band change along $k_z$ in Bernal graphite [92–95], parabolic energy dispersions due to quantum confinement in graphene nanoribbons [96, 97], Dirac cone in monolayer graphene [50, 98], multi-cones in AA graphene [99, 100], two pairs of parabolic dispersions in AB graphene [101–104] and composted band structure in twisted bilayer graphene [99], trilayer ABA graphene [101, 103, 104] and ABC-stacked trilayer graphene [101, 103, 104]. The electronic properties of these systems are attributed to the effects of various sizes, number of layers, stacking symmetry and chemical adsorption of adsorbed atoms or molecules. However,

ARPES measurements are not suitable for testing the quasiparticle properties of single-walled carbon nanotubes. It should be noted that the measured scattering rate of quasiparticles gives information about the quasiparticle dynamics to support theoretical predictions. The dynamics of quasiparticles is very useful for understanding the broadening mechanism of ultrafast optical absorption spectra [105]. Regarding monolayer doped germanene, experimental studies are still needed to clarify the main characteristics of the Coulomb decay rate and the de-excitation mechanism. In short, the quasiparticle decay rates contrast sharply with one another among the aforementioned graphene-related materials, indicating that the geometric shapes and dimensions play a critical role in the many-body effect.

## 3.1 Dynamic and Static Dielectric Functions in Layered Systems

E–e interactions play critical roles in the fundamental properties of quasiparticles, such as Coulomb excitations and decays, dynamic plasmon-assisted charge density waves and Friedel oscillations. These popular topics in condensed matter physics are explained in detail in Chapters 3–5 of this book. In order to cover all the important mechanisms of the Hamiltonian, the RPA and self-energy methods need to be strongly modified to be consistent with the tight-binding model based on layers and sublattices, in which the band-structure effects dominate their excitation/de-excitation phenomena [30, 82, 106–116]. The most important point is the electric polarization function tensor of sublattice index, which has never been presented in previous studies [30, 107–110].

In the presence of an external Coulomb potential in the system, the electrons will redistribute and create an induced potential to screen this perturbation [29]. The bare and screened response functions are thoroughly explained by the dynamically inelastic scattering of electrons. Within the linear response, the dimensionless dielectric function in $(\mathbf{q}, \omega)$ is defined as the ratio of the bare potential and the effective potential

$$\epsilon(\mathbf{q},\omega) = \lim_{V^{\mathrm{ex}}\to 0} \frac{V^{\mathrm{ex}}(\mathbf{q},\omega)}{V^{\mathrm{eff}}(\mathbf{q},\omega)}, \tag{1}$$

which can also be presented by the charge density and the longitudinal electric field, namely $\rho^{\mathrm{ex}}/\rho^{\mathrm{tot}}$ and $D_l/E_l$[117]. The bare polarization function ($P$) is used to describe the induced charged density in linear response to the effective Coulomb potential. Using the Poisson equations and the self-consistent field approach, the induced Coulomb potential is calculated as the product of the induced charged density and the bare Coulomb potential. As a result, the dielectric function is expressed as

$$\epsilon(q,\phi,\omega) = \epsilon_0 - V_q P(q,\phi,\omega), \tag{2}$$

where $\epsilon_0(= 2.4)$ is the background dielectric constant due to $\sigma$-electron excitations, and $V_q = 2\pi e^2/q$ for the 2D bare Coulomb potential [111].

Taking the band-structure effects into account, the bare polarization function is given by

$$P(q,\phi,\omega) = \sum_{h,h'=c,v} \langle \mathbf{k};h|e^{-i\mathbf{q}\cdot\mathbf{r}}|\mathbf{k}+\mathbf{q};h'\rangle \times \frac{f(E^{h'}(\mathbf{k}+\mathbf{q})) - f(E^h(\mathbf{k}))}{E^{h'}(\mathbf{k}+\mathbf{q}) - E^h(\mathbf{k}) - (\omega + i\Gamma)}, \tag{3}$$

where $h = v/c$ stands for the valence/conduction state. The Fermi–Dirac distribution is adjusted according to intrinsic and extrinsic cases at a finite temperature. The transferred momentum and frequency are conserved during the e–e Coulomb interaction, which is a necessary condition for describing the electronic excitations. The expressed dielectric function is also suitable to describe monolayer silicene and germanene with buckled honeycomb lattices; however, the electronic states are modified by the significant spin–orbital interactions, resulting in remarkable differences in rich and diversified electronic excitations [106, 114, 118].

The dielectric function of $N$-layer graphene is more complicated [82]. The perturbed Coulomb potentials and induced charges arise from all the layers; that is, the effective potential needs to simultaneously consider the external and induced Coulomb potentials due to each layer in the modified RPA. The stacking configuration and the number of layers determine the full band structure that can provide rich and unique electronic excitations in the $(\mathbf{q},\omega)$-phase diagrams. When an electron beam is incident on a graphene plane, the potential is assumed to be uniform on each layer, so the $\pi$ electrons on different layers experience similar bare Coulomb potentials. Specifically, the excited electrons and holes in each excitation pair due to the Coulomb perturbation frequently occur on distinct layers. The Feynman diagram of the Coulomb excitations is depicted in Fig. 3.2. Characterized by the Dyson equation, the effective Coulomb potential for the two electrons on the $l$th and $l'$th layers is expressed as

$$\epsilon_0 V^{\mathrm{eff}}_{ll'}(\mathbf{q},\omega) = V_{ll'}(\mathbf{q}) + \sum_{mm'} V_{lm}(\mathbf{q}) P^{(1)}_{mm'}(\mathbf{q},\omega) V^{\mathrm{eff}}_{m'l'}(\mathbf{q},\omega), \tag{4}$$

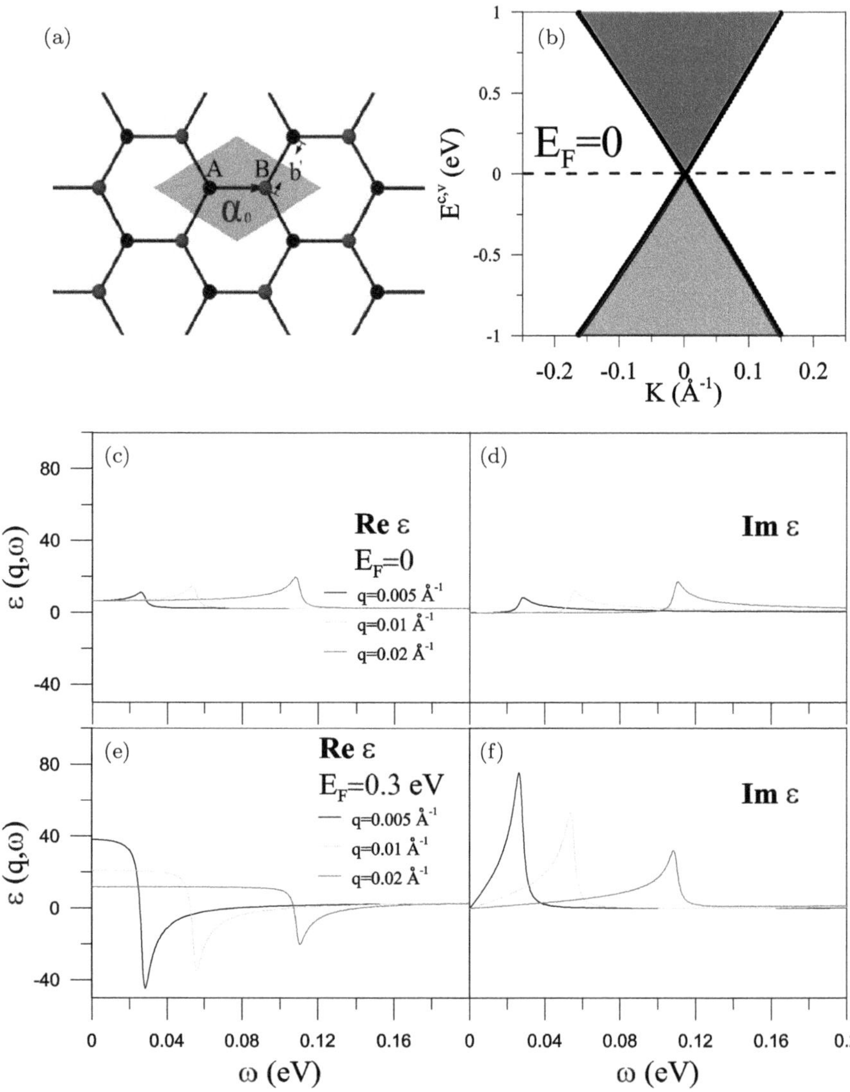


Fig. 3.1 (a) Honeycomb lattice of graphene. (b) The monolayer Dirac cone at the K point. (c) and (d) The dielectric function for various **q** under $E_F = 0\,\text{eV}$ and $E_F = 0.3\,\text{eV}$.

in which each term is labeled by layer indexes and clearly reveals the Coulomb potential, induced charge density and response function associated with any two layers. The first term is also useful in understanding the Coulomb decay rates in layered systems. The bilayer-like bare response

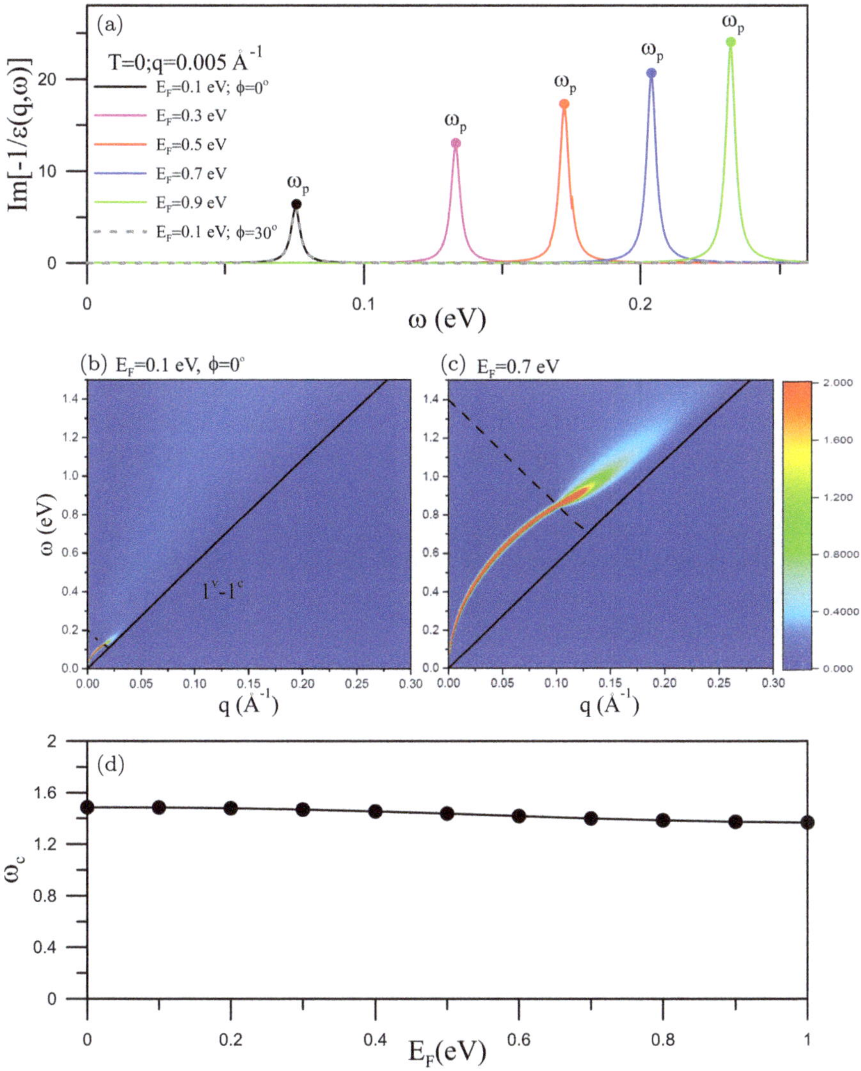


Fig. 3.2 (a) The energy loss function spectra of monolayer graphene for $\mathbf{q} = 0.005$ Å$^{-1}$ under different Fermi levels, where $\omega_p$ indicates the plasmon energy. Phase diagram of monolayer graphene at (b) $E_F = 0.1\,\text{eV}$ and (d) $E_F = 0.7\,\text{eV}$. The solid and dashed black curves, respectively, correspond to the boundaries of intraband and interband excitations. (d) The variation of the characteristic frequency with different Fermi levels.

function for the RPA bubble diagram is given by

$$P^{(1)}_{mm'}(\mathbf{q},\omega) = 2\sum_{k}\sum_{nn'}\sum_{h,h'=c,v}\left(\sum_{i} u^{h}_{nmi}(\mathbf{k})u^{h'*}_{n'm'i}(\mathbf{k}+\mathbf{q})\right)$$
$$\times\left(\sum_{i'} u^{h*}_{nmi'}(\mathbf{k})u^{h'}_{n'm'i'}(\mathbf{k}+\mathbf{q})\right)$$
$$\times\frac{f(E^{h}_{n}(\mathbf{k})) - f(E^{h'}_{n'}(\mathbf{k}+\mathbf{q}))}{E^{h}_{n}(\mathbf{k}) - E^{h'}_{n'}(\mathbf{k}+\mathbf{q}) + \hbar\omega + i\Gamma}. \tag{5}$$

Any electronic excitation that satisfies the Pauli exclusion principle and the conservation of transferred momentum and frequency can be decomposed into a number of layer-dependent contributions by analyzing their wave functions. This conceptual analysis performed on the tight-binding functions critically matches the layer-dependent Coulomb potential. This indicates that the modified RPA could be generalized to layered systems, even in the presence of external magnetic and electric fields. Based on the linear relationship between bare and screened potentials, we can write the layer-dependent dielectric function as

$$\epsilon_{ll'}(\mathbf{q},\omega) = \epsilon_0\delta_{ll'} - \sum_{m} V_{lm}(\mathbf{q})P_{m,l'}(\mathbf{q},\omega), \tag{6}$$

showing that

$$\sum_{l''}\epsilon_{ll''}(\mathbf{q},\omega)V^{\text{eff}}_{l''l'}(\mathbf{q},\omega) = V_{ll'}(\mathbf{q}). \tag{7}$$

The zero points of the dielectric tensor in $\omega$ space are available for understanding the energies of plasmon modes, while their intensities can't be determined. In inelastic experimental measurements, the observed energy loss spectrum is associated with $\mathbf{q}$ and $\omega$, which need to be defined in further formulas for describing the collective excitations. In this chapter, we focus on the inelastic scattering rate of the quasiparticles that transfer momentum and energy to the system. The derivation of the formula is given in the following sections. Moreover, the plasmon-assisted oscillations of the dynamical charge density will be discussed in detail in Chapter 4.

More interestingly, the static charge screening phenomena are largely diversified by the atomic binding in condensed matter systems. For the particular case of $\omega = 0$, the screening length is characterized by the static effective Coulomb potential between two charges, which is obtained by the Fourier transform from the momentum-space domain to the real-space domain [29]. The static screening effect is closely related to the Fermi momentum and density of free carriers in layered systems. Taking monolayer graphene as an example, the Fermi level is located at the Dirac point (the conduction/valence cone) in an intrinsic (extrinsic) case, so it belongs to a zero-gap semiconductor (a metal with the free electron/hole density roughly proportional to the square of the Fermi momentum). The static dielectric function used to describe the screening ability is zero/finite under the long-wavelength limit ($q \to 0$) for the intrinsic/extrinsic graphene, which will affect the long-range Coulomb behavior away from impurities [107, 119]. As a result, the effective Coulomb potential is predicted to remain inverse distance decay in the former case. However, for an electron-doped graphene, the e–e interaction close to the charged impurity will rapidly decrease and cause a long-distance oscillation of the charge density on the surface. This static screening phenomenon is the well-known Friedel oscillation, which comes from the singularity in the static dielectric function of the wavevector at $2k_F$ (evaluated from an isotropic Fermi surface). The free carrier density and the Fermi surface, or the energy gap, are expected to play critical roles in the unusual screening behaviors. Whether the long-range effective Coulomb interactions decay quicker than the inverse of the characteristic momentum is the standard criterion for charge screening ability. This fascinating Friedel oscillation is the focus of Chapter 6. Further studies demonstrate interesting phenomena in various systems, including other important intrinsic factors, such as chirality, dimensionality, layer number, stacking configuration and extrinsic factors, e.g., temperature and external fields.

## 3.2 Lifetime of Quasiparticles

The inelastic scattering rate of the probing electrons that transfer the specific momentum and frequency $(\mathbf{q}, \omega)$ to the system can be delicately evaluated using the Born approximation [29]. Appropriately and reliably characterizing the experimental energy loss spectra, the dielectric function tensor in layered systems is given as follows using the Dyson equation [29, 83]:

$$\mathbf{Im}[-1/\epsilon] \equiv \sum_{l} \mathbf{Im}\Big[-V_{ll}^{\text{eff}}(\mathbf{q}, \omega)\Big] \Big/ \Big(\sum_{lm} V_{lm}(q)/N\Big), \tag{8}$$

which takes the average of all the external Coulomb potentials on the different layers. The dimensionless loss function is utilized to explore the various plasmon modes; the entire methodology developed in the framework of modified RPA is suitable for layered condensed matter systems, such as layered graphene, germanene, phosphorene, tinene, silicene, antimonene and bismuthene.

The unified theoretical framework of the properties of single particles and many particles is very useful for a thorough understanding of different quasiparticle excitations and decays. For example, the calculations in graphene-related systems stressfully exploring the e–h excitation region and plasma mode under the effects of the lattice symmetry, number of layers, stacking configuration, temperature, magnetic field, doping, etc. [83]. In principle, high-resolution ARPES is a powerful technique for testing all the information about quasiparticles. The e–e interactions, orbital hybridizations, hopping integrals and external fields are unified to calculate the self-energy of the quasiparticle, which can be used to characterize its particular evolution under various scattering mechanisms of these effects. At low temperatures, the RPA self-energy, dominated by the e–e Coulomb interactions, is given by the Matsubara Green's function [29]:

$$\Sigma(\mathbf{k},h,ik_n) = -\frac{1}{\beta}\sum_{\mathbf{q},h',i\omega_m} V^{\text{eff}}(\mathbf{k},h,h';\mathbf{q},i\omega_m)G^{(0)}(\mathbf{k}+\mathbf{q},h',ik_n+i\omega_m), \tag{9}$$

where $G^{(0)}$ indicates the noninteracting Matsubara Green's function, $\beta = (k_B T)^{-1}$, the complex fermion frequency $ik_n = i(2n+1)/\pi/\beta$ and the complex boson frequency $i\omega_m = i2m\pi/\beta$. The screened Coulomb potential is expressed as

$$\begin{aligned} V^{\text{eff}}(\mathbf{k},h,h';\mathbf{q},i\omega_m) &= V(\mathbf{k},\mathbf{q},h,h')/\epsilon(\mathbf{q},i\omega_m) \\ &= V_q|\langle h',\mathbf{k}+\mathbf{q}|e^{i\vec{q}\cdot\vec{r}}|h,\mathbf{k}\rangle|^2/[\epsilon(\mathbf{q},i\omega_n)], \end{aligned} \tag{10}$$

where the effects of band structures and the intraband and interband de-excitation channels are in the dielectric function. The real and imaginary parts of the self-energy can be used to define the energy shifts and widths of the quasiparticle.

Under the analytic continuation $ik_n \to E^h(\mathbf{k})$, the Coulomb decay rate of the $|\mathbf{k},h\rangle$ state is characterized as

$$\begin{aligned}\frac{-1}{2\tau(\mathbf{k},h)} &= \frac{-1}{2\tau_e(\mathbf{k},h)} + \frac{-1}{2\tau_h(\mathbf{k},h)} \\ &= 2\sum_{\mathbf{q},h'} Im[-V^{\mathrm{eff}}(\mathbf{k},h,h';\mathbf{q},\omega_{\mathrm{de}})] \\ &\quad \times \{n_B(-\omega_{\mathrm{de}})[1-n_F(E^{h'}(\mathbf{k}+\mathbf{q}))] - [n_F(E^{h'}(\mathbf{k}+\mathbf{q}))]\}, \quad (11)\end{aligned}$$

where $n_B$ and $n_F$ indicate the Bose–Einstein and Fermi–Dirac distribution functions, respectively, and the summation takes over all available de-excitions from $|\mathbf{k},h\rangle$ $|\mathbf{k}+\mathbf{q},h'\rangle$, satisfying the Pauli exclusion principle and the conservation of energy and momentum (for detailed calculations, see Ref. [35]). In the above equation, the decay rate can be divided into electron decay and hole decay (indexed by subscripts $e$ and $h$, respectively) for the $|\mathbf{k},h\rangle$ state with the de-excitation/decay energy $\omega_{\mathrm{de}} = E^h(\mathbf{k}) - E^{h'}(\mathbf{k}+\mathbf{q})$ above or below the Fermi level. The different excitation patterns in momentum and frequency space become the available de-excitation channels for the Coulomb decay rate of quasiparticles, leading to very sensitive dependencies on the Dirac points, Fermi level and strong anisotropy.

The Coulomb decay rates have the following expression for the excited electrons and holes at zero temperature:

$$\begin{aligned}\frac{1}{\tau_e(\mathbf{k},h)} + \frac{1}{\tau_h(\mathbf{k},h)} &= -2\sum_{\mathbf{q},h'} Im[-V^{\mathrm{eff}}(\mathbf{k},h,h';\mathbf{q},\omega_{\mathrm{de}})] \\ &\quad \times [-\Theta(\omega_{\mathrm{de}})\Theta(E^{h'}(\mathbf{k}+\mathbf{q}) - E_F)) \\ &\quad + [\Theta(-\omega_{\mathrm{de}})\Theta(E_F - E^{h'}(\mathbf{k}+\mathbf{q}))], \quad (12)\end{aligned}$$

where $E_F$ is the Fermi level and $\Theta$ is the Heaviside step function that limits the available de-excitation channels. The derived Coulomb decay rate is just twice the energy width of the quasiparticle state. The theoretical framework of the modified self-energy equations could be applied using the layer-projection method for few-layer graphene-related systems and further generalized to cover spin–orbital effects in silicene and germanene [106, 114].

The interesting charge screening behaviors are investigated in this book, e.g., dynamic charge density waves assisted by plasmons, beating Friedel oscillations, dipole- or monopole-like induced charge distributions, and the complicated wavevector dependencies of Coulomb scattering. In addition, a large number of numerical calculation requirements related to wave functions have been overcome by the consistent framework. This theoretical framework is generalized to cover more quasiparticle phenomena such as chemical modification, spin configuration in two-particle interactions and inelastic scattering related to phonons. The unified framework depends on an understanding of lattice symmetry and scattering mechanisms.

In addition to the modified RPA self-energies, the Fermi golden rule could be utilized to fully explore the Coulomb decay rates of quasiparticle states, the excited electrons or holes, in systems such as carbon nanotubes and graphenes. The e–e inelastic scattering rate at any temperature $T$, which occurs from a specific state with wavevector $\mathbf{k}$ and band index $h$ to another state with $\mathbf{k}+\mathbf{q}$ and $h'$, is expressed as

$$\frac{1}{\tau_e(\mathbf{k},h)} = 2\pi \sum_{\mathbf{q},h',h'',h'''} f(E^{h'}(\mathbf{k})[1-f(E^{h'''}(\mathbf{k}-\mathbf{q}))][1-f(E^{h''}(\mathbf{k}+\mathbf{q}))]$$
$$\times \left|V^{\mathrm{eff}}(\mathbf{k},\mathbf{q},h,,h',h'',h''')\right|^2$$
$$\times\, \delta(E^{h'''}(\mathbf{k}-\mathbf{q}) + E^{h''}(\mathbf{k}+\mathbf{q}) - E^{h'}(\mathbf{k}) - E^{h}(\mathbf{k})), \qquad (13)$$

where $f(E^{h'}(\mathbf{k}))$ are the Fermi–Dirac distribution probabilities, and the electron scattering satisfies the Pauli exclusion principle. The available channel numbers during the scattering obey the conservation of momentum and energy, and the scattering strength after charge screening is evaluated using the effective Coulomb potential. The most critical factor is the effective Coulomb potential calculated from the matrix elements of the bare Coulomb potential and the dielectric tensor from RPA, which is expressed in Eq. (5). A similar formula for a quasiparticle hole under Coulomb scattering is given by

$$\frac{1}{\tau_h(\mathbf{k},h)} = 2\pi \sum_{\mathbf{k},\mathbf{q},h',h'',h'''} [1-f(E^{h'}(\mathbf{k}))][f(E^{h'''}(\mathbf{k}-\mathbf{q}))][f(E^{h''}(\mathbf{k}+\mathbf{q}))]$$
$$\times \left|V^{\mathrm{eff}}(\mathbf{k},\mathbf{q},h,,h',h'',h''')\right|^2$$
$$\times\, \delta(-E^{h'''}(\mathbf{k}-\mathbf{q}) - E^{h''}(\mathbf{k}+\mathbf{q}) + E^{h'}(\mathbf{k}) + E^{h}(\mathbf{k})). \qquad (14)$$

These two equations are very useful in calculating the quasiparticle decay rate of the unoccupied conduction bands and the occupied valence ones. For layered materials, the use of layer-decomposed wave function, Coulomb interaction and electrical polarization are completely introduced in the theoretical framework, and the formula of quasiparticle decay rate can be generalized. This is an applicable and reliable method to obtain the decay rate in multilayer graphenes. We also made a comparison with double-walled carbon nanotubes.

## 3.3 Dynamic Screening and Elementary Excitation in Multilayer Graphenes

The dielectric function ($\epsilon$) and the dimensionless energy loss function ($\mathrm{Im}[-1/\epsilon]$), which, respectively, represent the bare and screened response ability of charged carriers under a Coulomb field perturbation, are critical to fully understanding the intrinsic excitation/decay properties of condensed matter materials. In general, the imaginary part of $\epsilon$ and the prominent peaks in $\mathrm{Im}[-1/\epsilon]$, respectively, correspond to the single-particle and collective excitations (e–h excitations and plasmon modes). $\epsilon$ is strictly defined for bulk graphites [120], monolayer graphene [120] and cylindrical carbon nanotubes [121] since such systems possess good translational symmetry, respectively, in the 3D, 2D and 1D spaces. That is, the bare Coulomb e–e interactions, mainly determined by the transferred momentum ($\mathbf{q}$), exhibit well-behaved/dimension-determined forms. On the other hand, the longitudinal dielectric function is expressed as a tensor for the layered graphenes. The layer-projection method, which is closely related to the Coulomb scattering of the initial and final electronic states, will be introduced to deal with the electric polarizations. As a result, $\epsilon$ is a layer-dependent tensor function with double indices ($\epsilon_{ll'}$, where $l$ is the layer index), in which the band-structure effects on the excitation spectra and Coulomb matrix elements are fully taken into account in the theoretical model. For example, $N$-layer graphene possesses $N$ pairs of valence and conduction bands due to the carbon $2p_z$ orbitals, and the sublattice-decomposed wave functions are utilized to accurately evaluate the layer-dependent $P_{ll'}$ or $\epsilon_{ll'}$. Under the Born approximation, the effective energy loss function is characterized by the inelastic scattering probability of the incident electron beam. It has been very successful in understanding the diverse electronic excitation spectra in layered systems, e.g., AA- [113, 122–124], AB- [113, 125] and ABC-stacked graphenes [82]. The magneto-plasmon is discussed in this book through the development of the modified RPA and the generalized Peierls tight-binding model [126, 127]. It is very suitable for studying the inter-landau-level (inter-LL), single-particle excitations and magneto-plasmon modes [115]. Moreover, electronic excitation spectra are also quite efficient decay channels for the inelastic Coulomb scatterings, with strong effects on the energy widths of the quasiparticle states (the excited electrons or holes). The Coulomb decay rates are evaluated using the self-energy method in the detailed derivations. They will show certain important differences among monolayer graphene, silicene and germanene with the electron/hole doping. As to the experimental side, EELS and inelastic X-ray scattering (IXS) are two very powerful techniques for providing Coulomb excitation phenomena. Their recent developments [128, 129] and measurements on the screened response functions of graphene-related materials [130, 131] are explored in detail. Also, the quasiparticle energy spectra obtained using the measurements through ARPES are useful in determining the Coulomb decay rates [50, 132, 133].

For monolayer graphene depicted in Fig. 3.3(a), the tunability of the electronic properties can be achieved by means of either n-type or p-type electrical doping using electrical [24, 25] or chemical [26–28] methods. A previous study shows that graphene could be used to investigate the quasiparticle properties under n- or p-type dopings, such as the $E_F$-dependent energy spectra and lifetimes [7]. The plasmon excitations have been widely studied within the framework of the RPA. There are several relevant papers that focus on the plasmons in monolayer and few-layer graphenes and adopt the rigid shift model to adjust the Fermi level in the doping case.

Figure 3.3(b) clearly shows a Dirac cone band structure at the K/K′ point. The dispersion of the energy range of interest is described by $E^{c,v}(\mathbf{k}) = \pm v_F|\mathbf{k}|$, which is isotropic with respect to the K/K′ point. An analytic formula of the Bloch wave function is used to derive the dielectric function under a small momentum-transfer limit (details in Ref. [111]). The free electrons/holes can produce intra-$\pi$-band transitions (Fig. 3.3(b)) between the same conduction/valence bands in the Dirac cone, resulting in an excitation channel that is very different from the original one in the Coulomb excitation. The transitions in the $\pi$ band dominate the characteristics of the dielectric loss function (Figs. 3.3(c) and 3.3(f)), where their $\mathbf{q}$ evolution is shown in doped and undoped graphenes.

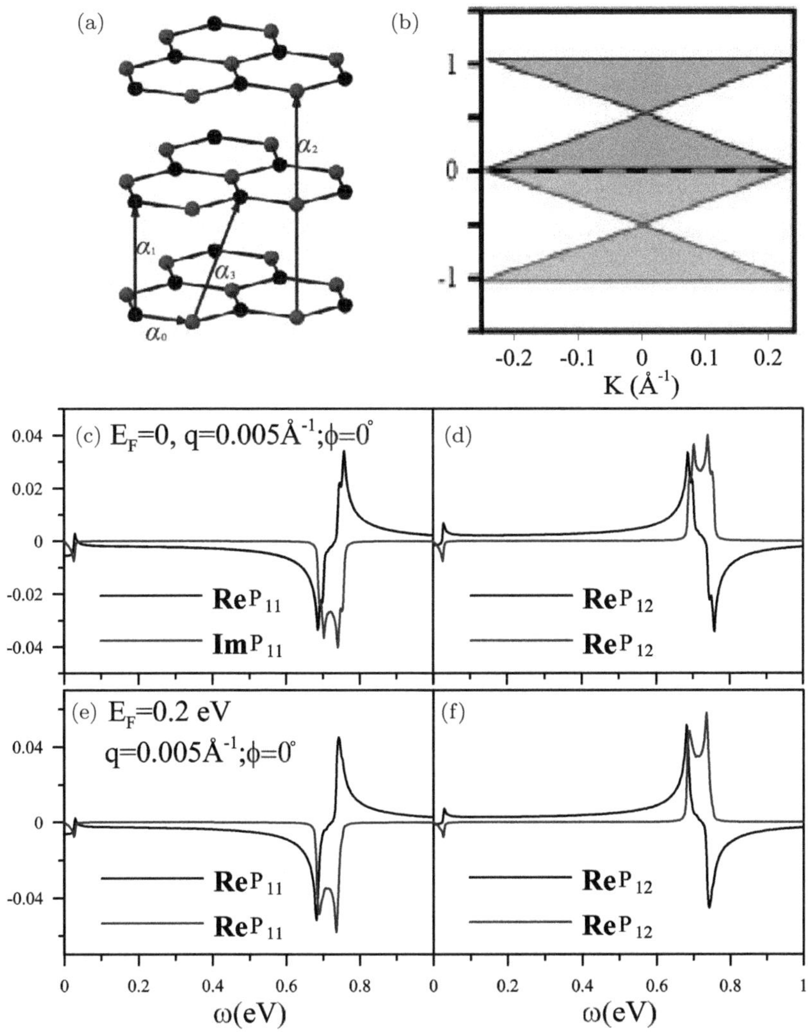


Fig. 3.3 (a) Structure of AA-stacked graphene. (b) The multilayer Dirac cone at the K point. (c) and (d) The dielectric function of bilayer AA-stacked graphene for various **q** under $E_F = 0\,\text{eV}$ and $E_F = 0.3\,\text{eV}$.

Under pristine condition, interband excitations are the only available excitation channel. The singularity in the dielectric function appears at the threshold frequency of single-particle excitations, i.e, $\omega_{\text{sp}} = v_F q$. A delicate expression for the divergence of the dielectric function, $\propto \sqrt{v_F q - \omega}$

and $\propto \sqrt{\omega - v_F q}$, is given for the imaginary and real parts of the dielectric function [111]. Both intraband and interband excitations contribute to electronic excitation with an increase in the density of free carriers, and the former gradually dominates at a sufficiently high Fermi level. The two zeros of Re$\epsilon$ can be interpreted as collective excitation; however, Landau damping determines the intensity of the excitation.

For the cases of different electron dopings, the change in the acoustic plasma mode under different **q** is explained (Fig. 3.4). The plasmon frequency $\omega_p$ is enhanced by the Fermi level, indicating that due to the higher electron density, the coherent oscillations of free conduction electrons are relatively easy to observe by EELS measurements [128]. The electron energy loss spectrum shows an excitation phase diagram on a large area $(\mathbf{q}, \omega)$, where the acoustic plasma mode with the group velocity $d\omega_p = dq$ is prominent in the region enclosed by the boundaries of the single-particle excitation channels. Plasmons usually occur in a specified region of the phase diagram where the plasmon dispersions don't coincide with the continuum e–h spectrum. Plasmon refers to the propagation of electron density waves, the wavevector is **q**, and the group velocity is $d\omega_q/dq$. Similar to 2D electron gas systems, the plasmon frequency is described by the dispersion relation of $\sqrt{q}$. Based on the linear dispersion, the two boundaries are described by the expressions $v_F q - 2v_F k_F$ and $2v_F k_F - v_F q$, respectively, due to the intraband and interband transitions. It should be noted that single-particle excitations are not allowed there. Moreover, unlike the cases in 3D and 2D electron gases [29], the coexistence of plasmon and e–h excitations is not allowed. Beyond that region, the plasma excitation is remarkably destroyed, called Landau damping, and the response function becomes dominated by single-particle excitations. As mentioned above, the plasmon frequency is estimated as $\omega_p \propto \sqrt{n_f q}$ by calculating the frequency of the second zero point in the real-part dielectric function (Figs. 3.3(c) and 3.3(e)). The dispersion relation is the same as that derived for 2D electron gas systems, although the electronic states in these two systems are described by different energy dispersions [9]. This is because they both exhibit the same excitation frequency dependence on the long-wavelength limit, which might explain the intrinsic excitation phenomena that are qualitatively similar among metallic 3D, 2D and 1D condensed matter systems, such as the lifetime of quasiparticles [134]. As a result, the characteristic frequency $\omega_c$ gradually becomes the integration of the collective excitation frequency with an increasing $E_F$. As for multilayer graphenes, there are more intra- and

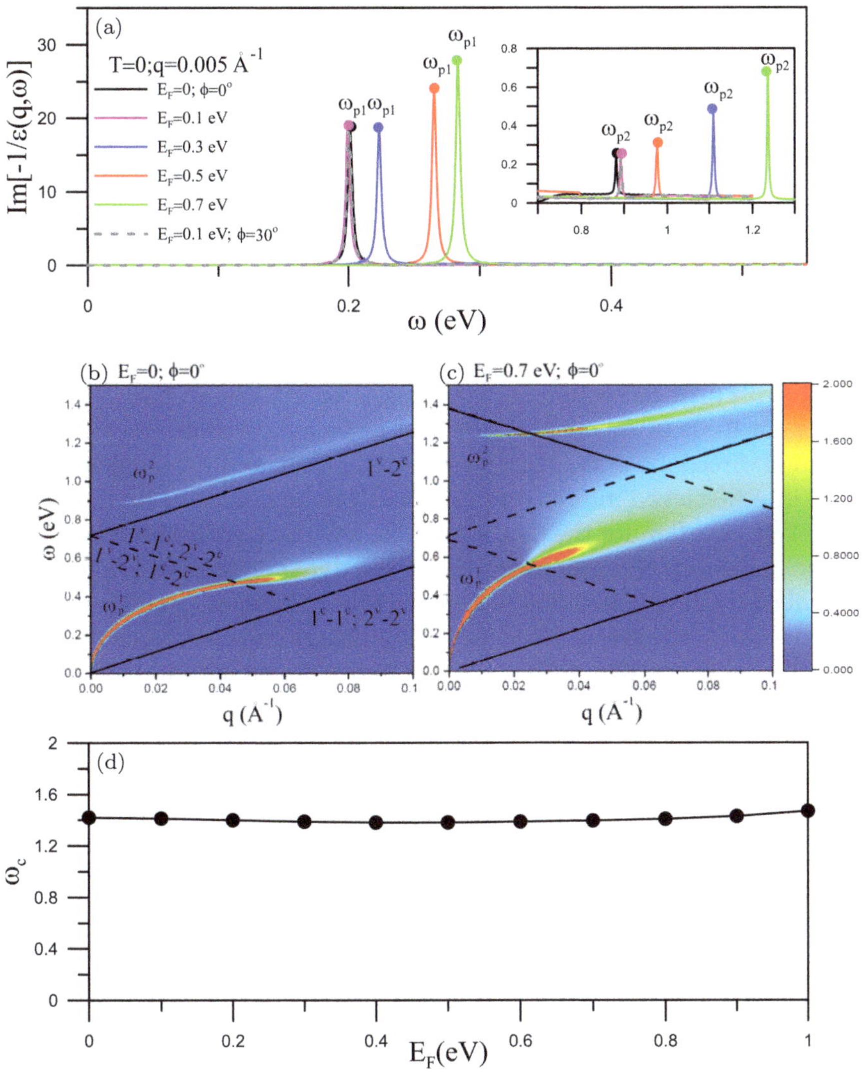


Fig. 3.4 (a) The energy loss function spectra of AA-stacked bilayer graphene for $\mathbf{q} = 0.005$ Å$^{-1}$ under different Fermi levels, where $\omega_p$ indicates the plasmon energy. Phase diagram of AA-stacked bilayer graphene at (b) $E_F = 0.1\,\text{eV}$ and (d) $E_F = 0.7\,\text{eV}$. The solid and dashed black curves, respectively, correspond to the boundaries of intraband and interband excitations. (d) The variation of the characteristic frequency with different Fermi levels.

interband plasmon oscillations. In short, the dimensionless energy loss spectra and the plasmon waves vary due to internal and external parameters, such as the stacking configuration of the metallic layered graphenes, the number of layers, free carrier density, gate voltage, magnetic field and temperature.

The AA-stacking configuration is highly symmetric, formed by perpendicularly stacking graphene sheets along the $\hat{z}$ direction. The geometric structure is depicted on the top-left side in Fig. 3.3(a), where $\alpha_1$, $\alpha_2$, $\alpha_3$ and $\alpha_4$ are four representative tight-binding parameters for the calculation of the electronic band structure. As shown in Fig. 3.3(b), the band structure consists of two monolayer-like Dirac cones located at the K/K′ point, with their respective Dirac points given by $-\alpha_1$ and $\alpha_1$. This indicates that the loss function is deduced to be almost isotropic along the different directions of the transferred momentum, and the single-particle state in AA-stacked multilayer graphenes can be described by the massless Dirac equation for the case where the low-lying states are retained in the semi-metallic AA graphenes [113]. The dynamic response function is described by the polarization functions of the intralayer $P_{ll}$ and interlayer $P_{l2}$; they are, respectively, responses to the charge correlations on the same layer and between two different layers under dynamic Coulomb interactions. In the undoped case, there are excitation channels resulting from transitions between the intra-Dirac cone and inter-Dirac cone structures. Given a transferred momentum $\mathbf{q}$, the former are available for both Dirac cones because they possess free electrons or holes when the Fermi level is at zero energy. This leads to a peak created at $\omega \sim 0.02\,\text{eV}$ in the screened response function, which is expressed as $1/\sqrt{\omega^2 - \omega_{\text{sp}}^2}$ (Figs. 3.3(c) and 3.3(d)), where $\omega_{\text{sp}}$ is the single-particle excitation energy corresponding to the excitation of the Fermi momentum states for either of the two Dirac cones. The latter channel triggers the other prominent structure at $E \sim 0.7\,\text{eV}$, whose intensity is relatively stronger due to the higher density of states and the symmetric relationship of the wave functions [113]. When the Fermi level is increased to $E_F = 0.2\,\text{eV}$, the response function maintains a similar behavior over all $\omega$, reflecting the low-lying resemblance to the linear dispersions of the Dirac cone. Furthermore, the two polarization functions are expected to induce two modes for the collective excitations, depending on the relationship between the oscillation phases.

We first review the energy loss functions of the AA-stacked bilayer graphene, as shown in Fig. 3.4(a) for various $E_F$ values. The Fermi level determines the interplay between the intrapair and interpair excitations. For any $E_F$, there are two intrinsic plasmon peaks, labeled by $\omega_p^1$ and $\omega_p^2$ with solid dots of different colors. The first peak is attributed to an acoustic mode, which comes from the interplay of two intrapair channels with in-phase oscillations between the two layers. In contrast, the channel of the two interpair excitations triggers the optical peak of $\omega_p^2$. As a typical combination of two oscillation modes, the intensity of the in-phase mode is much stronger than that of the out-of-phase mode, which indicates that the collective plasmons of the multi-Dirac cone are very different from those described in monolayer graphene. In short, the phase relationship between two polarizations can provide the ideal means of defining the composite of plasmon modes.

The $(\mathbf{q}, \omega)$-phase diagram of the energy loss function could provide complete information about the single-particle and collective excitations, as clearly shown in Figs. 3.4(c) and 3.16(d). The former spreads over the whole spectrum except for a triangular region near $q \longrightarrow 0$ and $\omega \longrightarrow 0$ and a large region below the first solid line according to the shape of the Fermi surface. Two various plasmon modes are displayed in the $(q, \omega)$ phase diagram for a doped AA-stacked bilayer graphene under different $E_F$ values (Figs. 3.4(b) and 3.4(c)). Collective modes dominate the excitation spectrum over the range where the continuum spectrum of e–h pairs is weak or even vanishes; that is, the plasmons can survive under weak Landau dampings. As clearly demonstrated in the left-bottom corner, the strong acoustic mode propagates with a group velocity of $d\omega_q/dq$ and a wavelength of $2\pi/q$ for $E_F = 0\,\text{eV}$ and $E_F = 0.2\,\text{eV}$. The other plasmon branch with a finite frequency at $q \rightarrow 0$ is ascribed to the optical mode from the interpair excitation channels. The collective excitations are significantly damped when dispersing into the continuum region of the single-particle excitations. With the change in the doping level, the characteristic frequency $\omega_c$ represents the different degrees of the combination of the two plasmon modes, as shown in Fig. 3.4(c). As a result, it might be hard to describe the composite wave oscillation by the sole $\omega_c$ at a given $E_F$. It should be noted that the acoustic branch can be triggered for other stacking configurations, which are usually accompanied by a wide range of Landau dampings because the low-energy bands are closer to the Fermi level as compared to those of the AA-stacking configuration. Moreover, the optical plasmon branches are restricted to the AA configuration, in which the separated Dirac points can give enough space in the excitation spectrum to hold these plasmon modes. The potential wave propagation

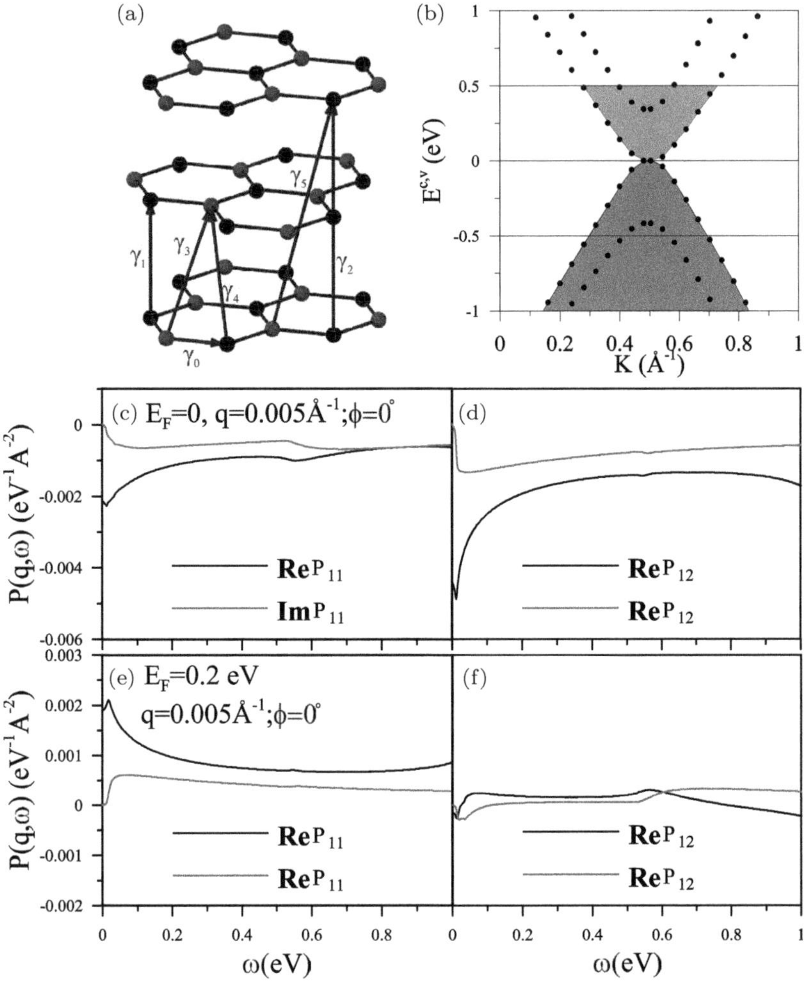


Fig. 3.5 (a) Structure of AB-stacked graphene. (b) The two pais of parabolic bands at the K point. (c) and (d) The dielectric function of bilayer AB-stacked graphene for various **q** under $E_F = 0\,\text{eV}$ and $E_F = 0.3\,\text{eV}$.

in the domains of $r$ and $t$ is expected to exhibit the dynamic charge screening from single-particle excitations and the acoustic and optical plasmon excitations.

The AB-stacking configuration is clearly shown in Fig. 3.5(a), where the two layers shift relative to each other by one C–C bond length ($b$) along the armchair direction ($\hat{x}$). The tight-binding wave functions are built from the atomic wave functions of the four carbon atoms in a primitive unit cell. All the significant electron hoppings are included in the tight-binding model, i.e., the nearest-neighbor intralayer hopping integral ($\gamma_0 = -3.12\,\text{eV}$), three neighboring-layer hopping integrals ($\gamma_1 = 0.38\,\text{eV}$, $\gamma_3 = 0.28\,\text{eV}$ and $\gamma_4 = 0.12\,\text{eV}$), and the chemical environment difference between A and B sublattices ($\gamma_6 = -0.0366\,\text{eV}$). The details of the expression of the Hamiltonian matrix can be found in Ref. [113]. The contributions due to the $2p_z$ orbitals of carbon atoms under the influence of the atomic interactions among carbons and the chemical environment are sufficient for calculating the low-energy bands and electronic excitations in the bilayer graphene. As shown in Fig. 3.5(b), the band structure consists of two pairs of parabolic bands located at the K/K′ point. The quadratic dispersion relation is expected to cause the 2D acoustic plasma to exist in a narrower space. The same condition can also be applied to the second plasmon mode.

At $E_F = 0$, the bare response function exhibits a special structure at the specific frequency for the excitations related to the band-edge states or the Fermi-momentum states. Except for the intrapair transitions, the transitions between different pairs of energy bands are forbidden due to the antisymmetry of the wave functions of the initial and final states. As for pristine trilayer ABA stacking, the available Coulomb excitation channels lead to the special structures appearing at low frequencies, ~0.02 and ~0.5 eV for $\mathbf{q} = 0.005\,\text{Å}^{-1}$, as shown in Figs. 3.5(c) and 3.5(d). When the Fermi level is increased to $E_F = 0.2\,\text{eV}$, there are more free carriers in conduction bands, which construct a bigger Fermi surface. Both excitation channels create a strong response at the same frequency below 0.1 eV. The divergent structure is reversed against the $y$-axis, and the special structure at ~0.5 eV is weakened because the occupied conduction electrons suppress the interband excitations through the Pauli exclusion principle. Electronic excitations are drastically changed during the variation of the Fermi level.

The energy loss function is used to define the plasmon modes and realize the evolution under the Landau dampings, which depends on the doping level and the magnitude of the transferred momentum. The energy loss spectra are rarely dependent on the direction of the transferred momentum because of the nearly isotropic energy bands at low energies. As shown in Fig. 3.6(a), the relatively stronger plasmon peak is observed for an $E_F$ higher than 0.2 eV, in addition to the weak one appearing at $\omega$ near 0.5 eV. In a pristine system, plasmon modes are completely

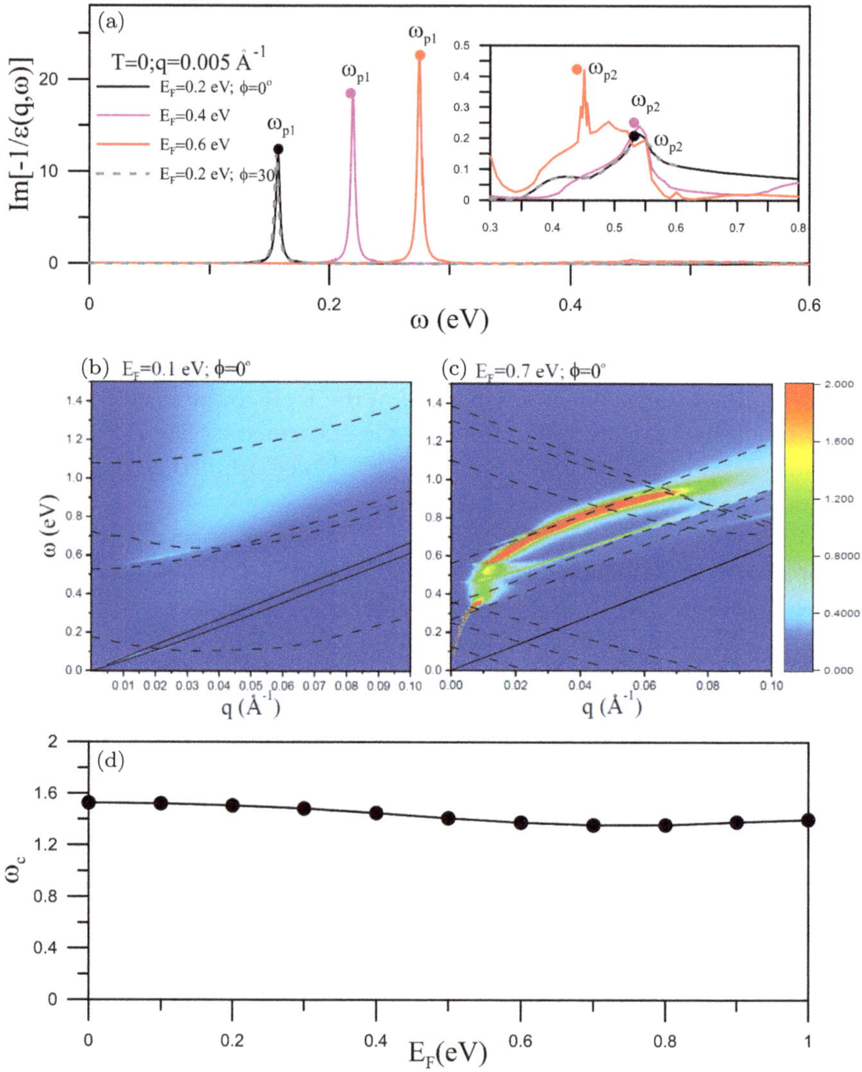


Fig. 3.6 (a) The energy loss function spectra of AB-stacked bilayer graphene for $\mathbf{q} = 0.005$ Å$^{-1}$ under different Fermi levels, where $\omega_{p1}$ and $\omega_{p2}$ indicate the acoustic and optical plasmon energies. Phase diagram of AB-stacked bilayer graphene at (b) $E_F = 0.1\,\text{eV}$ and (d) $E_F = 0.7\,\text{eV}$. The solid and dashed black curves, respectively, correspond to the boundaries of intraband and interband excitations. (d) The variation of the characteristic frequency with different Fermi levels.

suppressed by the e–h excitations. There are too few free carriers in conduction bands to make them excite collectively. Partially filled conduction electrons can create two prominent peaks with either in-phase or out-of phase oscillations in the excitation spectra. The lower-frequency plasmon at low frequencies has a rather strong intensity as compared to the higher one as a result of the in-phase acoustic plasmon mode of all the free conduction carriers. However, the second plasmon mode arising from the deeper interband excitations is obscure even under heavy doping. The $(\mathbf{q}, \omega)$-phase diagram of the energy loss function could provide complete information about the single-particle and collective excitations, as shown in Figs. 3.6(c) and 3.6(d). At $E_F = 0$, there are no prominent plasmon modes throughout the whole range of the $(\mathbf{q}, \omega)$ phase space. This is partly because of a few conduction electrons and partly because of the strong e–h Landau dampings. Once we compare the cases between $E_F = 0\,\text{eV}$ and $E_F = 0.7\,\text{eV}$, we find that the number and intensity of the e–h excitations are increased, and the e–h boundaries are drastically changed with different Fermi surfaces. The first plasmon mode, with a $\sqrt{q}$-dependent frequency, extends to a range of high momentum and gradually decay and disappear at a certain critical momentum. The second plasmon mode is identified as an optical mode, while the intensity is always weak at any Fermi level. The excitation spectrum directly reflects the dynamic response on the propagation behavior of plasma waves in real space.

Under the influence of dynamic e–e interactions, the trilayer graphene exhibits a rich and unique Coulomb excitation spectrum that is highly dependent on the stacked structure. It is claimed that under heavily doped conditions, a 2D $\sqrt{q}$-dependent plasmon mode appears in the layered graphene system. In addition, each system has three plasmon modes: one is an acoustic mode, and the other two are optical. However, its intensity is drastically changed by interband excitations varying with different Fermi levels, while the phase diagram of the energy loss function is almost irrelevant with the change in the direction of $\mathbf{q}$. For AA trilayer excitation spectra (Figs. 3.7(a) and 3.7(b)), three plasmon peaks are revealed at low frequencies, of which the acoustic plasmon from intrapair excitations with in-phase charge oscillations is the most prominent. The $(\mathbf{q}, \omega)$-phase diagrams could provide complete information about the single-particle and collective excitations in the system. The plasmon energy is higher than the single-particle excitation energy and is affected by Landau dampings. In addition, the other two optical plasmon modes, respectively, come from the channels between two neighboring Dirac cones and between the

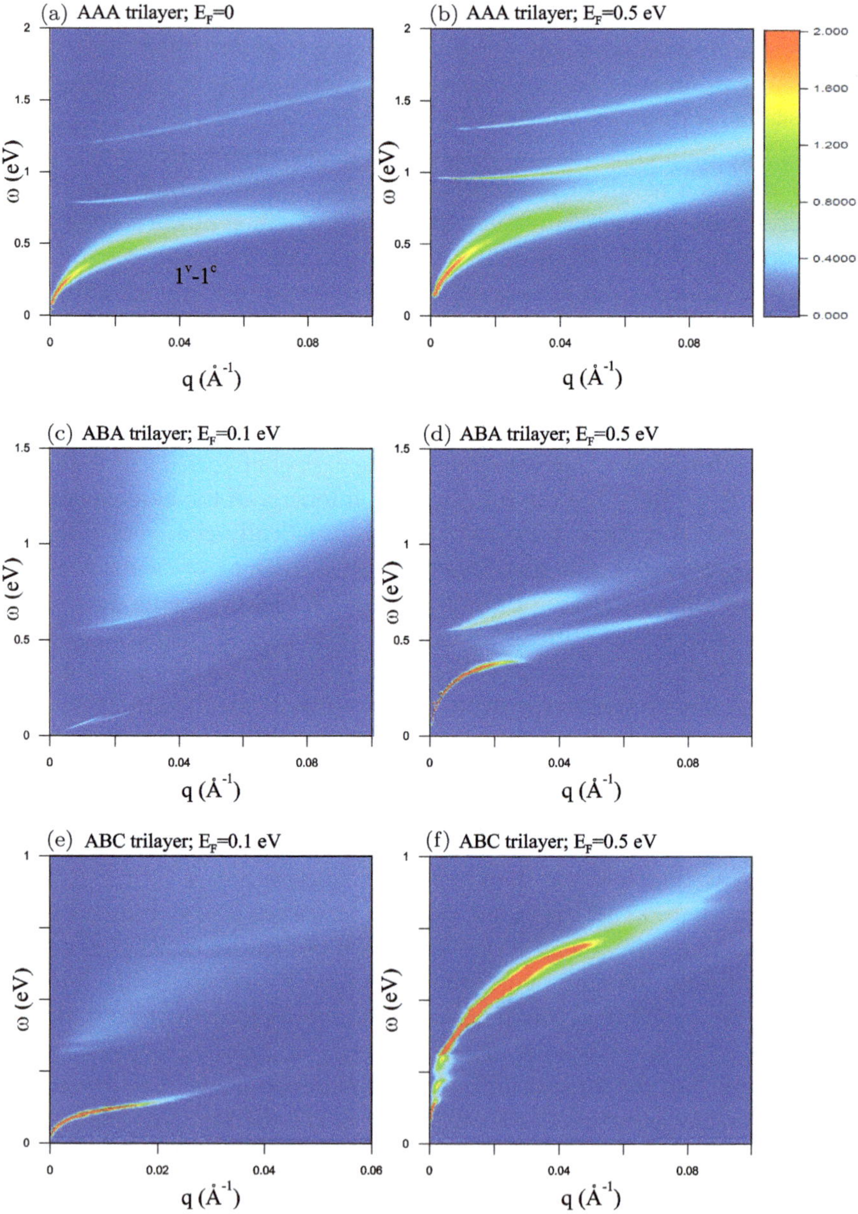


Fig. 3.7 Phase diagram of trilayer graphenes with (a) and (b) AAA, (c) and (d) ABA, and (e) and (f) ABC stacking configurations.

next nearest-neighbor interpairs. Their charge oscillations are out of phase with the two outermost layers or are related to each layer separately. In most cases, plasmons and single-particle excitations coexist in AA-stacked graphenes due to the multi-Dirac cone structure. When the Fermi level is increased, the induced free carriers enhance the single-particle excitations as well as the corresponding plasmons. As the frequency of various plasmon modes is consistent with the frequency of single-particle excitations, their intensity will decrease, resulting in the Landau damping of the plasmon. The mechanism to control the abnormal frequency region is mainly determined by the lowest interband channel. Moreover, the acoustic modes under a fixed $\mathbf{q}$ value have similar excitation energies, which indicates the existence of a multi-Dirac cone structure with a modified linear slope. The branches of the optical plasmons are limited to the AA-stacking configuration because the large spacing between the Dirac cones provides sufficient energy and momentum transfer space for various plasma modes.

For undoped AA-stacked bilayer graphene, the intrinsic plasmon mode is absent due to a lack of an intraband channel that does not create some $(\mathbf{q}, \omega)$ Coulomb interactions. When the free carriers are increased, all the single-particle excitations and the related boundaries are drastically changed by the different Fermi surfaces. As previously verified in the 2D electron gas, each doped system can display the strongest acoustic plasma, with a relative frequency of $\sqrt{q}$ under the limit of long wavelength [111]. Such a mode will gradually decay as $\mathbf{q}$ increases and disappear at the critical momentum $q_c$, which increases as the doping level increases. Due to the finite frequency at $\mathbf{q} \to 0$, the second plasma associated with the middle interband channel is identified as an optical mode. However, there are only two plasmon modes at low frequencies under low doping, e.g., $E_F = 0.1\,\text{eV}$ in Fig. 3.7(c). The suppression of the third plasmon mode is due to the strong Landau dampings caused by interband excitations at the K/K$'$ point. The second plasmon frequency is concave parabolic and has no simple relationship with $\mathbf{q}$. Landau damping also greatly reduces the strength. As the doping level increases (for example, 0.5 eV in Fig. 3.7(d)), this plasmon mode is easily observed. In addition, for the Fermi level passing through the highest conduction band, a third plasmon mode appears between the first and second modes because the highest interband channel is triggered with a sufficiently high energy to avoid interband Landau dampings.

Various plasmon modes of the trilayer ABC-stacked graphene display rich and unique behaviors of collective excitations, as shown in Figs. 3.7(e) and 3.7(f). The plasmon mode usually appears in several specified domains of the phase diagram under weak excitations between $\pi_1^v \rightarrow \pi_1^c$ partially flat subbands. The biggest difference from the previous two systems is the intrinsic acoustic plasmon mode, which has a frequency of up to 0.6 eV. In particular, this mode exhibits a linear dependence on $\mathbf{q}$ and is assigned to the collective excitation mode of surface local states with high state density to prevent the coupling of excitations between interbands. The intrinsic acoustic mode at zero temperature is dedicated to multilayer graphene systems with specific ABC-stacking configurations. On the other hand, the interband single-particle excitation produces strong Landau damping near the plasmon excitation energy. After passing through this region, plasmons similar to the acoustic dispersion can be formed. With an increase in $E_F$, two plasmons with different dispersion relations are shown in the case of $E_F = 0.1$ eV in Fig. 3.7(e). Under the influence of different degrees of Landau damping in the specified domain, they behave as acoustic and optical modes in the low and medium $(\mathbf{q}, \omega)$ regions without continuous single-particle excitation. When the collective excitation channel is switched from interband to intraband, the former deviates from the original linear dispersion, which indicates the influence of band structure and doping carrier density. Nevertheless, if more subbands are partially occupied, the plasmon mode will change drastically. At $E_F = 0.5$ eV (Fig. 3.7(f)), the acoustic plasma is enhanced because the intraband collective channel gradually becomes dominant in the excitation spectrum. According to the band effects, Landau damping is very strong for the induced interband single-particle excitations, for example, in the region of $0.2 \leq \omega \leq 0.3$ eV. This results in a weak plasmon mode with concave or convex dispersion, which indicates the robust Landau dampings coming from the particular partially flat and slowly dispersive Sombrero subbands.

In summary, Coulomb excitation is greatly diversified due to stacking configuration and the number of layers. Monolayer graphene with a linear Dirac cone only exhibits interband excitation without carrier doping. Therefore, 2D acoustic plasma does not exist because the system is a zero-gap semiconductor with zero DOS at $E_F = 0$. However, additional intraband excitation and acoustic plasmon can survive at finite temperatures and with limited electron/hole doping, where the latter experiences severe

Landau damping due to the high intensity of the interband e–h pair [113]. For pristine bilayer AA stacking, the two isolated Dirac cones each have enough free carriers to produce two plasmon modes together, namely acoustic and optical modes, whose frequency and intensity are not sensitive to doping effects [113]. On the contrary, bilayer AB graphene is an almost electrically neutral system and cannot induce intrinsic plasmons at room temperature. However, as the population of conduction electrons in the valence and conduction bands near the Fermi level increases, 2D plasma can usually be observed by increasing the Fermi level. On the other hand, there are rich and unique excitation behaviors in trilayer AAA, ABA and ABC stacked graphenes. The three stacking configurations with different symmetries show their specific three-plasmon modes. The e–h excitation boundary defined by different Fermi surface or band-edge states becomes more complicated. According to research, one acoustic and two optical plasmon modes are, respectively, related to the intraband excitation of conduction electrons and the interband excitation of valence electrons. It is worth noting that the first two are intrinsic acoustic plasmons; their characteristic conversion under different $E_F$ is expected to have a great impact on the lifetime of quasiparticles near the Fermi level and the propagating plasmon waves (will be discussed in Chapter 5).

Generally, the energy loss function, Landau damping and plasma mode in three-layer ABA graphene do not show proper decomposition into monolayer and bilayer excitation phenomena, even for any doping condition. Although the single-particle excitations in the undoped system are consistent with the perturbation concept, it cannot be understood from their linear superposition that the screening response functions from monolayer and bilayer parts are calculated independently. High-resolution EELS is one of the most powerful experimental techniques that can be used to study Coulomb excitation in widely studied 2D layered systems, such as graphene, silicene, germanene, tinene and phosphorene. EELS measurements on monolayer and fewlayer graphenes confirmed acoustic, $\pi$ and $\pi + \sigma$ plasmons. The latter two plasmons are observed at frequencies higher than 4.8 and 14.5 eV, and the frequency increases with the number of layers [36, 90]. In addition, in terms of acoustic plasmon dispersion below 1 eV, considerable interband Landau damping has been observed at larger momentum [36, 90]. However, there is still a lack of experimental identification of the multi-mode acoustic modes predicted in few-layer systems. Experimental measurements are very useful for thoroughly understanding various Coulomb excitation and decay phenomena. Furthermore, the viewpoint of energy band theory provides complete information about all excitation behaviors.

## 3.4 Coulomb Decay Rates

Quasiparticle properties of multilayer graphenes vary with the stacking configuration and the number of layers, leading to the fact that the decay spectra of AA-, AB- and ABC-stacked graphenes are sharply in contrast with each other. It can be immediately recognized in any system that the e–h pair or collective excitation can show a rapid decay rate of quasiparticles. Regarding the plasmon resonance generated by acoustic and optical plasmon modes, their influence on Coulomb scattering is related to Landau damping. From the perspective of the Coulomb decay rate of quasiparticles under the composite effect of many-body interactions, the roles played by the multi-plasmon modes, e–h excitations and multi-band structure are discussed in detail. The predictions can be verified by ARPES observations of the valence states.

### 3.4.1 *Coulomb decay rate in monolayer graphene*

Monolayer graphene exhibits the feature-rich Coulomb decay of Dirac quasiparticles in condensed matter. The linear Dirac bands are isotropic for low doping ($E_F \leq 1\,\text{eV}$), causing Coulomb scattering to have similar behavior along different directions in the Brillouin zone. Fermi energy and free carrier density determine the main characteristics of the elementary excitations that dominate the Coulomb decay channels. Here, we only consider the case of electron doping since the Dirac cone is symmetric about $E_F = 0$. The excited electrons/holes could decay into conduction/valence band states at zero temperature only via intraband single-particle excitations when the graphene is natural. However, the increase in electron doping produces intraband e–h excitations and plasmons, of which the latter provide a very effective decay channel and significantly increase the decay rate. The de-excitation channels can be categorized into four kinds according to the Dirac cone structure, as shown in Figs. 3.8(a), 3.8(c), 3.9(a) and 3.9(c). In particular, the $\mathbf{q}$ and $\omega$ regions transferred during the de-excitation process mapped in the phase space of the excitation spectrum, as shown in Figs. 3.8(b), 3.8(d), 3.9(b) and 3.9(d), are straightforward to

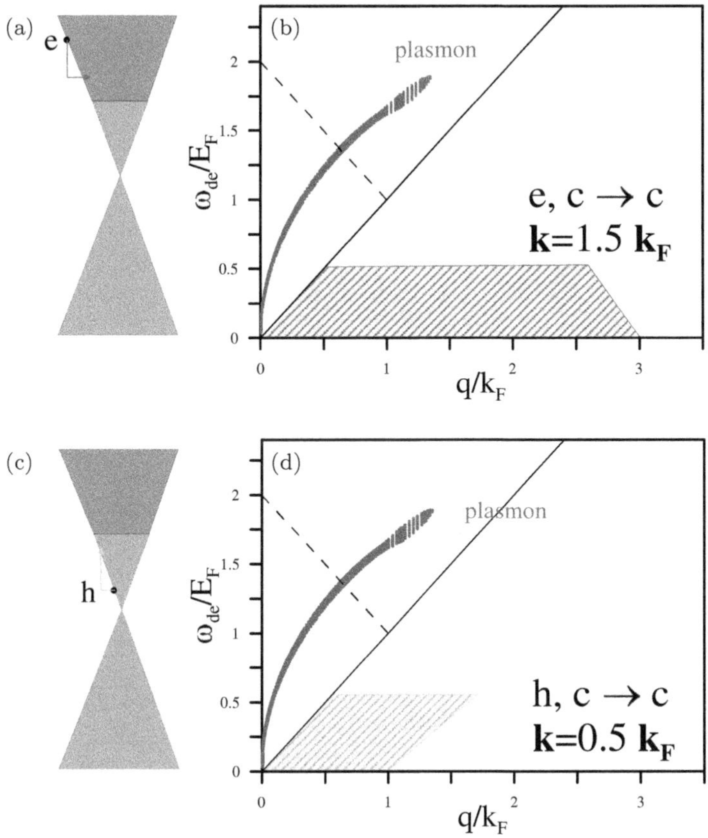


Fig. 3.8 In a monolayer graphene, the de-excitation process of (a) an excited conduction electron with $k = 1.5\,k_F$ and (c) an excited conduction hole with $k = 0.5\,k_F$; the corresponding spaces of transferred $\mathbf{q}$ and $\omega$ are shown in (d) and (d), respectively. The acoustic plasmon mode appears in the area drawn in red.

check whether the excited state can decay into either e–h pairs or plasmons. It can be clearly seen from these figures that the decay of intraband and interband processes corresponds to the regions of $\omega_{\text{de}} > v_F q$ and $\omega_{\text{de}} < v_F q$, respectively, while the decay of plasmon emission is only in the former region. In the first case, for the excited conduction electrons with wavevector $\mathbf{k} > k_F$, the Coulomb de-excitation process is depicted in Fig. 3.8(a). It can decay into the final state above the Fermi level. This process belongs to the intraband channel, where the available final states give the transferred $(q, \omega)$ space, as indicated by the pink

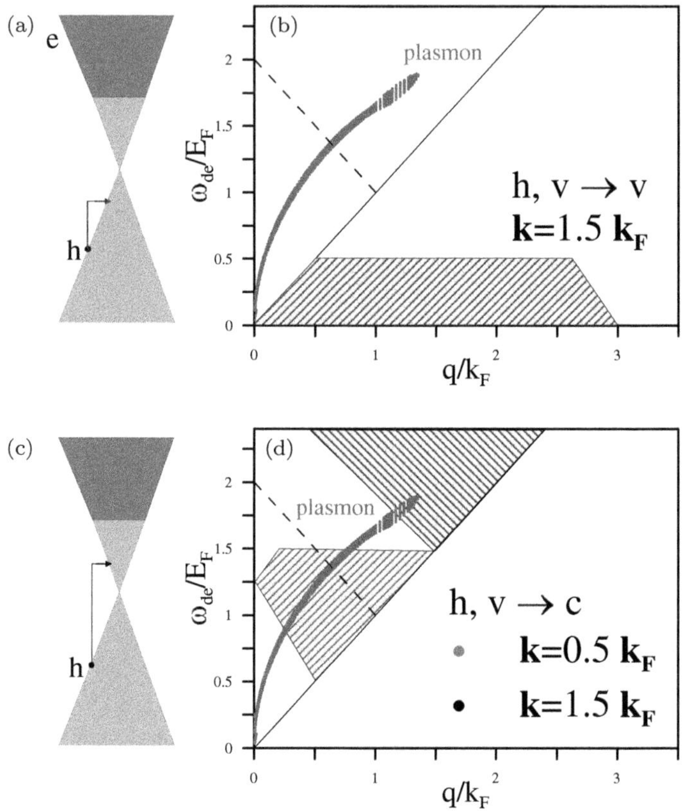


Fig. 3.9 The same plot as in Fig. 3.8 but showing the two de-excitation processes of a excited valence hole.

trapezoid in Fig. 3.8(b). This area is located below the excitation boundary of the intraband, drawn according to the Pauli exclusion principle, and satisfies the conservation of energy and momentum. Conduction electrons decay into e–h pairs through intraband channels. Therefore, their decay rate increases with the increase in state energy $E^c = v_F|\mathbf{k}|$. For the excited hole above the Dirac point, Fig. 3.8(c) shows the second de-excitation channel. They can easily decay into the upper region of the Dirac cone through many intraband e–h excitations. The $\mathbf{q}$ and $\omega$ transferred during Coulomb scattering are indicated by the parallelogram in Fig. 3.8(d), where the area is much larger than the area for states just above the Fermi level. On the other hand, the de-excitations of valence holes are divided into two kinds of decay processes: the lower or upper part of the Dirac cone (the third and fourth channels), as shown in Figs. 3.9(a) and 3.9(c). The former is composed of intraband e–h excitations, and the area in the $(q, \omega)$ domain is the same as that of the first channel at large **k**. However, for excited holes with small **k** (just below the Dirac point), there are few available de-excitations with sufficient momentum transfer via e–h pairs. This implies that the scattering rate is very weak at the Dirac point and only passes through a few channels. In particular, the fourth kind of channel in the plasmon resonance region is the most important and effective channel in the de-excitation process of conduction electrons. Regarding the plasmon resonance in the $k = 0.5\ k_F$ and $k = 1.5\ k_F$ states, although for the latter the acoustic plasmon mode is damped by the interband excitation (after the dotted line), they still contribute the most to the decay rate.

These de-excitation channels determine the decay rate of the quasiparticle states of $(\mathbf{k}, h)$ shown in Fig. 3.10. The colors in Figs. 3.8, 3.9 and 3.10 are consistent. At zero temperature, the excited electrons and holes have a decay rate symmetric about the natural Dirac point. The decay rate is set to zero at the Fermi level, corresponding to the discontinuous Fermi–Dirac distribution. Regarding the excited conduction electrons near the Fermi level, their Coulomb decay rate increases monotonically with $E^c - E_F$, roughly proportional to $(E^c - E_F)^2 ln|E^c - E_F|$ for $E^c \lesssim 1.5E_F$, because the intraband process is the only channel that makes a major contribution to the de-excitation process (the pink curves in Figs. 3.10(b) and 3.10(c)). Such energy dependence is a characteristic of 2D electron gas, and it is also applicable to the excited states dominated by the first three de-excitation channels described above [7, 29]. Previous work on graphite intercalation compounds has also reported the same decay rates [120]. Similar results are observed in doped silicene and germanene [116]. These systems have the common characteristics of Coulomb scattering because, when $E^{c,v} \to E_F$, the de-excitation energy is linearly proportional to **q** in the long-wavelength range and is reflected in the following phase space. It should be noted that $1/\tau$ at the Dirac point is finite because the distorted linear dispersion introduces additional channels with sufficient momentum transfer through the e–h pair, and its local minimum also implies the robustness of the Dirac point. The Coulomb decay rate $1/\tau$ shows isotropic behavior along the $KM$ and $K\Gamma$ directions for $E^{c,v} \lesssim 0.5\,\text{eV}$ under $E_F = 0.1\,\text{eV}$. Nevertheless, for high-energy conduction electrons, anisotropic $1/\tau$ is reflected by the band structure that deviates from linear dispersion; $1/\tau$ is mainly determined by the number of de-excitations for **q** around the distorted Dirac cone. High-energy electronic states have

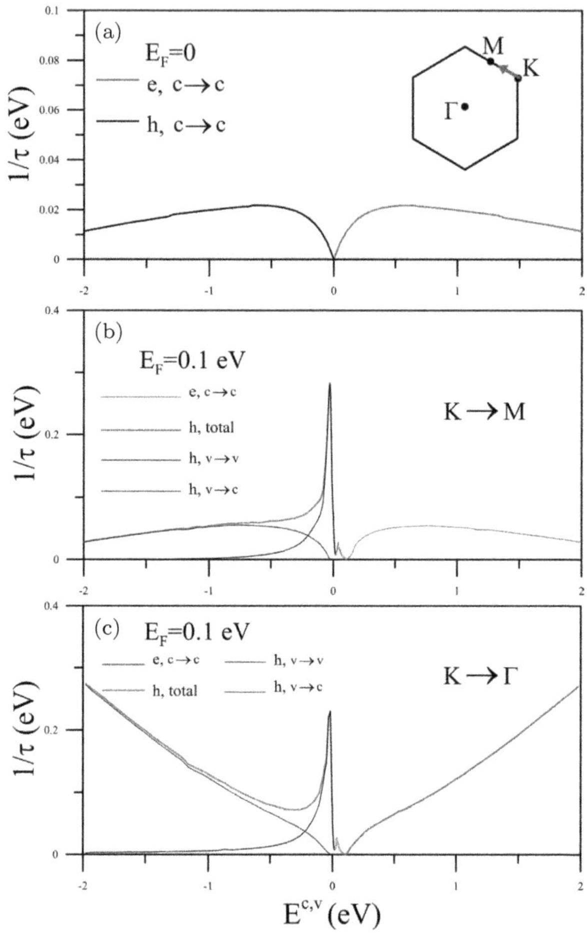


Fig. 3.10 The energy-dependent Coulomb decay rates of the quasiparticle states in monolayer graphene at $E_F = 0\,\text{eV}$ and $E_F = 0.1\,\text{eV}$ along the $KM$ and $K\Gamma$ directions. The inset in (a) shows the first Brillouin zone.

stronger energy dispersion along $K\Gamma$ than in $KM$, resulting in their relatively large de-excitation energy range in the former direction at a large rotational movement. A similar situation will also be seen in valence hole decay.

What's more noteworthy is the de-excitation by the plasmon emission, which is revealed by the decay rate of the valence holes (black curve in Figs. 3.10, 3.11 and 3.12) being very different from that of the conduction

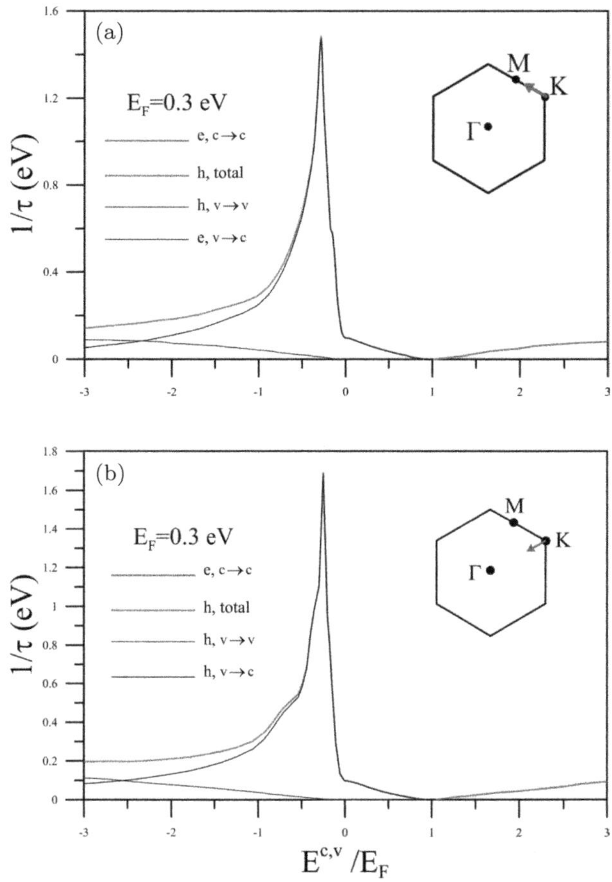


Fig. 3.11 The energy-dependent Coulomb decay rates of the quasiparticle states in monolayer graphene along the high-symmetry directions: (a) $KM$ and (b) $K\Gamma$ under $E_F = 0.3\,\text{eV}$. The insets show the first Brillouin zone.

electrons. The valence state slightly below the Dirac point has a significant decay rate as a consequence of the fourth decay process (Figs. 3.9(c) and 3.9(d)) that provides an efficient de-excitation channel through the undamped acoustic plasma mode. The plasmon de-excitations almost disappear for the deeper valence states for $E^v < -1.5E_F$ according to the phase diagram in Fig. 3.9(d). In other words, $1/\tau$ of the deeper states is dominated by the e–h excitations, reflecting their symmetry with respect to the Dirac point, as shown by the pink and orange curves in Fig. 3.10. The focus of interest is, in particular, on the plasmon resonance region of

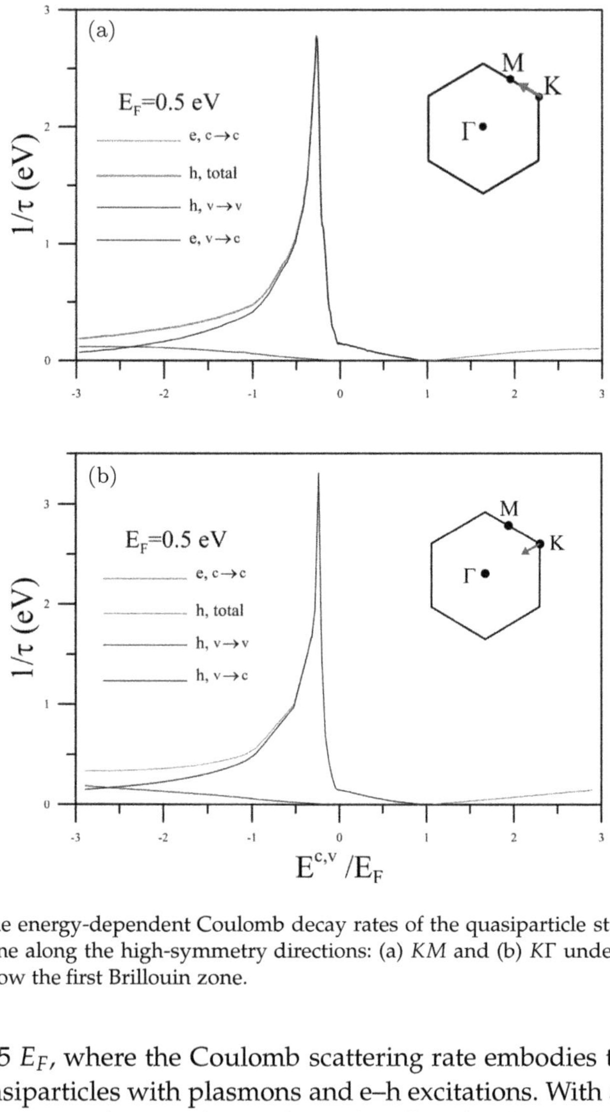


Fig. 3.12 The energy-dependent Coulomb decay rates of the quasiparticle states in monolayer graphene along the high-symmetry directions: (a) $KM$ and (b) $K\Gamma$ under $E_F = 0.5\,\text{eV}$. The insets show the first Brillouin zone.

$|E^{c,v}| < 1.5\ E_F$, where the Coulomb scattering rate embodies the interaction of quasiparticles with plasmons and e–h excitations. With an increase in valence energy, the two decay channels affect the Coulomb decay rate at the same time, and the dominance of channels is worthy of further discussion under different electron dopings. The usable range of the acoustic plasma mode increases and then rapidly decreases with the monotonically increasing single-event effect. The turning point is about 0.5 $E_F$, exactly corresponding to the position of the peak of $1/\tau$ as a function of energy. Quasiparticles have fast Coulomb decay with more plasmon de-excitation channels to produce a wider Lorentzian peak in the momentum distribution curve in the ARPES measurement. This result is easier to measure for $E_F = 0.3$ and 0.5 eV, as shown in Figs. 3.11 and 3.12. The physical pictures of both the e–h and electron–plasmon couplings are necessary to understand the dynamics of Dirac quasiparticles in related materials, such as layered Group-IV materials [135, 136], the semimetal ZrSiSe [137], $Bi_2Te_3$ [138] and 3D HgTe [139, 140].

### 3.4.2 *AA-stacked bilayer graphene*

AA bilayer graphene has more kinds of de-excitation processes and richer Coulomb decay features. The two Dirac bands isotropic at low energies ($E_F \leq 1$ eV) dominate the Coulomb scattering of quasiparticles in the Brillouin zone. The de-excitation channels associated with intraband e–h excitations and plasmons are as displayed in Figs. 3.13 and 3.14, of which the focus is on the latter, which is very effective. It is clear from Fig. 3.13 that, in addition to each cone, the Coulomb decay should be considered with the contribution of the two Dirac cones. In the pristine case of AA-stacked graphene, the displacements of the two pairs of Dirac cones relative to the Fermi energy lead to corresponding equivalent electron and hole dopings with $E_F = 0.4$ eV, respectively. An excited conduction electron can decay into the final state of either the upper or lower cone. The two de-excitation processes are supposed to be equally important. The area of $(q, \omega)$ in the intra-cone is located between the acoustic and optical plasmon modes according to the Pauli exclusion principle and the conservation of energy and momentum. Meanwhile, the inter-cone de-excitations also play an equally important role as compared to the former one. The e–h channel is also very important between the intra-cone and inter-cone processes. The available de-excitations for excited holes just below the Dirac point can be significantly enhanced by the inter-cone channel, which makes the Coulomb decay at the Dirac point of bilayer graphene faster than that of monolayer graphene. In the case of electron doping, the two Dirac cones reflect different doping concentrations and different de-excitation channels, as shown in Fig. 3.14. This implies that the excitation of two different Dirac cone quasiparticles can be modulated by the Fermi surface, and the interplay between them is expected to cause drastic changes in the Coulomb decay rate. Regarding the plasmon resonance with the optical plasmon mode, the deeper excited electrons or holes will have a faster decay rate, accompanied by the effects of the inter-cone

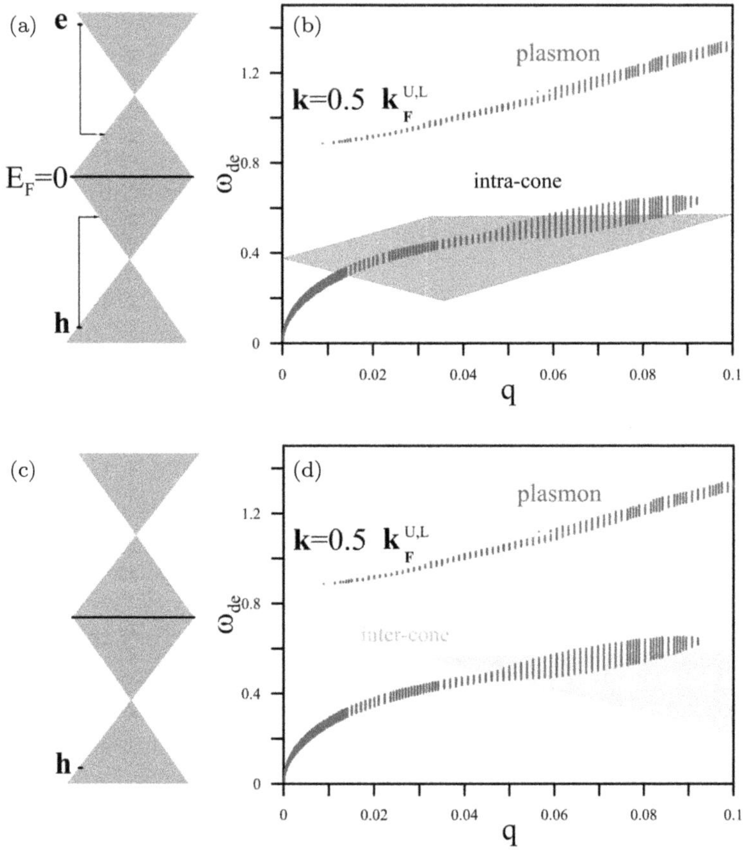


Fig. 3.13 The de-excitation process and the corresponding space of transferred **q** and $\omega$ in AA-stacked bilayer graphene at $E_F = 0\,\text{eV}$ for (a) and (b) intra-cone channel and (c) and (d) inter-cone channel. $k_F^{U,L}$ indicates the effective Fermi momentum for the upper and lower cones, respectively. The two plasmon modes appear in the area drawn in red.

and Landau damping. The result will be fully presented in the spectra of Coulomb excitations and decays of AA-stacked graphene.

At $E_F = 0$, the decay rate of the upper Dirac cone is symmetric to that of the lower cone with respect to $E_F = 0$, as shown in Fig. 3.15. That is, the former is shown as the effective electron doping of $E_F^{\text{eff}} = 0.4\,\text{eV}$. The latter corresponds to hole doping of $E_F^{\text{eff}} = -0.4\,\text{eV}$ in the same way. $1/\tau$ of the excited state near the Fermi level is $(E^{c,v} - E_F^{\text{eff}})^2 ln|E^{c,v} - E_F^{\text{eff}}|$, which is approximately the same as the contribution of the majority of the intraband decay process. The four kinds of decay rates of the intra-cone

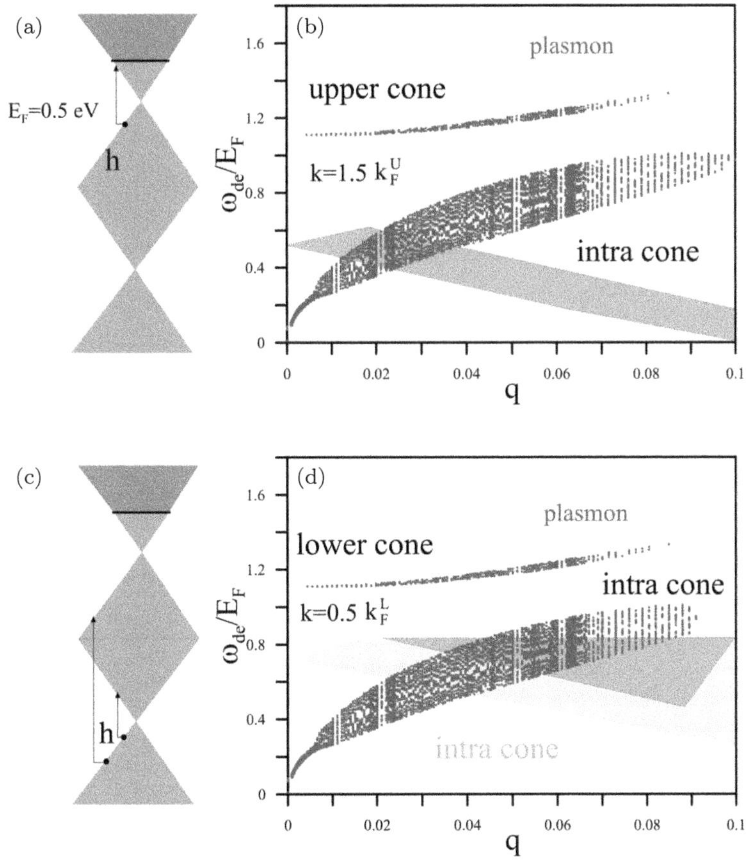


Fig. 3.14 The same plot as in Fig. 3.13 but showing the de-excitation processes at $E_F = 0.5\,\mathrm{eV}$.

de-excitation known are $1/\tau_{e,c\to c}$, $1/\tau_{h,v\to v}$, $1/\tau_{h,v\to c}$ and $1/\tau_{h,c\to c}$, which are similar to those of monolayer graphene, while the inter-cone plays an important role in the de-excitation process, especially for the state of the wavevector $>k_F^{\text{eff}}$ in the upper and lower cones. The Dirac point of a single Dirac cone no longer corresponds to the minimum value of the decay rate. In addition, the plasmon-resonance peak becomes slightly wider and is approximately located at the same value of the wavevector. For the lower cone, the excited valence holes of the wavevector $>k_F^{\text{eff}}$ can decay through the intra-cone and inter-cone decay processes together, as the acoustic plasmon mode provides a very effective decay channel. Unlike

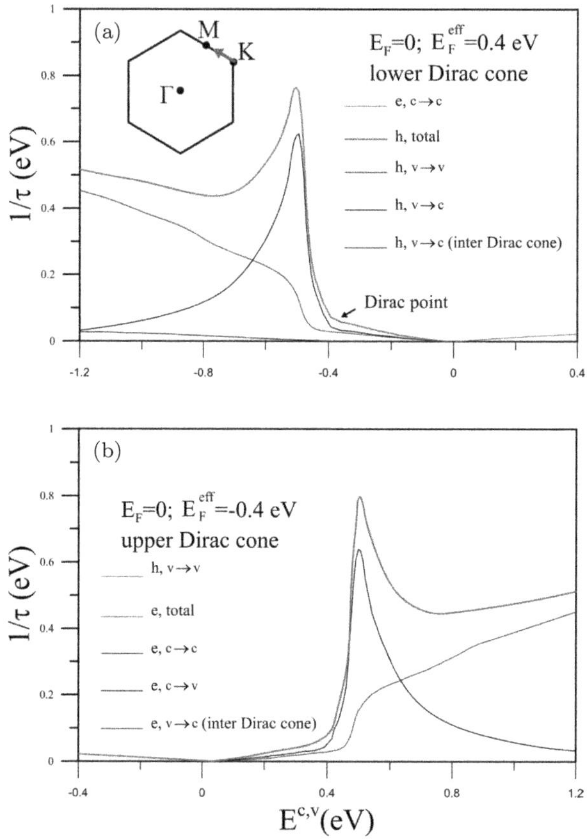


Fig. 3.15 The energy-dependent Coulomb decay rates of the quasiparticle states in (a) the lower Dirac cone and (b) the upper Dirac cone of AA-stacked bilayer graphene at $E_F = 0.5$ eV along the $KM$ direction. The inset shows the first Brillouin zone of AA-stacked bilayer graphene.

the former, whose main effect is below 1.5 $k_F$, the latter's effect increases as $k$ becomes larger. This phenomenon is mainly caused by two almost parallel conduction bands. Even when the momentum transfer is small, the inter-cone transitions between them will cause a significant decay of plasma emission. The same reasoning applies to the hole decay of the upper Dirac cone with hole doping regarding the plasmon emission.

When $E_F$ changes, the Coulomb decay rate behaves differently between the states of the two Dirac cones. As shown in Fig. 3.16, $1/\tau_e$ for the lower cone increases overall as the available frequency of the

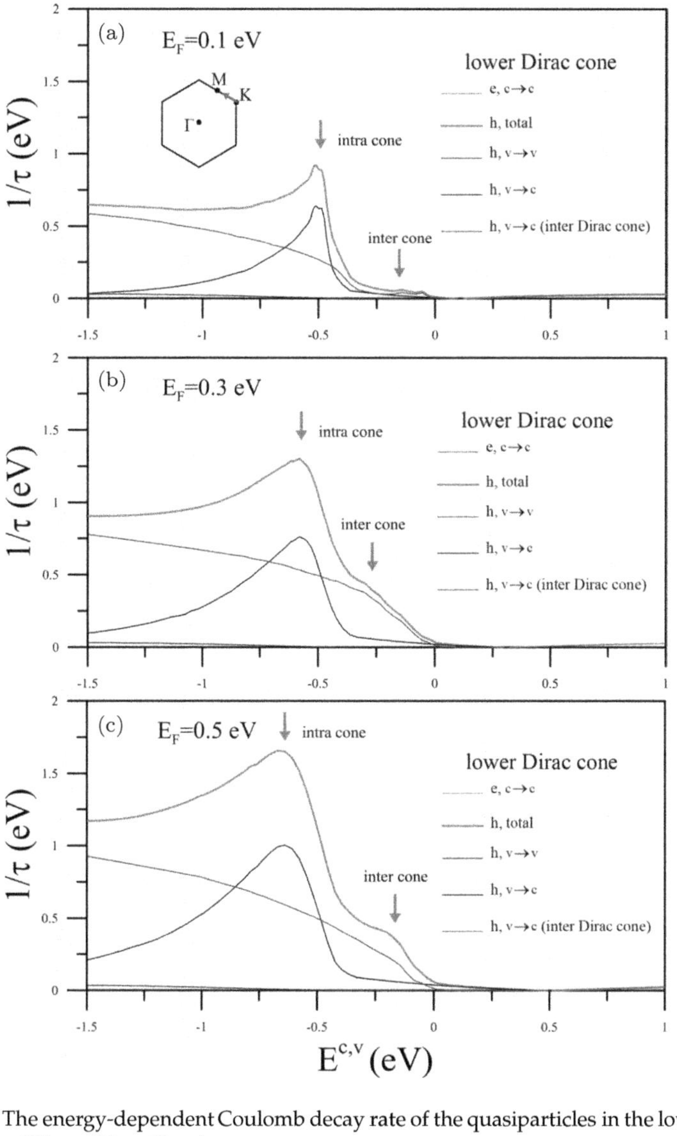


Fig. 3.16 The energy-dependent Coulomb decay rate of the quasiparticles in the lower Dirac cone under different Fermi levels.

de-excitation channel becomes wider, and it largely becomes faster for the states with plasmon emission as $E_F$ changes from zero to 0.5 eV. There are two peaks in the spectrum of the Coulomb decay rate, indicated by the orange arrows. The higher of the two is from the emission of an acoustic

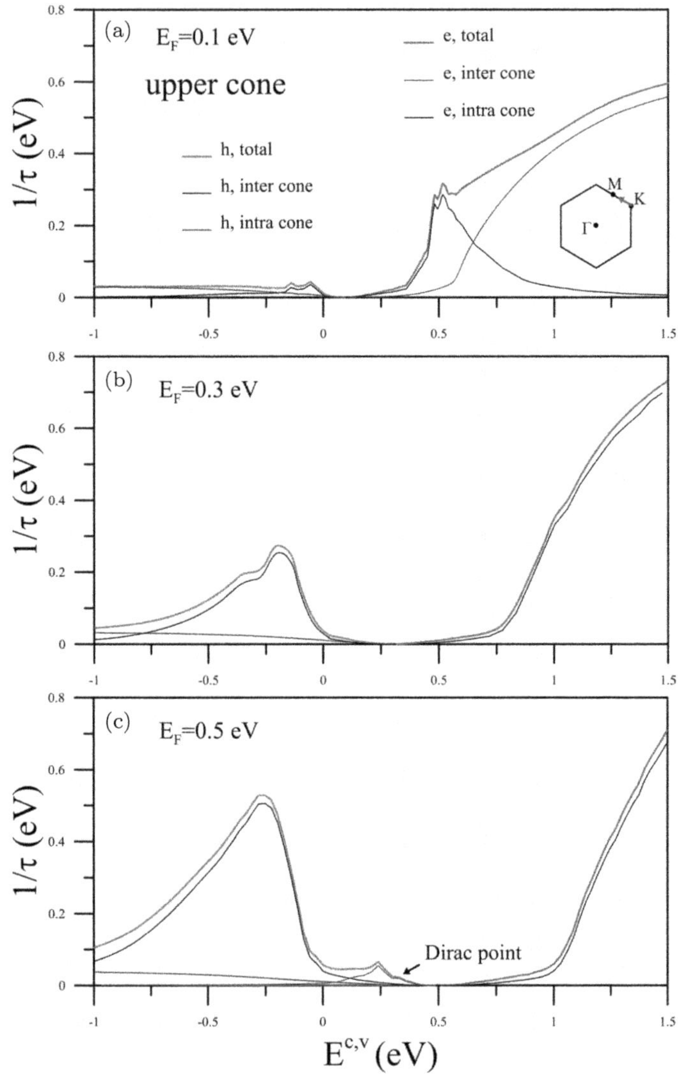


Fig. 3.17 The energy-dependent Coulomb decay rate of the quasiparticles in the upper Dirac cone under different Fermi levels.

plasmon that occurs in the transition within the cone, and the lower one is decayed by the emission of an optical plasmon that occurs in the transition between the cones. On the other hand, the upper cone shows a non-monotonic change with $E_F$, as shown in Fig. 3.17. When $E_F$ is changed from 0 to 0.1 eV, $1/\tau_e$ slows down, while $1/\tau_h$ becomes slightly faster near the Fermi energy, which may be triggered by the acoustic plasmon. However, at $E_F = 0.3$ eV, a plasmon-resonance peak occurs at $1/\tau_h(E^v)$ as a result of the inter-cone excitation, which can trigger the efficient plasmon channel to de-excite the hole under a small momentum transfer. As the $E_F$ increases further, the two plasmon-resonance peaks of $1/\tau_h(E^v)$ mean that the electrical property changes from n-type to p-type effectively for the upper Dirac cone. In conclusion, each plasmon mode makes an important contribution to the Coulomb decay rate of quasiparticles. This phenomenon may be observed in all multilayer AA graphenes because of each of their prominent plasmon modes.

### 3.4.3 *AB-stacked bilayer graphenes and differences with trilayer graphenes*

In AB bilayer graphenes, the two pairs of parabolic bands at the K point dominate the Coulomb scattering of excited states. In the pristine case of AB-stacked graphene, the Coulomb excitation and decay occur only through the single-particle excitation mechanism; the higher the energy, the faster the decay rate. The same mechanism applies to both pairs of parabolic bands. The de-excitation channels are discussed under electron doping, as displayed in Figs. 3.18 and 3.19 for $E_F = 0.1$ and 0.5 eV, respectively. The conduction and valence excited states can decay utilizing the interband e–h excitations, the acoustic plasmon mode and the intraband e–h excitations in both intrapair and interpair transitions. The former is important in both AA and AB graphenes, while the latter is a more effective de-excitation channel under the same momentum transfer as AB graphene, because the plasma energy of the optical mode is lower than the intrapair excitation energy. It is clear from the two plasmon modes shown in the figures that their distribution is getting closer and closer to each other, extending from $\omega = 0$ to $\omega = 1.0$ eV. This will result in transitions with a large range of momentum transfer being used in the plasmon emission. The excitation valence holes of $k = 0.2\,k_F$ and $k = 0.5\,k_F$ decay using different plasmon modes. Therefore, it can be expected that this transition will result in an abnormal feature of the decay rate spectrum.

At $E_F = 0$, the decay rate for excited conduction electrons or valence holes in the two pairs of parabolic bands is displayed along the $KM$ and $K\Gamma$ directions, as shown in Fig. 3.20. The internal transition mechanism causes the energy dependence of the decay rate. In the first conduction/valence

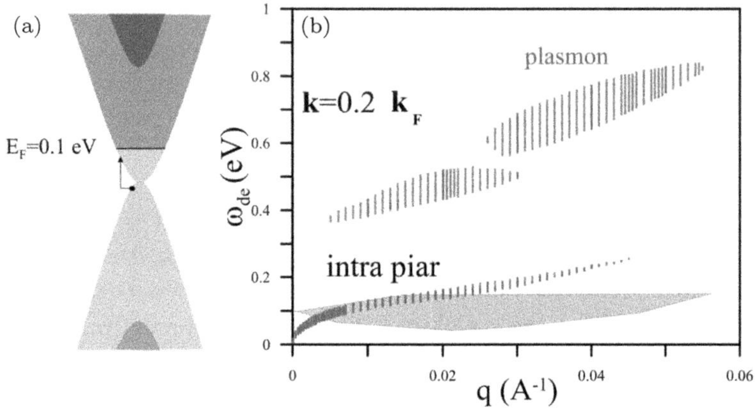


Fig. 3.18 In AB-stacked bilayer graphene, the de-excitation process and the corresponding space of transferred **q** and $\omega$ for the intra-cone channel at $k = 0.2\ k_F$ and $k = 1.0\ k_F$ under $E_F = 0.1$ eV. The two plasmon modes appear in the area drawn in red.

band, excited electrons/holes near the Fermi level mainly decay to empty states in the same band. In Fig. 3.20, $1/\tau_e(E^c)$ and $1/\tau_h(E^v)$ are symmetric with respect to zero energy and isotropic along the $KM$ and $K\Gamma$ directions. In the second conduction/valence band, the de-excitation can occur in the same band or between different bands; however, the higher de-excitation energy of the latter accounts for the anisotropic $1/\tau_{e,h}$. In the case of n-type doping, due to collective excitations, $1/\tau_e$ and $1/\tau_h$ are asymmetric, as shown in Fig. 3.21, where the contribution of each channel is shown in Figs. 3.21(b) and 3.21(c). By using the plasmon de-excitation channel (black curve), quasiparticles close to the Fermi level decay quickly. The remaining

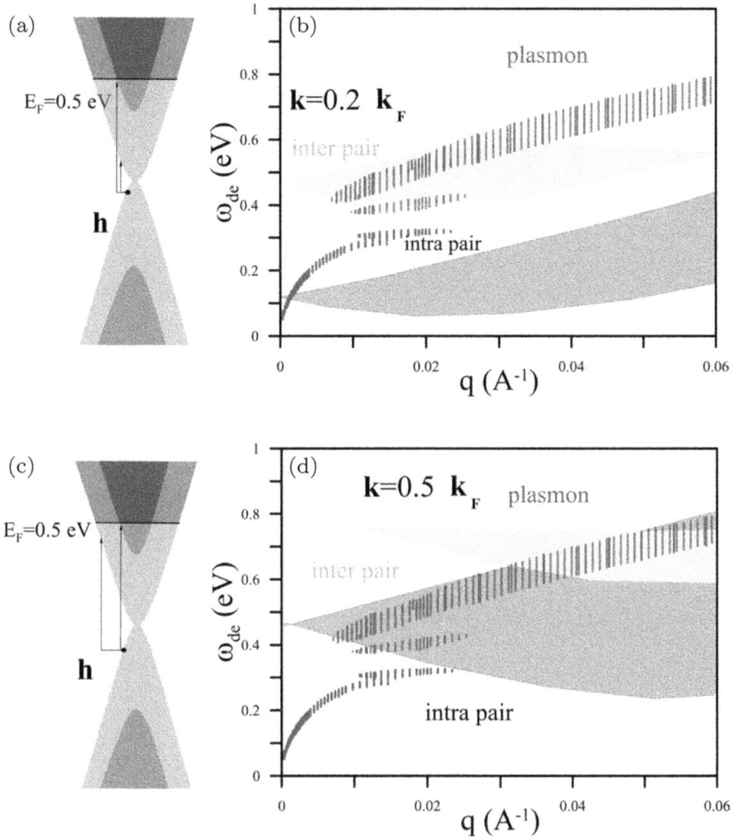


Fig. 3.19 The same plot as in Fig. 3.18 but showing the inter-cone de-excitation processes at $E_F = 0.5\,\text{eV}$.

excited states that use single-particle excitations have been discussed in detail by analyzing the decay mechanism under various momentum transfers. One can intuitively understand the results shown in Fig. 3.21(c).

The decay rate near the Fermi surface can better reflect the dynamics of quasiparticles under e–e interactions; its dependence on energy is worth studying under different doping conditions. When the electron concentration changes, the energy range of the plasmon-emission decay rate gradually increases. As a manifestation of the quasiparticle band structure, it explores the different degrees of decays by e–h pairs and plasmons in the intrapair and interpair de-excitation channels. In addition, both plasmon modes can participate in the de-excitation process of momentum

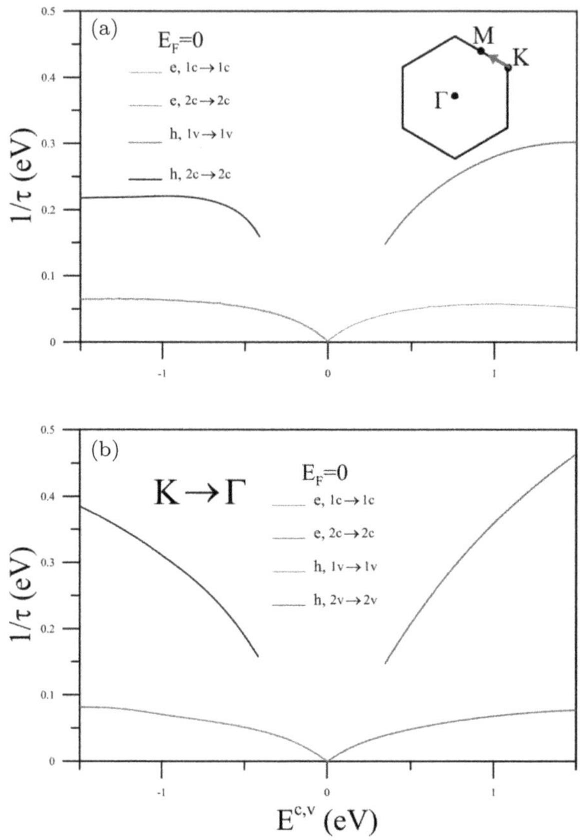


Fig. 3.20 The energy-dependent Coulomb decay rates of the quasiparticle states in AB-stacked bilayer graphene at $E_F = 0$ along the (a) $KM$ and (b) $K\Gamma$ directions. The inset in (a) shows the first Brillouin zone of AB-stacked bilayer graphene.

transfer according to the electron concentration to construct the effective Coulomb decay rates of $1/\tau_e(E^{c,v})$ and $1/\tau_h(E^{c,v})$. At $E_F = 0.1\,\text{eV}$ in Fig. 3.22(a), the three peak structures of $1/\tau_h$ near $E^v = -E_F$ originate from the undamped and damped collective excitations of the acoustic plasmon, which indirectly reveals the role of the Landau damping on plasmon excitations and the influence on the Coulomb decay rate. Incidentally, the interpair and intrapair de-excitations use acoustic and optical plasmon modes, respectively, to dominate the energy range of $1/\tau_h$ above and below $2E_F$. The former mechanism, however, is dominant at $E_F = 0.3\,\text{eV}$ in Fig. 3.22(b) because the latter is not enough to generate

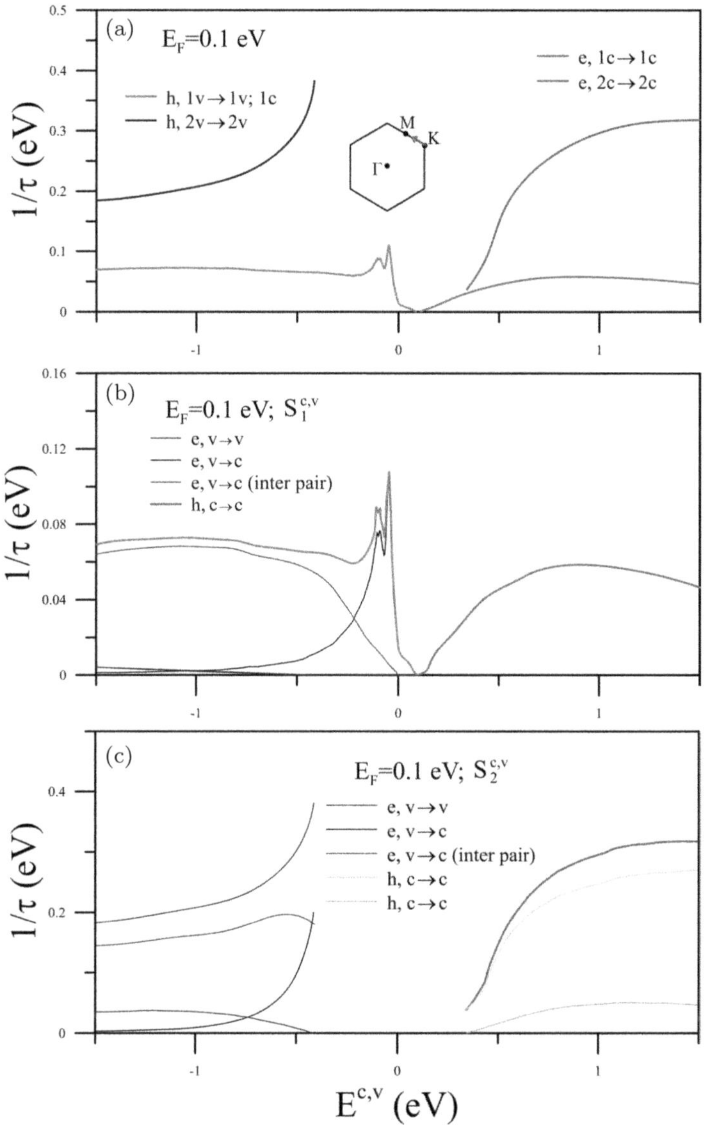


Fig. 3.21 (a) The energy-dependent Coulomb decay rate of the quasiparticle states in AB-stacked bilayer graphene at $E_F = 0.1\,\text{eV}$ along the *KM* direction. For the (b) first pair and (c) second pair parabolic bands, the separation contribution of each de-excitation channel is shown.

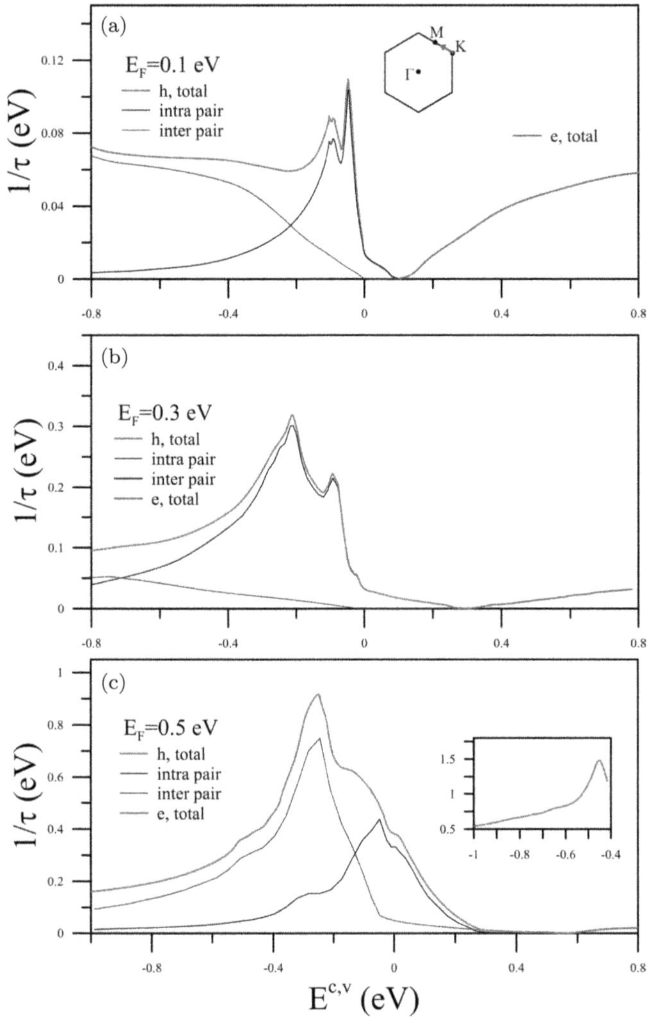


Fig. 3.22 The Coulomb decay rate of the first pair of the parabolic band in AB-stacked bilayer graphene under different Fermi levels.

the energy required for plasmon resonance in this momentum range. The electron gas behavior is derived from the purple curve. The extreme level of the plasmon de-excitation process is considered in the case where a certain part of the electron population highly occupies two conduction bands, e.g., $E_F = 0.5\,\text{eV}$ in Fig. 3.22(c). The excited states reveal the significant coupling between the electron and the plasmon.

The Coulomb decay rate of multilayer graphenes varies with the stacking configuration and the number of layers. The decay spectra of AA-, AB- and ABC-stacked graphenes sharply contrast with each other because of their distinctive quasiparticle excitation behavior, as shown in Figs. 3.23–3.25. The anisotropy of the interpair Coulomb decay rate is revealed in the $k$ space for any stacking configuration in the range close to

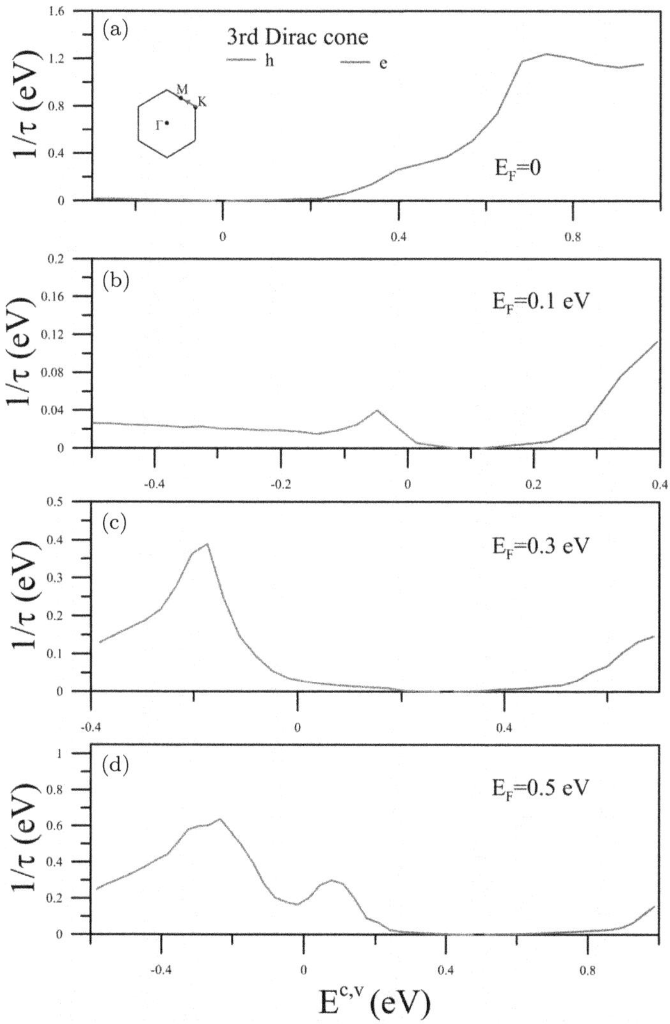


Fig. 3.23 The Coulomb decay rate for the highest Dirac cone in trilayer AAA graphene under different Fermi levels.

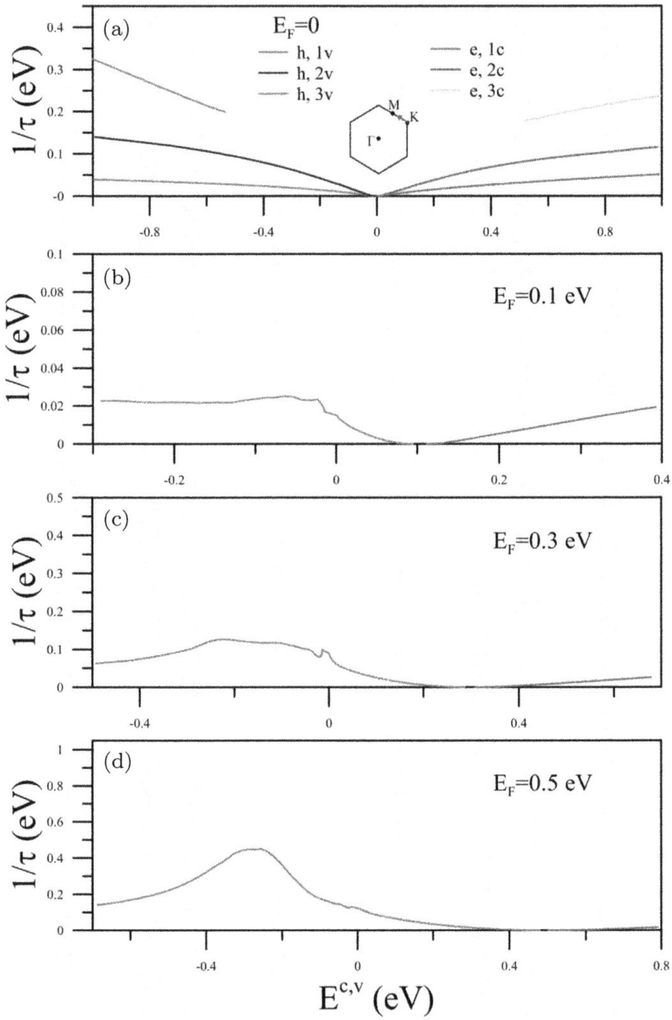


Fig. 3.24 The Coulomb decay rate for the lowest pair of the parabolic bands in trilayer ABA graphene under different Fermi levels.

and far from the Fermi level. As observed in conventional 2D systems, it can be clearly recognized in the three systems that the prominent acoustic plasmon mode manifests a fast decay rate of quasiparticles but with a clear energy dependence on electron concentration according to the stacked structure. Regarding the plasmon resonance played by the optical plasma mode, this effect on Coulomb scattering is reflected in a deeper excited

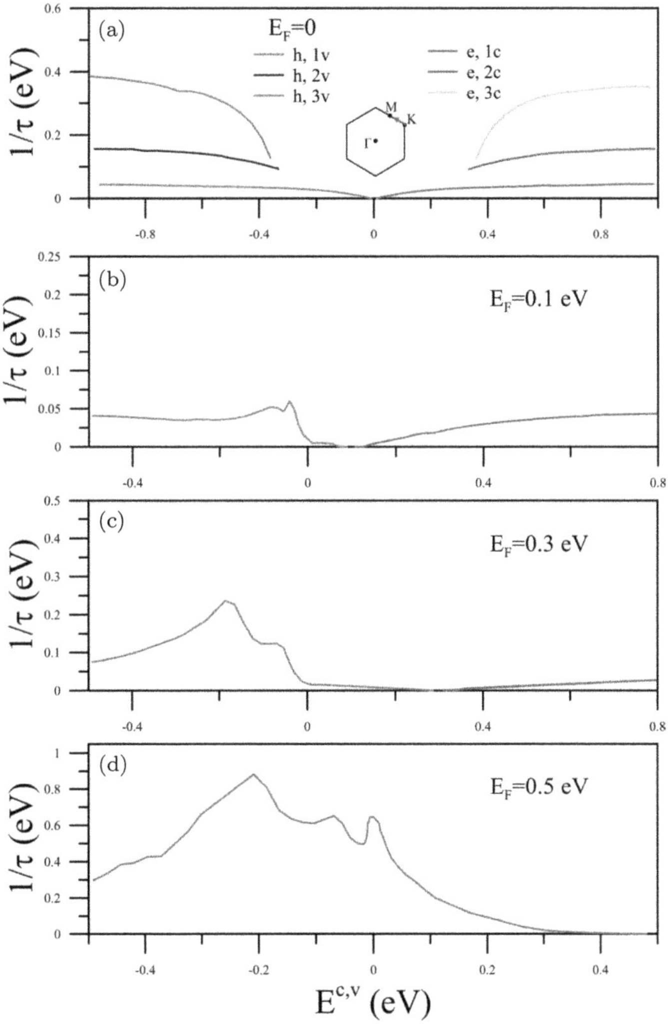


Fig. 3.25 The Coulomb decay rate for the lowest pair of the partially flat bands in trilayer ABC graphene under different Fermi levels.

state, but it may be suppressed by Landau damping and cannot show its importance. The three pairs of Dirac cone structures cause its AA stacking to have the most prominent plasmon mode and Coulomb decay rate among the three systems. The position of the peak in the Coulomb scattering spectrum indicates the hybrid nature of the respective Dirac cone. The plasmon-resonance peak can appear in electron or hole decay, even if AA graphene is doped with electrons. Because the prominent plasmon modes do not interfere with each other, this phenomenon can be observed in all multilayer AA graphenes and verified by ARPES observations of the valence states. In conclusion, the role played by the multi-plasmon modes can be fully presented in the spectra of Coulomb excitations and decays for AA-stacked graphene. In AB-stacked trilayer graphene, the quasiparticle decay rate under moderate doping cannot be traced from the composite effect of monolayer and bilayer graphenes, unlike the interpretation for the band structure. This is due to the fact that the optical plasmon mode is obscure in the case of more stacked layers. This leads to the fact that experimental measurements must cover quasiparticles with relatively larger energy and momentum ranges to verify the characteristics of AB stacking. On the other hand, the symmetric property of quasipartice behavior with respect to the K point might be strongly modified under ABC stacking, mainly owing to the surface-localized energy bands. The hybridization of the bulk energy bands around the K point [125] can enrich and diversify the Coulomb decay rates in ABC-stacked graphene because of the induced additional de-excitation channels for the middle-energy decay rate.

As is well known, the configuration dominates the symmetry of the crystal lattice, thus diversifying the basic properties of layered and coaxial condensed matter [80, 112, 118, 126]. Most importantly, the theoretical framework has been used to combine various intrinsic interactions and external perturbation factors and thoroughly explore the resulting basic properties, such as optical properties [112, 126], magneto-electronic properties [112, 118, 126] and Coulomb excitations [83]. The research involves emergent low-dimensional materials, such as Group-IV and Group-V mono-element materials and $MoS_2$. The development of this framework involves the combination of RPA and layer-decomposition self-energy methods to study the static and dynamic properties of quasiparticles under many-body interactions. For example, the Coulomb decay rates of quasiparticles in a multilayer graphene system with a stacked configuration, such as AA, AB or ABC, included in this chapter are appropriate for an in-depth study of the basic de-excitation characteristics under this development model. It has been shown that dimensions can dictate the electronic properties of materials with honeycomb lattices, thereby greatly enriching Coulomb excitation and quasiparticle characteristics and explaining the phenomena of many-body effects. The Dirac point in pristine graphene is electrically neutral. By changing the doping level, it is suitable for qualitatively and quantitatively discussing the quasiparticle picture in terms of single-particle and collective behavior. The inelastic Coulomb scattering can provide an abnormal temperature and wavevector correlation for quasiparticle coupling through the efficient channels of e–h excitation in the low-frequency band, the 2D acoustic plasmon and the interband e–h excitation. The latest experimental techniques, such as femtosecond time-resolved photoelectrons [141, 142], absorption and fluorescence spectroscopy [143–147], are performed on electronic states with prominent van Hove singularities to show the quasiparticle properties. In general, ARPES can be used to directly measure the energy and lifetime of quasiparticles; however, due to the photoelectric effect, the measurement is limited to occupying valence band states. The predicted many-body effect can be checked directly from the spectral properties in the high-resolution ARPES measurements.

## References


[1] Poterba J.M. and Summers L.H. 1987. Finite lifetimes and the effects of budget deficits on national saving. *J. Monet. Econ.* **20** 369–391.

[2] Tokuda S., Souma S., Segawa K., Takahashi T., Ando Y. *et al.* 2021. Unveiling quasiparticle dynamics of topological insulators through Bayesian modelling. *J. Phys. Commun.* **4** 170.

[3] Sessi P., Fan F.R., Kuster F., Manna K., Schroter, N.B.M. *et al.* 2020. Handedness-dependent quasiparticle interference in the two enantiomers of the topological chiral semimetal PdGa. *Nat. Commun.* **11** 3507.

[4] Hong T., Qiu Y., Matsumoto M., Tennant D.A., Coester K. *et al.* 2017. Field induced spontaneous quasiparticle decay and renormalization of quasiparticle dispersion in a quantum antiferromagnet. *Nat. Commun.* **8** 15148.

[5] Oldwig von R. and Jonas S.Z. 1961. Effect of electron exchange on the dispersion relation of plasmons. *Phys. Rev.* **121** 941.

[6] Nils E.D. and Robert van L. 2005. Self-consistent solution of the Dyson equation for atoms and molecules within a conserving approximation. *J. Chem. Phys.* **122** 164102.

[7] Giuliani G.F. and Quinn J.J. 1982. Lifetime of a quasiparticle in a two-dimensional electron gas. *Phys. Rev. B* **26** 4421.

[8] Caruso F., Novko D. and Draxl C. 2018. Phonon-assisted damping of plasmons in three- and two-dimensional metals. *Phys. Rev. B* **97** 205118.

[9] Frank S. 2018. Polarizability of a two-dimensional electron gas. *Phys. Rev. Lett.* **18** 546.

[10] Lin M.F. and Shung K.W.-K. 1993. Elementary excitations in cylindrical tubules. *Phys. Rev. B* **47** 6617–6624.

[11] Northrup J.E., Hybertsen M.S. and Louie S.G. 1989. Quasiparticle excitation spectrum for nearly-free-electron metals. *Phys. Rev. B* **39** 8198.

[12] Hubbard J. 1958. The description of collective motions in terms of many-body perturbation theory. II. The correlation energy of a free-electron gas. *Proc. R. Soc. Lond.* **14** 336–352.

[13] Singwi K.S., Tosi M.P., Land R.H. and Sjolander A. 1968. Electron correlations at metallic densities. *Phys. Rev.* **176** 589.

[14] Nozieres P. and Pines D. 1958. Correlation energy of a free electron gas. *Phys. Rev.* **111** 442.

[15] Rossi E. and Sarma S.D. 2008. Ground state of fraphene in the presence of random charged impurities. *Phys. Rev. Lett.* **101** 166803.

[16] Kim D., Li Q., Syers P., Butch N.P., Paglione J., Sarma S.D. and Fuhrer M.S. 2012. Intrinsic electron-phonon resistivity of $Bi_2Se_3$ in the topological regime. *Phys. Rev. Lett.* **109** 166801.

[17] Ericson T. 1958. Level density of degenerate Fermi systems. *Nucl. Phys. News* **8** 265–283.

[18] Chubukov A.V., Galitski V.M. and Yakovenko V.M. 2005. Quantum critical behavior near a density-wave instability in an isotropic Fermi liquid. *Phys. Rev. Lett.* **94** 046404.

[19] Csiba T., Kriza G. and Janossy A. 1989. Propagation of charge-density wave voltage noise along a blue-bronze, $Rb_{0.3}MoO_3$ crystal. *EPL* **9** 163.

[20] Bulutay C. and Tanatar B. 2002. Spin-dependent analysis of two-dimensional electron liquids. *Phys. Rev. B* **65** 195116.

[21] Hwang E.H. and Sarma S.D. 2008. Screening, Kohn Anomaly, Friedel oscillation, and RKKY interaction in bilayer graphene. *Phys. Rev. Lett.* **101** 156802.

[22] Kwon Y., Ceperley D.M. and Martin R.M. 1998. Effects of backflow correlation in the three-dimensional electron gas: Quantum Monte Carlo study. *Phys. Rev. B* **58** 6800.

[23] Driver K.P., Cohen R.E., Wu Z., Militzer B., Lopez Rios P., Towler M.D., Needs R.J. and Wilkins J.W. 2010. Quantum Monte Carlo computations of phase stability, equations of state, and elasticity of high-pressure silica. *Proc. Natl. Acad. Sci. USA* **107** 9519–9524.

[24] Geim A.K. and Novoselov K.S. 2007. The rise of graphene. *Nat. Mater.* **6** 183.

[25] Novoselov K.S., Geim A.K., Morozov S.V., Jiang D., Zhang Y. *et al.* 2004. Electric field effect in atomically thin carbon films. *Science* **306** 666.

[26] Schedin F., Geim A.K., Morozov S.V., Hill E.W., Blake P. *et al.* 2007. Detection of individual gas molecules adsorbed on graphene. *Nat. Mater.* **6** 652.

[27] Gierz I., Riedl C., Starke U., Ast C.R. and Kern K. 2008. Atomic hole doping of graphene. *Nano Lett.* **8** 4603.

[28] Shin S.Y., Kim N.D., Kim J.G., Kim K.S., Noh D.Y. *et al.* 2011. Control of the pi plasmon in a single layer graphene by charge doping. *App. Phys. Lett.* **99** 082110.

[29] Mahan G.D. 1990. *Many-Particle Physics* (New York: Plenum).

[30] Hwang E.H. and Das Sarma S. 2007. Dielectric function, screening, and plasmons in two-dimensional graphene. *Phys. Rev. B* **75** 205418.

[31] Wunsch B., Stauber T., Sols F. and Guinea F. 2006. Dynamical polarization of graphene at finite doping. *New J. Phys.* **8** 318.

[32] Borghi G., Polini M., Asgari R. and MacDonald A.H. 2009. Dynamical response functions and collective modes of bilayer graphene. *Phys. Rev. B* **80** 241402.

[33] Sensarma R., Hwang E.H. and Das Sarma S. 2010. Dynamic screening and low-energy collective modes in bilayer graphene. *Phys. Rev. B* **82** 195428.

[34] Zhu J.J., Badalyan S.M. and Peeters F.M. 2013. Plasmonic excitations in Coulomb-coupled N-layer graphene structures. *Phys. Rev. B* **87** 085401.

[35] Ho J.H., Chang C.P., Chen R.B. and Lin M.F. 2006. Electronic excitations of the multilayered graphite. *Phys. Lett. A* **357** 401.

[36] Shin S.Y., Hwang C.G., Sung S.J., Kim N.D., Kim H.S. *et al.* 2011. Observation of intrinsic intraband pi-plasmon excitation of a single-layer graphene. *Phys. Rev. B* **83** 161403.

[37] Sanchez-Barriga J., Varykhalov A., Springholz G., Steiner H., Kirchschlager R. *et al.* 2020. Nonmagnetic band gap at the Dirac point of the magnetic topological insulator $(Bi_{1-x}Mn_x)_2Se_3$. *Nat. Commun.* **7** 10559.

[38] Souma S., Komatsu M., Nomura M., Sato T., Takayama A. *et al.* 2012. Spin polarization of gapped dirac surface states near the topological phase transition in $TlBi(S_{1-x}Se_x)_2$. *Phys. Rev. Lett.* **109** 186804.

[39] Lv B.Q., Qian T. and Ding H. 2019. Angle-resolved photoemission spectroscopy and its application to topological materials. *Nat. Rev. Phys.* **1** 609.

[40] Renz R. and Schulz H.-J. 1983. The decay of infrared luminescence in II-VI compound semiconductors doped by 3d transition elements. *J. Phys. C: Solid State Phys.* **16** 4917.

[41] Katayama-Yoshida H. and Sato K. 2003. Materials design for semiconductor spintronics by ab initio electronic-structure calculation. *Physica B Condens. Matter* **327** 337.

[42] King P.D.C., McKeown Walker S., Tamai A., de la Torre A., Eknapakul T. *et al.* 2014. Quasiparticle dynamics and spin-orbital texture of the $SrTiO_3$ two-dimensional electron gas. *Nat. Commun.* **5** 3414.

[43] Riddoch F.A. and Ridley B.K. 1983. On the scattering of electrons by polar optical phonons in quasi-2D quantum wells. *J. Phys. C: Solid State Phys.* **16** 6971.

[44] Feldmann J., Peter G., Gobel E.O., Dawson P., Moore K., Foxon C. and Elliott R.J. 1988. Linewidth dependence of radiative exciton lifetimes in quantum wells. *Phys. Rev. Lett.* **60** 243.

[45] Ulstrup S., Giusca C.E., Miwa J.A., Sanders C.E. and Browning A. 2019. Nanoscale mapping of quasiparticle band alignment. *Nat. Commun.* **10** 3283.

[46] Read A.J., Needs R.J., Nash K.J., Canham L.T., Calcott P.D.J. and Qteish A. 1993. First-principles calculations of the electronic properties of silicon quantum wires. *Phys. Rev. Lett.* **69** 1232.

[47] Zhensheng T., Cong C., Tibor S., Mark K., Manos M. *et al.* 2016. Direct time-domain observation of attosecond final-state lifetimes in photoemission from solids. *Science* **353** 62.

[48] Hofmann M.S., Gluckert J.T., Noe J., Bourjau C. and Dehmel R. 2013. Bright, long-lived and coherent excitons in carbon nanotube quantum dots. *Nat. Nanotechnol.* **8** 502.
[49] Chen S.C., Shyu F.L., Lue C.S. and Lin M.F. 2006. Dielectric screening for carbon nanotubes in a gating electric field. *Physica E* **32** 577.
[50] Bostwick A., Ohta T., Seyller T., Horn K. and Rotenberg E. 2007. Quasiparticle dynamics in graphene. *Nat. Phys.* **3** 36.
[51] Xu S., Cao J., Miller C.C., Mantell D.A., Miller R.J.D. *et al.* 1996. Energy dependence of electron lifetime in graphite observed with femtosecond photoemission spectroscopy. *Phys. Rev. Lett.* **76** 483.
[52] Moos G., Gahl C., Fasel R., Wolf M. and Hertel T. 2001. Anisotropy of quasiparticle lifetimes and the role of disorder in graphite from ultrafast time-resolved photoemission spectroscopy. *Phys. Rev. Lett.* **87** 267402.
[53] Lin M.F., Huang C.S. and Chuu D.S. 1997. Plasmons in graphite and stage-1 graphite intercalation compounds. *Phys. Rev. B* **55** 13961.
[54] Lin M.F. and Shung K.W.-K. 1996. The self-energy of electrons in graphite intercalation compounds. *Phys. Rev. B* **53** 1109.
[55] Lin M.F. and Shung K.W.-K. 1994. Plasmons and optical properties of carbon nanotubes. *Phys. Rev. B* **50** 17744.
[56] Lin M.F. and Shung K.W.-K. 1997. Optical and magneto-optical properties of carbon nanotube bundles. *J. Phys. Soc. Jpn.* **66** 3294.
[57] Gusynin V.P. and Sharapov S.G. 2006. Transport of Dirac quasiparticles in graphene: Hall and optical conductivities. *Phys. Rev. B* **73** 245411.
[58] Tougaard S. and Chorkendorff I. 1987. Differential inelastic electron scattering cross sections from experimental reflection electron-energy-loss spectra: Application to background removal in electron spectroscopy. *Phys. Rev. B* **35** 6570.
[59] Rowe J.E. and Ibach H. 1973. Surface-dtate transitions of silicon in electron energy-loss spectra. *Phys. Rev. Lett.* **31** 102.
[60] Park C.H., Giustino F., Spataru C.D., Cohen M.L. and Louie S.G. 2009. Angle-resolved photoemission spectra of graphene from first-principles calculations. *Nano Lett.* **9** 4234.
[61] Rosato V., Guillope M. and Legrand B. 1989. Thermodynamical and structural properties of f.c.c. transition metals using a simple tight-binding model. *Philos. Mag.* **59** 321.
[62] Dresselhaus G. 1956. Effective mass approximation for excitons. *J. Phys. Chem. Solids* **1** 14.
[63] Xiong W., Xia C., Wang T., Peng Y. and Jia Y. 2016. Strain and spin-orbital coupling effects on electronic structures and magnetism of semi-hydrogenated stanene. *J. Phys. Chem. C* **120** 10622.
[64] Chiu Y.H., Ho J.H., Chang C.P., Chuu D.S. and Lin M.F. 2008. Low-frequency magneto-optical excitations of a graphene monolayer: Peierls tight-binding model and gradient approximation calculation. *Phys. Rev. B* **78** 245411.
[65] Chiu Y.H., Ho J.H., Ho Y.H., Chuu D.S. and Lin M.F. 2008. Effects of a modulated electric field on the optical absorption spectra in a single-layer graphene. *J. Nanosci. Nanotechnol.* **9** 6579.

[66] Panchakarla L.S., Subrahmanyam K.S., Saha S.K., Govindaraj A., Krishnamurthy H.R., Waghmare U.V. *et al.* 2009. Synthesis, structure, and properties of boron and nitrogend doped graphene. *Adv. Mater.* **21** 4726.
[67] Martins T.B., Miwa R.H., da Silva A.J.R. and Fazzio A. 2007. Electronic and transport properties of boron-doped graphene nanoribbons. *Phys. Rev. Lett.* **98** 196803.
[68] Shih P.H., Chiu C.W., Wu J.Y., Do T.N. and Lin M.F. 2018. Coulomb scattering rates of excited states in monolayer electron-doped germanene. *Phys. Rev. B* **97** 195302.
[69] Starodub E., Bostwick A., Moreschini L., Nie S., El Gabaly F., McCarty K.F. and Rotenberg E. 2011. In-plane orientation effects on the electronic structure, stability, and Raman scattering of monolayer graphene on Ir(111). *Phys. Rev. B* **83** 125428.
[70] Gierz I. 2017. Probing carrier dynamics in photo-excited graphene with time-resolved ARPES. *J. Electron Spectros. Relat. Phenomena* **219** 53.
[71] Weber C.P., Gedik N., Moore J.E., Orenstein J. and Stephens J. 2005. Observation of spin Coulomb drag in a two-dimensional electron gas. *Nature* **437** 1330.
[72] Norman M.R., Eschrig M., Kaminski A. and Campuzano J.C. 2001. Momentum distribution curves in the superconducting state. *Phys. Rev. B* **64** 184508.
[73] Chiu C.W. and Lin M.F. 2007. Electronic excitations and deexcitations in narrow-gap carbon nanotubes. *Nanotechnology* **18** 435401.
[74] Chiu C.W. and Lin M.F. 2008. Temperature-dependent Coulomb scattering rates of Fermi-momentum states in metallic carbon nanotubes. *Physica E* **40** 2720.
[75] Chiu C.W., Shyu F.L., Chang C.P., Chuu D.S. and Lin M.F. 2006. Coulomb scattering rates of excited carriers in moderate-gap carbon nanotubes. *Phys. Rev. B* **73** 235407.
[76] Chiu C.W., Shyu F.L., Chang C.P., Chen R.B., Chuu D.S. and Lin M.F. 2004. Electron lifetime in armchair carbon nanotubes. *J. Phys. Soc. Jpn.* **73** 2936.
[77] Chiu C.W., Chiu Y.H., Shyu F.L., Chang C.P., Chuu D.S. and Lin M.F. 2005. Temperature-dependent carrier dynamics in metallic carbon nanotubes. *Phys. Lett. A* **346** 347.
[78] Ho J.H., Chiu C.W. and Lin M.F. 2006. Temperature-dependent self-energy of single-walled carbon nantubes. *Physica E* **35** 212.
[79] Ho G.W., Ho Y.H., Li T.S., Chang C.P. and Lin M.F. 2006. Band structure and absorption spectrum of double-walled zigzag carbon nanotubes in an electric field. *Carbon* **44** 2323.
[80] Ho Y.H., Ho G.W., Chen S.C., Ho J.H. and Lin M.F. 2007. Low-frequency excitation spectra in double-walled armchair carbon nanotubes. *Phys. Rev. B* **76** 115422.
[81] Jiang H., Feng X.-Q., Huang Y., Hwang K.C. and Wu P.D. 2004. Defect nucleation in carbon nanotubes under tension and torsion: Stone- Wales transformation. *Comput. Methods Appl. Mech. Eng.* **193** 3419.
[82] Lin C.Y., Lee M.H. and Lin M.F. 2018. Coulomb excitations in ABC-stacked trilayer graphene. *Phys. Rev. B* **98** 041408(R).

[83] Lin C.Y., Wu J.Y., Chiu C.W. and Lin M.F. 2019. *Coulomb Excitations and Decays in Graphene-Related Systems* (CRC Press).
[84] Hermanson J. 1977. Final-state symmetry and polarization effects in angle-resolved photoemission spectroscopy. *Solid State Commun.* **22** 9.
[85] Ohta T., Bostwick A., McChesney J.L., Seyller T., Horn K. *et al.* 2007. Interlayer interaction and electronic screening in multilayer graphene investigated with angle-resolved photoemission spectroscopy. *Phys. Rev. Lett.* **98** 206802.
[86] Jin W., Yeh P.C., Zaki N., Zhang D., Sadowski J.T. *et al.* 2013. Direct measurement of the thickness-dependent electronic band structure of MoS2 using angle-resolved photoemission spectroscopy. *Phys. Rev. Lett.* **111** 106801.
[87] Ibach H. and Mills D.L. 2013. *Electron Energy Loss Spectroscopy and Surface Vibrations* (Academic Press).
[88] Politano A., Chiarello G., Formoso V., Agostino R.G. and Colavita E. 2006. Plasmon of Shockley surface states in Cu(111): A high-resolution electron energy loss spectroscopy study. *Phys. Rev. B* **74** 081401.
[89] Politano A., Formoso V., Agostino R.G., Colavita E. and Chiarello G. 2007. Influence of CO adsorption on the alkali-substrate bond studied by high-resolution electron energy loss spectroscopy. *Phys. Rev. B* **76** 233403.
[90] Lu J., Loh K.P., Huang H., Chen W. and Wee A.T. 2009. Plasmon dispersion on epitaxial graphene studied using high-resolution electron energy-loss spectroscopy. *Phys. Rev. B* **80** 113410.
[91] Ulstrup S., Giusca C.E., Miwa J.A., Sanders C.E., Browning A. *et al.* 2019. Nanoscale mapping of quasiparticle band alignment. *Nat. Commun.* **10** 3283.
[92] Gruneis A., Attaccalite C., Pichler T., Zabolotnyy V. and Shiozawa H. 2008. Electron-electron correlation in graphite: A combined angle-resolved photoemission and first-principles study. *Phys. Rev. Lett.* **100** 037601.
[93] Zhou S.Y., Gweon G.-H., Graf J., Fedorov A.V., Spataru C.D. *et al.* First direct observation of Dirac fermions in graphite. *Nat. Phys.* **2** 595.
[94] Cheng C.M., Hsu C.J., Peng J.L., Chen C.H., Yuh J.Y. *et al.* 2015. Tight-binding parameters of graphite determined with angle-resolved photoemission spectra. *Appl. Surf. Sci.* **354** 229.
[95] Leem C.S., Kim C., Park S.R., Kim M.K., Choi H.J. *et al.* 2009. High-resolution angle-resolved photoemission studies of quasiparticle dynamics in graphite. *Phys. Rev. B* **79** 125438.
[96] Ruffieux P., Cai J., Plumb N.C., Patthey L., Prezzi D. *et al.* 2012. Electronic structure of atomically precise graphene nanoribbons. *ACS Nano* **6** 6930.
[97] Sugawara K., Sato T., Souma S., Takahashi T. and Suematsu H. 2006. Fermi surface and edge-localized states in graphite studied by high-resolution angle-resolved photoemission spectroscopy. *Phys. Rev. B* **73** 045124.
[98] Siegel D.A., Regan W., Fedorov A.V., Zettl A. and Lanzara A. 2013. Charge-carrier screening in single-layer graphene. *Phys. Rev. Lett.* **110** 146802.
[99] Kim K.S., Walter A.L., Moreschini L., Seyller T., Horn K.E. *et al.* 2013. Coexisting massive and massless Dirac fermions in symmetry-broken bilayer graphene. *Nat. Mater.* **12** 887.

[100] Papagno M., Rusponi S., Sheverdyaeva P.M., Vlaic S., Etzkorn M. *et al.* 2012. Large band gap opening between graphene Dirac cones induced by Na adsorption onto an Ir superlattice. *ACS Nano* **6** 199.
[101] Ohta T., Bostwick A. and McChesney J.L. 2007. Interlayer interaction and electronic screening in multilayer graphene investigated with angle-resolved photoemission spectroscopy. *Phys. Rev. Lett.* **98** 206802.
[102] Moreau E., Godey S., Wallart X., Razado-Colambo I., Avila J. *et al.* 2013. Highresolution angle-resolved photoemission spectroscopy study of monolayer and bilayer graphene on the C-face of SiC. *Phys. Rev. B* **88** 075406.
[103] Coletti C., Forti S., Principi A., Emtsev K.V., Zakharov A.A. *et al.* 2013. Revealing the electronic band structure of trilayer graphene on SiC: An angle-resolved photoemission study. *Phys. Rev. B* **88** 155439.
[104] Sugawara K., Yamamura N., Matsuda K., Norimatsu W., Kusunoki M. *et al.* 2018. Selective fabrication of free-standing ABA and ABC trilayer graphene with/without Dirac-cone energy bands. *NPG Asia Mater.* **10** 466.
[105] Schomacker K.T. and Champion P.M. 1986. Investigations of spectral broadening mechanisms in biomolecules: Cytochrome-c. *J. Chem. Phys.* **84** 5314.
[106] Wu, J.Y., Chen, S.C. and Lin, M.F. 2014. Temperature-dependent Coulomb excitations in silicene. *New J. Phys.* **16** 125002.
[107] Wunsch B., Stauber T., Sols F. and Guinea F. 2006. Dynamical polarization of graphene at finite doping. *New J. Phys.* **8** 318.
[108] Borghi G., Polini M., Asgari R. and MacDonald A.H. 2009. Dynamical response functions and collective modes of bilayer graphene. *Phys. Rev. B* **80** 241402.
[109] Sensarma R., Hwang E.H. and Das Sarma S. 2010. Dynamic screening and low-energy collective modes in bilayer graphene. *Phys. Rev. B* **82** 195428.
[110] Zhu J.-J., Badalyan S.M. and Peeters F.M. 2013. Plasmonic excitations in Coulomb-coupled N-layer graphene structures. *Phys. Rev. B* **87** 085401.
[111] Shung K.W.-K. 1986. Dielectric function and plasmon structure of stage-1 intercalated graphite. *Phys. Rev. B* **34** 979.
[112] Lin C.Y., Chen R.B., Ho Y.H. and Lin M.F. 2018. *Electronic and Optical Properties of Graphite-Related Systems* (CRC Press).
[113] Ho J.H., Lu C.L., Hwang C.C., Chang C.P. and Lin M.F. 2006. Coulomb excitations in AA- and AB-stacked bilayer graphites. *Phys. Rev. B* **74** 085406.
[114] Shih P.H., Chiu Y.H., Wu J.Y., Shyu F.L. and Lin M.F. 2017. Coulomb excitations of monolayer germanene. *Sci. Rep.* **7** 40600.
[115] Wu J.Y., Chen S.C., Roslyak O., Gumbs G. and Lin M.F. 2011. Plasma excitations in graphene: Their spectral intensity and temperature dependence in magnetic field. *ACS Nano* **5** 1026.
[116] Shih P.H., Chiu C.W., Wu J.Y., Do T.N. and Lin M.F. 2018. Coulomb scattering rates of excited states in germanene. *Phys. Rev. B* **97** 195302.
[117] Kittel C. 2001. *Introduction to Solid State Physics*, 8th edn. (London: Wiley).
[118] Lin C.Y., Do T.N., Wu J.Y., Shih P.H., Lin S.Y., *et al.* 2019. *Diverse Quantization Phenomena in Emergent Layered Materials* (CRC Press).

[119] Lin C.Y., Yang C.H., Chiu C.W., Chung H.C., Lin S.Y. and Lin M.F. Many-particle interactions in carbon nanotubes. IOP Concise Physics (San Raefel, CA, USA: Morgan & Claypool Publishers).
[120] Shung K.W.-K. 1986. Lifetime effects in low-stage intercalated graphite systems. *Phys. Rev. B* **34** 2.
[121] Lin M.F. and Chuu D.S. 1997. Elementary excitations in a carbon nanotube. *J. Phys. Soc. Jpn.* **66** 757.
[122] Chuang Y.C., Wu J.Y. and Lin M.F. 2012. Analytical calculations on low-frequency excitations in AA-stacked bilayer graphene. *J. Phys. Soc. Jpn.* **81** 124713.
[123] Lin M.F., Chuang Y.C. and Wu J.Y. 2012. Electrically tunable plasma excitations in AA-stacked multilayer graphene. *Phys. Rev. B* **86** 125434.
[124] Chuang Y.C., Wu J.Y. and Lin M.F. 2013. Electric-field-induced plasmon in AA-stacked bilayer graphene. *Ann. Phys.* **339** 298.
[125] Ho J.H., Chang C.P. and Lin M.F. 2006. Electronic excitations of the multilayered graphite. *Phys. Lett. A* **352** 446.
[126] Lin C.Y., Do T.N., Huang Y.K. and Lin M.F. 2017. *Electronic and Optical Properties of Graphene in Magnetic and Electric Fields.* IOP Concise Physics (San Raefel, CA, USA: Morgan & Claypool Publishers).
[127] Chen S.C., Wu J.Y., Lin C.Y. and Lin M.F. 2017. *Theory of Magnetoelectric Properties of 2D Systems.* IOP Concise Physics (San Raefel, CA, USA: Morgan & Claypool Publishers).
[128] Egerton R.F. 1996. *Electron Energy-Loss Spectroscopy in the Electron Microscope,* Vol. 370 (New York, London: Plenum).
[129] Schülke W. 2007. *Electron Dynamics by Inelastic x-Ray Scattering* (Oxford: Oxford University Press).
[130] Zeppenfeld K. 1969. Wavelength dependence and spatial dispersion of the dielectric constant in graphite by electron spectroscopy. *Opt. Com.* **1** 119.
[131] Zeppenfeld K. 1971. Nonvertical interband transitions in graphite by intrinsic electron scattering. *Z. Phys.* **243** 229.
[132] Valla T., Camacho J., Pan Z.-H., Fedorov A.V., Walters A.C. *et al.* 2009. Anisotropic electron-phonon coupling and dynamical nesting on the graphene sheets in superconducting $CaC_6$ using angle-resolved photoemission spectroscopy. *Phys. Rev. Lett.* **102** 107007.
[133] Fedorov A.V., Verbitskiy N.I., Haberer D., Struzzi C. and Petaccia L. 2014. Observation of a universal donor-dependent vibrational mode in graphene. *Nat. Commun.* **5** 3257.
[134] Lin M.F. and Shyu F.L. 2000. Temperature-induced plasmons in a graphite sheet. *J. Phys. Soc. Jpn.* **69** 607.
[135] Li Z.Q., Henriksen E.A., Jiang Z., Hao Z. and Martin M.C. 2008. Dirac charge dynamics in graphene by infrared spectroscopy. *Nat. Phys.* **4** 532.
[136] Matthes L., Pulci O. and Bechstedt F. 2013. Massive Dirac quasiparticles in the optical absorbance of graphene, silicene, germanene, and tinene. *J. Condens. Matter Phys.* **25** 395305.

[137] Gatti G., Crepaldi A., Puppin M., Tancogne-Dejean N. and Xian L. 2020. Light-induced renormalization of the Dirac quasiparticles in the nodal-line semimetal ZrSiSe. *Phys. Rev. Lett.* **125** 076401.

[138] Miao L., Wang Z.F., Ming W.M., Yao M.Y. and Wang M.X. 2013. Quasiparticle dynamics in reshaped helical Dirac cone of topological insulators. *PNAS* **110** 2758.

[139] Liu C., Bian G., Chang T.R., Wang K. and Xu S.Y. 2020. Tunable spin helical Dirac quasiparticles on the surface of three-dimensional HgTe. *Phys. Rev. B* **92** 115436.

[140] Jost A., Bendias M., Bottcher J., Hankiewicz E. and Brune C. 2017. Electron-hole asymmetry of the topological surface states in strained HgTe. *PNAS* **114** 3381.

[141] Hertel T. and Moos G. 2000. Electron-phonon interaction in single-wall carbon nanotubes: A time-domain study. *Phys. Rev. Lett.* **84** 5002.

[142] Zamkov M., Woody N., Shan B., Chang Z. and Richard P. 2005. Lifetime of charge carriers in multiwalled nanotubes. *Phys. Rev. Lett.* **94** 056803.

[143] Lauret J.-S., Voisin C., Cassabois G., Delalande C., Roussignol P. *et al.* 2003. Ultrafast carrier dynamics in single-wall carbon nanotubes. *Phys. Rev. Lett.* **90** 057404.

[144] Lauret J.-S., Voisina C., Cassabois G., Tignon J., Delalande C. *et al.* 2004. Third-order optical nonlinearities of carbon nanotubes in the femtosecond regime. *Appl. Phys. Lett.* **85** 3572.

[145] Korovyanko O.J., Sheng C.-X., Vardeny Z.V., Dalton A.B. and Baughman R.H. 2004. Ultrafast spectroscopy of excitons in single-walled carbon nanotubes. *Phys. Rev. Lett.* **92** 017403.

[146] Ostojic G.N., Zaric S., Kono J., Strano M.S., Moore V.C. *et al.* 2004. Interband recombination dynamics in resonantly excited single-walled carbon nanotubes. *Phys. Rev. Lett.* **92** 117402.

[147] Huang L., Pedrosa H.N. and Krauss T.D. 2004. Ultrafast ground-state recovery of single-walled carbon nanotubes. *Phys. Rev. Lett.* **93** 017403.

Chapter 4

# MULTI-MODE FRIEDEL OSCILLATIONS AND STATIC CHARGE SCREENINGS IN LAYERED SYSTEMS

Chiun-Yan Lin, Chih-Wei Chiu, Ching-Hong Ho, Hsien-Ching Chung, Lu-Yao Wang, and Ming-Fa Lin

Static charge screening is a fundamental topic in physics that has been extensively studied [1–30]. It refers to the way in which valence charges in condensed matter systems fluctuate and polarize to screen the external Coulomb field caused by the presence of impurities. This screening is mediated by long-range interactions between electrons and is responsible for the basic phenomena of electromagnetic theories [29, 30]. The screening ability of a material can be characterized by its dielectric function, which is defined as the ratio of the external field to the total Coulomb field, including the perturbed and induced parts [29, 30]. In classical electrodynamics, this quantity is macroscopic and is called the dielectric constant, which provides a macroscopic averaged response over all frequencies. This perspective is appropriate for a comprehensive understanding of the polarization of electrons in semiconductors or insulators. However, dealing with quantum states from a microscopic viewpoint is challenging. Electric polarization in quantum systems involves transitions of different electronic states and is closely related to the momentum and frequency of the external electric field. When free carriers in n-type or p-type metals are disturbed by impurities, the screening behavior varies depending on the electronic states' momentum and frequency, which respond to external fields differently. Moreover, due to the long-range Coulomb interaction, a quantum phenomenon called Friedel oscillation appears. In this chapter, we present a theoretical framework that accurately calculates the dielectric function in metallic, narrow-gap, middle-gap and semi-metallic systems. We thoroughly illustrate the charge screening in the static limit of $\omega \rightarrow, 0$. Using the Fourier transform, we obtain the total Coulomb potential and induced charge density in **r**-space, demonstrating the effective screening length and the long-range behavior of Friedel oscillation.

Some theoretical quantum models are devoted to exploring the unusual charge screening behaviors of 3D free carriers in metals due to short- and long-range Coulomb fields. The most common of these is the Thomas–Fermi model [29, 31], which semiclassically studies the local Fermi electron gas and the slow variation in the real-space Coulomb potentials. The dielectric function is reliable only for a sufficiently high density of conduction carriers under the long-wavelength limit. This model shows that the effective Coulomb potential is rapidly decayed away from the impurity using the Yukawa potential form under the static screening effect. The tail of the Yukawa potential could survive for a long range in any semiconducting system, while the screening ability is much stronger in metals with a relatively larger Fermi momentum [31]. The effective screening length is defined as being inversely proportional to the square root of the Fermi momentum, beyond which the potential is largely reduced. For example, the effective screening length is $\sim$1.5 Å for the bcc sodium with a density of $r_s = 3.96$ in units of Bohr radius, equivalent to one electron for a sphere of radius $\sim$2 Å. However, the Thomas–Fermi model is a semiclassical theory for many-body systems. Using this approach, the Friedel oscillation of the electron density cannot be generated at a distance far greater than the Thomas–Fermi screening length because the Thomas–Fermi model fails to examine the singularity of the dielectric function across the Fermi surface, which is the origin of the Friedel oscillation. This issue has been successfully solved by several important many-body models for calculating the dielectric function, such as random-phase approximation, Hubbard, Singwi–Sjolander, and Singwi *et al.* [29]. More interestingly, Friedel oscillations occur for all the metallic systems when the free charges are influenced by the potential associated with impurities. The charge density distribution close to the impurity exhibits oscillatory charge screening behaviors at sufficiently long distances. This quantum phenomenon is mainly related to the excitations near the Fermi surface. It should be noted that the simultaneous existence of electron and hole clouds is based on the conservation of the total induced charges. The long-range Friedel oscillations are predicted to vary with Coulomb potentials and Fermi surfaces in different dimensions. The wavelength of such oscillations is related to the characteristic wavevector $q = 2k_F$ in the metal. For example, the 3D, 2D and 1D electron gas systems present the oscillatory forms of $\sin(2k_F r)/r^3$, $\sin(2k_F r)/r^2$ and $\sin(2k_F r)/r$, respectively [1, 3, 29, 31]. The effective screening length and long-range behavior of the Friedel oscillation are clearly displayed in the Fourier transform of the total Coulomb potential and induced charge density in **r**-space. This quantum phenomenon arises from the long-range Coulomb interaction and is characterized by oscillations that describe the behavior of the effective Coulomb potential at long distances, rather than the typical $1/r$ decay.

Friedel explained that when a charged impurity is introduced into a 3D simple metal, the screening behavior can cause the Coulomb field to change over a long range, resulting in oscillations of the charge density [32]. This phenomenon, known as Friedel oscillation, is mainly due to the singularity in the static dielectric function at the transferred momentum of $q = 2k_F$. This singularity arises from the abrupt change in the occupied or unoccupied density of states at the Fermi level. As a result, in metals, these Friedel oscillations are the primary mechanism for the long-range electron–electron Coulomb interactions. It was clarified by Ruderman and Kittel [33] that a similar mechanism of charge screening is the reason for the magnetic interaction between two nuclei induced by the Fermi surface. There have been many experiments [27, 34–37] and theoretical [38–43] studies verifying this phenomenon. Nevertheless, there are only a few theoretical predictions about Friedel oscillations caused by non-spherical Fermi surfaces [23, 25, 44–47]. It is difficult to directly measure the Friedel oscillation caused by impurities on the surface of a solid system. However, the Friedel oscillation can be directly verified by measuring the surface charge density and the distribution of Coulomb potential on the surface of 2D materials through scanning tunneling microscopy (STM). On the Be(0001) surface, STM observed the Friedel oscillation with a wavevector $2k_F$, but the amplitude is enhanced due to many-body effects [2]. Besides, the fine STM clearly reveals the anisotropic screening charge wave on the Be($10\bar{1}0$) near the surface defects and steps as a result of a non-circular Fermi line. Meanwhile, it is very interesting that a highly anisotropic 2D electron gas appears near the surface of black phosphorous through potassium adsorption atoms [48], which directly illustrates the effect of anisotropic charge screening. The unique relationship between Friedel oscillation, Berry phase and energy dispersion has been successfully clarified in the graphene system by studying the long-range behavior of the charge screening [26–28].

Previous works on graphite intercalation compounds provide more information on the essential physical properties of quasiparticles caused by screening phenomena. The dielectric function, calculated using the electronic states of the Dirac cone in the superlattice structure [49–52], clearly explains a significant layer dependence of the phenomena different from those in 2D/layered electron gases [3, 29], such as the strong screening ability of free carriers due to the intraband transition [51], the greatly reduced screening length by the interband transition [51] and the Friedel oscillation phenomenon at long distances [52]. In addition, the long-range Coulomb interactions give rise to the formation of magnetic ordering in graphite and are relevant for the intrinsic origin of the magnetic phenomena [53–55]. Based on the Born approximation, the relaxation time of quasiparticles due to elastic scatterings by the long-range potential created Friedel oscillations can be directly calculated. The relaxation time near the Fermi level can be used to verify the transport measurements on the stage-1 graphite alkali-intercalation compounds, such as the measurements of residual resistivities of $C_8K$, $C_8Rb$ and $C_8Cs$ [56] using a small amount of charged impurities. The Coulomb interactions between impurities and electrons are expected to dominate carrier transport and residual resistivity at extremely low temperatures [29, 51]. Such a relaxation time may entail several excitation channels during scatterings. The concept of quasiparticles can link the screening and transport behavior in layered graphene systems, which may be an interesting research topic.

The longitudinal polarization of a carbon nanotube is highly dependent on its radius and chirality, which determine its electric properties [57–59]. Experimental measurements of the longitudinal dielectric properties can distinguish between metallic and semiconducting carbon nanotubes [62]. The ability to screen the long-range Coulomb field depends on the free carrier density and impurity position [58, 59]. The range of Friedel oscillations in metallic carbon nanotubes can be controlled by the magnitude and position of the impurity [59]. Short-wavelength Friedel oscillations are more obvious at long distances for smaller single-walled armchair nanotubes, while semiconducting ones exhibit the opposite behavior due to their larger energy gap at smaller radii. Close proximity of charged impurities to the tubular surface results in significant short-wavelength Friedel oscillations in both single-walled and multi-walled armchair nanotubes [60]. Different oscillation behaviors, such as period, amplitude and coherence, are predicted on neighboring surfaces of multi-walled cylinders. In 1D quantum wires, electron–electron interactions cause inhomogeneity of the Fermi state, leading to the anisotropic scattering and beating behavior of Friedel oscillations [61]. The composite effect in double-walled nanotubes is more intriguing than in single-walled nanotubes. Therefore, the three typical types of coaxial armchair systems are promising candidates for studying Friedel charge oscillations in low-dimensional van der Waals materials, in which the band structure is a hybridization caused by interlayer atomic interactions.

The energy subbands of the 1D cylindrical systems are affected by transverse quantum confinement. Specifically, the response of a perpendicular electric field is fully screened in carbon nanotubes and predominantly depends on the tube radius [62, 63]. The $sp^3$ tight-binding model has been used to evaluate the static dielectric function, of which the component of $L = 1$ for the $\pi$ electrons represents the screening ability at the long-wavelength limit ($q \to 0$). The transverse dielectric screening is enhanced with the increasing free carrier density in carbon nanotubes; it mainly depends on the nanotube radius, while in contrast, the longitudinal part depends on the details of the radius and chirality. The transverse electric polarization of carbon nanotubes is independent of the nanotube radius and chirality angle and corresponds to a solid cylindrical material with a dielectric constant of about 10 [63]. The predicted result is consistent with the near-dc polarization of individual carbon nanotubes measured using high-resolution scanning force microscopy technology [15, 21]. The effects of the charge screening can be used to gain insight into other physical properties related to impurities. For example, the calculated effective potential can be further used to study impurity levels [22], intrinsic band gap [64, 65], magnetic ordering [66–71], electron correlation [72–77], residual resistivity [16–21, 78], electrodynamics of charged particles [79–82], exciton effects [82–84, 86, 87], etc.

Graphene-related systems, such as 1D carbon nanotubes, 2D layered graphenes and 3D graphite intercalated compounds, have completely different static screening phenomena in terms of different dimensions. The critical mechanisms for the dielectric functions are greatly diversified by the various dimensionalities, lattice symmetries and carrier densities, resulting in exemplary manifestations of the intrinsic interactions between electrons, such as the bare Coulomb potentials, energy dispersions of conduction bands and free carrier densities. Using 3D, 2D and 1D electronic gas model yields poor quantitative predictions of these systems, even failing to calculate the multi-mode oscillations and screening length. For the doped graphene systems, intraband and interband excitations contribute to four kinds of Coulomb scatterings from intravalley and intervalley processes, generating singularities of the static electric polarizations at certain critical moment transfers. The Friedel oscillations display beating phenomena as a combination of several normal modes, of which the $q = 2k_F$ component is expected to be the strongest oscillation with a short wavelength scaled as $\pi/k_F$. The few-layer doped graphene system will be an outstanding candidate for the first experimental and theoretical verification of the beating phenomenon due to electron and hole density fluctuations. Under the long-wavelength limit, the couplings of charge oscillations on different sheets of multilayer systems exhibit rich, static coherent oscillations in real space. The strong response of charge screenings clearly illustrates the discontinuity of electron distribution at the Fermi surface for three types of double-walled carbon nanotubes and few-layer graphenes under the doped cases.

This chapter is focused on the 2D Friedel oscillations in doped monolayer graphene and the AA- and AB-stacked bilayer graphenes. All the critical factors, including the intralayer and interlayer orbital hybridizations in planar or cylindrical lattices and the Coulomb interactions between impurities and electrons, are unified in the theoretical framework (details in Chapter 2). The effective Coulomb potentials and induced charge densities are evaluated using the self-consistent Dyson formula. It is worthwhile to systematically study the dominant mechanism of electron scattering among the layered graphene system, 1D carbon nanotubes and electron gas, especially for effective screening length and long-range charge oscillation. The subject of this chapter, the real-space Friedel oscillation, is straightforwardly derived from the standard scheme of the Fourier transform. Fortunately, real-space observation of screened potential and Friedel oscillation have been achieved in 2D systems by examining screening charge density oscillations scattered around defects or impurities using scanning tunneling spectroscopy (STS) [24, 26, 88–91]. Direct examination of various phenomena caused by the Coulomb field through high-resolution STM/STS tools has become an urgent issue.

## 4.1 Beating Phenomenon of Multi-mode Friedel Oscillation in Doped Monolayer Graphenes

The dielectric function in graphenes is mainly contributed by the $\pi$ electrons that come from the $2p_z$-orbital bondings. The $\pi$-electronic states start from the chemical potential and have a wide energy width of $\sim$10 eV around the Fermi level. Obviously, they dominate the static charge screenings and show a significant response to the Coulomb perturbation. On the other hand, the very strong $\sigma$ bonding mainly constructs the honeycomb lattice structure of graphene, which only generates deep energy states that provide background charge screening. In this chapter, the background dielectric constant, $\epsilon_0$, is evaluated as 2.4 using the fitting parameters of Bernal graphite from reflectance measurements [49]. The dielectric function for the screening ability of $\pi$ electrons is expressed as $\epsilon\left[\mathbf{q},\omega\right] = \epsilon_0 - V(q)\chi\left(\mathbf{q},\omega\right)$, derived from random-phase approximations (details in Chapter 2). The research based on the effective-mass model has studied the screening effect of a single impurity within the linear and parabolic dispersion regimes for monolayer and bilayer graphenes [92], which is only practical when dealing with excitations in a narrow energy range near the Fermi level. In contrast, the theoretical framework of quasiparticles elaborated in Chapter 2 can allow for the extension of studies to energies beyond the low-energy range.

The Coulomb scattering of electrons occurs within the same valley of Dirac cone structures or between the two different valleys. These two kinds of available scattering processes correspond to intravalley and intervalley channels under smaller and larger momentum transfers, respectively. In the undoped case, transitions only occur between interbands, while intraband transitions further participate in the case of doping. The static dielectric function involves the four transition channels included in the Coulomb matrix elements based on the energy bands of monolayer graphene and the symmetric/antisymmetric superposition of the A- and B-sublattice tight-binding functions. Due to the nearly isotropic energy spectrum, the anisotropy of the electron density in real space can be ignored.

The significant features of the static dielectric function are mainly determined by the doping-related Fermi surface. The static dielectric function $\chi\left[q,\phi = 0°,\omega = 0\right]$ is clearly illustrated in Fig. 4.1; $q$ extends to the

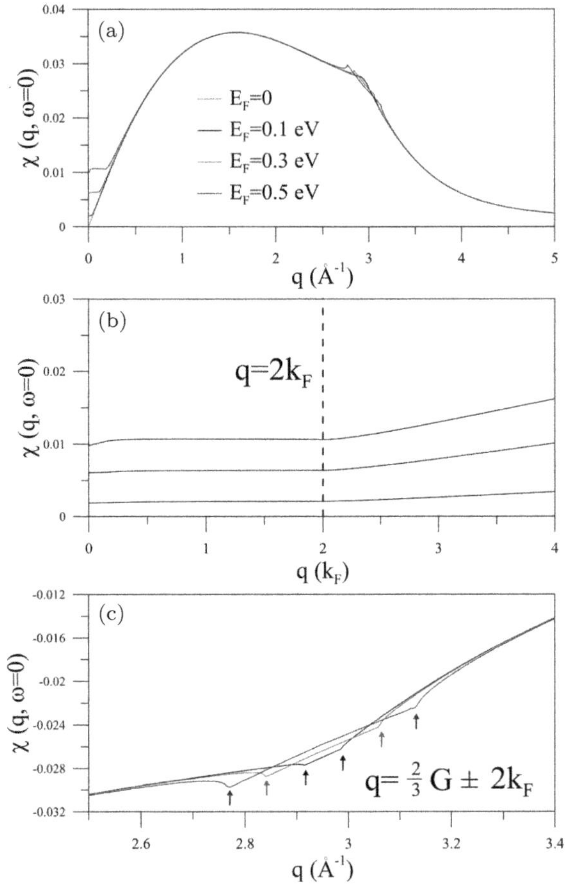


Fig. 4.1 (a) Static polarization function $\chi(q, \omega = 0)$ of monolayer graphene under different $E_F$ and at zero temperature. (b) and (c) show the zoomed-in views of $\chi(q, \omega = 0)$ for the intravalley and intervalley transitions close to $q \sim 2k_F$ and $q \sim \frac{2}{3}G \pm 2k_F$.

entire Brillouin zone to cover the whole band screening. The Friedel oscillation originates from a singularity in the dielectric function caused by discontinuous excitation across the Fermi surface. For a 2D electron gas, the singularity appears at the wavevector equal to $2k_F$, and both the initial and final states of the excitation are on the Fermi surface. However, the position and number of singularities in the graphene system vary with the doping level. As for pristine monolayer graphene, $\chi[q]$ vanishes at the long-wavelength limit, indicating the zero-gap semiconducting behavior at the (neutral) Dirac point. When the electron doping is gradually enhanced, $\chi\left[q \to 0\right]$ increases from zero to a finite value. This result shows enhanced screening ability with the increment in the Fermi level. This situation is easier to see in the Fourier transform (Figs. 4.2 and 4.3). For the excitation process within the Dirac cone, the critical momentum transfer of the dielectric singularity is $q_c = 2k_F$ (Fig. 4.1(b)), as previously observed in metallic systems [59]. But for scatterings between the K- and K′ valleys, two additional critical values are induced at $2G/3 - 2k_F$ and $2G/3 + 2k_F$, as indicated by colored arrows in Fig. 4.1(c). When the initial and final states are connected by the critical momenta at zero temperature, the induced singularity in the dielectric function dominates the static electron–electron Coulomb scatterings and leads to unusual charge screening behaviors. This is due to the discontinuity of the Fermi–Dirac distribution at the Fermi level. The response function clearly shows a tilted step structure in the vicinity of the critical points, $q_c$, where their first and second derivatives are discontinuous, consistent with the Fermi–Dirac distribution. Multi-critical momentum can produce beating oscillations formed by the linear combination of Friedel oscillations of multiple modes (discussed using Figs. 4.2 and 4.3 later). The charge screening phenomena are expected to be observed in 2D layered materials with two equivalent valleys, such as silicene [93, 94], germanene [95, 96] and tinene [97], or the anisotropic Fermi surface in silicene and germanene induced by uniaxial strains [98, 99], periodic potentials [100] or adatom chemisorption [101, 102].

In the undoped case, $V^{\text{eff}}$ is purely described by the non-oscillating behavior at a small distance from the impurity, and the long-distance Friedel oscillation is suppressed (the red curve in Fig. 4.2(a)). The n-type doped monolayer graphene exhibits metallic screening behavior, as shown by the effective Coulomb potential $V^{\text{eff}}$ in Fig. 4.2 and the induced charge density $\rho^{\text{in}}(r)$ in Fig. 4.3. $V^{\text{eff}}$ exhibits the Yukawa potential form $\exp(-r/r_c)/r$ within a short distance range close to the charged impurities, where the screening length $r_c$ is scaled by the value of $\pi/k_F$, e.g., $0.5\pi/k_F$ for $E_F = 0.1$ eV. The screening effect is obviously stronger than that in the original situation (red curve in Fig. 4.2). The non-oscillating part stems from the $q \longrightarrow 0$ behavior of the screening effect, which is the same as the Thomas–Fermi approximation for 2D and 3D metals with high charge density [29]. However, at $E_F = 0.1$ eV, $V^{\text{eff}}$ exhibits a Friedel oscillation with the 2D EG mode for distances away from the effective screening length. The oscillation period estimated as $\pi/k_F$ is attributed to the singularity in the dielectric function at $q_c = k_F$. It is clear from the figure

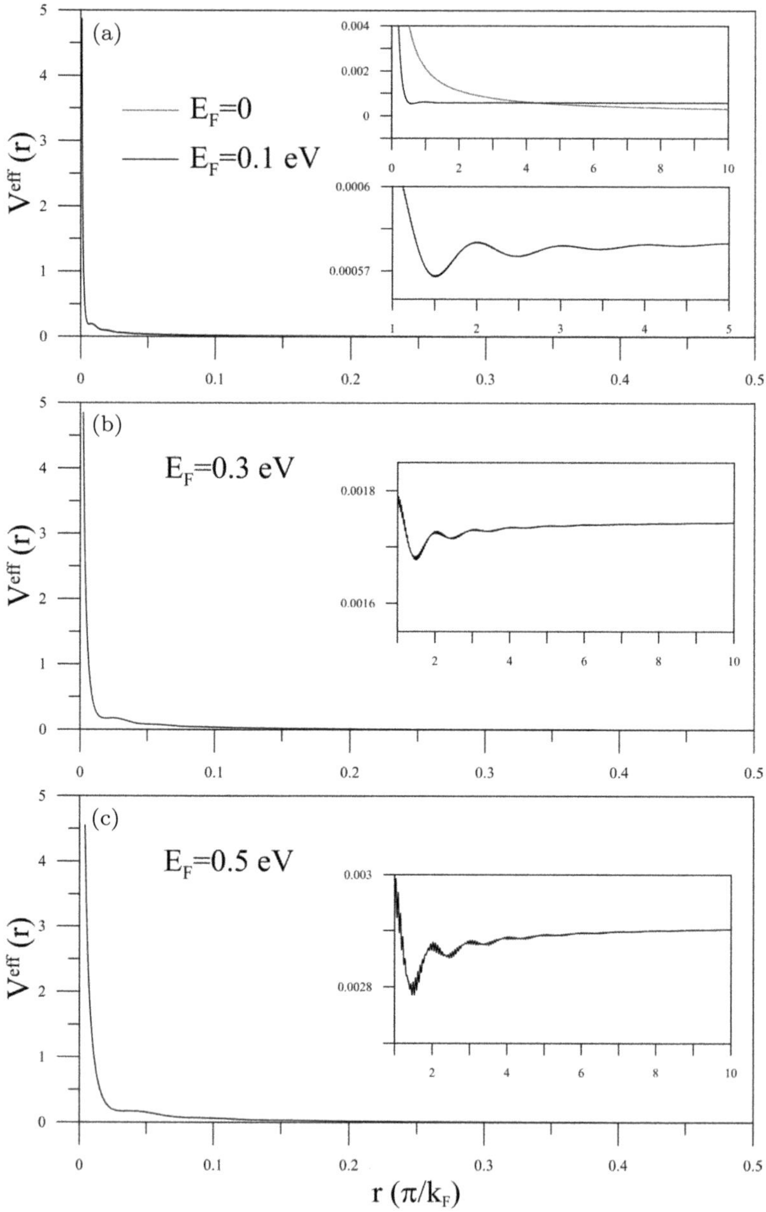


Fig. 4.2 (a) The effective Coulomb potential $V^{\text{eff}}(r)$ of monolayer graphene under a charged impurity at $r = 0$ and at zero temperature.

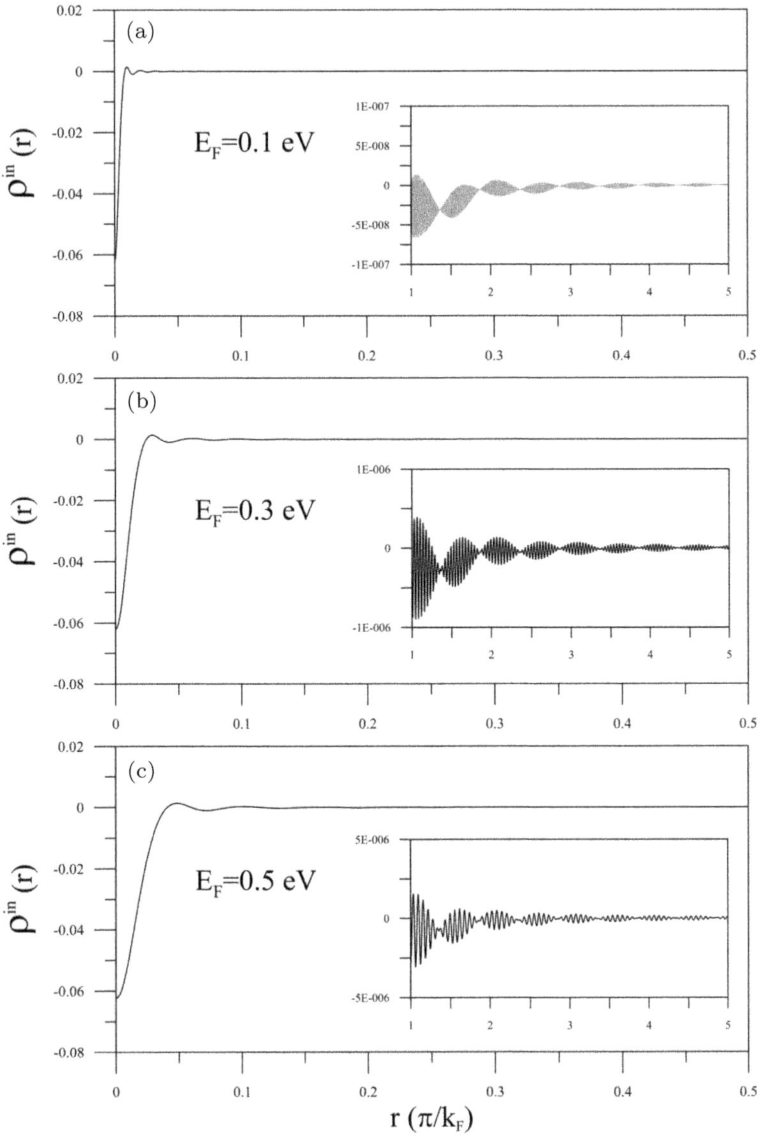


Fig. 4.3 Induced charge density distributions of monolayer graphene corresponding to a charged impurity at $r = 0$ under different $E_F$ and at zero temperature. The inserts show the charge density oscillations at long distance.

that the decline of $V^{\text{eff}}$ is faster than $1/r^2$ in the traditional 2D EG. This implies the singularity of the second derivative of the dielectric function. Moreover, the effective potential oscillates with a relatively long period as $E_F$ increases as a result of the singular $q_c$ being greater than $2k_F$, e.g., $E_F = 0.3\,\text{eV}$ and $E_F = 0.5\,\text{eV}$. This unexpected behavior is attributed to the excitation of a large energy range beyond the linear dispersion near the Fermi level. The biggest difference in the Friedel oscillations between monolayer graphene and 2D EG is that the former exhibits obvious beating behavior and consists of three spatial periods: $2k_F$, $2G/3 + 2k_F$ and $2G/3 - 2k_F$, constituting a beating periodic function. The multi-mode Friedel oscillations can survive in doped few-layer graphene systems with various stacking configurations, such as AA- and AB-stacked graphenes, making them an excellent candidate for experimental verification of the beating screening phenomenon caused by charge fluctuations. In Fig. 4.3, the induced charge density in the real space is significantly fluctuated by the Coulomb field generated by impurities and also exhibits an oscillating behavior. In other words, more charges are induced and distributed around the impurity to screen the applied Coulomb field. The beating behavior of the induced charge density is also observed due to the combination of multi-modes of the Friedel oscillation, displaying the charge fluctuations due to the composite periods of $\pi/k_F$, $\pi/(G/3 - k_F)$ and $\pi/(G/3 + k_F)$, as demonstrated by the insets in Fig. 4.3. The first from intravalley scattering will dominate the static dielectric behavior under low electron doping. However, with the increase in free carriers in the conduction band, intervalley scattering becomes important.

Cylindrical carbon nanotubes are a kind of analog material used to check Friedel oscillations in graphene-related systems with different dimensions. The screening behaviors of carbon nanotubes are strongly affected by temperature, nanotube chirality and radius, as well as the coupling between tubes in multi-walled nanotubes. Metallic carbon nanotubes can effectively screen the Coulomb field caused by charged impurities, while semiconductor carbon nanotubes cannot. At the long-wavelength limit, the dielectric function for any single-walled and multi-walled armchair carbon nanotubes is finite at $q \to 0$ due to the reflection symmetry with respect to the plane passing through the cylinder axis. Therefore, the short-wavelength screening is prominent and varies with the nanotube radius. Armchair carbon nanotubes, such as (7, 7) and (14, 14), exhibit an intriguing behavior, i.e., strong and rapid Friedel oscillations. The long-distance periodic electron cloud comes from the first-order singularity at the wavevector of $q = 2k_F$ and $L = 0$. When the charged impurities are close enough to the surface of the cylinder, an oscillating carrier cloud will appear. It shows complexity through the significant Friedel oscillations along the tube axis and the dependence on the azimuth angle. This topic can be simplified by using the average value over the azimuth angles, that is, the electron–electron scattering of $L = 0$ will be the only contribution to impurity screening. This unusual phenomenon is relatively easy to observe in smaller armchair nanotubes; they can survive at room temperature but may disappear beyond 1200 K. In addition, long-distance Friedel oscillations also exist in all zigzag carbon nanotubes, but the oscillation period is relatively small because their electric polarization functions have a singular structure at $q = G$. Friedel oscillations may survive in double-walled carbon nanotubes and few-layer graphenes. These materials are excellent candidates for the first experimental test of the beating phenomena in the fluctuation of the carrier clouds coupling in layered systems.

## 4.2 Stacking-Enriched Friedel Modes and Static Screenings in AA-Stacked Bilayer Graphenes

The dielectric function in the AA-stacked bilayer graphene is mainly contributed by the two Dirac cones vertically located at the K and K′ points. The intravalley and intervalley excitations can occur within the same Dirac cone or between two different cones. However, with the clarification of monolayer graphene, the former dominate the intensity and oscillation behavior of the effective potential and the induced charge density. While the latter cause only a minor modification of the intensity, they also cause extra modes of the Friedel oscillations. The static polarization function can be divided into intralayer and interlayer responses as a function of $q$, expressed as $\chi_{11}(q)$ and $\chi_{12}(q)$, respectively, where the subscript number represents the layer number, and their changes under different Fermi levels are shown in Fig. 4.4. The two static polarization functions, $\chi_{11}[q, \phi = 0°, \omega = 0]$ and $\chi_{12}[q, \phi = 0°, \omega = 0]$, are clearly illustrated for $q$ extending to the entire Brillouin zone, just like in monolayer graphene. Also, the anisotropy of the screening effects in real space is ignored.

When the electron doping is gradually enhanced, the increase in $\chi_{11}$ and $\chi_{12}$ at the long-wavelength limit indicates that the screening ability

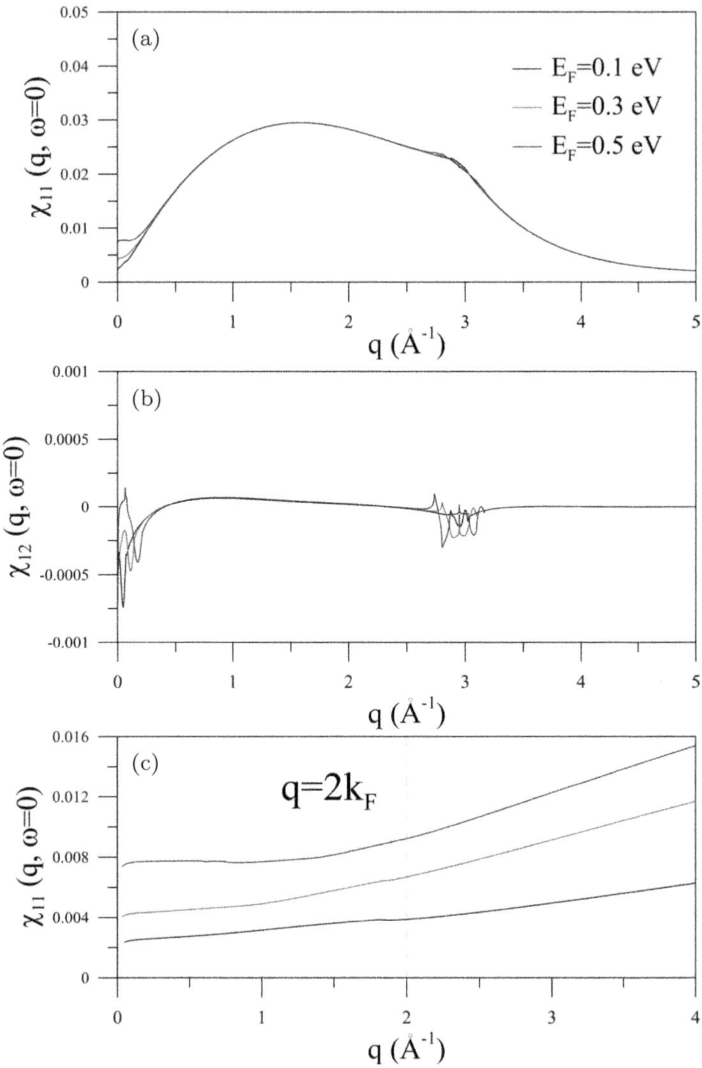


Fig. 4.4 Momentum-dependent static electric polarization function of a AA-stacked bilayer graphene under different $E^F$ and at zero temperature: (a) intralayer $\chi^{11}$ and (b) interlayer $\chi^{11}$. A charged impurity is located at $r = 0$ on the first layer. (c) Zoomed-in view of $\chi^{11}$ in the vicinity of $q = 2k_F$.

at a higher Fermi level is enhanced as a fast-declining screened Coulomb potential in the Fourier transform. The singularities in the static polarization function are mainly contributed by intravalley and intervalley excitations, given by $q_c \rightarrow 0.1\,\text{Å}^{-1}$ $(2k_F)$ and $q_c \rightarrow 3.0\,\text{Å}^{-1}$ $(2G/3 - 2k_F$ $2G/3 + 2k_F)$, respectively. These singular behaviors might manifest themselves in the periods and intensities of the multi-mode Friedel oscillations. The two intravalley channels individually cause a divergence of the first derivative of the polarization function. One of the critical momentum transfers is $q_c$ $2k_F$ and the other is less than $2k_F$, where the tilted step structures clearly shown near $q_c$ and $k_F$ are defined as the maximum momentum on the Fermi surface (Fig. 4.4(c)). The combined effect of all singular polarizabilities may make the stacked system more likely to produce multi-mode Friedel oscillations.

The metallic screening behaviors on two different graphene sheets are shown by the screened Coulomb potential in Figs. 4.5 and 4.6. It should be remembered that the charge impurity is situated on the first layer. The characteristics of the effective Coulomb field on the two graphene sheets are similar for long distances, but they are completely different at short distances and in the Friedel oscillation region. $V_{11}^{\text{eff}}$ exhibits the Yukawa potential form $exp(-r/r_c)/r$ within a short distance range close to the charged impurity, while it declines at a slower rate than the potential screened by an impurity in monolayer graphene. This is mainly due to the different energy band structures between these two systems that the divergence of the dielectric singularity at the critical momentum is relatively weak in AA-stacked bilayer graphene. In the first and second graphene layers, the oscillation intensity is clearly visible in the first two periods $(\pi/k_F\ 3\pi/k_F)$ of the Friedel oscillation, while the beating behavior is vaguely visible only for $V_{11}^{\text{eff}}$. The screening effect in multi-layer graphene is highly dependent on the relative position of the single impurity and each layer. At long distances, $V_{11}^{\text{eff}}$ and $V_{22}^{\text{eff}}$ both display the non-oscillating potential stemming from the long-wavelength limit of the screening effect, which is unclear even if the electron doping increases (Fig. 4.6). The disappearance of the Friedel oscillation means that the excitations related to the Fermi surface undergo a drastic change (Fig. 4.7). It is interpreted that the dielectric singularity becomes weaker under the influence of the intervalley excitations, which is the main reason for the above phenomenon. In short, the screening effect of Dirac quasiparticles is completely different between monolayer and AA-stacked graphenes.

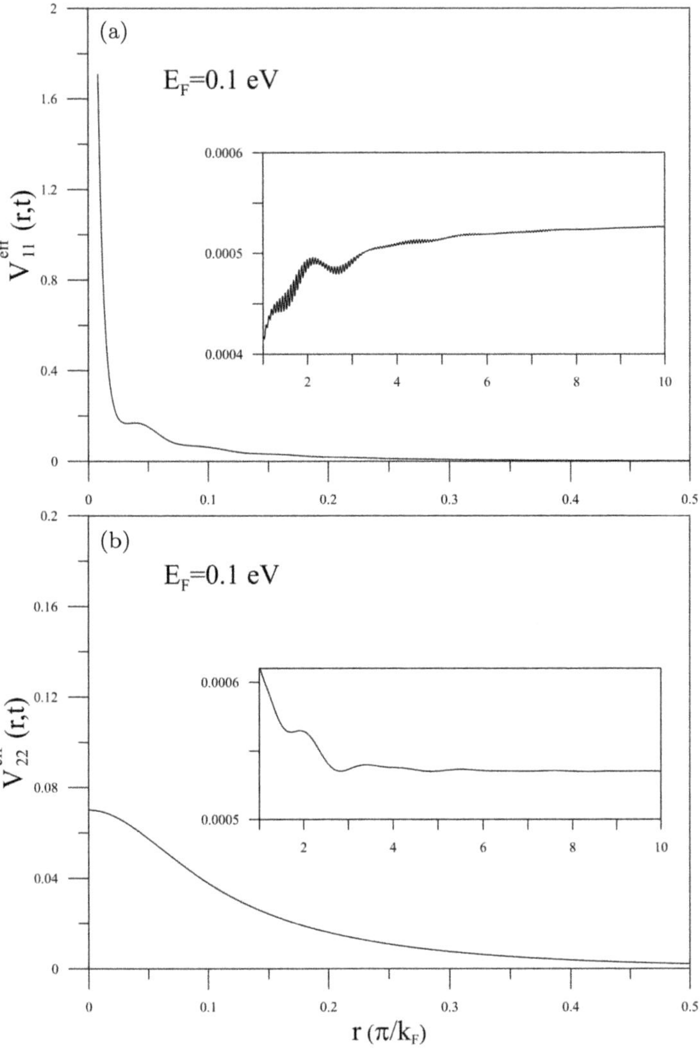


Fig. 4.5 Effective potentials on (a) the first layer $V_{11}^{\text{eff}}$ and (b) the second layer $V_{22}^{\text{eff}}$ for a AA-stacked bilayer graphene under $E^F = 0.1\,\text{eV}$ and at zero temperature. A charged impurity is located at $r = 0$ on the first layer. The inserts show $V_{11}^{\text{eff}}$ and $V_{22}^{\text{eff}}$ at long distance.

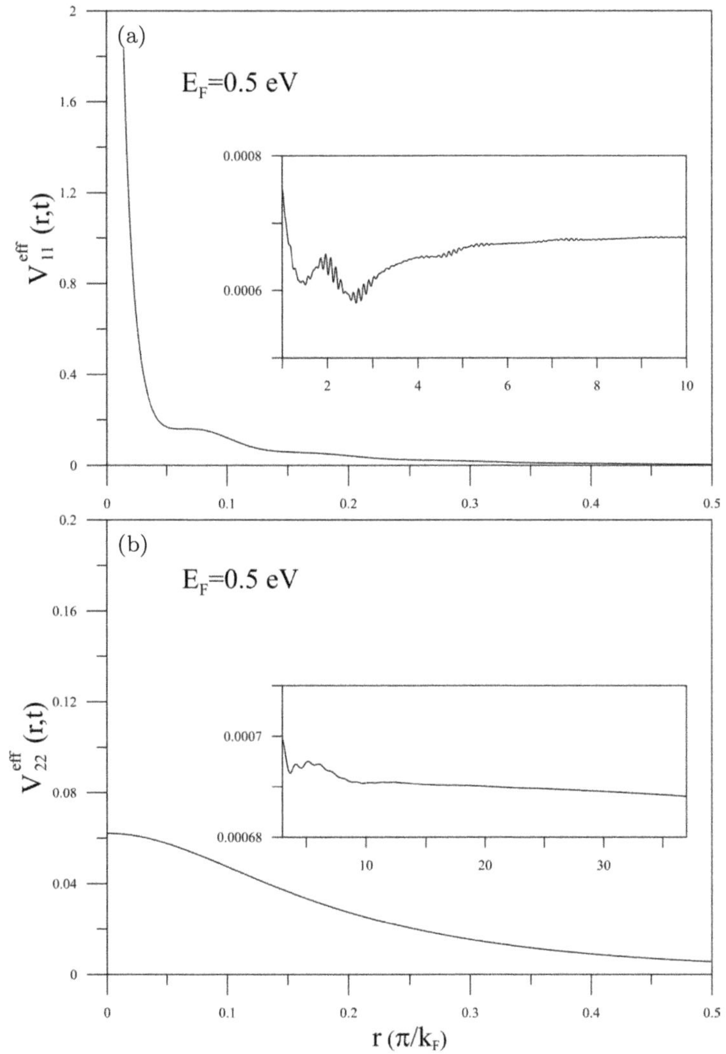


Fig. 4.6 Effective potentials on (a) the first layer $V_{11}^{\text{eff}}$ and (b) the second layer $V_{22}^{\text{eff}}$ for a AA-stacked bilayer graphene under $E^F = 0.5\,\text{eV}$ and at zero temperature. A charged impurity is located at $r = 0$ on the first layer. The inserts show $V_{11}^{\text{eff}}$ and $V_{22}^{\text{eff}}$ at long distance.

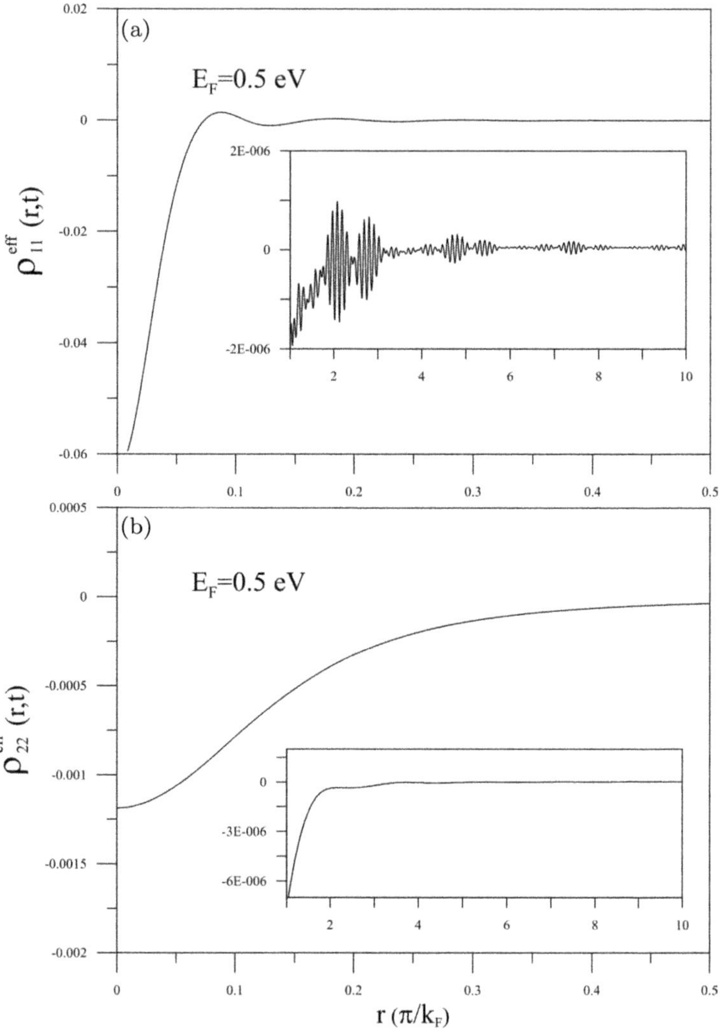


Fig. 4.7 Induced charge density on (a) the first layer $\rho_{11}^{\mathrm{in}}$ and (b) the second layer $\rho_{22}^{\mathrm{in}}$ for a AA-stacked bilayer graphene under $E^F = 0.5\,\mathrm{eV}$ and at zero temperature. A charged impurity is located at $r = 0$ on the first layer. The inserts show $\rho_{11}^{\mathrm{in}}$ and $\rho_{22}^{\mathrm{in}}$ at long distance.

## 4.3 Peculiar Friedel Oscillations in AB-Stacked Bilayer Graphenes

The static polarization function tensor, $\chi$, in AB-stacked bilayer graphenes is mainly contributed by the two pairs of parabolic bands around the K and K$'$ points, as shown in Fig. 4.8. The intravalley excitations can occur within the two pairs of energy bands and dominate the intensity and oscillation behavior of the effective potential and induced charge density, while the main contribution of the intervalley ones causes additional Friedel oscillation modes. These characteristics are reflected in two independent components, $\chi_{11}(q, \omega = 0)$ and $\chi_{22}(q, \omega = 0)$, of which the screening effects from the singularities of $d\chi/dq$ around $2k_F$ and their Fourier transform are worthy of further discussion. This number of singularity is the same as the number of energy bands spanned by the Fermi surface, for example, one singular point when $E_F = 0.1$ and $0.3$ eV, and two when $E_F = 0.5$ eV, as indicated by arrows in Fig. 4.8(c). The screening phenomenon is easier to control through the Fermi energy in the AB-stacking configuration, especially for the Friedel oscillation.

In the AB-stacked bilayer graphene, the phenomenon of conduction electrons screening impurities placed on the first layer is described by the effective Coulomb potential on two different graphene sheets, as shown in Figs. 4.9 and 4.10. The screening effects are completely different between the AB- and AA-stacked graphenes. The Yukawa potential decay $\exp(-r/r_c)/r$ with the Friedel oscillation is demonstrated on $V_{11}^{\text{eff}}$, but not on $V_{22}^{\text{eff}}$, which is mainly determined by the impurity position, a feature enabling the multilayer graphene to screen a single impurity immersed in it. It is very interesting that for $E_F = 0.1$ eV, the static screening phenomenon of the two graphene layers both display well-behaved Friedel oscillations. Moreover, very different from the AA stacking, the induced charge density only oscillates in a single mode of $q_c = 2k_F$ because in the AB stacking configuration, other dielectric singularities are inhibited by the special selection rule for the transition of electronic states. As explained above, under appropriate electron doping, the transition of conduction electrons between different bands is not allowed, which relatively causes the singularity in the dielectric function to become prominent, so the Friedel oscillation is also similar. As a consequence, the Coulomb field screening effect at $E_F = 0.5$ eV in Fig. 4.10 can be proved by the interband transitions for the same reason. Furthermore, the static screening effect in multilayer graphene systems can be controlled by the amount and location

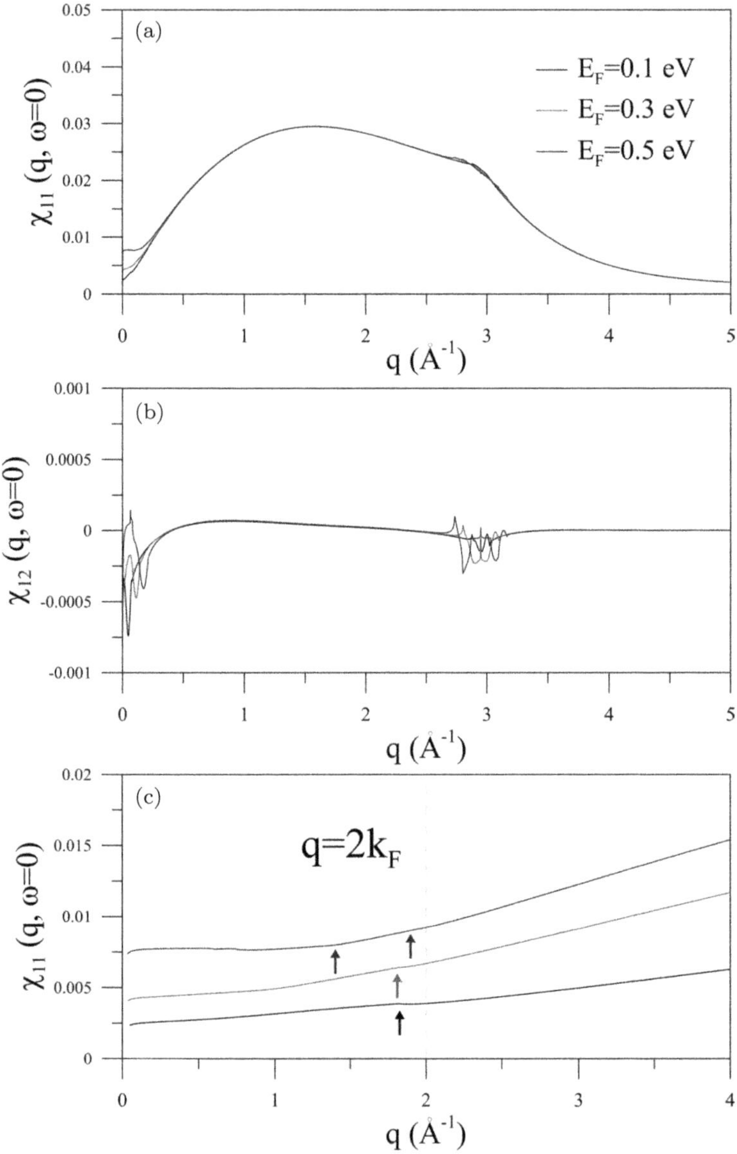


Fig. 4.8 Momentum-dependent static electric polarization function of a AB-stacked bilayer graphene under different $E^F$ and at zero temperature: (a) intralayer $\chi^{11}$ and (b) interlayer $\chi^{11}$. A charged impurity is located at $r = 0$ on the first layer. (c) Zoomed-in view of $\chi^{11}$ in the vicinity of $q = 2k_F$.

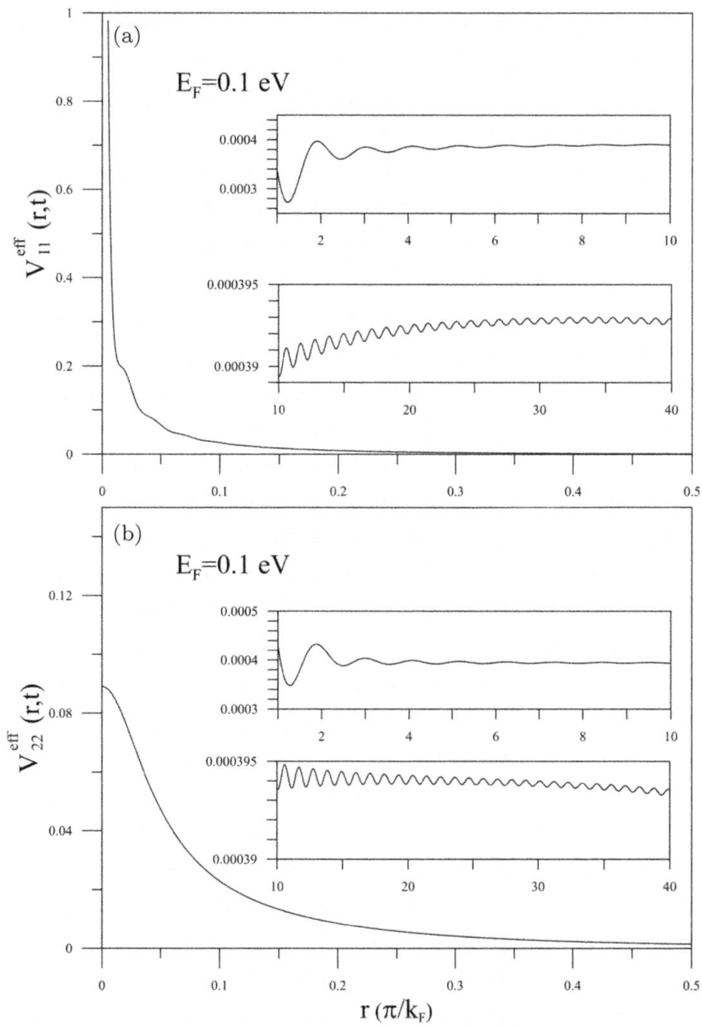


Fig. 4.9 Effective potentials on (a) the first layer $V_{11}^{\rm eff}$ and (b) the second layer $V_{22}^{\rm eff}$ for a AB-stacked bilayer graphene under $E^F = 0.1$ eV and at zero temperature. A charged impurity is located at $r = 0$ on the first layer. The inserts show $V_{11}^{\rm eff}$ and $V_{22}^{\rm eff}$ at long distance.

of impurities. The interference between each layer will cause the quasiparticles near the Fermi surface to be scattered by the overall Friedel potential and play a key role in related physical properties. The interference between each layer will cause the overall Friedel oscillation in terms of the effective potential and the induced charge density, which is expected

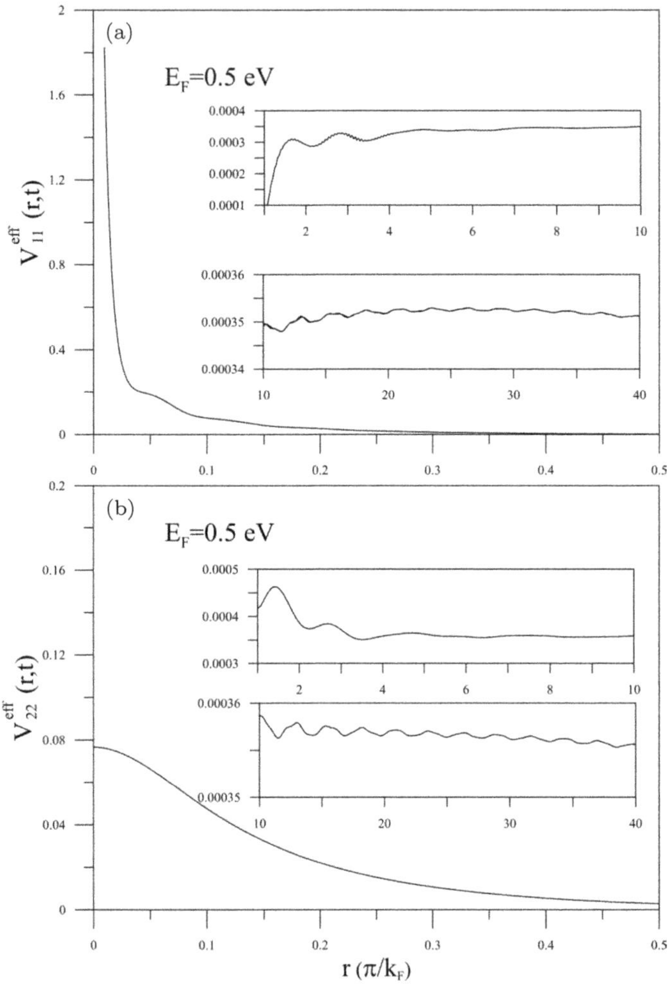


Fig. 4.10 Effective potentials on (a) the first layer $V_{11}^{\text{eff}}$ and (b) the second layer $V_{22}^{\text{eff}}$ for a AB-stacked bilayer graphene under $E^F = 0.5\,\text{eV}$ and at zero temperature. A charged impurity is located at $r = 0$ on the first layer. The inserts show $V_{11}^{\text{eff}}$ and $V_{22}^{\text{eff}}$ at long distance.

to play a key role in the behavior of quasiparticles near the Fermi surface and its related physical properties.

The spatial fluctuation of the charge density distribution can provide a variety of phenomena of essential quasiparticles under the influence of Coulomb screenings. Theoretical study [59] and high-resolution STM clearly illustrate this aspect [6, 7]; these phenomena include dipole screening in semiconductors [103], long-range Friedel oscillations in metals of composite Fermi surface [6, 7, 13, 104, 106], the giant Friedel oscillations due to the many-body effect [2], the limit of Friedel oscillation in heterostructures [107] and the topological Berry phase in monolayer graphene [26, 108], bilayer graphene [28] and quasi-1D photonic insulators [109]. The latest STM surface morphology and its Fourier transform image are very successful in verifying the theoretical predictions of the electronic properties and can even be used as an effective tool to clarify the characteristics of quantum topologies. For example, the charge screening of a 2D electron gas from the Coulomb field of charged impurities is expected to show an effective potential energy, and the decline form and oscillation period of the induced charge density are given by $1/r^2$ and $\pi/k_F$, respectively. This prediction of $1/r^2$ decaying and the period $\pi/k_F$ mediated by a 2DES was confirmed by STM observations on three adsorbate/substrate materials: Cu/Cu[111], Co/Cu[111] and Co/Ag[111][8]. The observations of the charge density distribution around the potassium chain on black phosphorus clearly show anisotropic screening effects of a non-circular Fermi line [10]. The average Coulomb potential is substantially consistent with the energy dispersion modified by the dimensionality and anisotropy of the charge screening. Most importantly, by studying the charge density waves scattered around defects or impurities, STM measurements can be very powerful for real-space observations of Friedel oscillations. The Friedel oscillations on the beryllium (0001)[2] and ($10\bar{1}0$)[24] surfaces, respectively, show strong isotropic and anisotropic screening phenomena near the surface defects. The orientation of the metal surface verifies the unique Fermi line and surface states. These results can be further compared with ARPES measurements. Specifically, the charge density wave measured through STM serves as a powerful observable to provide insight into quasiparticle properties, such as chirality [110] and the Berry phase [26, 28, 108]. The number of additional wavefronts observed in the dislocation of the charge density wave in monolayer graphene is direct evidence of the Berry phase in real space.

The STM and its Fourier transform images are very suitable and reliable for thoroughly investigating the real-space charge fluctuation and wave function interference in low-dimensional emergent materials. The prediction of electrostatic charge screening in 2D graphenes exhibits diverse Coulomb perturbations. The induced charge density and effective Coulomb potential clearly illustrate the unique phenomenon of metallic screenings, and they cover abundant characteristics, such as the screening

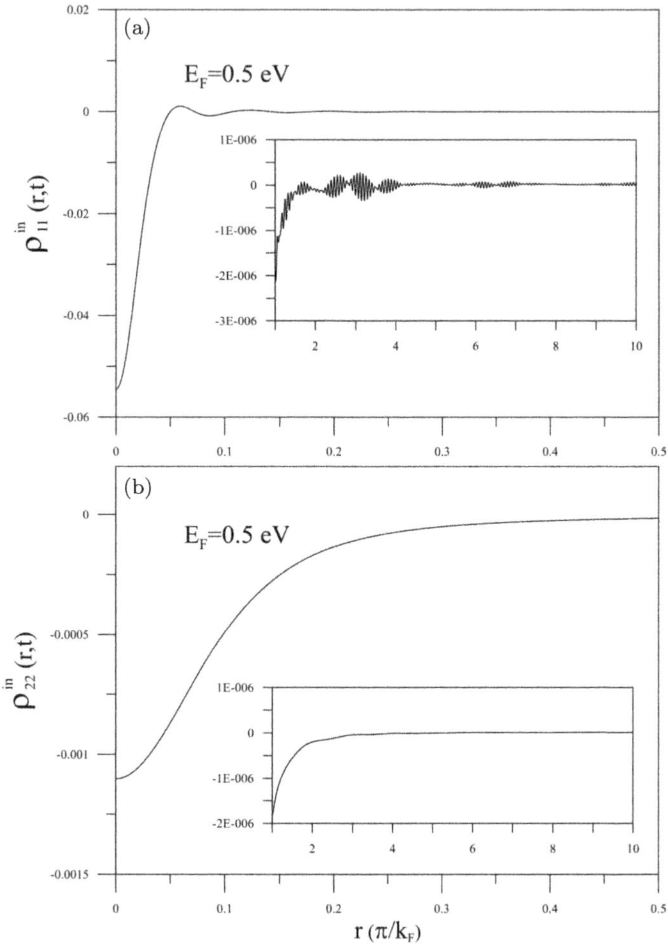


Fig. 4.11 Induced charge density on (a) the first layer $\rho_{11}^{\text{in}}$ and (b) the second layer $\rho_{22}^{\text{in}}$ for a AA-stacked bilayer graphene under $E^F = 0.5\,\text{eV}$ and at zero temperature. A charged impurity is located at $r = 0$ on the first layer. The inserts show $\rho_{11}^{\text{in}}$ and $\rho_{22}^{\text{in}}$ at long distance.

length of the Yukawa Coulomb potential, the rapid charge fluctuations related to the impurity position, single-mode or multi-mode Friedel oscillation, beating screening behavior and non-oscillating behavior at short and long distances for monolayer and single-mode Friedel oscillation in AB-stacked bilayer systems. The composite Friedel oscillations of the unusual beating phenomenon are closely related to the interplay between the intravalley and intervalley Coulomb interactions on the Fermi states.

The single-mode oscillation in AB bilayer graphenes represents its special selection rules related to electronic transitions. The theoretical calculations and their experimental verification can provide direct evidence for the stacking-enriched Coulomb screening phenomenon of the electronic states in multilayer graphenes.

## References

[1] Hu G.Y. and O'Connell R.F. 1997. Dielectric response of a quasi-one-dimensional electron system. *J. Phys. Condens. Matter.* **2** 9381.

[2] Sprunger P.T., Petersen L., Plummer E.W., Laegsgaard E. and Besenbacher F. 1997. Giant Friedel oscillations on the beryllium(0001) surface. *Science* **275** 1764.

[3] Stern F. 1967. Polarizability of a two-dimensional electron gas. *Phys. Rev. Lett.* **18** 546.

[4] Crommie M.F., Lutz C.P. and Eigler D.M. 1993. Imaging standing waves in a two-dimensional electron gas. *Nature* **363** 524.

[5] Min H., Hwang E.H. and Das Sarma S. 2012. Polarizability and screening in chiral multilayer graphene. *Phys. Rev. B* **86** 081402(R).

[6] Petersen L., Sprunger P.T., Hofmann P., Laegsgaard E., Briner B.G. *et al.* 2000. Direct imaging of the two-dimensional Fermi contour: Fourier-transform STM. *Phys. Rev. B* **57** R6858.

[7] Petersen L., Laitenberger P., Laegsgaard E. and Besenbacher F. 1998. Screening waves from steps and defects on Cu(111) and Au(111) imaged with STM: Contribution from bulk electrons. *Phys. Rev. B* **58** 7361.

[8] Knorr N., Brune H., Epple M., Hirstein A., Schneider M.A. *et al.* 2002. Long-range adsorbate interactions mediated by a two-dimensional electron gas. *Phys. Rev. B* **65** 115420.

[9] Pereira V.M., Nilsson J. and Castro Neto A.H. 2007. Coulomb impurity problem in graphene. *Phys. Rev. Lett.* **99** 166802.

[10] Kiraly B., Knol E.J., Volckaert K., Biswas D., Rudenko A.N. *et al.* 2019. Anisotropic two-dimensional screening at the surface of black phosphorus. *Phys. Rev. Lett.* **123** 216403.

[11] Hwang E.H. and Das Sarma S. 2008. Screening, Kohn anomaly, Friedel oscillation, and RKKY interaction in bilayer graphene. *Phys. Rev. Lett.* **101** 156802.

[12] Lu C.K. 1997. Friedel oscillation near a van Hove singularity in two-dimensional Dirac materials. *J. Phys. Condens. Matter* **28** 065001.

[13] Bacsi A. and Virosztek A. 2002. Local density of states and Friedel oscillations in graphene. *Phys. Rev. B* **82** 193405.

[14] Dutreix C. and Katsnelson M.I. 2016. Friedel oscillations at the surfaces of rhombohedral N-layer graphene. *Phys. Rev. B* **93** 035413.

[15] Zou Y.L., Song J.T., Bai C.X. and Chang K. 2016. Impurity scattering and Friedel oscillations in monolayer black phosphorus. *Phys. Rev. B* **94** 035431.

[16] Yosida Y. 1999. Electrical resistance in sheets composed of multiwalled carbon nanotubes. *J. Phys. Chem. Solids* **60** 1.

[17] Pal S.K., Son Y., Borca-Tasciuc T., Borca-Tasciuc D.-A., Kar S. *et al.* 2011. Thermal and electrical transport along MWCNT arrays grown on Inconel substrates. *J. Mater. Res.* **23** 2099.

[18] Behnam A., Guo J. and Ural A. 2011. Effects of nanotube alignment and measurement direction on percolation resistivity in single-walled carbon nanotube films. *J. Appl. Phys.* **102** 044313.

[19] Chiodarelli N., Masahito S., Kashiwagi Y., Li Y. and Arstila K. 2011. Measuring the electrical resistivity and contact resistance of vertical carbon nanotube bundles for application as interconnects. *Nanotechnology* **22** 8.

[20] Behnam A., Noriega L. and Choi Y. 2006. Resistivity scaling in single-walled carbon nanotube films patterned to submicron dimensions. *Appl. Phys. Lett.* **89** 093107.

[21] Cheianov V.V. and Fal'ko V.I. 2006. Friedel oscillations, impurity scattering, and temperature dependence of resistivity in graphene. *Phys. Rev. Lett.* **97** 226801.

[22] Yi J.Y. and Bernholc J. 1993. Atomic structure and doping of microtubules. *Phys. Rev. B* **47** 1708.

[23] Sadhukhan K. and Agarwal A. 2017. Anisotropic plasmons, Friedel oscillations, and screening in 8-P mmn borophene. *Phys. Rev. B* **96** 035410.

[24] Hofmann P., Briner B.G., Doering M., Rust H.P., Plummer E.W. *et al.* 1997. Anisotropic two-dimensional Friedel oscillations. *Phys. Rev. Lett.* **79** 265.

[25] Kiraly B., Knol E.J., Volckaert K., Biswas D., Rudenko A.N. *et al.* 2016. Anisotropic density fluctuations, plasmons, and Friedel oscillations in nodal line semimetal. *New J. Phys.* **18** 043010.

[26] Dutreix C., Gonzalez-Herrero H., Brihuega I., Katsnelson M.I., Chapelier C. *et al.* 2019. Measuring the Berry phase of graphene from wavefront dislocations in Friedel oscillations. *Nature* **574** 219.

[27] Khan M.A., Chang P.H., Ghimire N., Bretz-Sullivan T.M., Bhattacharya A. *et al.* 2020. Fermi surface topology and nontrivial Berry phase in the flat-band semimetal Pd3Pb. *Phys. Rev. B* **101** 245113.

[28] Zhang Y., Su Y. and He L. 2020. Local Berry phase signatures of bilayer graphene in intervalley quantum interference. *Phys. Rev. Lett.* **125** 116804.

[29] Mahan G.D. 1990. *Many-Particle Physics* (New York: Plenum).

[30] Jackson J.D. 1975. *Classical Electrodynamics* (New York: Wiley).

[31] Grosso G. and Parravicini G. 2014. *Solid State Physics*, 2nd ed. (Barakhamba: Reed Elsevier India Private Limited).

[32] Friedel J. 1958. Metallic alloys. *Il Nuovo Cimento* **7** 287–311.

[33] Ruderman M.A. and Kittel C. 1954. Indirect exchange coupling of nuclear magnetic moments by conduction electrons. *Phys. Rev.* **96** 99.

[34] Norman M.R., Ding H., Randeria M., Campuzano J.C., Yokoya T. *et al.* 1998. Destruction of the Fermi surface underdoped high-T-c superconductors. *Nature* **392** 6672.

[35] Klotz J., Butcher T.A., Forster T., Hornung J., Sheikin I. *et al.* 2020. Fermi surface investigation of the noncentrosymmetric superconductor alpha-PdBi. *Phys. Rev. B* **101** 235139.
[36] Elatresh S.F., Hossain M.T., Bhowmick T., Grockowiak A.D., Cai W.Z. *et al.* 2020. Fermi surface studies of the low-temperature structure of sodium. *Phys. Rev. B* **101** 220103.
[37] Khan M.A., Chang P.H., Ghimire N., Bretz-Sullivan T.M., Bhattacharya A. *et al.* 2020. Three-dimensional Fermi surface and small effective masses in Mo8Ga41. *App. Phys. Lett.* **116** 202601.
[38] Luttinger J.M. 1960. Fermi surface and some simple qauilibrium properties of a system of interacting fermimons. *Phys. Rev. B* **119** 1153.
[39] Forster T., Kraft I., Sheikin I., Bianchi A.D., Wosnitza J. *et al.* 2019. Fermi surface of $LaFe_2P_2$-a detailed density functional study. *J. Phys. Condens.* **32** 025503.
[40] Gong S.S., Zheng W.N., Lee M., Lu Y.M., Sheng D.N. *et al.* 2019. Chiral spin liquid with spinon Fermi surfaces in the spin-1/2 triangular Heisenberg model. *Phys. Rev. B* **100** 241111.
[41] Xu Y.J., Sheng Y.T. and Yang Y.F. 2019. Quasi-two-dimensional Fermi surfaces and unitary spin-triplet pairing in the heavy fermion superconductor $UTe_2$. *Phys. Rev. Lett.* **123** 217002.
[42] Efremov D.V., Shtyk A., Rost A.W., Chamon C., Mackenzie A.P. *et al.* 2019. Multicritical Fermi surface topological transitions. *Phys. Rev. Lett.* **123** 207202.
[43] Battilomo R., Scopigno N. and Ortix C. 2019. Berry curvature dipole in strained graphene: A Fermi surface warping effect. *Phys. Rev. Lett.* **123** 196403.
[44] Roth L.M., Zeiger H.J. and Kaplan T.A. 1966. Generalization of the Ruderman-Kittel-Kasuya-Yosida interaction for nonspherical Fermi surfaces. *Phys. Rev.* **149** 519.
[45] Mahan G.D. 1995. Friedel oscillation on a square lattice. *Int. J. Mod. Phys. B* **9** 1327.
[46] Farajollahpour T., Khamouei S., Shateri S.S. and Phirouznia A. 2018. Anisotropic Friedel oscillations in graphene-like materials: The Dirac point approximation in wave-number dependent quantities revisited. *Sci. Rep.* **8** 2667.
[47] Herman F. and Schrieffer R. 1992. Generalized Ruderman-Kittel-Kasuya-Yosida theory of oscillatory exchange coupling in magnetic multilayers. *Phys. Rev. B* **46** 5806.
[48] Kiraly B., Knol E.J., Volckaert K., Biswas D., Rudenko A.N., Prishchenko D.A. *et al.* 2019. Anisotropic two-dimensional screening at the surface of black phosphorus. *Phys. Rev. Lett.* **123** 216403.
[49] Shung K.W.-K. 1986. Dielectric function and plasmon structure of stage-1 intercalated graphite. *Phys. Rev. B* **34** 979.
[50] Shung K.W.-K. 1986. Lifetime effects in low-stage intercalated graphite systems. *Phys. Rev. B* **34** 1264.

[51] DiVincenzo D.P. and Mele E.J. 1984. Self-consistent effective-mass theory for intralayer screening in graphite intercalation compounds. *Phys. Rev. B* **29** 1685.
[52] Gomez-Santos G. and Stauber T. 2011. Measurable lattice effects on the charge and magnetic response in graphene. *Phys. Rev. Lett.* **106** 045504.
[53] Esquinazi P., Spemann D., Höhne R., Setzer A., Han K.-H. *et al.* 2003. Magnetic ordering by proton irradiation in graphite. *Phys. Rev. Lett.* **91** 227201.
[54] Esquinazi P., Setzer A., Höhne R., Semmelhack C., Kopelevich Y. *et al.* 2002. Ferromagnetism in oriented graphite samples. *Phys. Rev. B* **66** 024429.
[55] Iye Y., McNeil L.E. and Dresselhaus G. 1984. Effect of impurities on the electronic phase transition in graphite in the magnetic quantum limit. *Phys. Rev. B* **30** 7009.
[56] Potter M.E., Johnson W.D. and Fischer J.E. 1981. Transport properties of alkali metal-graphite intercalation compounds. *Solid State Commun.* **37** 713.
[57] Sasaki K. 2002. Charge screening effect in metallic carbon nanotubes. *Phys. Rev. B* **65** 195412.
[58] Léonard F. 2002. Dielectric response of semiconducting carbon nanotubes. *Appl. Phys. Lett.* **81** 4835.
[59] Lin M.F. and Chuu D.S. 1997. Impurity screening in carbon nanotubes. *Phys. Rev. B* **56** 4996.
[60] Inaba T., Morimoto T., Yamazaki S., Okazaki T. 2021. Electron scattering by Friedel oscillations in carbon nanotubes. *Nano Res.*
[61] Urban D.F. and Komnik A. 2008. Interaction-induced beats of Friedel oscillations in quantum wires. *Phys. Rev. Lett.* **100** 146602.
[62] Lu W., Wang D. and Chen L. 2007. Near-static dielectric polarization of individual carbon nanotubes. *Nano Lett.* **7** 2729.
[63] Chen S.C., Hseih W.C. and Lin M.F. 2005. Charge screening of single-walled carbon nanotubes in a uniform transverse electric field. *Phys. Rev. B* **72** 193412.
[64] Mitchell J.S., Daniel R.M., Neda L., Vikram V.D. and Ethan D.M. 2005. Universal interaction-driven gap in metallic carbon nanotubes. *Phys. Rev. B* **97** 035445.
[65] Aspitarte L., McCulley D.R., Bertoni A., Island J.O., Ostermann M. *et al.* 2017. Giant modulation of the electronic band gap of carbon nanotubes by dielectric screening. *Sci. Rep.* **7** 8828.
[66] Barzola-Quiquia J., Esquinazi P., Rothermel M., Spemann D., Butz T. *et al.* 2007. Experimental evidence for two-dimensional magnetic order in proton bombarded graphite. *Phys. Rev. B* **76** 161403(R).
[67] Fujita M., Wakabayashi K., Nakada K. and Kusakabe K. 1996. Peculiar localized state at zigzag graphite edge. *J. Phys. Soc. Jpn.* **65** 1920.
[68] Červenka J., Katsnelson M.I. and Flipse C.F.J. 2009. Room-temperature ferromagnetism in graphite driven by two-dimensional networks of point defects. *Nat. Phys.* **5** 840.
[69] Gregory S. 1978. Magnetic susceptibility of oxygen adsorbed on graphite. *Phys. Rev. Lett.* **40** 723.

[70] Ohldag H., Esquinazi P., Arenholz E., Spemann D., Rothermel M. *et al.* 2010. The role of hydrogen in room-temperature ferromagnetism at graphite surfaces. *New J. Phys.* **12** 123012.
[71] Yang X., Xia H., Qin X., Li W., Dai Y. *et al.* 2009. Correlation between the vacancy defects and ferromagnetism in graphite. *Carbon* **47** 1399.
[72] Tans S.J., Devoret M.H., Groeneveld R.J.A. and Dekker C. 1998. Electron-electron correlations in carbon nanotubes. *Nature* **394** 761.
[73] Kang N., Hu J.S., Kong W.J., Lu L., Zhang D.L. *et al.* 2002. Consistent picture of strong electron correlation from magnetoresistance and tunneling conductance measurements in multiwall carbon nanotubes. *Phys. Rev. B* **666** 241403(R).
[74] Bockrath M., Cobden D.H., Lu J., Rinzler A.G., Smalley R.E. *et al.* 1999. Luttinger-liquid behaviour in carbon nanotubes. *Nature* **397** 598.
[75] Deilmann T., Drüppel M. and Rohlfing M. 2019. Three-particle correlation from a many-body perspective: Trions in a carbon nanotube. *Phys. Rev. Lett.* **116** 196804.
[76] Oreg Y., Byczuk K. and Halp B.I. 2000. Spin configurations of a carbon nanotube in a nonuniform external potential. *Phys. Rev. Lett.* **85** 365.
[77] Egger R. and Grabert H. 1995. Friedel oscillations for interacting fermions in one dimension. *Phys. Rev. Lett.* **75** 3505.
[78] Wang Z., Luo M., Yan D., Ying H. and Li W. 1995. Electronic structure and transport properties of fullerene nanotubules. *Phys. Rev. B* **51** 13833.
[79] Klimov V.V. and Letokhov V.S. 1996. Hard X-radiation emitted by a charged particle moving in a carbon nanotube. *Phys. Lett. A* **222** 424.
[80] Zhukova E.S., Grebenko A.K., Bubis A.V., Prokhorov A.S., Belyanchikov M.A. *et al.* 2017. Terahertz-infrared electrodynamics of single-wall carbon nanotube films. *Nanotechnology* **28** 44.
[81] Viennot J.J., Delbecq M.R., Dartiailh M.C., Cottet A. and Kontos T. 2011. Out-of-equilibrium charge dynamics in a hybrid circuit quantum electrodynamics architecture. *Phys. Rev. B* **89** 165404.
[82] Crochet J.J., Sau J.D., Duque J.G., Doorn S.K. and Cohen M.L. 2011. Electrodynamic and excitonic intertube interactions in semiconducting carbon nanotube aggregates. *ACS Nano* **5** 2611.
[83] Mohite A.D., Gopinath P., Shah H.M. and Alphenaar B.W. 2008. Exciton dissociation and Stark effect in the carbon nanotube photocurrent spectrum. *Nano Lett.* **8** 142.
[84] Matsuda K., Inoue T., Murakami Y., Maruyama S. and Kanemitsu Y. 2008. Exciton dephasing and multiexciton recombinations in a single carbon nanotube. *Phys. Rev. B* **77** 033406.
[85] Graf A., Held M., Zakharko Y., Tropf L., Gather M.C. *et al.* 2017. Electrical pumping and tuning of exciton-polaritons in carbon nanotube microcavities. *Nat. Mater.* **16** 911.
[86] Ichida M., Mizuno S., Tani K., Saito Y., Nakamura A. *et al.* 1999. Exciton effects of optical transitions in single-wall carbon nanotubes. *J. Phys. Soc. Jpn.* **68** 3131.

[87] Varsano D., Sorella S., Sangalli D., Barborini M., Corni S. *et al.* 2017. Carbon nanotubes as excitonic insulators. *Nat. Commun.* **8** 1461.

[88] Simon L., Bena C., Vonau F., Cranney M. and Aubel D. 2011. Fourier-transform scanning tunnelling spectroscopy: The possibility to obtain constant-energy maps and band dispersion using a local measurement. *J. Phys. D Appl. Phys.* **44** 46.

[89] Lee J., Eggert S., Kim H., Kahng S.-J., Shinohara H. *et al.* 2004. Real space imaging of one-dimensional standing waves: Direct evidence for a Luttinger liquid. *Phys. Rev. Lett.* **93** 166403.

[90] Hasegawa Y., Ono M., Nishigata Y., Nishio T. and Eguchi T. 2007. Real-space observation of screened potential and Friedel oscillation by scanning tunneling spectroscopy. *J. Phys. Conf. Ser.* **61** 399.

[91] Hosur P. 2012. Friedel oscillations due to Fermi arcs in Weyl semimetals. *Phys. Rev. B* **86** 195102.

[92] Bena C. 2008. Effect of a single localized impurity on the local density of states in monolayer and bilayer graphene. *Phys. Rev. Lett.* **100** 076601.

[93] Tabert C.J. and Nicol E.J. 2014. Dynamical polarization function, plasmons, and screening in silicene and other buckled honeycomb lattices. *Phys. Rev. B* **89** 195410.

[94] Ezawa M. 2012. A topological insulator and helical zero mode in silicene under an inhomogeneous electric field. *New J. Phys.* **14** 033003.

[95] Li L., Lu S.Z., Pan J., Qin Z., Wang Y. *et al.* 2014. Buckled germanene formation on Pt(111). *Adv. Mater.* **26** 4820.

[96] Derivaz M., Dentel D., Stephan R., Hanf M., Mehdaoui A. *et al.* 2015. Continuous germanene layer on Al(111). *Nano Lett.* **15** 2510.

[97] Zhu F.F., Chen W.J., Xu Y., Gao C.L., Guan D.D. *et al.* 2015. Epitaxial growth of two-dimensional stanene. *Nat. Mater.* **14** 1020.

[98] Yan J.A., Gao S.P., Stein R. and Coard G. 2015. Tuning the electronic structure of silicene and germanene by biaxial strain and electric field. *Phys. Rev. B* **91** 245403.

[99] Qin R., Zhu W., Zhang Y. and Deng X. 2014. Uniaxial strain-induced mechanical and electronic property modulation of silicene. *Nano Res. Lett.* **9** 521.

[100] Lu W.T., Li Y.F. and Tian H.Y. 2018. Spin- and valley-dependent electronic structure in silicene under periodic potentials. *Nano Res. Lett.* **13** 84.

[101] van den Broek B., Houssa M., Scalise E., Pourtois G. and Afanas'ev V.V. 2014. First-principles electronic functionalization of silicene and germaneneby adatom chemisorption. *App. Surf. Sci.* **291** 104.

[102] Liu G., Liu S.B., Xu B., Ouyang C.Y., Song H.Y. *et al.* 2015. Multiple Dirac points and hydrogenation-induced magnetism of germanene layer on Al (111) surface. *J. Phys. Chem. Lett.* **6** 4936.

[103] Liu W.H., Luo J.W., Li S.S. and Wang L.W. 2019. Impurity diffusion induced dynamic electron donors in semiconductors. *Phys. Rev. B* **100** 165203.

[104] Wunsch B., Stauber T., Sols F. and Guinea F. 2006. Dynamical polarization of graphene at finite doping. *New J. Phys.* **100** 165203.

[105] Bácsi Á. and Virosztek A. 2010. Local density of states and Friedel oscillations in graphene. *Phys. Rev. B* **82** 193405.
[106] Kanisawa K., Butcher M.J., Yamaguchi H. and Hirayama Y. 2001. Imaging of Friedel oscillation patterns of two-dimensionally accumulated electrons at epitaxially grown InAs(111)A surfaces. *Phys. Rev. Lett.* **86** 3384.
[107] Zhang R.F., Argon A.S. and Veprek S. 2009. Friedel oscillations are limiting the strength of superhard nanocomposites and heterostructures. *Phys. Rev. Lett.* **102** 015503.
[108] Zhang S.H., Yang J., Shao D.F., Wu Z. and Yang W. 2021. Robust wavefront dislocations of Friedel oscillations in gapped graphene. *Phys. Rev. B* **103** L161407.
[109] Dutreix C., Bellec M., Delplace P. and Mortessagne F. 2021. Wavefront dislocations reveal the topology of quasi-1D photonic insulators. *Nat. Commun.* **12** 3571.
[110] Brihuega I., Mallet P., Bena C., Bose S., Michaelis C. *et al.* 2021. Quasiparticle chirality in epitaxial graphene probed at the nanometer scale. *Phys. Rev. Lett.* **101** 206802.

Chapter 5

# PROPAGATION OF LOW-FREQUENCY PLASMONS AND e–h EXCITATIONS

Chiun-Yan Lin, Ching-Hong Ho, Lu-Yao Wang, Chih-Wei Chiu, Hsien-Ching Chung, and Ming-Fa Lin

Over the past year, research on the dynamic behavior of quasiparticles using ultrafast spectroscopy has become a major topic in solid-state physics [1–12]. By utilizing ultrafast laser pulses, researchers have been able to generate and monitor the dynamic oscillations of electrons. The integration of ultrafast laser pulse technology with scanning tunneling microscopy (STM) has led to the development of a new microscopy technique that allows for the real-space imaging of transient carrier dynamics in nanostructures [10, 11]. Coherent nanoimaging and nanospectroscopy have also observed ultrafast coherent dynamics and related nanoscale phenomena with nanometer and femtosecond temporal resolutions [11, 13–15]. With the advent of near-field optical microscopes, there has been an increased interest in the real-space and time propagation of plasmons [13–16]. The Fourier transform STM provides a method to determine the Fermi surface directly by measuring standing-wave patterns around defects in a metal [17–19]. Efficiently converting the transient carrier response into measurable electrical signals is crucial for realizing the dynamic properties of quasiparticles governed by single-particle or collective excitations. Plasmon oscillations in conducting materials can produce large optical field confinement and enhancement, highlighting their potential for practical applications.

Up until now, there has been limited theoretical research on the dynamic charge oscillation phenomenon outside of electron gas systems [20–22]. In particular, a few studies have focused on the frequency response of external perturbations, such as a step potential. In the $(q, \omega)$ domain, the electric polarization is calculated within the linear response theory under random-phase approximation (RPA). The manifestations of static and dynamic screenings of charge carriers with an infinite frequency domain are obtained together. The screenings of electrons in a medium are based on the concept of a quasiparticle in quantum mechanics. This approach allows for the charge response to be decomposed into two parts: one caused by the plasmon mode and the other related to the electron–hole (e–h) pair based on a clear classification of collective and single-particle excitations. The division is one of the most important aspects of elementary excitations. By using the inverse Fourier transform, the $(q, \omega)$ domain is converted into the $(r, t)$ domain. The obtained position- and time-dependent screened charge density distributions might be suitable for investigating the creation of core-holes in simple metals in the limit of large photoelectron energies. A notable result is the shock wave, which consists of the static distribution surrounded by a ring related to plasmons with small transferred momenta. The concise picture is clarified for the electronic "shock wave," which propagates from the core hole and has a certain dispersion, and the group velocity is centered on the Fermi velocity. This will become clearer as the transient effects on the core-hole decay are discussed for a 2D electron gas [21] and a semi-infinite metal surface [22]. The charge oscillation time is evaluated to be about 0.1–1.0 fs.

The screening of charges in solid-state systems can be effectively characterized by using the dimensionless dielectric function or energy loss function [23]. This allows us to describe the overall screening ability of charges using a universal dimensionless physical quantity, enabling reliable comparisons between different condensed matter systems. On the other hand, the latter has been widely used to describe the collective excitations, plasmons, using fast electron beams or hard X-ray sources in the emergent 2D layered materials [24, 25] and 1D coaxial nanotubes [26], since the theory of RPA is generalized for the Coulomb correlation energy. The plasma wave is investigated through the Fourier components of the energy loss function or the screened charge density, demonstrating its propagation in the real-space and time domains. Choosing one of the two viewpoints to describe the dynamic Coulomb field phenomenon is worthy of systematic research to verify the consistency of theoretical and experimental results. In 2001, femtosecond spectroscopy analysis demonstrated that screening requires a finite time ($\approx$10 fs) to be established in GaAs in order to cause the collective behavior of electrons in the system [27]. In the previous chapter, we chose multilayer graphenes as a model system to explore their static screening. Here, we further consider the dynamic frequency response on a finite timescale. The duration of the external perturbation plays an important role in the excitation spectrum and plasma waves. The graphene systems are outstanding candidates for clearly illustrating the diverse carrier dynamics due to the diversity of stacked configurations. The RPA is modified to account for electron–electron (e–e) interactions and inter- and intralayer atomic interactions based on the theoretical framework of the generalized tight-binding model. With the development of the theoretical framework, the charge oscillation is expected to be reflected in the dynamic screening of graphenes. Apparently, the propagating waves related to the essential physical properties are required to be clarified thoroughly; they might be detected by high-speed spectroscopy with femtosecond ultrahigh resolutions.

The dynamic charge screenings are expected to be greatly diversified in the position and time domains according to the various band structures near the Fermi level. Previous studies have shown that electron concentration is a critical mechanism that affects the screening effect and its modulation in various systems, such as 1D carbon nanotubes [28, 29] and 2D few-layer graphene systems [30, 31], could be achieved through intrinsic factors, such as tube radius, chirality [32] and stacking configuration [33], and extrinsic factors, such as temperature and gate voltage [26]. It should be noted that all the aforementioned effects on the screened response functions are unified and simultaneously combined with the development of the theoretical framework (details in Chapter 3). The evolution of the effective potential and charge oscillation is investigated in detail in terms of their propagation in time and position domains. It reflects the dynamic screening resulting from the single-particle and collective excitations, which can be realized from the characteristics of the energy loss spectrum and might be verified by the high-resolution pump-probe femtosecond optical spectroscopies [8, 9]. For example, a low-frequency plasmon below 10 meV will collectively be excited as a coherent wave with a period of 400 fs, which has been directly observed by current experimental measurements in the time domain [34].

## 5.1 Dynamic Screenings in Low-dimensional Systems

The spatially induced charge distributions and the total Coulomb potentials using the RPA dielectric function are often simulated and discussed in electron gas systems [35–41, 43–46, 53]. Related experimental studies are available on this research topic [47–52, 54, 55]. Owing to the parabolic energy dispersion and the corresponding plane wave, the electric polarization of the electron gas is expressed as an analytic function in the linear response theory. The imaginary part of the dielectric function can be directly calculated using the Kubo formula, and the real part can be obtained through the Kramers–Kronig relations [23]. Alternatively, one can also define the Matsubara polarization function on the frequency complex plane and calculate the integral by analytical continuation to derive the dielectric function. See Refs. [21, 56, 57] for a detailed calculation. The derived formulas of the dielectric function are very useful in exploring the essential physical properties, e.g., the pair distribution functions [58–61], the exchange and correlation energies [45, 46, 61, 62], the Coulomb decay rates [43, 52, 53, 63–65], the modified effective masses and the renormalization constants [66–68]. In particular, the different dimensions from 1D to 3D play a critical role in electronic properties and e–e Coulomb interactions, enriching and diversifying the charge screening behaviors under many-particle interactions. In the context of this knowledge, we have established a theory to study the propagation of charge oscillations caused by the excitation of plasmon and electron holes in a layered system [64]. This model can also be applied to other van der Waals materials.

The characteristic length ($r_c$) and time ($t_c$) are kinematic quantities, which are very helpful to further understand the dynamic response of the charge oscillation. The former is defined as $r_c = \pi/k_F$, which is the wavelength of the Friedel oscillations in metallic systems. The latter is given by $t_c = 2\pi/\omega_c$, where $\omega_c$ is the overall spectral intensity and expressed as

$$\omega_c = \frac{\int_0^\infty \int_0^\infty \omega * Im[-1/\epsilon] * dq^2 d\omega}{\int_0^\infty \int_0^\infty Im[-1/\epsilon] * dq^2 d\omega}. \tag{1}$$

Specifically, the Fermi momentum is defined by the lowest conduction subband in the case of electron doping. $r_c$ and $t_c$ vary depending on how high the carrier concentration is in relation to the pristine case, i.e., they, respectively, reflect the static screening length and the energy scales of single-particle and collective excitations. In the case of high carrier doping, $t_c$ is dominated by the plasma frequency because the collective excitation overwhelms the single-particle excitation. $r_c$ and $t_c$ can serve as the characteristic scales for the screening phenomena in position and time domains where dominant patterns emerge; their definitions can be generalized to metallic solid-state systems.

As one of the most typical manifestations of the interaction between electrons, screening ability can be understood from the effective Coulomb potential, which is directly defined by the known RPA dielectric function in the $(q, \omega)$ domain. Its Fourier components can describe the Coulomb interaction as a function of position and time due to the dynamic charge oscillations under external perturbations. First, we calculate the case where an external point charge is suddenly created at the origin when $t = 0$. The Coulomb potential is given by

$$V^{\text{ex}}(\mathbf{r}, t) = \frac{e^2}{r}\theta(t), \tag{2}$$

where $\theta(t)$ is the Heaviside step function, describing the time dependence. The Fourier transform of the suddenly created potential in the momentum domain and frequency domain of a 2D system is expressed as

$$V^{\text{ex}}(\mathbf{q}, \omega) = \frac{4\pi e^2}{q^2}\frac{i}{\omega + i\delta} = \frac{4\pi e^2}{q^2}\frac{\delta + i\omega}{\omega^2 + \delta^2}, \tag{3}$$

where the infinitesimal positive quantity $\delta$ is set to maintain the material stability by digesting the increasing heat energy during the perturbation. In the above formula, the components of $\omega$ are decomposed into a static term and a dynamic term, namely the delta function $\delta(\omega)$ and a Gaussian function of $\omega$, respectively. This assumption is reliable only under the linear-response approximation. Using the definition of dielectric function, $\epsilon(\mathbf{q}, \omega) = V^{\text{ex}}(\mathbf{q}, \omega)/V^{\text{eff}}(\mathbf{q}, \omega)$, and the substitution into Eq. (2) yield the effective Coulomb potential in the real-space and time domains:

$$\begin{aligned} V^{\text{eff}}(\mathbf{r}, t) &= \frac{1}{(2\pi)^3}\int_0^{2\pi}\int_0^{\infty}\int_{-\infty}^{\infty} V^{\text{eff}}(\mathbf{q}, \omega)\exp[i(\mathbf{q}\cdot\mathbf{r} - wt)]qdqd\theta d\omega \\ &= \frac{e^2}{2\pi^2}\left[\int_0^{\infty} dq \frac{\sin(qr)}{qr}\frac{1}{\epsilon(q, 0)} + 2\int_0^{\infty}\int_0^{\infty} dqd\omega \right. \\ &\quad \left. \times \frac{\sin(qr)}{qr}\frac{\mathbf{Im}[\epsilon(q, \omega)]\cos(\omega t) + \mathbf{Re}[\epsilon(q, \omega)]\sin(\omega t)}{\omega\,|\epsilon(q, \omega)|^2}\right], \end{aligned} \tag{4}$$

where the integration on $\theta$ is easy to obtain, as the dielectric function is isotropic in the momentum space.

In the above equation, the Coulomb potential is a real function of $\mathbf{r}$ owing to the fact that $\mathbf{Re}[\epsilon(q,\omega)]$ and $\mathbf{Im}[\epsilon(q,\omega)]$ are even and odd functions of $\omega$, respectively. The first term is static, and the screening effect it causes has been discussed in the previous chapter. In connection with the linear response theory, the charge fluctuation that depends on $\omega$ is a focus of interest, in addition to the dynamic potential. The current work only explores very fast charge oscillations in coherent and non-coherent ways in relation to the energy loss spectrum. The results are also compared to the charge Coulomb potential that exponentially decays with the timescale of $t_c$, i.e., $V^{\text{ex}}(\mathbf{r},t) = \frac{e^2}{r}\exp(-\alpha t)\theta(t)$.

In addition, we replace the Heaviside step function with a rectangular function to reduce the perturbation time to a finite duration, that is, the Coulomb potential in the domains $t$ and $\omega$ are, respectively, given by

$$V^{\text{ex}}(\mathbf{r},t) = \begin{cases} \dfrac{e^2}{r} & \text{for } t \geq 0 \quad \text{and} \quad t \leq T_a \\ 0 & \text{for } t > T_a \quad \text{and} \quad 0 \end{cases} \tag{5}$$

and

$$V^{\text{ex}}(\mathbf{q},\omega) = \frac{4\pi e^2}{q^2} \times \left[\frac{\sin(\omega T_a)}{\omega} - i\frac{2\sin^2\left(\frac{\omega T_a}{2}\right)}{\omega}\right], \tag{6}$$

where $T_a$ stands for the duration of perturbation, which is expected to modulate the coherent plasmon waves in the dynamic response regime. When discussing $V^{\text{eff}}(\mathbf{r},t)$ and $\rho^{\text{in}}(\mathbf{r},t)$ in a dynamic manner, the energy transport in materials might become an emergent issue due to recent advances in the generation of ultrahigh speed spectroscopy [69].

The cases of 1D nanotubes and 2D and 3D electron gases are reviewed first, showing the diverse phenomena of static and dynamic charge screenings. Figures 5.1 and 5.2 clearly illustrate the position- and time-dependent effective Coulomb potentials at $E_F = 0.5$ eV, respectively, under the isotropic Coulomb interaction of the point charge being suddenly turned on at $t = 0$. When the Coulomb field produces a long-distance perturbation in the medium, the interaction can create an infinite number of intermediate excited states in a sufficient phase space. That is, based on the Pauli exclusion principle and conservation law, the potential divergence of small $q$ gives rise to countless ways of transferring energy and momentum for the inelastic Coulomb scatterings between carriers. This argument accounts for the transformation among elementary excitations

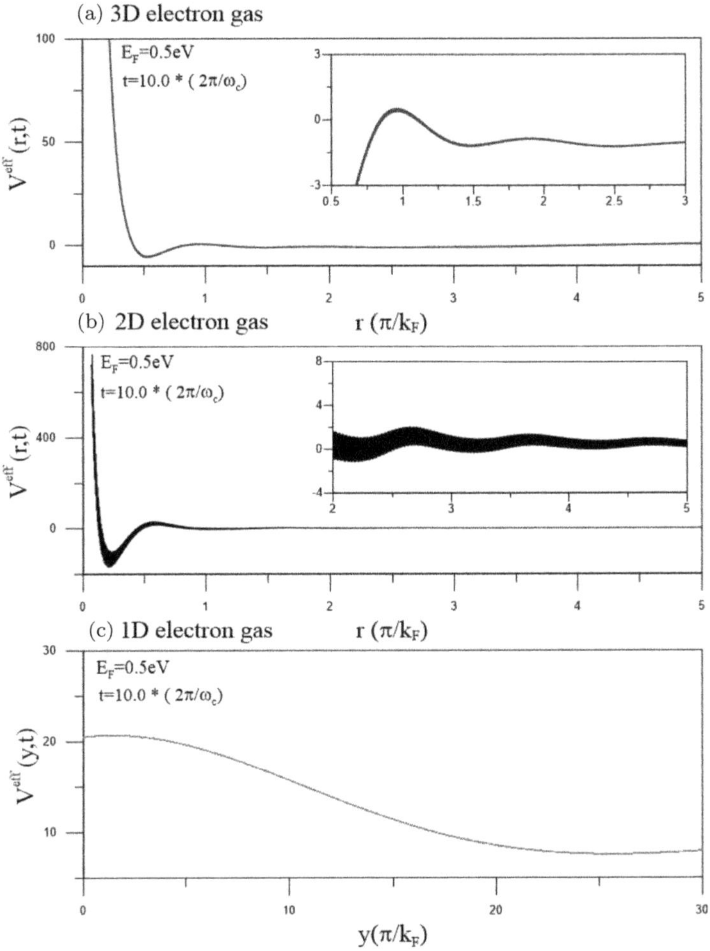


Fig. 5.1 Effective Coulomb potential as a function of the position at $t = 10 \times t_c$ for (a) 3D electron gas, (b) 2D electron gas and (c) 1D nanotube electron gas at $E_F = 0.5\,\text{eV}$ under the influence of a point charge generated suddenly at $r = 0$ and $t = 0$.

based on the RPA model. In the $(r, t)$ domain, the screening behavior in a higher-dimensional system is relatively easy to observe due to the high electron density. The charge oscillations could be regarded as propagating plasma waves, and their energy is dissipated due to Landau damping beyond a certain range. Obviously, $V^{\text{eff}}$ in the metallic system quickly drops off from the perturbed charge, as shown in Fig. 5.1, where the 3D and 2D electron gases exhibit a decrease as observed in the Yukawa

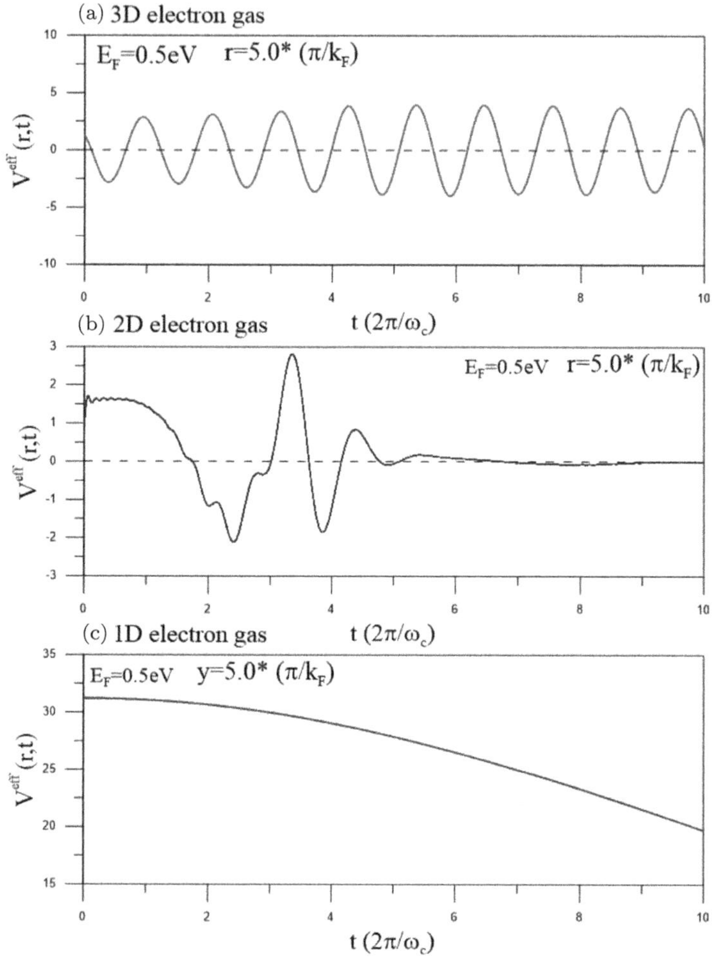


Fig. 5.2 Time variation of the effective Coulomb potential at $r = 5 \times \pi/k_F$ for (a) 3D electron gas, (b) 2D electron gas and (c) 1D nanotube electron gas at $E_F = 0.5\,\text{eV}$ under the influence of a point charge generated suddenly at $r = 0$ and $t = 0$.

potential. This screening is very effective over a very short length of a few $r_c$. The magnitude of the screening length is estimated to be of 1 Å [70]. More interestingly, the Friedel oscillations are also observed on a timescale of a sufficiently long distance (approximately of the order of $\pi/k_F$), which is a quantum phenomenon caused by the Fermi surface. For 3D and 2D electron gas systems with isotropic quadratic energy dispersion, the charges decay with $\sin(2k_F r)/r^3$ and $\sin(2k_F r)/r^2$ oscillatory functions [71, 71]. Charge conservation produces electron and hole charge variations and a consecutively enhanced electric field in a few Friedel oscillations with a characteristic length of $\pi/k_F$. For the 1D nanotube electron gas, $V^{\text{eff}}$ only shows observable monotonic dependence along the axial direction. Relevant studies have shown that only impurities on the tube surface can cause the Friedel oscillations [72].

Potential behavior in the $r$ domain can better show the difference in screenings between quasiparticles caused by dimensions, as shown in Fig. 5.2. The 3D electron gas oscillation behavior is the most significant, followed by 2D, and that in 1D is the weakest. This is because the plasma intensity decreases sequentially with the dimension. Refer to Ref. [73] to see the important differences between the momentum dependence of the optical and acoustic plasmon modes in the 0D–3D electron gas systems. The regular oscillation in 3D electron gas clearly displays a period of the featured characteristic time $t_c$. The abnormal oscillation revealed in the 2D case reflects the combination of oscillations with different plasmon frequencies. However, in the 1D case, only monotonic decays with distance are observed. Experimentally, short periods on the position and timescale would cause measurement difficulties. However, emergent materials with low free carrier densities, i.e., high $t_c$, have the potential to contribute to the aspects of dynamics for the experimental tests. Intrinsic interactions can be tuned via the duration of the perturbation, and the dynamics of electrons can be resolved due to the longer timescales. High-resolution STM observations may be a usable tool, with ultrahigh resolutions being needed/resolved in recent research [52].

Equations (3) and (5) provide ideal means for studying various forms of Coulomb perturbation through the linear superposition of point charge perturbations within the linear response theory. Those formulas are suitable for calculating the RPA response function. The dielectric responses of $N$-layer graphenes become more complicated because the perturbed Coulomb potential will induce charge distribution on all layers, thereby further screening the potential. That is, in the linear response theory, all external and induced Coulomb potentials of each layer need to be considered in a consistent manner. Fortunately, the RPA is generalized in the tight-binding model to incorporate the hopping integrals related to the electronic states and the Coulomb interactions together. Based on the characteristics of the layer-dependent wave functions, the dielectric response under the e–e Coulomb interactions displays layer-dependent properties; therefore, it varies according to the stacking configuration of the $N$-layer graphene. The development of the modified RPA is of great benefit to calculating the screened response functions in graphene systems because the atomic interactions and e–e interactions among all the layers are considered simultaneously. While the energy loss spectra of multilayer graphenes have been extensively investigated using the self-consistent Dyson equation [64], the rich and unique charge dynamics might be of great interest during the movement of the electron beam through the medium. This topic is worthy of systematic research on emerging materials.

## 5.2 Electron-Doped Monolayer Graphene

Previous works have shown that either n-type or p-type electrical doping is conveniently controlled by electrical and static electric gating. Both the electrical [74, 75] and chemical [76, 77, 80] dopings do not deteriorate the unique nature of graphene. The tunability of the electronic properties of monolayer graphene makes it an outstanding material to study the dynamics of Dirac quasiparticles, such as the quasiparticle energy spectrum and lifetime [63–65]. Relevant theoretical and experimental works have proved that the rigid shift model can explain the electronic properties in the case of doping. Under e–e interactions, the dielectric function with a small momentum transfer limit is studied by the analytic formulas of the Bloch wave function within an energy range of interest, as described by the isotropic linear dispersion $E^{c,v}(\mathbf{k}) = \pm v_F|\mathbf{k}|$. The intra-$\pi$-band transitions from free electrons/holes between the same conduction band/valence band within the Dirac cone give rise to excitation channels that are very distinct from the interband ones from valence electrons in undoped graphene [64]. As discussed in Chapter 3, the acoustic plasmon mode in 2D materials is enhanced by the Fermi level, as shown by the electron energy loss spectrum. Both of these excitations contribute to the plasmons, while the former dominates the strong plasmon mode and weak Landau damping with sufficiently high electron doping. As a result, the characteristic frequency $\omega_c$ gradually becomes the integration of the collective excitation frequency with the increasing $E_F$. How the plasmon resonance in energy reacts to the propagation of plasma waves in time and space domains is the subject of this chapter.

Doped monolayer graphene presents rich and unique screening behaviors, as clearly illustrated by the screened potential in the domains of $r$ and $t$ in Figs. 5.3 and 5.4. Important differences are observed between

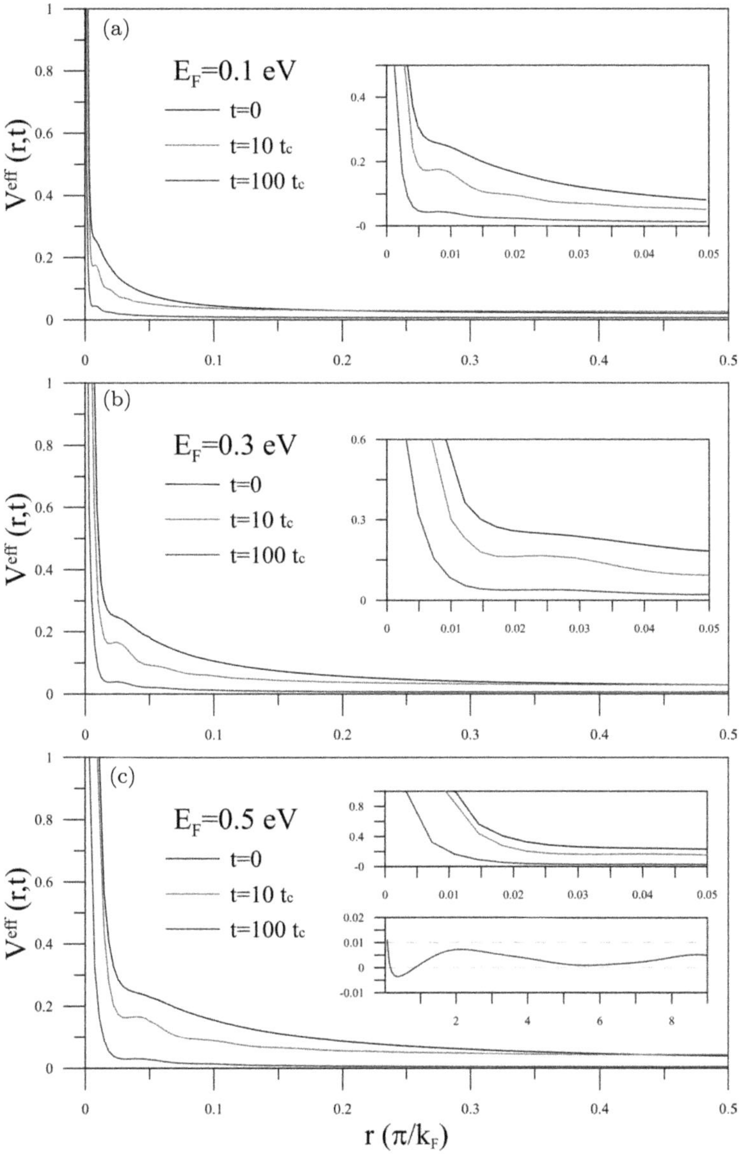


Fig. 5.3 Distance- and time-dependent effective Coulomb potential in the n-type doped graphene system perturbed by a point charge at $r = 0$ from $t = 0$: (a), (b) and (c), respectively, correspond to $E_F = 0.1\,\text{eV}$, $E_F = 0.3\,\text{eV}$ and $E_F = 0.5\,\text{eV}$ at different times after the perturbation.

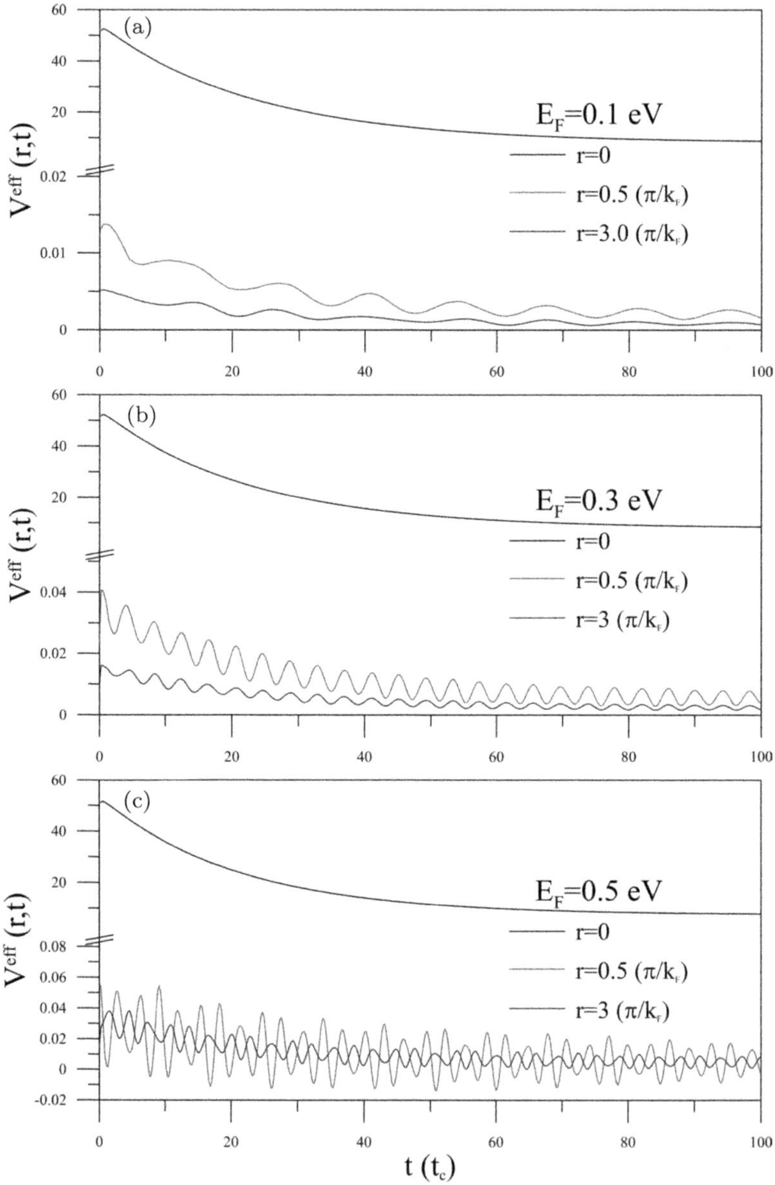


Fig. 5.4 Time variation of the effective Coulomb potential shown in Fig. 5.3.

graphene and 2D electron gas in terms of the screening length and oscillatory phenomena. The conduction and valence electrons are occupied at the K and K′ valleys simultaneously, thus leading to the intravalley and intervalley e–e Coulomb scatterings. However, the former effect is much stronger than the latter, so we only consider the former's contribution to dynamic screenings [63]. As shown in Fig. 5.3, $V^{\text{eff}}$ gradually declines from $r = 0$ and display fascinating Friedel oscillations that vary depending on $E_F$ and $t$. The screening ability in terms of the length scale $\pi/k_F$ is, of course, very sensitive to $t$ at different $E_F$, that is, the larger the $E_F$, the smaller the scale value and the stronger the screening effect. Regarding the time dependence (Fig. 5.4), one can observe a time-decaying plasmon wave that corresponds to the composition of the Friedel oscillations occurring away from the screening length, and it could survive for a long time in the cases of heavy density carriers. As mentioned in the previous section, the oscillatory phenomenon dependent on $t$ becomes regular with a period of $t_c$ at a high enough $E_F$, e.g., $E_F = 0.3\,\text{eV}$ and $E_F = 0.5\,\text{eV}$, where the plasmon oscillation predominates over the excitations. In addition, this result directly reflects the gradual energy loss of the plasmon wave that is dissipated by Landau damping. The induced charge density and effective potential have the same oscillating behavior and shielding phenomenon in the $r$ and $t$ domains for electron-doped graphene and clearly demonstrate their decay with $r$ and $t$, as shown in Figs. 5.5 and 5.6. The induced charge at the origin has a similar oscillation behavior over time; however, the charge decays faster at the same distance under high $E_F$, and the short-distance oscillation is more obvious, as depicted for $E_F = 0.5\,\text{eV}$ in Fig. 5.6(c). In the inset, the charge oscillates with a period of $t_c$. Graphene-related emerging materials have well-behaved lattice symmetry and are considered outstanding candidates for thoroughly exploring the diverse phenomena caused by charge density waves.

The duration and intensity of the perturbation play an important role in the effective potential and induced charge density of the system. When the external potential decays exponentially with $t$, the screening length and the plasmon wave propagating in time show drastic changes, as shown in Fig. 5.7. For the two cases of different external potentials, the time period of the plasmon wave remains the same because it originates from the same plasmon resonance at the specified Fermi level. Due to the decrease in $V^{\text{ex}}$, $V^{\text{eff}}$ is reduced more effectively with the decrease in the decaying factor $\alpha$, as shown in Fig. 5.7(a). The physical significance of the potential change emerges when $r \sim 0.03\ \pi/_F$, which means that

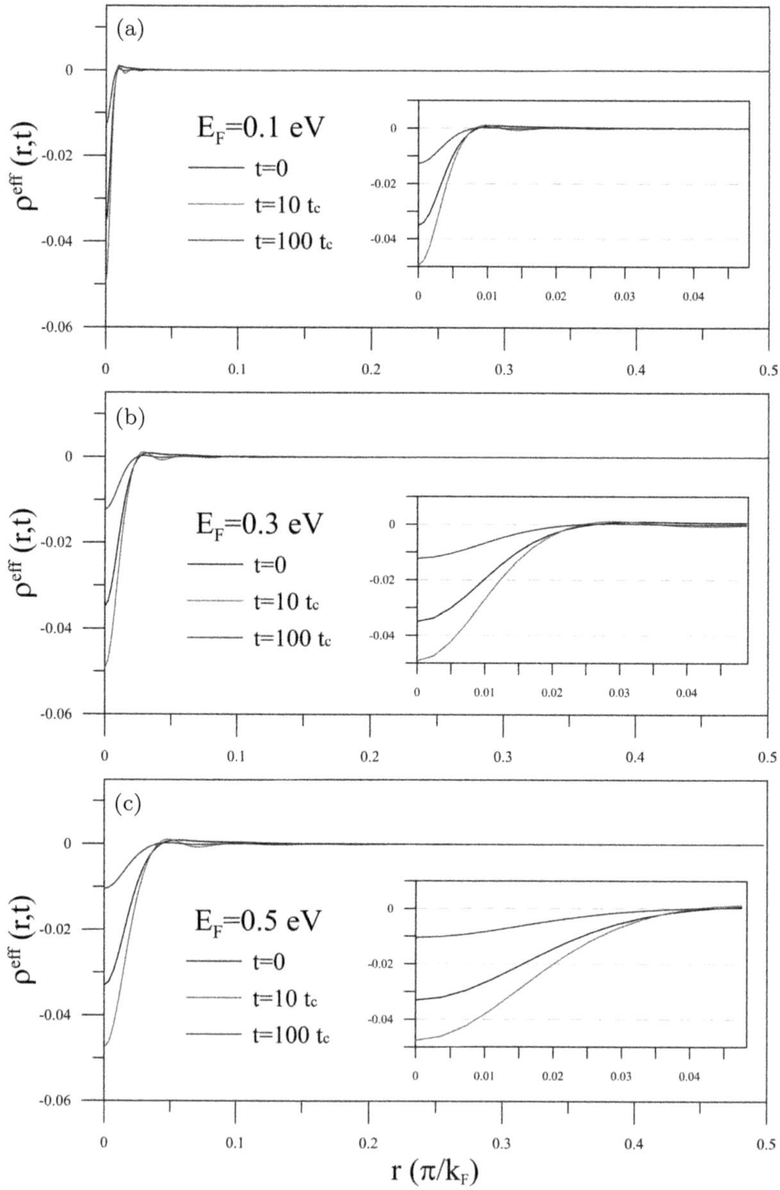


Fig. 5.5 Distribution of the induced charge density in the position space corresponding to the effective Coulomb potential shown in Fig. 5.3.

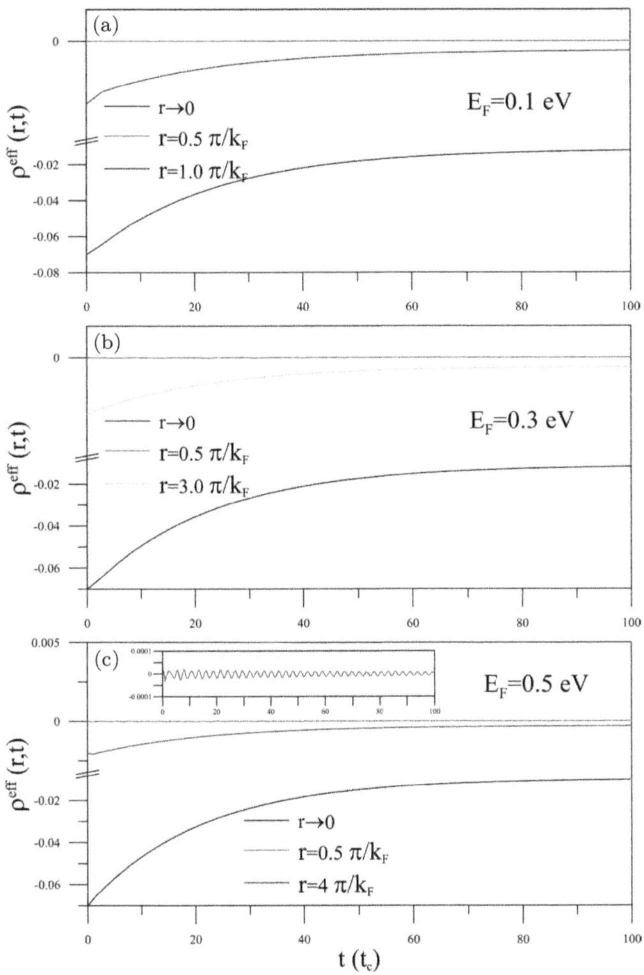


Fig. 5.6 Time variation of the induced charge density corresponding to the effective Coulomb potential shown in Fig. 5.4.

the oscillatory phenomenon may be enhanced in a certain region of the $(r, t)$ domain. It should be noted that the oscillation amplitudes at different positions show different time dependences.

Figures 5.7(b) and 5.7(c) show that the potential amplification at the position of $r = 0.5\pi/k_F$ can still be observed when $t$ expands to more than $100t_c$ while that at $r = \pi/k_F$ is greatly reduced. This indicates that the plasmon wave can survive for a long time at the moderate Fermi level corresponding to the weak Landau damping, which may make the dynamic

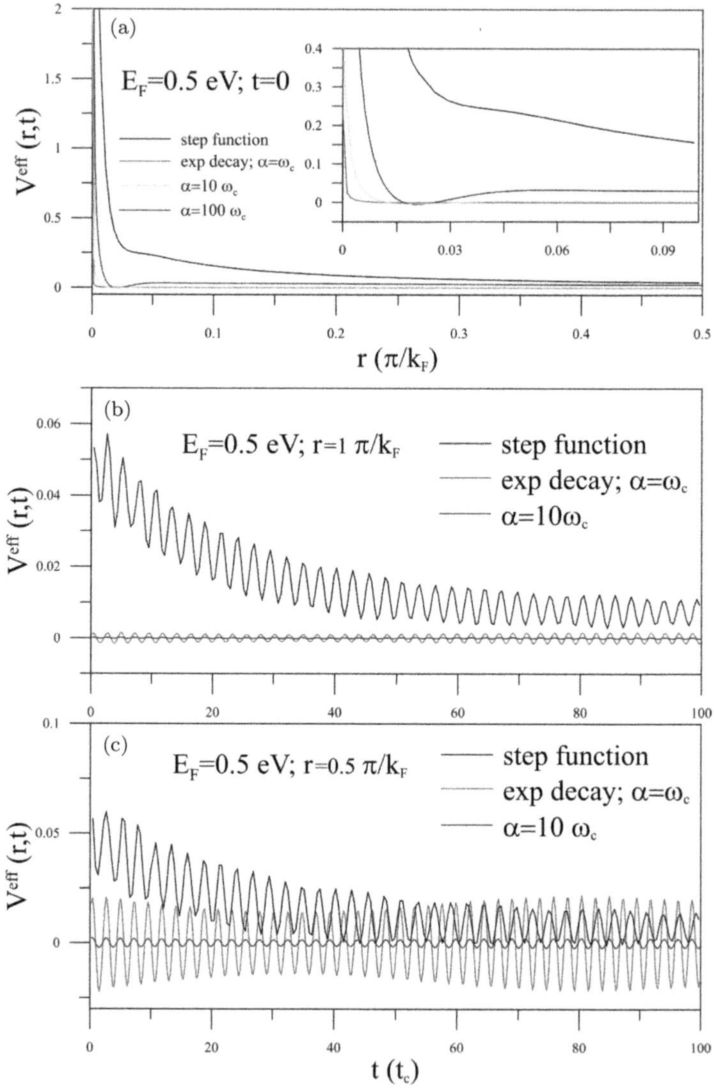


Fig. 5.7 Distance- and time-dependent effective potential in the n-type doped graphene system under perturbation of the charge Coulomb potential that exponentially decays with time.

phenomenon easier to observe. On the other hand, the rectangular potential in time will also cause a similar enhancement of the plasmon wave. As shown in Fig. 5.8, the amplitude of the potential vibration is significantly enlarged. In addition, with the shortening of the perturbed time,

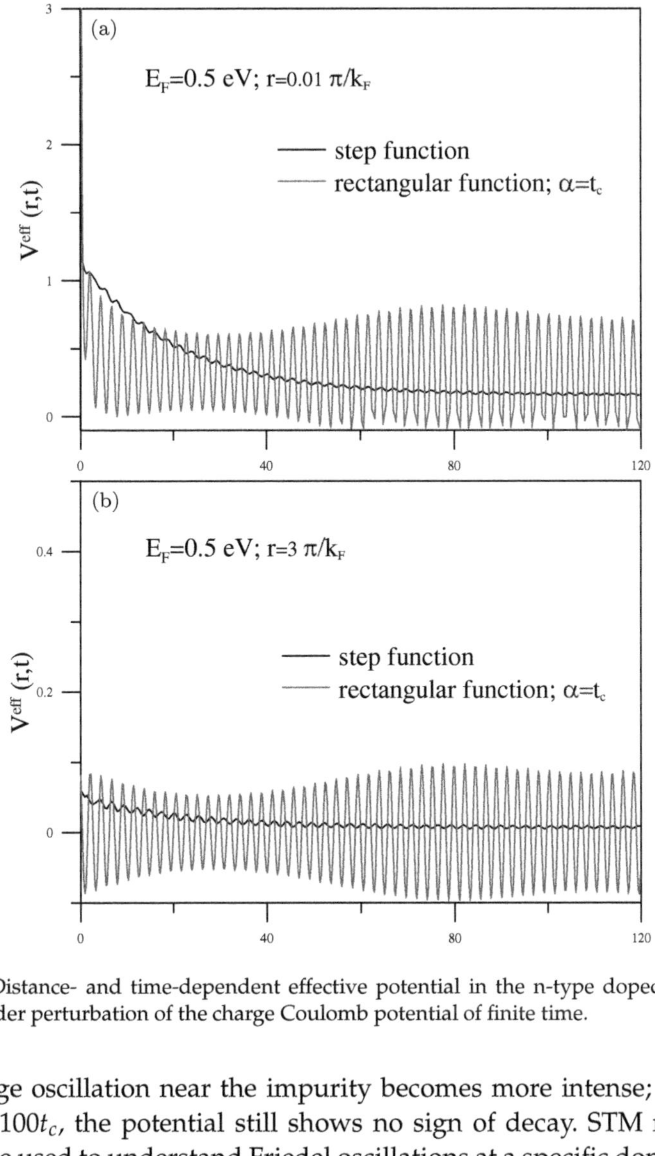


Fig. 5.8 Distance- and time-dependent effective potential in the n-type doped graphene system under perturbation of the charge Coulomb potential of finite time.

the charge oscillation near the impurity becomes more intense; even if $t$ exceeds $100t_c$, the potential still shows no sign of decay. STM measurements are used to understand Friedel oscillations at a specific doping level in the vicinity of the impurity. While STM observations of standing-wave patterns generated by defects provide information on the electronic properties of a system, which is fully understood, how to check and identify their dynamics under resonance and on long timescales may bring new challenges to high-resolution STM measurements [18, 52, 79].

Theoretical studies have shown that the excitation spectrum of graphene exhibits a low-frequency acoustic plasma mode, in addition to the intermediate frequency $\pi$ plasma $\omega_p > 5$ eV [63]. The predictions have been verified by electron energy loss spectroscopy (EELS) and inelastic X-ray spectroscopy, which are two very powerful techniques that provide the Coulomb excitation spectrum [80–96, 111]. The prominent Coulomb interaction in graphene sheets clearly indicates that it is relatively easy to observe plasmon modes at low frequencies compared to other solid-state systems [97, 98], while for Fermi levels lower than 2.0 eV, it is difficult to observe coherent charge oscillations in the EELS measurements [97, 98]. Nevertheless, the ultrafast coherent dynamics, which come from the linear dispersion, and the associated nanoscale phenomena have been observed through coherent nanoimaging and nanospectroscopy with nanometer and femtosecond temporal resolutions [11, 13–15]. The dynamic effective potential and charge density caused by related single-particle and collective excitations can be observed with the advancement of broadband nanospectroscopy.

## 5.3 AA-Stacked Bilayer Graphenes

The geometric structure of the AA configuration is formed by stacking graphene sheets vertically along the $\hat{z}$ direction. As a result of the high symmetric structure, the band structure consists of multi monolayer-like Dirac cones located at the K/K′ point, with energy spacings caused by $\alpha_1$. The layer-dependent dynamic polarization function responds to the dynamic Coulomb interactions between electrons on the same layer or different layers. In the undoped case, free electrons or free holes lead to intrinsic composite plasmon modes, i.e., an acoustic and an optical collective excitation mode attributed to in-phase and out-of-phase charge coupling between two layers. Without doping, the acoustic branch can only be triggered in the AA configuration as well as the undamped optical plasmon one. Such two collective modes dominate the excitation spectrum over a wide area of the $(q, \omega)$ space due to the robust Dirac cone structure in the AA configuration. The potential wave propagation is expected to be dominated by the dynamic charge screening from single-particle excitations and the different degrees of the combination of acoustic and optical plasmon excitations.

When a point charge is suddenly created in one layer of a bilayer graphene sheet, the effective potentials on the two layers respond to the external one in a consistent manner, as defined in the linear response theory (the calculated results are shown in Figs. 5.9–5.11). The interlayer and

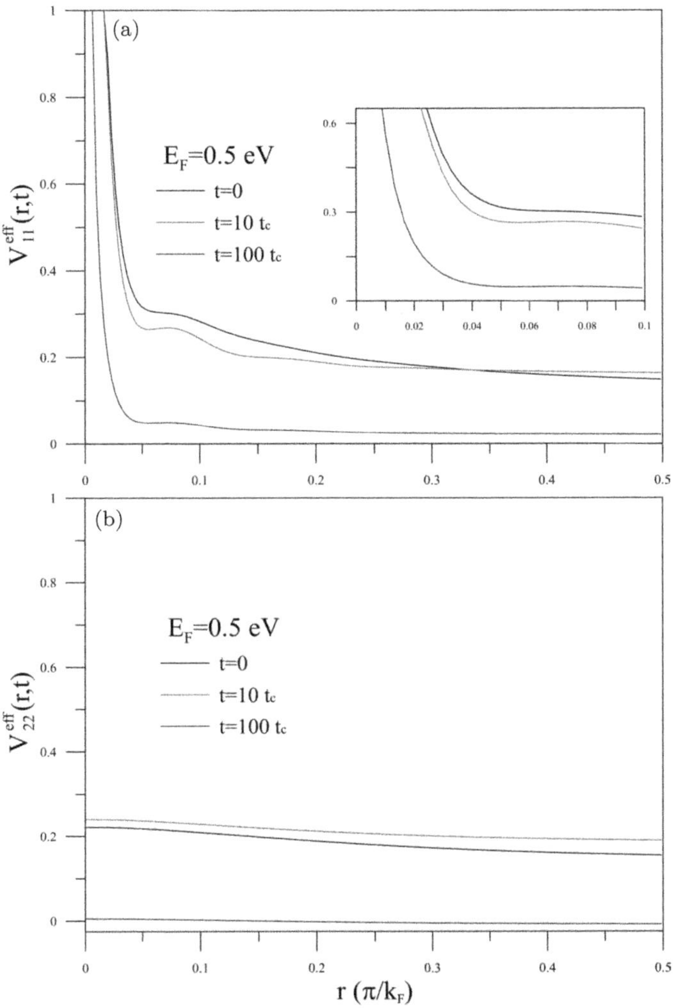


Fig. 5.9 Effective potentials on (a) the first layer $V_{11}^{\text{eff}}$ and (b) the second layer $V_{22}^{\text{eff}}$ as functions of $r$ at various $t$ in AA-stacked bilayer graphene at $E^F = 0.5\,\text{eV}$. A charged impurity is suddenly created at $r = 0$ and $t = 0$ on the first layer. The insert show the zoomed-in view of $V_{11}^{\text{eff}}$ at short distance.

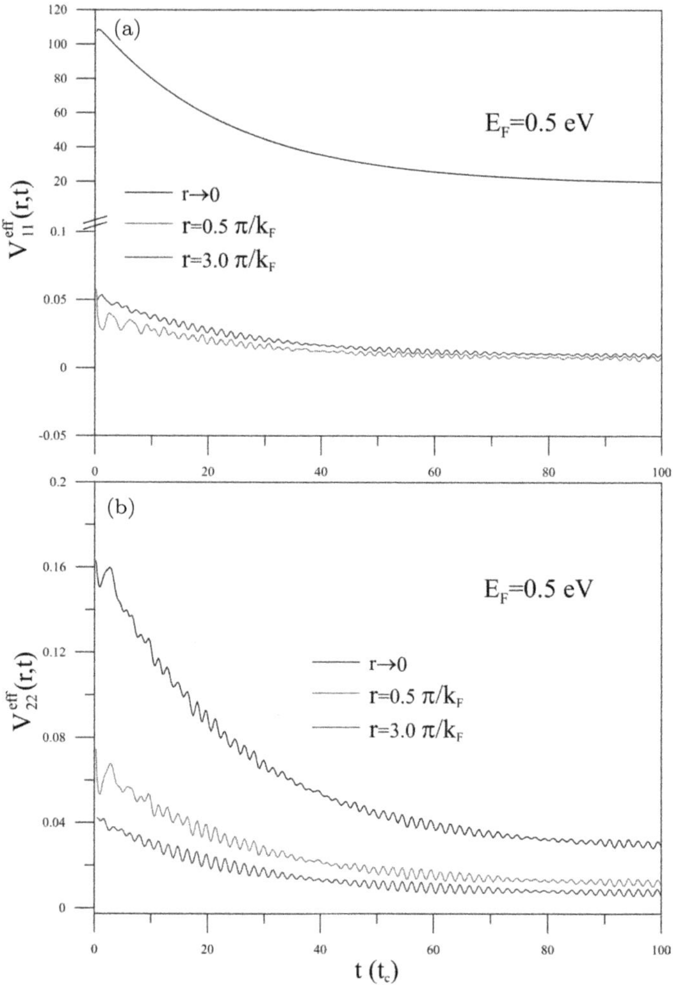


Fig. 5.10 Time variation of the effective Coulomb potential shown in Fig. 5.9.

interlayer components of the dynamic response function are responsible for the full screening effect in layered systems. If the charge location is set on the first layer, the effective potential is expressed as $V_{11}^{\text{eff}}(r,t)$, which shows the screening phenomenon in the $r$ space similar to that of monolayer graphene, such as the Friedel oscillation and the screening length. As to the effective potential on the second layer, $V_{22}^{\text{eff}}(r,t)$, it behaves much differently from $V_{11}^{\text{eff}}(r,t)$. The potential decreases monotonically with the

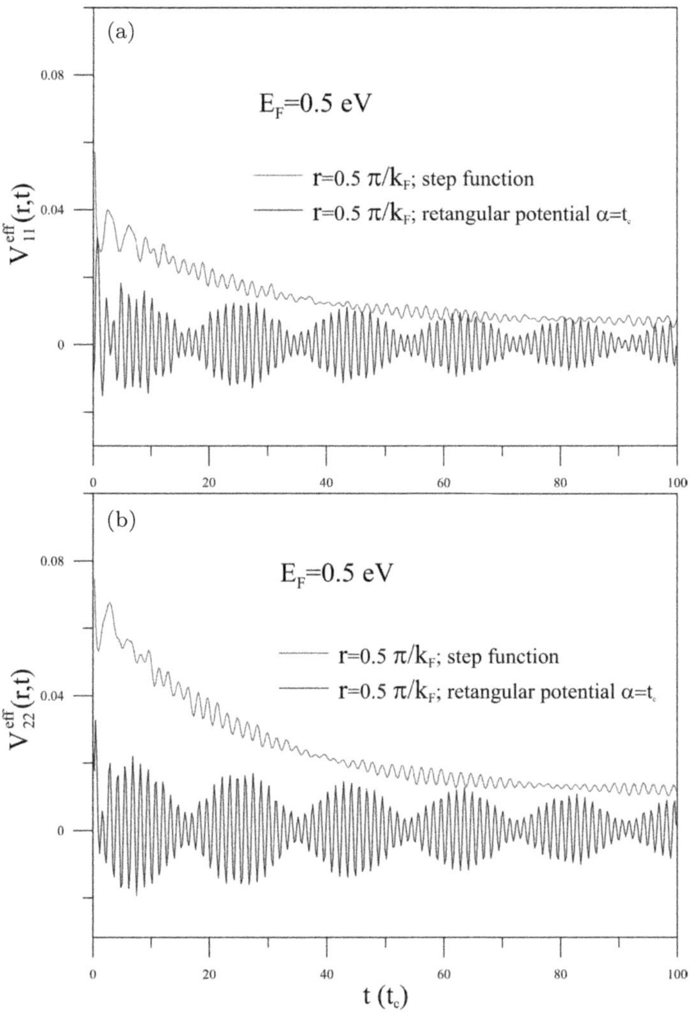


Fig. 5.11 Time-dependent effective potential on the two layers in the n-type AA-stacked bilayer graphene system under perturbation of the charge Coulomb potential of finite time.

increasing distance from the impurity. The Friedel oscillation is absent because the Coulomb interactions in the second layer are too weak to cause a significant response on the effective potential in the short $q$ limit. However, $V_{11}^{\text{eff}}(r,t)$ and $V_{22}^{\text{eff}}(r,t)$ depend on each other. The potential reduction of the first layer may be compensated by the second layer, and vice versa, in order to achieve the overall potential consistency. Compared with the monolayer graphene, the AA-stacked bilayer graphene equivalently exhibits stronger screening scaled with a larger Fermi momentum because there already exist free carriers within the two Dirac cones in the pristine case. Both $V_{11}^{\text{eff}}(r,t)$ and $V_{22}^{\text{eff}}(r,t)$ belong to a combination of the Fourier transform of single-particle and coherent charge excitations. At a sufficiently high Fermi level, $E_F = 0.5\,\text{eV}$, both $V_{11}^{\text{eff}}$ and $V_{22}^{\text{eff}}$ reflect plasma wave oscillations in the time domain (Fig. 5.10); they are coherent and originate from resonance with the two plasmon modes described in Figs. 5.3–5.4.

## 5.4 AB-Stacked Bilayer Graphenes

The AB-stacking configuration is commonly observed in natural bulk graphite, which belongs to mirror-stacking symmetry, where the two layers shift relative to each other by one C–C bond length ($b$) along the armchair direction ($\hat{x}$). When a point charge is suddenly created in one layer of a bilayer graphene sheet, the two pairs of $\pi$-parabolic bands in the AB-stacked bilayer graphene play a critical role in the propagation of the low-energy plasmon waves on graphene sheets, as shown in Figs. 3.12–3.14. $V_{11}^{\text{eff}}$ and $V_{22}^{\text{eff}}$ are responsible for the full screening effect of the interlayer and interlayer dynamic response in layered systems, reflecting distinct energies and wave functions for the electronic states. They depend on each other to achieve the consistency of the overall effective potential on the different layers. For the impurity charge on the first layer, the Friedel oscillation of the effective potential, $V_{11}^{\text{eff}}$, is clearly demonstrated for distances beyond the Thomas–Fermi screening length (Fig. 5.12(a)), accompanied by periodic oscillations in time (Fig. 5.13(a)); however, the screening phenomenon in the $r$ and $t$ domains reflects the screening effect between different stackings. The effective potential on the second layer, $V_{22}^{\text{eff}}$, reflects the screening of the free carriers on the second layer with increasing distance from the impurity. Both the cases of AA and AB stackings exhibit a monotonic decrease without the Friedel oscillations as a function of $r$, and the oscillation behavior is absent in the $r$ space. The AB-stacked bilayer graphene exhibits a stronger metallic screening ability than the AA-stacked one due to the quadratic energy dispersion of the low-lying energy electronic states, as shown by the shorter screening length. Moreover, due to the lack of interpair excitation channels, the Friedel oscillation of the AB-stacked graphene in the $r$ or $t$ space is

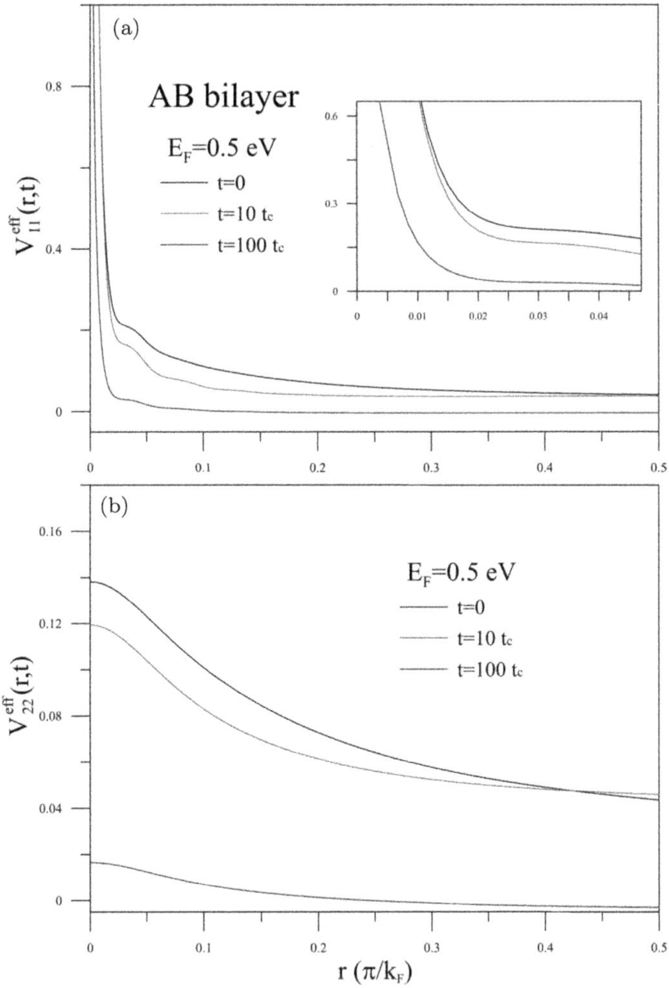


Fig. 5.12 Effective potentials on (a) the first layer $V_{11}^{\text{eff}}$ and (b) the second layer $V_{22}^{\text{eff}}$ as functions of $r$ at various $t$ in AB-stacked bilayer graphene at $E^F = 0.5\,\text{eV}$. A charged impurity is suddenly created at $r = 0$ and $t = 0$ on the first layer. The insert show the zoomed-in view of $V_{11}^{\text{eff}}$ at short distance.

more pronounced than that of the AA-stacked graphene. At a sufficiently high Fermi level, $E_F = 0.5\,\text{eV}$, the plasmon oscillations of $V_{11}^{\text{eff}}$ and $V_{22}^{\text{eff}}$ both reflect the resonance mode with the acoustic plasmon described in Chapter 3. The charge almost oscillates in space with a fixed period at a single frequency scaled by $\omega_c$. The aforementioned characteristics indicate

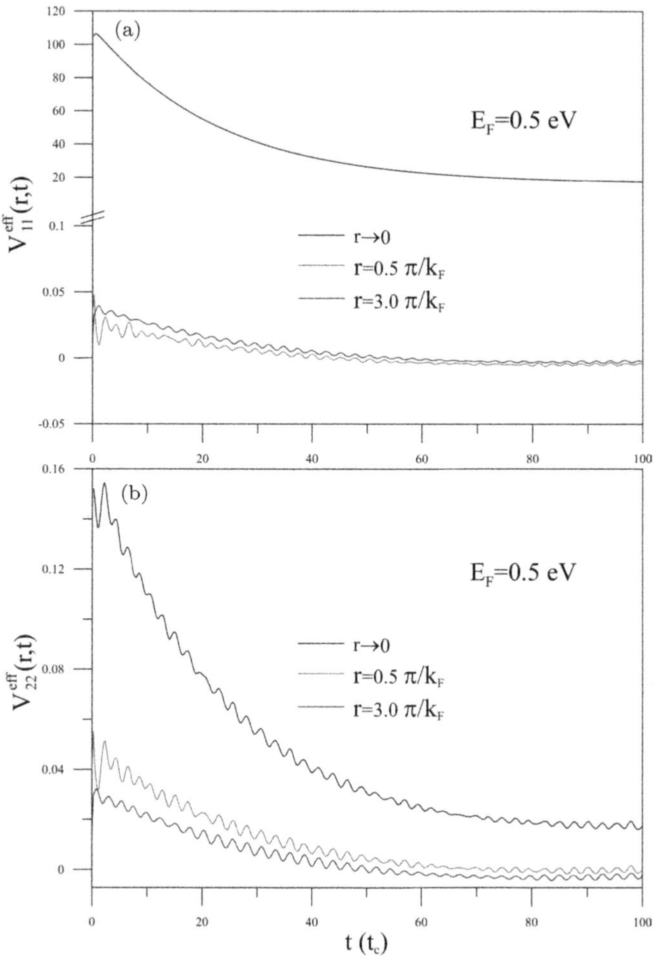


Fig. 5.13 Time variation of the effective Coulomb potential shown in Fig. 5.12.

that in the layered system, the stacking configuration, in contrast to the carrier concentration, predominates the difference in the screening phenomenon between the AA and AB systems.

The 3D, 2D and 1D electron gases, doped graphene and carbon nanotubes have some important similarities and differences in low-frequency elementary excitations and the resonant dynamic plasmon waves. Obviously, in the above five systems, the main features of charge screening are very diverse. For any small magnitude of transferred momentum, the

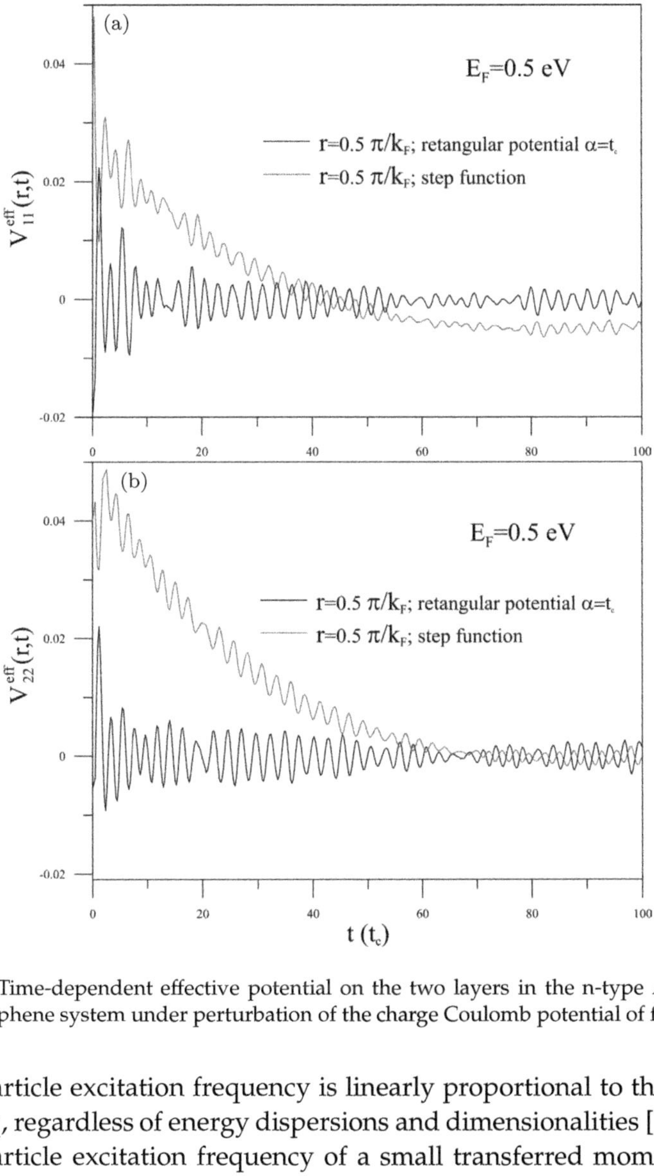


Fig. 5.14 Time-dependent effective potential on the two layers in the n-type AB-stacked bilayer graphene system under perturbation of the charge Coulomb potential of finite time.

single-particle excitation frequency is linearly proportional to the magnitude of **q**, regardless of energy dispersions and dimensionalities [106]. The single-particle excitation frequency of a small transferred momentum is linearly proportional to $|\mathbf{q}|$, independent of energy dispersion and dimensionality [106]. Generally, a substantial reduction in the real part of the dielectric function reflects the screening effect. When the real part is close to zero and there is no e–h pair excitation, coherent oscillation will occur. At $q \to 0$, the square of the excitation energy, $\omega_0^2$, is proportional to the free carrier density. However, the energy dispersions of low-energy subbands vary with dimensionality and quantum confinement. For example, 1D electron gas and carbon nanotubes show larger vacant regions among these five systems, owing mainly to the presence of only two discrete Fermi momenta in each lower-energy conduction subband. This leads to different coherent frequency dependences on $q$ in the above five systems, which are given by a constant value plus $Cq^2$, $\sqrt{q}$, $q\sqrt{|ln(qr)|}$, $\sqrt{q}$ and $q\sqrt{|ln(qr)|}$, respectively [43]. These collective excitations experience Landau damping when $q$ exceeds the critical momentum. They are caused by intra-$\pi$-band excitation, except for the 1D system, in which the cause is inter-$\pi$-band e–h excitation.

The plasmon damping in electron gas systems exceeding the critical momentum is subjected to intra-$\pi$-band e–h excitation, while that in graphene systems is caused by interband excitation. The intensity of the plasmon peak in graphene-related systems is lower than that derived from the 3D, 2D and 1D nanotube gas models, and there is no simple description to explain them. This shows that at the zero point of the real part of the dielectric function (corresponding to the collective condition), the intraband and interband single-particle excitations will further affect the main characteristics of the plasmon mode, such as frequency and intensity. When the electron doping is sufficiently high in these 1D systems, EELS confirms the predicted inter-$\pi$-band plasmon modes at high frequencies due to the population of electrons in the several $\pi$-valence subbands under the quantum confinement in cylindrical structures [99]. The latest measurement of the energy loss spectrum has successfully verified the low-energy acoustic plasmons in doped 3D semiconductors and metals [100–103, 107–109], 2D quantum wells [104, 110, 111], 1D quantum wires [105] and graphene systems [80–91], except for the experimental verification of 1D carbon nanotube positions. However, what is observed is only a single mode, and the experimental verification of other modes is helpful to indirectly infer the transmission behavior of plasma waves.

## References


[1] Cocker T.L., Peller D., Yu P., Repp J. and Huber R. 2016. Tracking the ultrafast motion of a single molecule by femtosecond orbital imaging. *Nature* **539** 263.

[2] Eisele M., Cocker T.L., Huber M.A., Plankl M., Viti L. *et al.* 2014. Ultrafast multi-terahertz nano-spectroscopy with sub-cycle temporal resolution. *Nat. Photonics* **8** 841.

[3] Sola I.J., Mével E., Elouga L., Constant E., Strelkov V. *et al.* 2006. Controlling attosecond electron dynamics by phase-stabilized polarization gating. *Nat. Phys.* **2** 319.
[4] Cavalieri A.L., Müller N., Uphues T., Yakovlev V.S., Baltuška A. *et al.* 2003. Attosecond control of electronic processes by intense light fields. *Nature* **421** 611.
[5] Baltuška A., Udem T., Uiberacker M., Hentschel M., Goulielmakis E. *et al.* 2007. Attosecond spectroscopy in condensed matter. *Nature* **449** 1029.
[6] Zahn D., Hildebrandt P.-N., Vasileiadis T., Windsor Y.W., Qi Y. *et al.* 2007. Anisotropic nonequilibrium lattice dynamics of black phosphorus. *Nano Lett.* **20** 3728.
[7] Bao J.M., Pfeiffer L.N., West K.W. and Merlin R. 2007. Ultrafast dynamic control of spin and charge density oscillations in a GaAs quantum well. *Phys. Rev. Lett.* **92** 236601.
[8] Roskos H.G., Thomson M.D., Kreß M. and Loffler T. 2007. Broadband THz emission from gas plasmas induced by femtosecond optical pulses: From fundamentals to applications. *Laser & Photon. Rev.* **4** 349.
[9] Margetic V., Pakulev A., Stockhaus A., Bolshov M., Niemax K. *et al.* 2000. A comparison of nanosecond and femtosecond laser-induced plasma spectroscopy of brass samples. *Spectrochim. Acta B* **55** 31771.
[10] Cocker T.L., Jelic V., Gupta M., Molesky S.J. and Jacob A.J. 2013. An ultrafast terahertz scanning tunnelling microscope. *Nat. Photonics* **7** 620.
[11] Terada Y., Yoshida S., Takeuchi O. and Shigekawa H. 2010. Real-space imaging of transient carrier dynamics by nanoscale pump-probe microscopy. *Nat. Photonics* **4** 869.
[12] Jiang T., Kravtsov V., Tokman M., Belyanin A. and Raschke M.B. 2019. Ultrafast coherent nonlinear nanooptics and nanoimaging of graphene. *Nat. Nanotechnol.* **14** 838.
[13] Fei Z., Rodin A.S., Andreev G.O., Bao W., McLeod A.S. *et al.* 2012. Gate-tuning of graphene plasmons revealed by infrared nano-imaging. *Nature* **487** 82.
[14] Ma W., Alonso-González P., Li S., Nikitin A.Y., Yuan J. *et al.* 2018. In-plane anisotropic and ultra-low-loss polaritons in a natural van der Waals crystal. *Nature* **562** 557.
[15] Ni G.X., McLeod A.S., Sun Z., Wang L., Xiong L. *et al.* 2018. Fundamental limits to graphene plasmonics. *Nature* **557** 530.
[16] Wagner M., Fei Z., McLeod A.S., Rodin A.S. *et al.* 2018. Ultrafast and nanoscale plasmonic phenomena in exfoliated graphene revealed by infrared pump-probe nanoscopy. *Nano Lett.* **14** 894.
[17] Riechers K., Hueck K., Luick N., Lompe T., Moritz H. *et al.* 2017. Detecting Friedel oscillations in ultracold Fermi gases. *Eur. Phys. J. D* **71** 232.
[18] Simon L., Bena C., Vonau F., Cranney M., Aubel D. *et al.* 2011. Fourier-transform scanning tunnelling spectroscopy: The possibility to obtain constant-energy maps and band dispersion using a local measurement. *J. Phys. D Appl. Phys.* **44** 464010.

[19] Goda K. and Jalali B. 2013. Dispersive Fourier transformation for fast continuous single-shot measurements. *Nat. Photonics* **7** 102.
[20] Canright G.S. 1988. Time-dependent screening in the electron gas. *Phys. Rev. B* **38** 1647.
[21] Stern F. 1967. Polarizability of a two-dimensional electron gas. *Phys. Rev. Lett.* **18** 546.
[22] Bagchi A. 1977. Transverse dielectric response of a semi-infinite metal: Surface effect. *Phys. Rev. B* **15** 3060.
[23] Kittel C. 2001. *Introduction to Solid State Physics*, 8th ed. (London: Wiley).
[24] Shyu F.L. and Lin M.F. 2000. Plasmons and optical properties of semimetal graphite. *J. Phys. Soc. Jpn.* **69** 3781.
[25] Chiu C.W., Lin M.F. and Shyu F.L. 2011. Electronic specific heat of nanographite ribbons. *Physica E* **11** 356.
[26] Yannouleas C., Bogachek E.N. and Landman U. 1996. Collective excitations of multishell carbon microstructures: Multishell fullerenes and coaxial nanotubes. *Phys. Rev. B* **53** 10225.
[27] Huber R., Tauser F., Brodschelm A., Bichler M., Abstreiter G. *et al.* 2001. How many-particle interactions develop after ultrafast excitation of an electron-hole plasma. *Nature* **414** 286.
[28] Shyu F.L. and Lin M.F. 2002. Electronic and optical properties of narrow-gap carbon nanotubes. *J. Phys. Soc. Jpn.* **71** 1820.
[29] Lin M.F. and Shung K.W.-K. 1994. Plasmons and optical properties of carbon nanotubes. *Phys. Rev. B* **50** 17744.
[30] Lu C.L., Chang C.P., Huang Y.C., Chen R.B. and Lin M.F. 2006. Influence of an electric field on the optical properties of few-layer graphene with AB stacking. *Phys. Rev. B* **73** 144427.
[31] Chang C.P., Wang J., Lu C.L., Huang Y.C., Lin M.F. and Chen R.B. 2008. Optical properties of simple hexagonal and rhombohedral few-layer graphenes in an electric field. *J. Appl. Phys.* **103** 103109.
[32] Thurakitseree T., Kramberger C., Kumamoto A., Chiashi S., Einarsson E. and Maruyama S. 2013. Reversible diameter modulation of single-walled carbon nanotubes by acetonitrile-containing feedstock. *ACS Nano* **7** 2205.
[33] Puretzky A.A., Liang L., Li X., Xiao K., Wang K., Mahjouri-Samani M. *et al.* 2015. Low-frequency Raman fingerprints of two-dimensional metal dichalcogenide layer stacking configurations. *ACS Nano* **9** 6333.
[34] Knap W., Dyakonov M., Coquillat D., Teppe F., Dyakonova N., Lusakowski J. *et al.* 2009. Field effect transistors for terahertz detection: Physics and first imaging applications. *J. Infrared Millim. Terahertz Waves* **30** 1319.
[35] Caruso F., Novko D. and Draxl C. 2018. Phonon-assisted damping of plasmons in three- and two-dimensional metals. *Phys. Rev. B* **97** 205118.
[36] Torres J.A., Pascual J.I. and Saenz J.J. 1994. Theory of conduction through narrow constrictions in a three-dimensional electron gas. *Phys. Rev. B* **49** 16581.
[37] Halperin B.I. 1987. Possible states for a three-dimensional electron gas in a strong magnetic field. *Jpn. J. Appl. Phys.* **26** 1913.

[38] Perdew J.P. and Wang Y. 1992. Accurate and simple analytic representation of the electron-gas correlation energy. *Phys. Rev. B* **45** 13244.
[39] Ohmura Y. and Matsudaira N. 1992. Influence of Coulomb correlation on the scattering of X-ray by an electron gas. I. RPA approximation. *J. Phys. Soc. Jpn.* **19** 1355.
[40] Wang Y. and Perdew J.P. 1991. Correlation hole of the spin-polarized electron gas, with exact small-wave-vector and high-density scaling. *Phys. Rev. B* **44** 13298.
[41] Yasuhara H. 1974. Electron correlations and dielectric function of electron gas at metallic densities. *J. Phys. Soc. Jpn.* **26** 361.
[42] Zheng L. and MacDonald A.H. 1994. Correlation in double-layer two-dimensional electron-gas systems: Singwi-Tosi-Land-Sjölander theory at B=0. *Phys. Rev. B* **49** 5522.
[43] Giuliani G.F. and Quinn J.J. 1982. Lifetime of a quasiparticle in a two-dimensional electron gas. *Phys. Rev. B* **26** 4421.
[44] Nozières P. and Pines D. 1958. Correlation energy of a free electron gas. *Phys. Rev* **111** 4421.
[45] Dobson J.F. and Wang J. 2000. Energy-optimized local exchange-correlation kernel for the electron gas: Application to van der Waals forces. *Phys. Rev. B* **62** 10038.
[46] Czachor A., Holas A., Sharma S.R. and Singwi K.S. 1982. Dynamical correlations in a two-dimensional electron gas: First-order perturbation theory. *Phys. Rev. B* **25** 2144.
[47] van Wees B.J., van Houten H., Beenakker C.W.J., Williamson J.G., Kouwenhoven L.P. *et al.* 1988. Quantized conductance of point contacts in a two-dimensional electron gas. *Phys. Rev. Lett.* **60** 848.
[48] Naik G.V., Shalaev V.M. and Boltasseva A. 2013. Alternative plasmonic materials: Beyond gold and silver. *Adv. Mater.* **25** 3264.
[49] Brown S.A., Young J.F., Brum J.A., Hawrylak P. and Wasilewski Z. 1996. Evolution of the interband absorption threshold with the density of a two-dimensional electron gas. *Phys. Rev. B* **54** R11082(R).
[50] Pinczuk A., Schmitt-Rink S., Danan G., Valladares J.P., Pfeiffer L.N. *et al.* 1989. Large exchange interactions in the electron gas of GaAs quantum wells. *Phys. Rev. Lett.* **63** 1633.
[51] Huotari S., Soininen J.A., Pylkkänen T., Hämäläinen K., Issolah A. *et al.* 2010. Momentum distribution and renormalization factor in sodium and the electron gas. *Phys. Rev. Lett.* **105** 086403.
[52] Bürgi L., Jeandupeux O., Brune H., Kern K. *et al.* 1999. Probing hot-electron dynamics at surfaces with a cold scanning tunneling microscope. *Phys. Rev. Lett.* **82** 22.
[53] Kliewer J., Berndt R., Chulkov E.V., Silkin V.M., Echenique P.M. *et al.* 1999. Dimensionality effects in the lifetime of surface states. *Science* **288** 1399.
[54] Millard I.S., Patel N.K., Foden C.L., Linfield E.H., Simmons M.Y. *et al.* 2010. Effect of finite quantum-well width on the compressibility of a two-dimensional electron gas. *Phys. Rev. B* **55** 6715.

[55] Lei X.L., Liu S.Y. *et al.* 2003. Radiation-induced magnetoresistance oscillation in a two-dimensional electron gas in Faraday geometry. *Phys. Rev. Lett.* **91** 226805.
[56] Lin M.F. and Shung K.W.-K. 1993. Elementary excitations in cylindrical tubules. *Phys. Rev. B* **47** 6617.
[57] Ehrenreich H. and Cohen M.H. 1959. Self-consistent field approach to the many-electron problem. *Phys. Rev.* **115** 786.
[58] Zerah G. and Hansen J.P. 1986. Self-consistent integral equations for fluid pair distribution functions: Another attempt. *J. Chem. Phys.* **84** 2336.
[59] Pini D., Stell G. and Wilding N.B. 1986. A liquid-state theory that remains successful in the critical region. *Mol. Phys.* **95** 483.
[60] Duh D.M. and Haymet A.D.J. 1986. A integral equation theory for uncharged liquids: The Lennard-Jones fluid and the bridge function. *J. Chem. Phys.* **103** 2625.
[61] Barrat J.-L., Hansen J.-P. and Pastore G. 1986. Factorization of the triplet direct correlation function in dense fluids. *Phys. Rev. Lett.* **58** 2075.
[62] Zhao Q., Morrison R.C. and Parr R.G. 1994. From electron densities to Kohn-Sham kinetic energies, orbital energies, exchange-correlation potentials, and exchange-correlation energies. *Phys. Rev. A* **50** 2138.
[63] Chiu C.W., Chung Y.L., Yang C.H., Liu C.T. and Lin C.Y. 2020. Coulomb decay rates in monolayer doped graphene. *RSC Adv.* **10** 2337.
[64] Lin C.Y., Wu J.Y., Chiu C.W. and Lin M.F. 2019. *Coulomb Excitations and Decays in Graphene-Related Systems* (CRC Press).
[65] Hwang E.H., Hu B.Y.K. and Das Sarma S. 2020. Inelastic carrier lifetime in graphene. *Phys. Rev. B* **76** 115434.
[66] Johnson K. and Zumino B. 1959. Gauge fependence of the wave-function renormalization constant in quantum electrodynamics. *Phys. Rev. Lett.* **3** 351.
[67] Johnson K., Baker M. and Willey R. 1964. Self-energy of the electron. *Phys. Rev.* **136** B1111.
[68] Baker M. and Johnson K. 1971. Asymptotic form of the electron propagator and the self-mass of the electron. *Phys. Rev. D* **3** 2516.
[69] Sprunger P.T., Petersen L., Plummer E.W., Lægsgaard E. and Besenbacher F. 1997. Giant friedel oscillations on the beryllium(0001) surface. *Science* **275** 1764.
[70] Khrapak S.A., Ivlev A.V., Morfill G.E. and Zhdanov S.K. 2003. Scattering in the attractive Yukawa potential in the limit of strong interaction. *Phys. Rev. Lett.* **90** 225002.
[71] Simion G.E. and Giuliani G.F. 2005. Friedel oscillations in a Fermi liquid. *Phys. Rev. B* **72** 045127.
[72] Lin M.F. and Chuu D.S. 1997. Impurity screening in carbon nanotubes. *Phys. Rev. B* **56** 4996.
[73] Mahan G.D. 1990. *Many-Particle Physics* (New York: Plenum).
[74] Geim A.K. and Novoselov K.S. 2007. The rise of graphene. *Nat. Mater.* **6** 183.
[75] Novoselov K.S., Geim A.K., Morozov S.V., Jiang D., Zhang Y. *et al.* 2004. Electric field effect in atomically thin carbon films. *Science* **306** 666.

[76] Schedin F., Geim A.K., Morozov S.V., Hill E.W., Blake P. *et al.* 2007. Detection of individual gas molecules adsorbed on graphene. *Nat. Mater.* **6** 652.
[77] Gierz I., Riedl C., Starke U., Ast C.R. and Kern K. 2008. Atomic hole doping of graphene. *Nano Lett.* **8** 4603.
[78] Rahnejat K.C., Howard C.A., Shuttleworth N.E., Schofield S.R., Iwaya K. *et al.* 2011. Charge density waves in the graphene sheets of the superconductor $CaC_6$. *Nat. Commun.* **2** 558.
[79] Ruckhofer A., Benedek G., Bremholm M., Ernst W.E., Tamtogl A. *et al.* 2022. Observation of Dirac charge-density waves in $Bi_2Te_2Se$. *Nanomater.* **12,** 504.
[80] Shin S.Y., Kim N.D., Kim J.G., Kim K.S., Noh D.Y. *et al.* 2011. Control of the pi plasmon in a single layer graphene by charge doping. *App. Phys. Lett.* **99** 082110.
[81] Zhang L., Schwertfager N., Cheiwchanchamnangij T., Lin X., Glans-Suzuki P.-A. *et al.* 2012. Electronic band structure of graphene from resonant soft x-ray spectroscopy: The role of core-hole effects. *Phys. Rev. B* **86** 245430.
[82] Hage F.S., Hardcastle T.P., Gjerding M.N., Kepaptsoglou D.M., Seabourne C.R. *et al.* 2018. Local plasmon engineering in doped graphene. *ACS Nano* **12** 1837.
[83] Jovanović V.B., Radović I., Borka D. and Mišković Z.L. 2011. High-energy plasmon spectroscopy of freestanding multilayer graphene. *Phys. Rev. B* **84** 155416.
[84] Luo X., Qiu T., Lu W. and Ni Z. 2013. Plasmons in graphene: Recent progress and applications. *Mater. Sci. Eng. R* **74** 351.
[85] Eberlein T., Bangert U., Nair R.R., Jones R., Gass M. *et al.* 2008. Plasmon spectroscopy of free-standing graphene films. *Phys. Rev. B* **77** 233406.
[86] Wachsmuth P., Hambach R., Kinyanjui M.K., Guzzo M., Benner G. *et al.* 2013. High-energy collective electronic excitations in free-standing single-layer graphene. *Phys. Rev. B* **88** 075433.
[87] Politanoa A., Radović I., Borkab D., Mišković Z.L., Yu H.K. *et al.* 2017. Dispersion and damping of the interband $\pi$ plasmon in graphene grown on Cu(111) foils. *Carbon* **114** 70.
[88] Liou S.C., Shie C.-S., Chen C.H., Breitwieser R., Pai W.W. *et al.* 2015. $\pi$-plasmon dispersion in free-standing graphene by momentum-resolved electron energy-loss spectroscopy. *Phys. Rev. B* **91** 045418.
[89] Lu J., Loh K.P., Huang H., Chen W. and Wee A.T.S. 2009. Plasmon dispersion on epitaxial graphene studied using high-resolution electron energy-loss spectroscopy. *Phys. Rev. B* **80** 113410.
[90] Nelson F.J., Idrobo J.-C., Fite J.D., Miskovic Z.L., Pennycook S.J. *et al.* 2014. Electronic excitations in graphene in the 1–50 eV range: The $\pi$ and $\pi + \sigma$ peaks are not plasmons. *Nano Lett.* **14** 3827.
[91] Politano A., Marino A.R., Formoso V., Farías D., Miranda R. *et al.* 2011. Evidence for acoustic-like plasmons on epitaxial graphene on Pt(111). *Phys. Rev. B* **84** 033401.

[92] Gao X., Burns C., Casa D., Upton M., Gog T. *et al.* 2011. Development of a graphite polarization analyzer for resonant inelastic x-ray scattering. *Rev. Sci. Instrum.* **82** 113108.
[93] Devereaux T.P. 2007. Inelastic light scattering from correlated electrons. *Rev. Mod. Phys.* **79** 175.
[94] Ament L.J.P., van Veenendaal M., Devereaux T.P., Hill J.P. and van den Brink J. 2001. Resonant inelastic x-ray scattering studies of elementary excitations. *Rev. Mod. Phys.* **83** 705.
[95] Hambach R., Giorgetti C., Hiraoka N., Cai Y.Q., Sottile F. *et al.* 2008. Anomalous angular dependence of the dynamic structure factor near Bragg reflections: Graphite. *Phys. Rev. Lett.* **101** 266406.
[96] Brink H.A., Barfels M.M.G., Burgner R.P. and Edwards B.N. 2003. A sub-50 meV spectrometer and energy filter for use in combination with 200 kV monochromated (S)TEMs. *Ultramicroscopy* **96** 367.
[97] Egerton R.F. 1996. *Electron Energy-Loss Spectroscopy in the Electron Microscope*, p. 370 (New York and London: Plenum).
[98] Pichler T., Knupfer M., Golden M.S., Fink J., Rinzler A. *et al.* 1998. Localized and delocalized electronic states in single-wall carbon nanotubes. *Phys. Rev. Lett.* **80** 4729.
[99] Tuan D. and Khanh N.Q. 2013. Plasmon modes of double-layer graphene at finite temperature. *Phys. E Low Dimens. Syst. Nanostruct.* **54** 267.
[100] Cox P.A., Egdell R.G., Goodenough J.B., Hamnett A. and Naish C.C. 1983. The metal-to-semiconductor transition in ternary ruthenium (IV) oxides: A study by electron spectroscopy. *J. Phys. C: Solid State Phys.* **16** 6221.
[101] Huotari S., Albergamo F., Vanko G., Verbeni R. and Monaco G. 2006. Resonant inelastic hard x-ray scattering with diced analyzer crystals and positionsensitive detectors. *Rev. Sci. Instrum.* **77** 053102.
[102] Kotani A. and Shin S. 2001. Resonant inelastic x-ray scattering spectra for electrons in solids. *Rev. Mod. Phys.* **73** 203.
[103] Mitterbauer C., Kothleitner G., Grogger W., Zandbergen H., Freitag B. *et al.* 2003. Electron energy-loss near-edge structures of 3d transition metal oxides recorded at high-energy resolution. *Ultramicroscopy* **96** 469.
[104] Blancon J.-C., Stier A.V., Tsai H., Nie W., Stoumpos C.C. *et al.* 2018. Scaling law for excitons in 2D perovskite quantum wells. *Nat. Commun.* **9** 2254.
[105] Barati M. and Aghajamali A. 2016. Near-infrared tunable narrow filter properties in a 1D photonic crystal containing semiconductor metamaterial photonic quantum-well defect. *Phys. E Low Dimens. Syst. Nanostruct.* **79** 20.
[106] Shung K.W.-K. 1986. Dielectric function and plasmon structure of stage-1 intercalated graphite. *Phys. Rev. B* **34** 979.
[107] Mohr M., Maultzsch J., Dobardžić E., Reich S., Milošević I. *et al.* 2007. Phonon dispersion of graphite by inelastic x-ray scattering, phonon dispersion of graphite by inelastic x-ray scattering. *Phys. Rev. B* **76** 035439.
[108] Kimura K., Matsuda K., Hiraoka N., Fukumaru T., Kajihara Y. *et al.* 2014. Inelastic x-ray scattering study of plasmon dispersions in solid and liquid Rb. *Phys. Rev. B* **89** 014206.

[109] Tirao G., Stutz G., Silkin V.M., Chulkov E.V. and Cusatis C. 2007. Plasmon excitation in beryllium: Inelastic x-ray scattering experiments and first-principles calculations. *J. Phys. Condens. Matter* **19** 046207.
[110] Galambosi S., Soininen J.A., Mattila A., Huotari S., Manninen S. *et al.* 2005. Inelastic x-ray scattering study of collective electron excitations in $MgB_2$. *Phys. Rev. B* **71** 060504.
[111] Diaconescu B., Pohl K., Vattuone L., Savio L., Hofmann P. *et al.* 2007. Low-energy acoustic plasmons at metal surfaces. *Nature* **448** 57.

Chapter 6

# OPTICAL SCATTERINGS IN LAYERED SYSTEMS

Chih-Wei Chiu, Lu-Yao Wang, Hsien-Ching Chung, Ching-Hong Ho, Chiun-Yan Lin, and Ming-Fa Lin

Layered graphene systems are among the mainstream materials [1–19] owing to their various crystal symmetries. Modifying the number of layers, stacking configuration and external fields can lead to diverse phenomena [3, 20], e.g., easy modulation of vertical optical transitions. There are a number of theoretical [20–39] and experimental studies [40–57] on few-layer graphene systems. Their main focus is on low-frequency absorption structures in the presence/absence of gate voltage and magnetic field. Specifically, the magneto-optical theory is developed through the generalized tight-binding model based on the two sublattices of each honeycomb crystal. Systematic investigations for diversified magnetic quantization have been done for 2D materials [20, 21], such as the unusual Landau levels and magneto-optical selection rules. This theoretical framework is useful in fully understanding the electronic [1, 2], magnetic [21, 40], optical [20–23, 40–44], transport [58] and Coulomb-excitation [59] properties under intrinsic atomic interactions and field perturbations. However, the Kubo formula is not enough for exploring the reflectance and transmittance spectra of thin films with prominent boundary effects. In this work, an optical scattering model is developed within the previous framework. The calculated results could be examined through high-resolution optical measurements [40–57].

The $\pi$-electronic states of $sp^2$-bonding systems [1–3] are responsible for the essential properties, such as in graphenes [1, 60], carbon nanotubes

[61–79], graphene nanoribbons [80–99], onions [100–106] and fullerenes [107–116]. More interestingly, such carriers dominate the dynamic/static screening responses under significant perturbation of an electromagnetic/Coulomb field [20, 59], as clearly revealed in the rich absorption spectra of few-layer graphene systems [40–42]. The inter-$\pi$-band transitions, which correspond to the occupied valence subbands and the unoccupied conduction ones, can absorb photon energies and thus induce an obvious decline in the propagating electromagnetic wave. The close relations between charges and fields are well characterized by the transverse dielectric function [117]. $\epsilon(\omega)$ of a multi-/few-layer graphene system will be evaluated under the generalized tight-binding model [59], which is suitable under perpendicular gate voltages and uniform magnetic fields.

In this work, a full optical theory is developed for the 2D layered graphene systems. Their band structures and wave functions are calculated within the generalized tight-binding model in the presence/absence of external fields [20, 59]. They are responsible for the dynamic charge screenings of an electromagnetic field through the evaluated dielectric function at the long-wavelength limit. The perturbation of an electric dipole moment is calculated using the gradient approximation [20, 21]. Through the direct combination of $\epsilon(\omega)$ and two boundary conditions at the upper and lower surfaces, the whole propagating path can be characterized by the optical excitation process inside the multi-layer honeycomb lattices and the continuous boundary scatterings of incidence, reflection and refraction. As a result, both reflectance and transmittance are expressed as analytic formulas, encompassing useful information about crystal symmetries [118] and the main features of electronic properties [20, 118, 119].

## 6.1 Theories of Optical Scatterings

For few-layer graphenes with AA, AB and ABC stackings (Figs. 6.1(c)–6.1(e)), the definition of the geometric structure and the development of the tight-binding model to calculate the $\pi$ bands are well established by previous works [128–131]. As for AAB-stacked systems of any layer number ($N_L$), the $(3N+1)$th and $(3N+2)$th ($(3N+1)$th and $(3N+3)$th) layers (where $I$ is a natural number), as clearly shown in Fig. 6.1(f), are arranged in the AA (AB) stacking. The interlayer distance and the C–C bond length are, respectively, $c = 3.35\,\text{Å}$ and $b' = 1.42\,\text{Å}$. There are $2N_L$ carbon atoms in a primitive unit cell. The zero-field Hamiltonian, which is built using the $2N_L$ tight-binding functions of the $2p_z$ orbitals, is dominated by the intralayer and interlayer atomic interactions $r_i$. There exist nine kinds of atom–atom interactions, which correspond to the nine atomic hopping integrals in the Hamiltonian matrix. $r_0 = -2.569\,\text{eV}$ represents the nearest-neighbor intralayer atomic interaction; $r_1 = -0.263\,\text{eV}$ and $r_2 = 0.032\,\text{eV}$ represent the interlayer atomic interactions between the $(3N+1)$th and $(3N+2)$th layers; $r_3 = -0.413\,\text{eV}$, $r_4 = -0.177\,\text{eV}$ and $r_5 = -0.119\,\text{eV}$ are associated with the interlayer atomic interactions between the $(3N+1)$th and $(3N+3)$th layers; $r_6 = -0.013\,\text{eV}$, $r_7 = -0.0177\,\text{eV}$ and $r_8 = -0.0119\,\text{eV}$ refer to the interlayer atomic interactions between the $(3N+1)$th and $(3N+3)$th layers. The hopping integrals $r_1$, $r_3$ and $r_6$ belong to the vertical interlayer atomic interactions, whereas all others are non-vertical. For $N_L$-layer AAB-stacked graphene, the C-$2p_z$-based Hamiltonian, the $2N_L \times 2N_L$ Hermitian matrix, could be expressed as

$$H_{l,m} = \begin{cases} r_0 f & \text{if } m = l+1,\ m = 6N+1\ \&\ 6N+3 \\ r_2 f^* & \text{if } m = l+1,\ m = 6N+2 \\ r_4 f^* & \text{if } m = l+1,\ m = 6N+4; \\ & \quad m = l+3,\ m = 6N+3 \\ r_0 f^* & \text{if } m = l+1,\ m = 6N+5 \\ r_4 f & \text{if } m = l+1,\ m = 6N+6; \\ & \quad m = l+3,\ m = 6N+5 \\ r_1 & \text{if } m = l+2,\ m = 6N+1\ \&\ 6N+2 \\ r_3 & \text{if } m = l+2,\ m = 6N+3\ \&\ 6N+5 \\ r_5 f & \text{if } m = l+2,\ m = 6N+4 \\ r_5 f^* & \text{if } m = l+2,\ m = 6N+6 \\ r_2 f & \text{if } m = l+3,\ m = 6N+1 \\ r_7 f^* & \text{if } m = l+3,\ m = 6N+2\ \&\ 6N+4; \\ & \quad m = l+5,\ m = 6N+1 \\ r_7 f & \text{if } m = l+3,\ m = 6N+6; \\ & \quad m = l+5,\ m = 6N+3\ \&\ 6N+5 \\ r_6 & \text{if } m = l+4, \\ & \quad m = 6N+1,\ 6N+3,\ 6N+4,\ 6N+5 \\ r_8 f & \text{if } m = l+4,\ m = 6N+2 \\ r_8 f^* & \text{if } m = l+4,\ m = 6N+6 \\ 0 & \text{otherwise,} \end{cases} \quad (1)$$

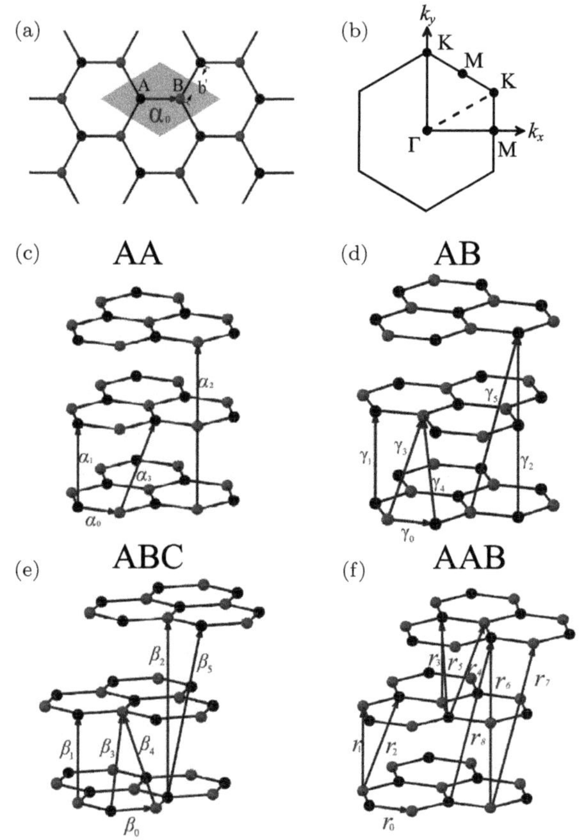


Fig. 6.1 (a) Honeycomb lattice structure of monolayer graphene. (b) The first Brillouin zone of the layered graphene with any stacking configuration. $K$, $M$ and $\Gamma$ are the symmetric points for the state energy. The geometric structures of (c) AA-, (d) AB-, (e) ABC- and (f) AAB-stacked graphenes. The two sublattices are indicated by black and red colors. $b'$ represents the carbon–carbon bond length. $\alpha$s, $\gamma$s, $\beta$s and $r$s are the atomic hopping integrals indicated by blue arrows.

where $f = \exp(ik_xb) + cos(k_y\sqrt{3}b/2)\exp(-ik_yb/2)$. Similar forms are revealed for AA, AB and ABC [120–123] but under quite different interlayer hopping integrals ($\alpha_i$s, $\gamma_i$s and $\beta_i$s) shown in Figs. 6.1(c)–6.1(e). The diverse band structures are easily modulated by the layer number and the stacking configuration. Apparently, these features strongly affect the optical excitation responses and thus the transmittance and reflectance.

When any condensed matter system is present in an electromagnetic field, all the charges very efficiently screen the external perturbations by creating an induced current density. Their electronic states are vertically excited from the occupied to the unoccupied ones through the quasiparticle absorption process [125]. That is, a valence electron absorbs one photon, becoming an excited conduction one. The electric dipole is responsible for driving all the optical transition channels. Through delicate calculations of its inner products, the imaginary-part dielectric function could be evaluated using the Fermi golden rule, which is expressed as

$$Im[\epsilon(\omega)] = \frac{8\pi e^2}{N_L} \sum_{h,h'} \int_{1stBZ} \frac{d\mathbf{k}}{(2\pi)^2} \frac{\left|\left\langle \Psi_{\mathbf{k}}^{h'} \middle| \frac{\hat{\mathbf{E}}\cdot\mathbf{P}}{m_e} \middle| \Psi_{\mathbf{k}}^{h} \right\rangle\right|^2}{\omega_{\mathrm{ex}}^2(\mathbf{k})} \times \frac{\left[n_F(E^{h'}(\mathbf{k})) - n_F(E^{h}(\mathbf{k}))\right]\delta}{[\omega_{\mathrm{ex}}(\mathbf{k}) - \omega]^2 - \delta^2}, \tag{2}$$

where $h$ ($h'$) represents the initial (final) state, $n_F(E(\mathbf{k}))$ is the Fermi–Dirac distribution function and $\delta$ ($= 0.01\,\mathrm{eV}$), the energy width due to various de-excitation mechanisms, is treated as a free parameter in the calculations. The excitation energy $\omega_{\mathrm{ex}}(\mathbf{k}) = E^{h'}(\mathbf{k}) - E^{h}(\mathbf{k})$ arises from the same and different subbands. Furthermore, the real-part dielectric function is calculated using the Kramers–Kronig relations (the principle-value integrations) [125]. Equation (2) covers the joint density of states and the velocity matrix elements, respectively, dominating the number of excitation channels and scattering amplitudes (optical selection rules) [20–23]. The gradient approximation is very convenient and reliable for resolving the second term, e.g., the successful evaluations in graphite intercalation compounds [126], carbon nanotubes [127] and few-layer graphenes [20, 21, 128, 129, 131]. This linear Kubo formula, which is directly combined with the generalized model, has been successfully used to explore the diverse magneto-optical excitations of 2D layered structures, e.g., the systematic investigations of magnetic quantization phenomena of layered Group-IV and Group-V materials [132].

Notably, the significant couplings of carrier charges and electromagnetic waves are rather sensitive to position and time. In general, their dynamic screening and propagating behaviors are dominated by the Helmholtz equation [125]. The linear energy dispersion relation in free space would become a nonlinear one of $\epsilon(\omega)\,\omega^2 = c^2k^2$ in any material. The real and imaginary parts of the dielectric function could lead to a reduction in the group velocity and spectrum intensity, mainly owing to the imaginary wavevector through the specific relations $k = [\omega/c]N(\omega)$

and $N(\omega) = \sqrt{\epsilon(\omega)} = n(\omega) + i\kappa(\omega)$ [125]. $n$ and $\kappa$, respectively, represent the refractive index and extinction coefficient. The absorption coefficient, $\alpha = [\kappa\,\omega/c]^{-1}$, is very useful in understanding the characteristic decay length under the efficient photon absorption of the valence quasiparticles. Apparently, layered graphene systems provide effective inter-$\pi$-band transitions in particle–field couplings (discussed in the following section), even dominating the continuous scatterings at the upper and lower boundaries. It should be noted that such $\pi$ valence electrons absorb the electromagnetic-field energy and then release it to a heat reservoir by gradually increasing the temperature. The energy transfer between them is an interesting open issue [133, 134].

An electromagnetic wave with $\mathbf{E_I} = E_I \exp[i(\frac{\omega}{c}z - \omega t)]\,\hat{x}$ and $\mathbf{B_I} = E_I \exp[i(\frac{\omega}{c}z - \omega t)]\,\hat{y}$ in a free space (a dielectric constant of $\epsilon_0$) is assumed to be normally incident on the upper surface of a finite-width sample, as clearly illustrated in Fig. 6.2. Electric and magnetic fields

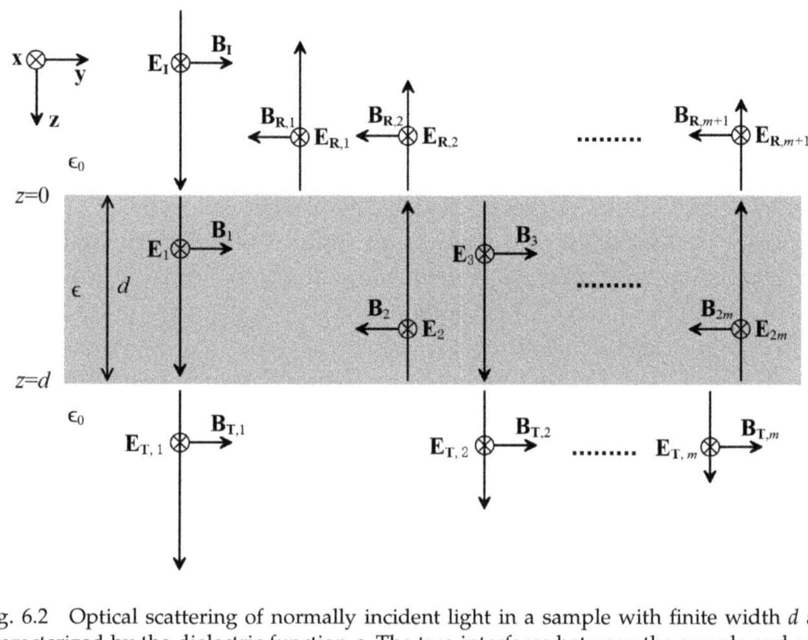


Fig. 6.2 Optical scattering of normally incident light in a sample with finite width $d$ and characterized by the dielectric function $\epsilon$. The two interfaces between the sample and environment are located in the $x$–$y$ plane at $z = 0$ and $z = d$. The reflection and transmission of an electromagnetic wave with the electric field **E** and magnetic field **B** are perpendicular to and lie in the incident plane, respectively. The subscripts I, R and T stand for the incident, reflected and transmitted waves, respectively; the Arabic numerals represent the number of times scattering occurred on the interface.

behave as

$$
\begin{aligned}
&\mathbf{E}_{\mathbf{R},j} = E_{\mathrm{R},j} \exp\left[i\left(-\frac{\omega}{c}z - \omega t\right)\right] \hat{x} \quad \& \\
&\mathbf{B}_{\mathbf{R},j} = -E_{\mathrm{R},j} \exp\left[i\left(-\frac{\omega}{c}z - \omega t\right)\right] \hat{y} && \text{for } z < 0, \\
&\mathbf{E}_j = E_j \exp\left\{i\left[(-1)^{j-1}\frac{\omega}{c}Nz - \omega t\right]\right\} \hat{x} \quad \& \\
&\mathbf{B}_j = (-1)^{j-1}E_j \exp\left\{i\left[(-1)^{j-1}\frac{\omega}{c}Nz - \omega t\right]\right\} \hat{y} && \text{for } 0 < z < d, \\
&\mathbf{E}_{\mathbf{T},j} = E_{\mathrm{T},j} \exp\left[i\left(\frac{\omega}{c}z - \omega t\right)\right] \hat{x} \quad \& \\
&\mathbf{B}_{\mathbf{T},j} = E_{\mathrm{T},j} \exp\left[i\left(\frac{\omega}{c}z - \omega t\right)\right] \hat{y} && \text{for } z > d.
\end{aligned}
\tag{3}
$$

Its propagation experiences the first incidence, reflection and transmission, indicating the significant coupling of fields and charges. Obviously, this scattering event is mainly determined by two independent boundary conditions, namely the continuities of tangential electric and magnetic fields:

$$
\begin{aligned}
&E_{\mathrm{I}} + \sum_j^{m+1} E_{\mathrm{R},j} = \sum_j^{2m} E_j \quad \& \\
&E_{\mathrm{I}} - \sum_j^{m+1} E_{\mathrm{R},j} = N\sum_j^{2m}(-1)^{j-1}E_j && \text{at } z = 0, \\
&\sum_j^{2m} E_j \exp\left[(-1)^{j-1}i\frac{\omega}{c}Nd\right] = \exp\left[i\frac{\omega}{c}d\right]\sum_j^{m} E_{\mathrm{T},j} \quad \& \\
&N\sum_j^{2m}(-1)^{j-1}E_j \exp\left[(-1)^{j-1}i\frac{\omega}{c}Nd\right] = \exp\left[i\frac{\omega}{c}d\right]\sum_j^{m} E_{\mathrm{T},j} && \text{at } z = d.
\end{aligned}
\tag{4}
$$

The transmitted wave will propagate along the $z$-direction, but with a reduced intensity caused by the Landau damping of the inter-$\pi$-band excitations [125]. After reaching the opposite surface, a similar scattering behavior arises there. A part of the wave can penetrate into the outside environment and then make a certain contribution to the transmitted field. The remaining part is directly reflected and continues its interaction with the whole valence charge. With the next scattering at the initial surface, the second contribution to the reflected field is achieved through the same boundary conditions. A series of reflections and transmissions, which, respectively, occur at the upper and lower surfaces, could be analytically calculated and expressed as

$$\frac{\sum_j^{m+1} E_{R,j}}{E_I} = \frac{[1-N^2(\omega)][1-e^{i2N(\omega)\frac{\omega}{c}d}]}{[1+N(\omega)]^2-[1-N(\omega)]^2e^{i2N(\omega)\frac{\omega}{c}d}}, \quad (5)$$

and

$$\frac{\sum_j^{m} E_{T,j}}{E_I} = \frac{4N(\omega)e^{-i[1-N(\omega)]\frac{\omega}{c}d}}{[1+N(\omega)]^2-[1-N(\omega)]^2e^{i2N(\omega)\frac{\omega}{c}d}}. \quad (6)$$

The total electric fields of the reflected (Eq. (5)) and transmitted (Eq. (6)) electromagnetic waves are very sensitive to the dielectric function and sample width ($d$). Their squares are the frequency-dependent reflectance and transmittance spectra ($R(\omega)$ and $T(\omega)$, respectively), which are directly examined by optical measurements [40–44].

## 6.2 Electronic Structures

The AA-, AB-, ABC- and AAB-stacked configurations are most common in 2D layered graphenes (Figs. 6.1(c)–6.1(f)). The $\pi$-electronic structures ($E^{c,v}(\mathbf{k})$) developed from the $C$-$2p_z$-based Hamiltonian are calculated for the monolayer system and the few-layer ones with the four kinds of stacked configurations (Fig. 6.3). $E^{c,v}(\mathbf{k})$ of monolayer graphene is briefly reviewed (blue dashed line in Fig. 6.3(a)). The conduction and valence bands are symmetric about Fermi energy ($E_F = 0$). The low-energy bands, with $|E^{c,v}| \leq 0.5$ eV, exhibit two linear bands intersecting at $E_F$; the Fermi-momentum state $\mathbf{k_F}$ is located at the $K$ point. The energy dispersions are isotropic near the $K$ point and described by $|E^{c,v}| = 3\alpha_0 b|\mathbf{k} - K|/2$, with a slope of $dE^{c,v}/dk = \pm 3\alpha_0 b/2$. The middle-energy states at $M$ belong to saddle points; their state energies are $E^{c,v}(M) = \pm\alpha_0$. For any stacked-configuration system, the $N_L$-layer graphenes have $N_L$ pairs of energy bands, but the symmetry between the conduction and valence bands does not remain. The middle energy still exhibits the saddle points at $M$ with the state energy. In AA-stacked few-layer graphenes (Figs. 6.3(a)–6.3(c)), for each pair, the energy dispersions are similar to those of monolayer graphene along $\Gamma \to M \to K \to \Gamma$ in the first Brillouin zone and a shift in the energy–wavevector space. The conduction- and valence-band states merge at the intersecting point $K$. The energy spacings (or the optical excitation energy), $E^s(\mathbf{k}) = E^c(\mathbf{k}) - E^v(\mathbf{k})$, of most states in the lower pair

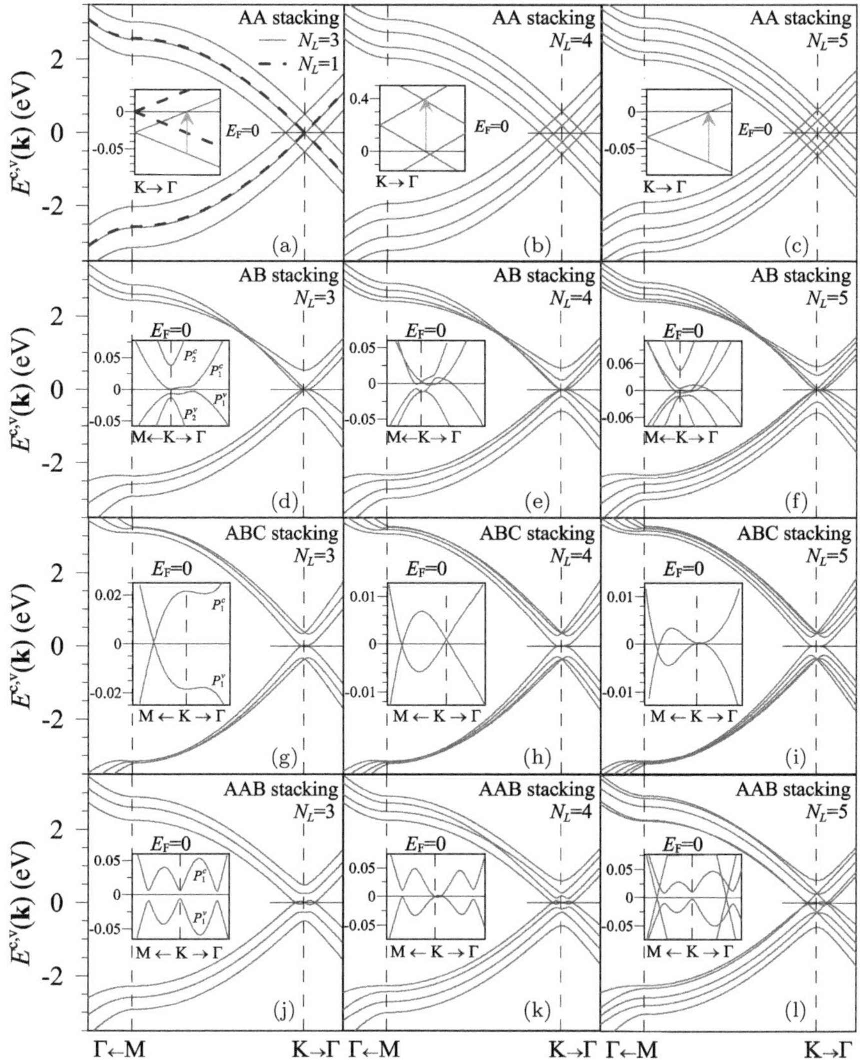


Fig. 6.3 Band structures are shown for (a) $N_L = 3$ and 1, (b) 4 and (c) 5 graphene with AA stacking along the symmetric points. (d)–(f), (g)–(i) and (j)–(l) are similar plots for graphenes with AB, ABC and AAB stacking, respectively. The insets show the detailed results near the Fermi energy, and the arrows in (a)–(c) point out the threshold energy for optical excitations.

are larger than those in the higher pair, for example, $E^c(M) - E^v(M)$ in the top (bottom) pair is the smallest (largest). At low energies (insets in Figs. 6.3(a)–6.3(c)), $E^s(\mathbf{k})$ increases from zero as $\mathbf{k}$ departs from $K$. However, both the initial and final states of the excitation located above (below) $E_F$ are forbidden; therefore, the threshold energy is $E^c(\mathbf{k_F}) - E^v(\mathbf{k_F})$ (indicated by arrows). Such threshold energy in odd-$N_L$ systems is within 0.1 eV for any $N_L$. On the other hand, the threshold energy in even-$N_L$ systems reduces quickly from 0.7 eV as $N_L$ increases from 2 and is close to that in odd-$N_L$ systems when $N_L \geq 14$ [128].

The stacking effect is clearly exhibited through the quite dissimilar low-energy bands of the AA-, AB-, ABC- and AAB-stacked graphenes (Figs. 6.3(d)–6.3(l)). The energy dispersions are completely modified by the interlayer interactions, which depend on the geometric structures (the layer numbers and stacking configurations). Apparently, most of the energy bands are parabolic, and anisotropic energy dispersions around the $K$ point are thus produced. For the AB stacking (Figs. 6.3(d)–6.3(f)), the $N_L = 3$ graphene is semiconducting, with a narrow indirect gap in the first pair of the conduction and valence bands ($P_1^c$ and $P_1^v$), and the others are metallic with band overlaps in the first and second pairs ($P_1^{c,v}$ and $P_2^{c,v}$) near or at $K$ (insets). Two or three pairs of the parabolic bands are farther away from $E_F$, and the band-edge states are exactly at $K$. For the ABC stacking (Figs. 6.3(g)–6.3(i)), $P_1^{c,v}$ of $N_L = 3, 4, 5$ graphene have a similar intersection at a certain wavevector along the $KM$ direction but, respectively, form a small energy spacing along the $K\Gamma$ direction, an intersection at $K$ and a tiny overlap near $K$ (insets). The other pairs are parabolic bands, with the band-edge states in the vicinity of the $K$ point. $\mathbf{k_b}$ values of $P_2^{c,v}$ ($P_3^{c,v}$) appear in the $K\Gamma$ ($KM$) direction, with the smallest (second smallest) $E^s(\mathbf{k_b})$ for $N_L = 3, 4, 5$, and those of $P_4^{c,v}$ ($P_4^{c,v}$ and $P_5^{c,v}$) are almost at $K$ for $N_L = 4$ ($N_L = 5$). Moreover, the state energies of the conduction or valence band at $M$, $E^c(M)$ or $E^v(M)$, are the closest to each other among all stacking configurations. As for AAB stacking (Figs. 6.3(j)–6.3(l)), $P_1^c$ ($P_1^v$) of few-layer graphenes exhibits the 3-V shape with non-monotonic wavevector dependence, i.e., there simultaneously exist three (two) local minima and two (three) maxima near $K$. The $N_L = 3$ ($N_L = 4$) graphene has a narrow gap (a tiny overlap) in such a pair, being a semiconductor (a metal). $P_1^c$ and $P_1^v$ also show an energy spacing for the $N_L = 5$ graphene, but the $N_L = 5$ one is metallic because of the two-pair linear bands crossing the Fermi level. The other pairs are parabolic or sombrero-shaped bands, with $K$ as the center for any $N_L$. Except for the AA stacking, those few-layer graphenes with different stacking configurations may be a metal or semiconductor, but all of them have the band-edge states of the conduction and valence bands that are very close to the Fermi level, denoted by $\mathbf{k}^c_{\mathbf{b0}}$ and $\mathbf{k}^v_{\mathbf{b0}}$ (insets in Figs. 6.3(d)–6.3(l))]. Those states become richer and more in number with the increase in $N_L$, which is reflected in the optical properties.

## 6.3 Excitation Spectra

The single-particle excitation (SPE) spectrum, $Im[\epsilon(\omega)]$, of AA-, AB-, ABC- and AAB-stacked graphenes shown for different $N_L$ values in Figs. 6.4(a)–6.4(d), is associated with the absorption coefficient. When the width of the graphene systems approaches the critical value of about $N_L > 50$ (black dashed curve and circle), the $N_L$ dependence of $Im[\epsilon(\omega)]$ vanishes, and the spectrum converges to that of 3D graphite. As regards the AA stacking, due to the symmetry of the wave function, the velocity matrix elements (the second term of Eq. (2)) equal zero if the initial and final states arise from the energy bands belonging to different pairs; therefore, only states in the same pair take part in the SPEs. The spectra exhibit complex structures in the middle ($5.01 < \omega < 5.27$ eV) (lower inset) and low frequency ($\omega < 1.47$ eV) ranges (upper inset); otherwise, they are identical among different $n_L$. At $n_L = 3$ (red curve in Fig. 6.4(a)), there are three peaks at $\omega = 5.05$, 5.14 and 5.23 eV, and two peaks at $\omega = 0.066$ and 1.07 eV. The former three peaks arise from the optical excitations of the top, center and bottom pairs at $M$, respectively, and the peak frequencies ($\omega_p$) are close to the $E^s(M)$ values. The latter two peaks are caused by the excitations of the center pair and the other pairs at $\mathbf{k_F}$, respectively, and $\omega_p$ values are close to the $E^s(\mathbf{k_F})$ values. The peak of $\omega = 1.07$ eV results from two excitation channels of the top and bottom pairs. However, the two excitation energies lie so close together that the two peaks are hardly distinguishable. With an increase in $n_L$, due to the existence of the $N_L$ pairs of energy bands, the number of excitation channels becomes $N_L$. Therefore, $N_L$ peaks exist in both the middle- and low-frequency ranges. Because of the characteristics of $E^s(\mathbf{k})$ among different pairs, a simple relationship between the $\omega_p$ values and the pairs can be established. For the middle-frequency range, the peak frequencies from low to high correspond to the excitation energies at the $M$ point from the top pair to the bottom pair. As for the low-frequency range, the highest $\omega_{ex}$ arises from the mixing of the excitations of the top and bottom pairs at $\mathbf{k_F}$. The lowest $\omega_p$, the threshold energy of the SPEs, corresponds to $E^s(\mathbf{k_F})$ of the pair with the smallest $E^{c(v)}(K)$. At odd $N_L$, the strongest peak at the smallest excitation energy $\omega_p \sim 0.1$ eV only appears in the odd-layer spectra. The fact that $\mathbf{k_F}$ of the center-pair energy band is so close to the $K$ point causes the small excitation energy. The strong Coulomb interactions due to this small excitation energy increase the peak strength. Moreover, since $|\mathbf{k_F} - K|$ in these few-layer systems are similar, the numbers of the states at $\mathbf{k_F}$ of the center pair are close to one another. Therefore, the density of states is inversely proportional to $n_L$, and this peak strength is roughly proportional to $1/n_L$.

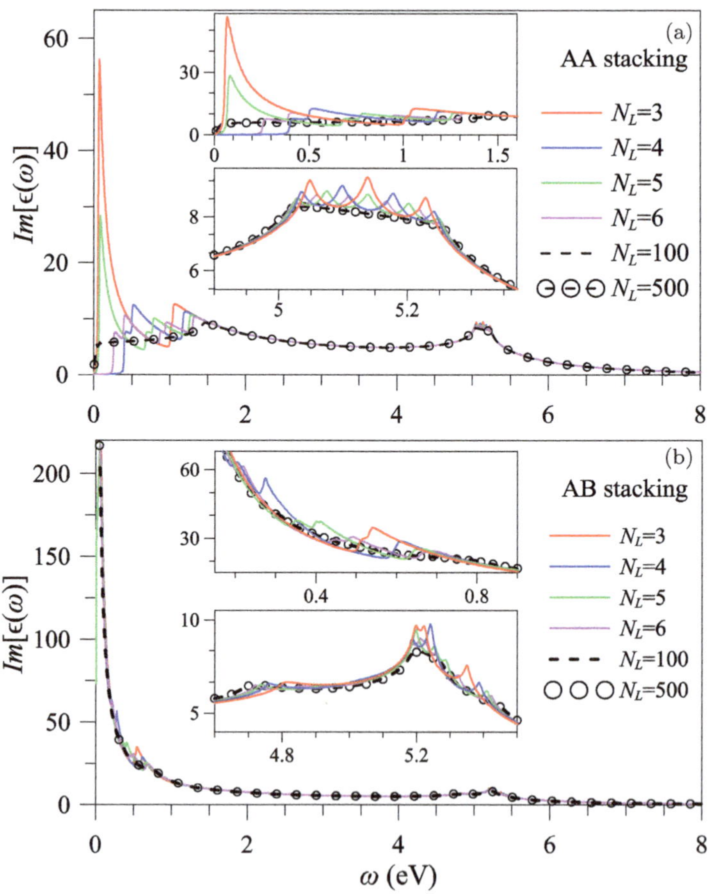


Fig. 6.4 (*Continued*)

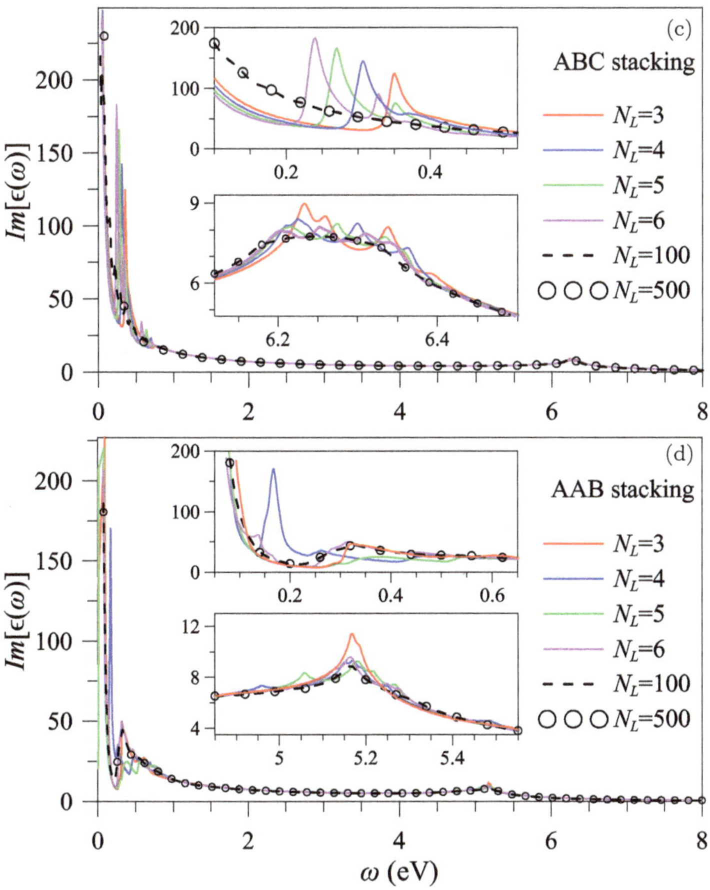


Fig. 6.4 The single-particle excitation spectra of graphenes with (a) AA, (b) AB, (c) ABC and (d) AAB stackings shown for $N_L = 3, 4, 5, 6, 100$ and 500. Each upper and lower insets show the detailed results in the low- and middle-frequency ranges.

Concerning the AB, ABC and AAB stackings, the $N_L$ dependence of the excitation spectra also exists in certain ranges of the low and middle frequencies for $N_L < 100$. However, it differs from the AA one in that the serious restrictions of the excitations only coming from the same-pair energy band are absent. The relation between the excitation channels and the velocity matrix elements is not simple. The largest value is very close to zero frequency in each spectrum (Figs. 6.4(b)–6.4(d)) since the initial and final states of the excitations are, respectively, located at or near $\mathbf{k}^{\mathbf{v}}_{\mathbf{b0}}$ and $\mathbf{k}^{\mathbf{c}}_{\mathbf{b0}}$ (insets in Figs. 6.3(d)–6.3(l)). Besides the largest, in the low-frequency region (upper inset in Fig. 6.4(b)), there is one shoulder at $\omega_p = 0.52\,\mathrm{eV}$ and one peak at $\omega_p = 0.54\,\mathrm{eV}$ for $n_L = 3$ AB-stacked graphene (red line), induced by SPEs from $\mathbf{k_b}$ of $P_3^v \to \mathbf{k_{b0}^c}$ and $\mathbf{k_{b0}^v} \to \mathbf{k_b}$ of $P_3^c$. At $N_L = 4$ ($N_L = 5$) (blue curve (green curve)), the spectrum shows three peaks at $\omega_p = 0.20$, 0.27 and 0.61 eV ($\omega_p = 0.35$, 0.41 and 0.67 eV) from $\mathbf{k_b}$ of $P_3^v \to \mathbf{k_{b0}^c}$, $\mathbf{k_{b0}^v} \to \mathbf{k_b}$ of $P_3^c$ and the mixing of $\mathbf{k_b}$ of $P_4^v \to \mathbf{k_{b0}^c}$ and $\mathbf{k_{b0}^v} \to \mathbf{k_b}$ of $P_4^c$, respectively. Because of $\mathbf{k_{b0}^c}$ and $\mathbf{k_{b0}^v}$ close to $E_F = 0$, there exist $N_L$ ($N_L - 1$) peaks at $\omega_p \sim$ the state energy of $\mathbf{k_b}$ of $P_j^c$ and $\omega_p \sim$ the negative state energy of $\mathbf{k_b}$ of $P_j^v$ in the low-frequency region, where $j > \frac{n_L}{2}$ ($j > \frac{n_L+1}{2}$) for even (odd) $N_L$. In the middle-frequency range (lower inset in Fig. 6.4(b)), the structures fully arise from SPEs at $M$. At $N_L = 3$ (red line), three peaks at $\omega_p = 5.19$, 5.22 and 5.35 eV and one shoulder at $\omega_p = 4.81\,\mathrm{eV}$ are, respectively, from $P_1^v \to P_1^c$, $P_2^v \to P_2^c$, $P_3^v \to P_3^c$ and $P_1^v \to P_3^c$. With an increase in $N_L$, due to the existence of $N_L$ pairs of energy bands, the number of peak structures becomes $n_L$, corresponding to $P_1^v \to P_1^c$, $P_2^v \to P_2^c, \ldots, P_{N_L}^v \to P_{N_L}^c$, and the shoulder structure appears at the smallest excitation energy of $P_1^v \to P_{N_L}^c$. For the $N_L = 3$ ABC-stacked graphene (Fig. 6.4(c)), there is one peak at $\omega_p = 0.32\,\mathrm{eV}$ (red curve in the upper inset), caused by the four kinds of excitation channels between $\mathbf{k_{b0}^c}$ ($\mathbf{k_{b0}^v}$) and $P_2^v$ ($P_3^v$) ($P_2^c$ ($P_3^c$)). For larger $N_L$, the intensity and frequency of this peak are stronger and lower because of the smaller band curvature and energy spacing (Figs. 6.3(g)–6.3(i)), respectively. The excitations between $\mathbf{k_b}$ of $P_4^{c,v}$ ($P_4^{c,v}$ and $P_5^{c,v}$) and $\mathbf{k_{b0}^{v,c}}$ induce one shoulder structure (one shoulder structure) at $\omega_p = 0.37$ ($\omega_p = 0.35$) eV for $N_L = 4$ ($N_L = 5$). Due to the more concentrated energy bands near the $M$ point, the $N_L$ dependence of the middle-frequency spectra is within a narrower range ($6.18 < \omega < 6.42\,\mathrm{eV}$). There exist three peaks and one shoulder at $\omega_p = 6.23$, 6.26, 6.34 and 6.39 eV, corresponding to the excitations of $P_1^v \to P_1^c$, $P_{2(3)}^v \to P_1^c$, $P_1^v \to P_{2(3)}^c$ and $P_1^v \to P_3^c$ at $M$, respectively. Those structures become more and more complicated as $N_L$ increases. As for the $N_L = 3$ AAB-stacked graphene (Fig. 6.4(d)), one wide peak at $\omega_p = 0.32\,\mathrm{eV}$ (red curve in the upper inset) arises from the excitations between $P_1^{c,v}$ of the 3-V-shaped bands and $P_2^{v,c}$ of the sombrero-shaped bands (Fig. 6.3(j)). This peak becomes sharp at $N_L = 4$ (blue curve) since the $\mathbf{k_b}$ values of $P_1^{c,v}$ and $P_2^{c,v}$ are aligned together, concentrating the excitation energy (Fig. 6.3(k)). It broadens to disappear with further increase in $N_L$. The spectrum of the $N_L = 3$ graphene exhibits the strongest peak in the middle-frequency range among the spectra of all $N_L$ since the allowed excitation energy of the $E^s(M)$ values of the three pairs of energy bands are almost the same. As $N_L$ increases, the single peak splits into more and becomes weaker, spreading out in the range of $4.8 < \omega < 5.9$ eV.

## 6.4 Transmittance and Reflectance

The quantum-confinement effects of optical excitations are clearly shown in layered graphene systems with AA, AB, ABC and AAB stackings. Reflectance and transmittance, as shown in Figs. 6.5–6.8, present strong frequency and layer dependences. More interestingly, the $\pi$-electronic excitations create prominent vertical transitions at the low- and middle-frequency ranges. The former and latter mainly arise from the initial and saddle-point $\pi$-electronic states near the $K$ and $M$ valleys, respectively, directly reflecting the main features of electronic energy spectra and wave functions (Fig. 6.3) [20]. Furthermore, their van Hove singularities are strengthened and widened by the greatly enhanced interlayer van der Waals interactions [20]. There exist several strong peaks in the valence/conduction density of states (not shown), or a quadratic number of obvious ones in the joint density of states during the vertical optical transitions. However, part of the transition channels are forbidden in the low-/middle-frequency electric-dipole excitations. This is responsible for the whole $\pi$-electronic optical excitation. It should be noted that the dielectric functions become hardly independent of sufficiently high layer numbers, e.g., those of $N_L \geq 50$ (Fig. 6.4). They account for the significant couplings between charges and waves and thus determine the diverse absorption, reflectance and transmittance behaviors.

The rich and unique optical properties deserve a closer examination, especially for cooperation or competition between $\pi$-electronic excitations and finite-width quantum confinement. Both $R(\omega)$ and $T(\omega)$ present the non-monotonous frequency dependencies in the large-fluctuation forms. Such behaviors mainly come from the strong response functions under the prominent structures (the pronounced joint van Hove singularities). As for the layer number of graphene systems, it provides more information about the decline in the propagating wave and its couplings with charges. The transmittance (reflectance) of the four kinds of stacked graphenes is shown for the fewer-layer set of $N_L = 3, 5, 10, 15$ and 20 and the more-layer set of $N_L = 50, 100, 200, 500$ and 1000 in Figs. 6.5 and 6.6 (Figs. 6.7 and 6.8).

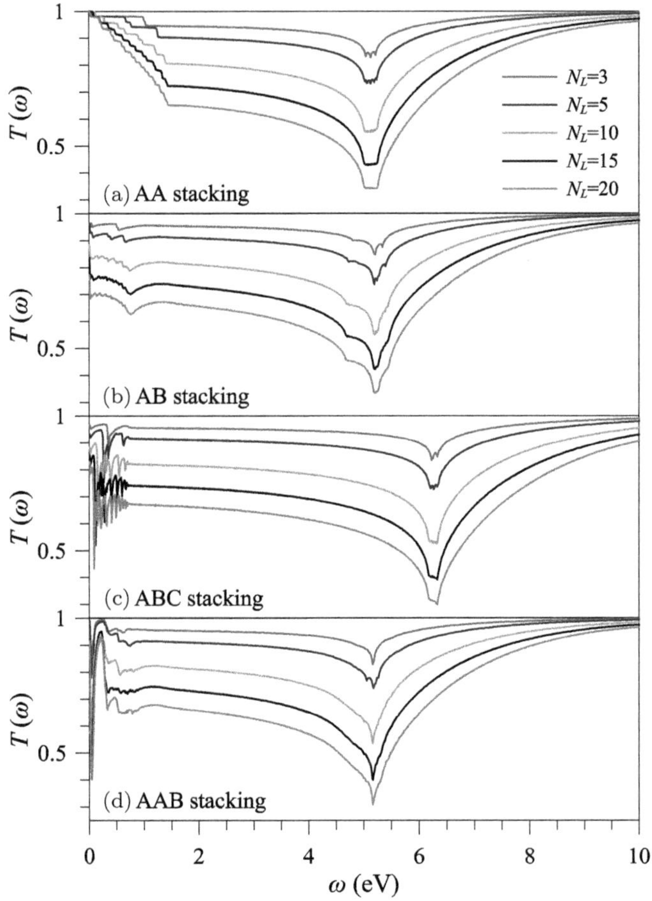


Fig. 6.5 Transmittance of graphenes with (a) AA, (b) AB, (c) ABC and (d) AAB stackings shown for $N_L = 3, 5, 10, 15$ and 20.

The two sets are separated by the critical layer ($N_L \sim 50$) of the excitation spectra, and both $T(\omega)$ and $R(\omega)$ only at the fewer-layer one reflect the $N_L$ dependence of the excitation spectra.

Transmittance exhibits an obvious reduction in the increment of $N_L$ (Figs. 6.5 and 6.6). It is almost vanishing within the whole frequency range, while the layer number is beyond 1000. Such results suggest a sufficiently high absorption coefficient (the observable imaginary part of the dielectric functions). Moreover, due to the material absorption, the sagging transmittance roughly corresponds to the prominent structures in the

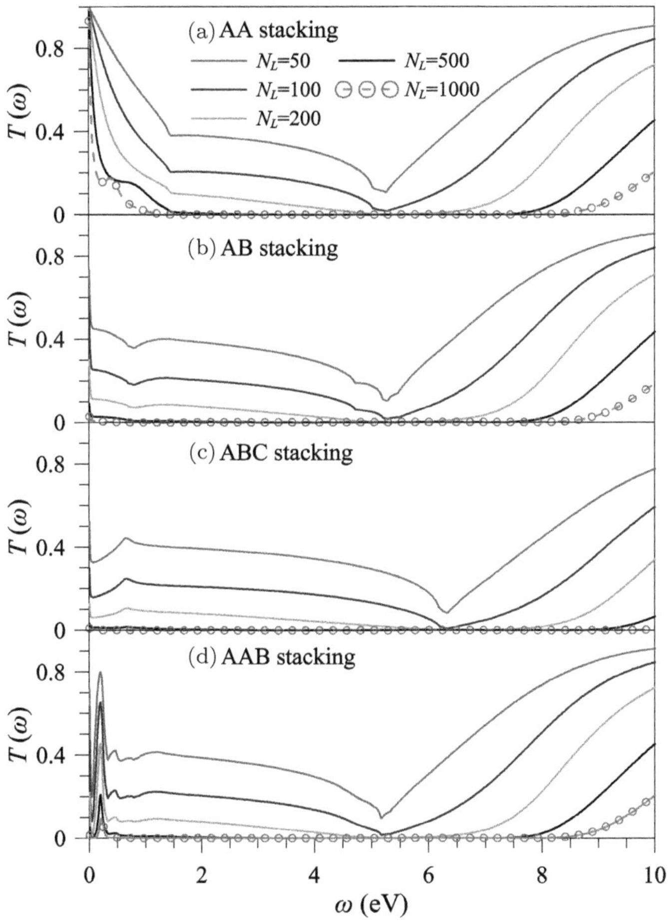


Fig. 6.6 Transmittance of graphenes with (a) AA, (b) AB, (c) ABC and (d) AAB stackings shown for $N_L = 50, 100, 200, 500$ and 1000.

excitation spectra. For the fewer-layer set (Fig. 6.5), the concentrated structures, steps, shoulders or peaks in the low- and middle-frequency $T(\omega)$ may be more complicated for larger $N_L$, reflecting the $N_L$-dependent excitation spectra and energy dispersions. However, for the more-layer set (Fig. 6.6), the reduction of $T(\omega)$ is purely due to the heavier consumption by the thicker graphene. On the other hand, the reflectance of the fewer-layer set grows as $N_L$ increases and also shows the $N_L$-dependence characters of the excitations and the energy bands (Fig. 6.7). For the more-layer set, reflectance might present the irregular/regular $N_L$ dependence

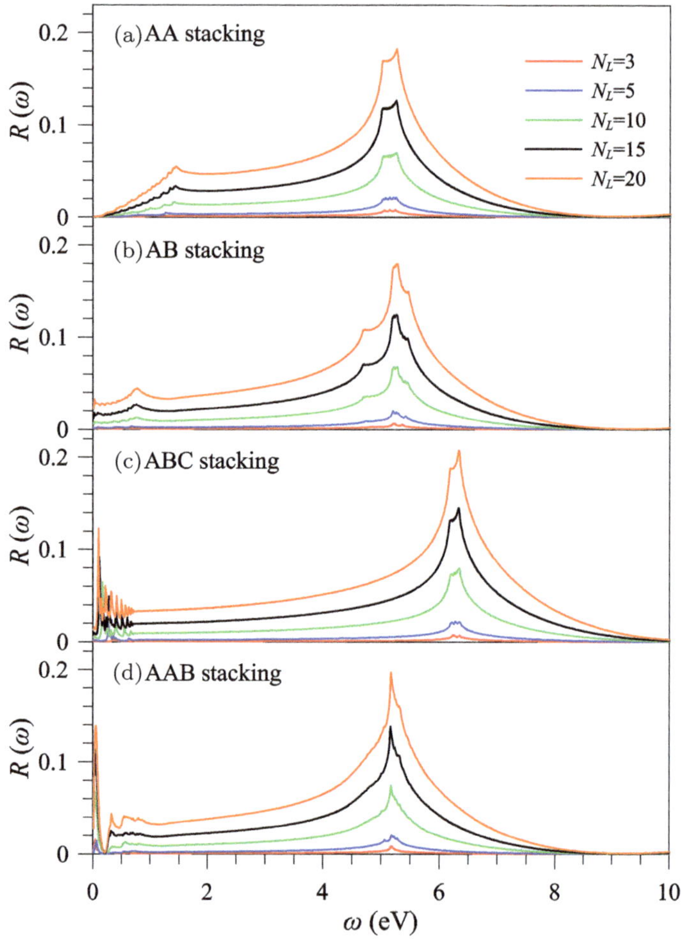


Fig. 6.7 Reflectance of graphenes with (a) AA, (b) AB, (c) ABC and (d) AAB stackings shown for $N_L = 3, 5, 10, 15$ and 20.

at $\omega > 0.2$ eV / others (Fig. 6.8). The unusual phenomenon of $R(\omega)$ should be closely related to the magnitude of the dielectric function. For example, $\epsilon(\omega)$ is about >100 and <10 at low and middle frequencies, respectively. Under the latter case, more optical scatterings would appear at two boundaries, supported by the slow energy dissipation of the propagating electromagnetic waves. This might lead to comparable $R(\omega)$ values among the different layer numbers or the absence of the monotonic thickness dependence.

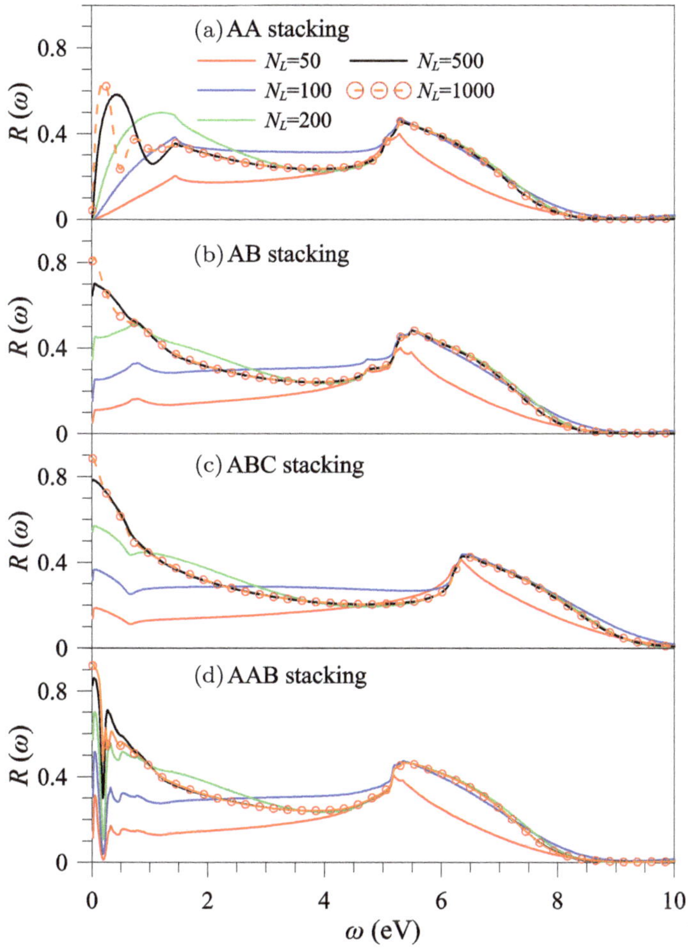


Fig. 6.8 Reflectance of graphenes with (a) AA, (b) AB, (c) ABC and (d) AAB stackings shown for $N_L = 50, 100, 200, 500$ and 1000.

In conclusion, it can be said that optical scatterings are fully explored for layered materials under the theoretical framework of quasiparticles [59], which is suitable for the rich and unique intrinsic interactions and the external fields. The fundamental properties are unified together and could be understood simultaneously. Specifically, layered graphene systems with AA, AB, ABC and AAB stackings clearly show the unusual reflectance and transmittance spectra through the quantum-size effects.

The strong frequency and thickness dependences illustrate the prominent $\pi$-electronic excitations and their significant couplings with electromagnetic waves.

## References


[1] Geim A.K. and Novoselov K.S. 2007. The rise of graphene. *Nat. Mater.* **6** 183.

[2] Castro Neto A.H., Guinea F., Peres N.M.R., Novoselov K.S. and Geim A.K. 2009. The electronic properties of graphene. *Rev. Mod. Phys.* **81** 109.

[3] Bae S. *et al.* 2010. Roll-to-roll production of 30-inch graphene films for transparent electrodes. *Nat. Nanotechnol.* **5** 574.

[4] Guinea F. 2007. Charge distribution and screening in layered graphene systems. *Phys. Rev. B* **75** 235433.

[5] Gutierrez F.A., Fernandes F.C.B., Rivas G.A. and Bueno P.R. 2017. Mesoscopic behaviour of multi-layered graphene: The meaning of supercapacitance revisited. *Phys. Chem. Chem. Phys.* **19** 6792–6806.

[6] Chen H.A., Hsin C.L., Huang Y.T., Tang M.L., Dhuey S., Cabrini S., Wu W.W. and Leone S.R. 2013. Measurement of interlayer screening length of layered graphene by plasmonic nanostructure resonances. *J. Phys. Chem. C* **117** 22211–22217.

[7] Englman R. and Vértesi T. 2008. Large Berry phases in layered graphene. *Phys. Rev. B* **78** 205311.

[8] Ansón-Casaos A., Puértolas J.A., Pascual F.J., Hernández-Ferrer J., Castell P., Benito A.M., Maser W.K. and Martínez M.T. 2014. The effect of gamma-irradiation on few-layered graphene materials. *Appl. Surf. Sci.* **301** 264–272.

[9] Masyukov M., Vozianova A., Grebenchukov A., Gubaidullina K., Zaitsev A. and Khodzitsky M. 2020. Optically tunable terahertz chiral metasurface based on multi-layered graphene. *Sci. Rep.* **10** 3157.

[10] Yuan G., Lin D., Wang Y., Huang X., Chen W., Xie X., Zong J., Yuan Q.Q., Zheng H., Wang D., Xu J., Li S.C., Zhang Y., Sun J., Xi X. and Gao L. 2020. Proton-assisted growth of ultra-flat graphene films. *Nature* **577** 204–208.

[11] Wang R., Shi M., Brewer B., Yang L., Zhang Y., Webb D.J., Li D. and Xu Y.Q. 2018. Ultrasensitive graphene optoelectronic probes for recording electrical activities of individual synapses. *Nano Lett.* **18** 5702–5708.

[12] Park H., Lim C., Lee C.J., Kang J., Kim J., Choi M. and Park H. 2018. Optimized poly(methyl methacrylate)-mediated graphene-transfer process for fabrication of high-quality graphene layer. *Nanotechnology* **29** 415303.

[13] Mendez J.P., Ramos F.A., Ortiz M. and Ariza M.P. 2018. Charge carrier transport across grain boundaries in graphene. *Acta. Mater.* **154** 199–206.

[14] Kang M., Kim J., Jang B., Chae Y., Kim J.H. and Ahn J.H. 2017. Graphene-based three-dimensional capacitive touch sensor for wearable electronics. *ACS Nano* **11** 7950–7957.

[15] Matruglio A., Nappini S., Naumenko D., Magnano E., Bondino F., Lazzarino M. and Zilio S.D. 2016. Contamination-free suspended graphene structures by a Ti-based transfer method. *Carbon* **103** 305–310.

[16] Jimenez-Villacorta F., Climent-Pascual E., Ramirez-Jimenez R., Sanchez-Marcos J., Prieto C. and de Andrés A. 2016. Graphene–ultrasmall silver nanoparticle interactions and their effect on electronic transport and Raman enhancement. *Carbon* **101** 305–314.
[17] Grushina A.L., Ki D.K., Koshino M., Nicolet A.A.L., Faugeras C., McCann E., Potemski M. and Morpurgo A.F. 2015. Insulating state in tetralayers reveals an even–odd interaction effect in multilayer graphene. *Nat. Commun.* **6** 6419.
[18] Tuan D.V., Ortmann F., Soriano D., Valenzuela S.O. and Roche S. 2014. Pseudospin-driven spin relaxation mechanism in graphene. *Nat. Phys.* **10** 857.
[19] Nandkishore R., Levitov L.S. and Chubukov A.V. 2012. Chiral superconductivity from repulsive interactions in doped graphene. *Nat. Phys.* **8** 158–163.
[20] Lin C.Y., Chen R.B., Ho Y.H. and Lin M.F. 2018. *Electronic and Optical Properties of Graphite-Related Systems* (CRC Press).
[21] Ho Y.-H., Chiu Y.-H., Lin D.-H., Chang C.-P. and Lin M.-F. 2010. Magneto-optical selection rules in bilayer Bernal graphene. *ACS Nano* **4** 1465–1472.
[22] Gusynin V.P., Sharapov S.G. and Carbotte J.P. 2007. Anomalous absorption line in the magneto-optical response of graphene. *Phys. Rev. Lett.* **98** 157402.
[23] Abergel D.S.L. and Fal'ko V. 2007. Optical and magneto-optical far-infrared properties of bilayer graphene. *Phys. Rev. B* **75** 155430.
[24] Shin J., Park Y., Chittari B.L., Sun J.H. and Jung J. 2021. Electron-hole asymmetry and band gaps of commensurate double moire patterns in twisted bilayer graphene on hexagonal boron nitride. *Phys. Rev. B* **103** 075423.
[25] Pack J., Russell B.J., Kapoor Y., Balgley J., Ahlers J., Taniguchi T., Watanabe K. and Henriksen E.A. 2020. Broken symmetries and Kohn's theorem in graphene cyclotron resonance. *Phys. Rev. X* **10** 041006.
[26] Ando T. 2019. Theory of weak-field magnetoresistance in bilayer graphene. *J. Phys. Soc. Jpn.* **88** 044707.
[27] Avetissian H.K. and Mkrtchian G.F. 2016. Coherent nonlinear optical response of graphene in the quantum Hall regime. *Phys. Rev. B* **94** 045419.
[28] Savostianova N.A. and Mikhailov S.A. 2015. Giant enhancement of the third harmonic in graphene integrated in a layered structure. *Appl. Phys. Lett.* **107** 181104.
[29] Carbotte J.P., LeBlanc J.P.F. and Ashby P.E.C. 2013. Impact of electron-phonon coupling on near-field optical spectra in graphene. *Phys. Rev. B* **87** 045405.
[30] Koshino M. 2013. Stacking-dependent optical absorption in multilayer graphene. *New J. Phys.* **15** 015010.
[31] Zhu J., Badalyan S.M. and Peeters F.M. 2012. Electron-phonon bound states in graphene in a perpendicular magnetic field. *Phys. Rev. Lett.* **109** 256602.
[32] Stauber T., Peres N.M.R. and Castro Neto A.H. 2008. Conductivity of suspended and non-suspended graphene at finite gate voltage. *Phys. Rev. B* **78** 085418.
[33] Mikhailov S.A. 2021. Nonperturbative quasiclassical theory of graphene photoconductivity. *Phys. Rev. B* **103** 245406.

[34] Fatima, Inerbaev T., Xia W. and Kilin D.S. 2021. Magnetic-field-driven electron dynamics in graphene. *J. Phys. Chem. Lett.* **12** 4749.

[35] Noro M. and Ando T. 2016. Weak-field Hall effect in graphene with long-range scatterers. *J. Phys. Soc. Jpn.* **85** 014708.

[36] Klimchitskaya G.L. and Mostepanenko V.M. 2016. Quantum electrodynamic approach to the conductivity of gapped graphene. *Phys. Rev. B* **94** 195405.

[37] Klimchitskaya G.L. and Mostepanenko V.M. 2016. Conductivity of pure graphene: Theoretical approach using the polarization tensor. *Phys. Rev. B* **93** 245419.

[38] Tahir M., Vasilopoulos P. and Peeters F.M. 2015. Magneto-optical transport properties of monolayer phosphorene. *Phys. Rev. B* **92** 045420.

[39] Orlita M., Crassee I., Faugeras C., Kuzmenko A.B., Fromm F., Ostler M., Seyller T., Martinez G., Polini M. and Potemski M. 2012. Classical to quantum crossover of the cyclotron resonance in graphene: A study of the strength of intraband absorption. *New J. Phys.* **14** 095008.

[40] Orlita M., Faugeras C., Plochocka P., Neugebauer P., Martinez G., Maude D.K., Barra A.-L., Sprinkle M., Berger C., de Heer W.A. and Potemski M. 2008. Approaching the dirac point in high-mobility multilayer epitaxial graphene. *Phys. Rev. Lett.* **101** 267601.

[41] Mak K.F., Shan J. and Heinz T.F. 2010. Electronic structure of few-layer graphene: Experimental demonstration of strong dependence on stacking sequence. *Phys. Rev. Lett.* **104** 176404.

[42] Wang F., Zhang Y., Tian C., Girit C., Zettl A., Crommie M. and Shen Y.R. 2008. Gate-variable optical transitions in graphene. *Science* **320** 206.

[43] Sensale-Rodriguez B., Yan R., Rafique S., Zhu M., Li W., Liang X., Gundlach D., Protasenko V., Kelly M.M., Jena D., Liu L. and Xing H.G. 2012. Extraordinary control of terahertz beam reflectance in graphene electro-absorption modulators. *Nano Lett.* **12** 4518–4522.

[44] Skulason H.S., Gaskell P.E. and Szkopek T. 2010. Optical reflection and transmission properties of exfoliated graphite from a graphene monolayer to several hundred graphene layers. *Nanotechnology* **21** 295709.

[45] Ju L., Wang L., Li X., Moon S., Ozerov M., Lu Z., Taniguchi T., Watanabe K., Mueller E., Zhang F., Smirnov D., Rana F. and McEuen P.L. 2020. Unconventional valley-dependent optical selection rules and landau level mixing in bilayer graphene. *Nat. Commun.* **11** 2941.

[46] Russell B.J., Zhou B., Taniguchi T., Watanabe K. and Henriksen E.A. 2018. Many-particle effects in the cyclotron resonance of encapsulated monolayer graphene. *Phys. Rev. Lett.* **120** 047401.

[47] Nakamura D., Saito H., Hibino H., Asano K. and Takeyama S. 2020. Quantum limit cyclotron resonance in monolayer epitaxial graphene in magnetic fields up to 560 T: The relativistic electron and hole asymmetry. *Phys. Rev. B* **101** 115420.

[48] Nedoliuk I.O., Hu S., Geim A.K., and Kuzmenko A.B. 2019. Colossal infrared and terahertz magneto-optical activity in a two-dimensional Dirac material. *Nat. Nanotechnol.* **14** 756.

[49] Massabeau S., Baillergeau M., Phuphachong T., Berger C., de Heer W.A., Dhillon S., Tignon J., de Vaulchier L.A., Ferreira R. and Mangeney J. 2017. Evidence of Fermi level pinning at the Dirac point in epitaxial multilayer graphene. *Phys. Rev. B* **95** 085311.

[50] Wendler F., Mittendorff M., König-Otto J.C., Brem S., Berger C., de Heer W.A., Böttger R., Schneider H., Helm M., Winnerl S. and Malic E. 2017. Symmetry-breaking supercollisions in Landau-quantized graphene. *Phys. Rev. Lett.* **119** 067405.

[51] König-Otto J.C., Mittendorff M., Winzer T., Kadi F., Malic E., Knorr A., Berger C., de Heer W.A., Pashkin A., Schneider H., Helm M. and Winnerl S. 2016. Slow noncollinear Coulomb scattering in the vicinity of the Dirac point in graphene. *Phys. Rev. Lett.* **117** 087401.

[52] Razado-Colambo I., Avila J., Chen C., Nys J.P., Wallart X., Asensio M.C. and Vignaud D. 2015. Probing the electronic properties of graphene on C-face SiC down to single domains by nanoresolved photoelectron spectroscopies. *Phys. Rev. B* **92** 035105.

[53] Mittendorff M., Wendler F., Malic E., Knorr A., Orlita M., Potemski M., Berger C., de Heer W.A., Schneider H., Helm M. and Winnerl S. 2015. Carrier dynamics in Landau-quantized graphene featuring strong Auger scattering. *Nat. Phys.* **11** 75.

[54] Malko D., Neiss C., Viñes F. and Görling A. 2012. Competition for graphene: Graphynes with direction-dependent Dirac cones. *Phys. Rev. Lett.* **108** 086804.

[55] Orlita M., Faugeras C., Grill R., Wysmolek A., Strupinski W., Berger C., de Heer W.A., Martinez G. and Potemski M. 2011. Carrier scattering from dynamical magnetoconductivity in quasineutral epitaxial graphene. *Phys. Rev. Lett.* **107** 216603.

[56] Bisti V.E. and Kirova N.N. 2011. Coulomb interaction and electron-hole asymmetry in cyclotron resonance of bilayer graphene in a high magnetic field. *Phys. Rev. B* **84** 155434.

[57] Crassee I., Levallois J., Walter A.L., Ostler M., Bostwick A., Rotenberg E., Seyller T., van der Marel D. and Kuzmenko A.B. 2011. Giant Faraday rotation in single- and multilayer graphene. *Nat. Phys.* **7** 48.

[58] Novoselov K.S., Geim A.K., Morozov S.V., Jiang D., Katsnelson M.I., Grigorieva I.V., Dubonos S.V. and Firsov A.A. 2005. Two-dimensional gas of massless Dirac fermions in graphene. *Nano Lett.* **438** 197.

[59] Lin C.Y., Wu J.Y., Chiu C.W. and Lin M.F. 2019. *Coulomb Excitations and Decays in Graphene-Related Systems* (CRC Press).

[60] Wallace P.R. 1946. The band theory of graphite. *Phys. Rev.* **71** 622.

[61] Iijima S. and Ichihashi T. 1993. Single-shell carbon nanotubes of 1-nm diameter. *Nature* **363** 603.

[62] Odom T.W., Huang J.-L., Kim P. and Lieber C.M. 2000. Structure and electronic properties of carbon nanotubes. *J. Phys. Chem. B* **104** 2794–2809.

[63] Saito R., Fujita M., Dresselhaus G. and Dresselhaus M.S. 1992. Electronic structure of chiral graphene tubules. *Appl. Phys. Lett.* **60** 2204.

[64] Arroyo-Gascón O., Fernández-Perea R., Morell E.S., Cabrillo C. and Chico L. 2020. One-dimensional Moiré superlattices and flat bands in collapsed chiral carbon nanotubes. *Nano Lett.* **20** 7588–7593.
[65] Fujii Y., Maruyama M., Cuong N.T. and Okada S. 2020. Pentadiamond: A hard carbon allotrope of a pentagonal network of $sp^2$ and $sp^3$ C atoms. *Phys. Rev. Lett.* **125** 016001.
[66] Yanagi K., Okada R., Ichinose Y., Yomogida Y., Katsutani F., Gao W. and Kono J. 2018. Intersubband plasmons in the quantum limit in gated and aligned carbon nanotubes. *Nat. Commun.* **9** 1121.
[67] Sato N., Tatsumi Y. and Saito R. 2017. Circular dichroism of single-wall carbon nanotubes. *Phys. Rev. B* **95** 155436.
[68] Santos H., Chico L., Alvarellos J.E. and Latgé A. 2017. Defect-enhanced Rashba spin-polarized currents in carbon nanotubes. *Phys. Rev. B* **96** 165401.
[69] Zhang F., Hou P.X., Liu C., Wang B.W., Jiang H., Chen M.L., Sun D.M., Li J.C., Cong H.T., Kauppinen E.I. and Cheng H.M. 2016. Growth of semiconducting single-wall carbon nanotubes with a narrow band-gap distribution. *Nat. Commun.* **7** 11160.
[70] Zhang D., Yang J., Hasdeo E.H., Liu C., Liu K., Saito R. and Li Y. 2016. Multiple electronic Raman scatterings in a single metallic carbon nanotube. *Phys. Rev. B* **93** 245428.
[71] Wei X., Tanaka T., Akizuki N., Miyauchi Y., Matsuda K., Ohfuchi M. and Kataura H. 2016. Single-chirality separation and optical properties of (5,4) single-wall carbon nanotubes. *J. Phys. Chem. C* **120** 10705–10710.
[72] Akizuki N., Aota S., Mouri S., Matsuda K. and Miyauchi Y. 2015. Efficient near-infrared up-conversion photoluminescence in carbon nanotubes. *Nat. Commun.* **6** 8920.
[73] Laird E.A., Kuemmeth F., Steele G.A., Grove-Rasmussen K., Nygård J., Flensberg K. and Kouwenhoven L.P. 2015. Quantum transport in carbon nanotubes. *Rev. Mod. Phys.* **87** 703–764.
[74] Gui H., Streit J.K., Fagan J.A., Hight Walker A.R., Zhou C. and Zheng M. 2015. Redox sorting of carbon nanotubes. *Nano Lett.* **15** 1642–1646.
[75] Saidi W.A. and Norman P. 2014. Spectroscopic signatures of topological and diatom-vacancy defects in single-walled carbon nanotubes. *Phys. Chem. Chem. Phys.* **16** 1479–1486.
[76] Zhao Y., Tang Y., Chen Y. and Star A. 2012. Corking carbon nanotube cups with gold nanoparticles. *ACS Nano* **6** 6912–6921.
[77] Zhao Y., Allen B.L. and Star A. 2011. Enzymatic degradation of multiwalled carbon nanotubes. *J. Phys. Chem. A* **115** 9536–9544.
[78] Umadevi D. and Narahari Sastry G. 2011. Quantum mechanical study of physisorption of nucleobases on carbon materials: Graphene versus carbon nanotubes. *J. Phys. Chem. Lett.* **2** 1572–1576.
[79] Lee H.W., Yoon Y., Park S., Oh J.H., Hong S., Liyanage L.S., Wang H., Morishita S., Patil N., Park Y.J., Park J.J., Spakowitz A., Galli G., Gygi F., Wong P.H.S., Tok J.B.H., Kim J.M. and Bao Z. 2011. Selective dispersion of high purity semiconducting single-walled carbon nanotubes with regioregular poly(3-alkylthiophene)s. *Nat. Commun.* **2** 541.

[80] Cai J. *et al.* 2010. Atomically precise bottom-up fabrication of graphene nanoribbons. *Nature* **466** 470.
[81] Nakada K., Fujita M., Dresselhaus G. and Dresselhaus M.S. 1996. Edge state in graphene ribbons: Nanometer size effect and edge shape dependence. *Phys. Rev. B* **54** 17954.
[82] Chung H.-C., Chang C.-P., Lin C.-Y. and Lin M.-F. 2016. Electronic and optical properties of graphene nanoribbons in external fields. *Phys. Chem. Chem. Phys.* **18** 7573–7616.
[83] Gani Y.S., Abergel D.S.L. and Rossi E. 2018. Electronic structure of graphene nanoribbons on hexagonal boron nitride. *Phys. Rev. B* **98** 205415.
[84] Jacobse P.H., Kimouche A., Gebraad T., Ervasti M.M., Thijssen J.M., Liljeroth P. and Swart I. 2017. Electronic components embedded in a single graphene nanoribbon. *Nat. Commun.* **8** 119.
[85] Jacobse P.H., Kimouche A., Gebraad T., Ervasti M.M., Thijssen J.M., Liljeroth P. and Swart I. 2017. Electronic components embedded in a single graphene nanoribbon. *Nat. Commun.* **8** 119.
[86] Nguyen T.T.N., de Vries N., Karakachian H., Gruschwitz M., Aprojanz J., Zakharov A.A., Polley C., Balasubramanian T., Starke U., Flipse C.F.J. and Tegenkamp C. 2021. Topological surface state in epitaxial zigzag graphene nanoribbons. *Nano Lett.* **21** 2876–2882.
[87] McCurdy R.D., Jacobse P.H., Piskun I., Veber G.C., Rizzo D.J., Zuzak R., Mutlu Z., Bokor J., Crommie M.F. and Fischer F.R. 2021. Synergetic bottom-up synthesis of graphene nanoribbons by matrix-assisted direct transfer. *J. Am. Chem. Soc.* **143** 4174–4178.
[88] Karakachian H., Nguyen T.T.N., Aprojanz J., Zakharov A.A., Yakimova R., Rosenzweig P., Polley C.M., Balasubramanian T., Tegenkamp C., Power S.R. and Starke U. 2020. One-dimensional confinement and width-dependent bandgap formation in epitaxial graphene nanoribbons. *Nat. Commun.* **11** 6380.
[89] Jolly A., Miao D., Daigle M. and Morin J.F. 2020. Emerging bottom-up strategies for the synthesis of graphene nanoribbons and related structures. *Angew. Chem. Int. Ed.* **59** 4624–4633.
[90] Pfeiffer M., Senkovskiy B.V., Haberer D., Fischer F.R., Yang F., Meerholz K., Ando Y., Grüneis A. and Lindfors K. 2018. Observation of room-temperature photoluminescence blinking in armchair-edge graphene nanoribbons. 2018 *Nano Lett.* **18** 7038–7044.
[91] Aprojanz J., Power S.R., Bampoulis P., Roche S., Jauho A.P., Zandvliet H.J.W., Zakharov A.A. and Tegenkamp C. 2018. Ballistic tracks in graphene nanoribbons. *Nat. Commun.* **9** 4426.
[92] Cortizo-Lacalle D., Mora-Fuentes J.P., Strutyński K., Saeki A., Melle-Franco M. and Mateo-Alonso A. 2018. Monodisperse N-doped graphene nanoribbons reaching 7.7 nanometers in length. *Angew. Chem. Int. Ed.* **57** 703–708.
[93] Mehdi Pour M., Lashkov A., Radocea A., Liu X., Sun T., Lipatov A., Korlacki R.A., Shekhirev M., Aluru N.R., Lyding J.W., Sysoev V. and Sinitskii A. 2017.

Laterally extended atomically precise graphene nanoribbons with improved electrical conductivity for efficient gas sensing. *Nat. Commun.* **8** 820.

[94] Ruffieux P., Wang S., Yang B., Sánchez-Sánchez C., Liu J., Dienel T., Talirz L., Shinde P., Pignedoli C.A., Passerone D., Dumslaff T., Feng X., Müllen K. and Fasel R. 2016. On-surface synthesis of graphene nanoribbons with zigzag edge topology. *Nature* **531** 489–492.

[95] Chen Y.C., Cao T., Chen C., Pedramrazi Z., Haberer D., de Oteyza D.G., Fischer F.R., Louie S.G. and Crommie M.F. 2015. Molecular bandgap engineering of bottom-up synthesized graphene nanoribbon heterojunctions. *Nat. Nanotechnol.* **10** 156–160.

[96] Kimouche A., Ervasti M.M., Drost R., Halonen S., Harju A., Joensuu P.M., Sainio J. and Liljeroth P. 2015. Ultra-narrow metallic armchair graphene nanoribbons. *Nat. Commun.* **6** 10177.

[97] Cai J., Pignedoli C.A., Talirz L., Ruffieux P., Söde H., Liang L., Meunier V., Berger R., Li R., Feng X., Müllen K. and Fasel R. 2014. Graphene nanoribbon heterojunctions. *Nat. Nanotechnol.* **9** 896–900.

[98] Vo T.H., Shekhirev M., Kunkel D.A., Morton M.D., Berglund E., Kong L., Wilson P.M., Dowben P.A., Enders A. and Sinitskii A. 2014. Large-scale solution synthesis of narrow graphene nanoribbons. *Nat. Commun.* **5** 3189.

[99] Jiao L., Wang X., Diankov G., Wang H. and Dai H. 2010. Facile synthesis of high-quality graphene nanoribbons. *Nat. Nanotechnol.* **5** 321–325.

[100] Pech D., Brunet M., Durou H., Huang P., Mochalin V., Gogotsi Y., Taberna P.-L. and Simon P. 2010. Ultrahigh-power micrometre-sized supercapacitors based on onion-like carbon. *Nat. Nanotechnol.* **5** 651.

[101] Tomita S., Sakurai T., Ohta H., Fujii M. and Hayashi S. 2001. Structure and electronic properties of carbon onions. *J. Chem. Phys.* **114** 7477.

[102] Wu Z.S., Parvez K., Feng X. and Müllen K. 2013. Graphene-based in-plane micro-supercapacitors with high power and energy densities. *Nat. Commun.* **4** 2487.

[103] Berman D., Narayanan B., Cherukara M.J., Sankaranarayanan S.K.R.S., Erdemir A., Zinovev A. and Sumant A.V. 2018. Operando tribochemical formation of onion-like-carbon leads to macroscale superlubricity. *Nat. Commun.* **9** 1164.

[104] Zeiger M., Jäckel N., Mochalin V.N. and Presser V. 2016. Review: Carbon onions for electrochemical energy storage. *J. Mater. Chem. A* **4** 3172–3196.

[105] Pech D., Brunet M., Durou H., Huang P., Mochalin V., Gogotsi Y., Taberna P.L. and Simon P. 2010. Ultrahigh-power micrometre-sized supercapacitors based on onion-like carbon. *Nat. Nanotechnol.* **5** 651–654.

[106] Van Aken K.L., McDonough J.K., Li S., Feng G., Chathoth S.M., Mamontov E., Fulvio P.F., Cummings P.T., Dai S. and Gogotsi Y. 2014. Effect of cation on diffusion coefficient of ionic liquids at onion-like carbon electrodes. *J. Phys. Condens. Matter* **26** 284104.

[107] Kroto H.W., Heath J.R., O'Brien S.C., Curl R.F. and Smalley R.E. 1985. $C_{60}$: Buckminsterfullerene. *Nature* **318** 162.

[108] Lin Y.-L. and Nori F. 1994. Electronic structure of single- and multiple-shell carbon fullerenes. *Phys. Rev. B* **49** 5020.

[109] Noël Y., De La Pierre M., Zicovich-Wilson C.M., Orlando R. and Dovesi R. 2014. Structural, electronic and energetic properties of giant icosahedral fullerenes up to C6000: Insights from an ab initio hybrid DFT study. *Phys. Chem. Chem. Phys.* **16** 13390.
[110] Pudlak M., Pincak R. and Osipov V.A. 2006. Low-energy electronic states in spheroidal fullerenes. *Phys. Rev. B* **74** 235435.
[111] Lin Y.L. and Nori F. 1996. Analytical application of the recursion and moments methods to the electronic structure of $C_{60}$: Exact solution for the $\pi$ and $\sigma$ states. *Phys. Rev. B* **53** 1641.
[112] Ando H. and Nakao Y. 2021. Quantum states of the endohedral fullerene $Li^{+}@C_{60}$ surrounded by anions: Energy decomposition analysis of nuclear wave functions. *Phys. Chem. Chem. Phys.* **23** 9785–9803.
[113] Changala P.B., Weichman M.L., Lee K.F., Fermann M.E. and Ye J. 2019. Rovibrational quantum state resolution of the $C_{60}$ fullerene. *Science* **363** 49–54.
[114] Moriyama R., Wu J.W.J., Nakano M., Ohshimo K. and Misaizu F. 2018. Small carbon nano-onions: An ion mobility mass spectrometric study. *J. Phys. Chem. C* **122** 5195.
[115] Wang Y., Díaz-Tendero S., Alcamí M. and Martín F. 2017. Generalized structural motif model for studying the thermodynamic stability of fullerenes: From $C_{60}$ to graphene passing through giant fullerenes. *Phys. Chem. Chem. Phys.* **19** 19646.
[116] Hedberg K., Hedberg L., Bethune D.S., Brown C.A., Dorn H.C., Johnson R.D., De Vries M. 1991. Bond lengths in free molecules of Buckminsterfullerene, $C_{60}$, from gas-phase electron diffraction. *Science* **254** 41041.
[117] Jackson J.D. 1975. *Classical Electrodynamics* (Wiley).
[118] Min H. and MacDonald A.H. 2008. Chiral decomposition in the electronic structure of graphene multilayers. *Phys. Rev. B* **77** 155416.
[119] Lin C.-Y., Wu J.-Y., Ou Y.-J., Chiu Y.-H. and Lin M.-F. 2015. Magnetoelectronic properties of multilayer graphenes. *Phys. Chem. Chem. Phys.* **17** 26008–26035.
[120] Charlier J.-C., Michenaud J.-P. and Lambin P. 1992. Tight-binding density of electronic states of pregraphitic carbon. *Phys. Rev. B* **46** 4540.
[121] McClure J.W. 1957. Band structure of graphite and de Haas-van Alphen effect. *Phys. Rev.* **108** 612.
[122] Nakao K. 1976. Landau level structure and magnetic breakthrough in graphite. *J. Phys. Soc. Jpn.* **40** 761.
[123] Charlier J.-C., Michenaud J.-P. and Gonze X. 1992. First-principles study of the electronic properties of simple hexagonal graphite. *Phys. Rev. B* **46** 4531.
[124] Do T.-N., Lin C.-Y., Lin Y.-P., Shih P.-H. and Lin M.-F. 2015. Configuration-enriched magneto-electronic spectra of AAB-stacked trilayer graphene. *Carbon* **94** 619–632.
[125] Dressel M. and Gruner G. 2002. *Electrodynamics of Solids: Optical Properties of Electrons in Matter* (Cambridge University Press).
[126] Blinowski J. *et al.* 1980. Band structure model and dynamical dielectric function in lowest stages of graphite acceptor compounds. *J. Phys. France* **41** 47–58.

[127] Lin M.F. and Shung Kenneth W.-K. 1994. Plasmons and optical properties of carbon nanotubes. *Phys. Rev. B* **50** 17744.
[128] Chiu C.-W., Chen S.-C., Huang Y.-C., Shyu F.-L. and Lin M.-F. 2013. Critical optical properties of AA-stacked multilayer graphenes. *Appl. Phys. Lett.* **103** 041907.
[129] Chiu C.-W., Ho Y.-H., Shyu F.-L. and Lin M.-F. 2014. Layer-enriched optical spectra of AB-stacked multilayer graphene. *Appl. Phys. Express* **7** 115102.
[130] Chang C.P., Wang J., Lu C.L., Huang Y.C., Lin M.F. and Chen R.B. 2008. Optical properties of simple hexagonal and rhombohedral few-layer graphenes in an electric field. *J. Appl. Phys.* **103** 103109.
[131] Chiu C.-W. and Chen R.-B. 2016. Influence of electric fields on absorption spectra of AAB-stacked trilayer graphene. *Appl. Phys. Express* **9** 065103.
[132] Lin C.Y., Do T.N., Wu J.Y., Shih P.H., Lin S.Y., Ho C.H. and Lin M.F. 2019. *Diverse Quantization Phenomena in Emergent Layered Materials* (CRC Press).
[133] Xiao S., Mei H., Han D., Dassios K.G. and Cheng L. 2017. Ultralight lamellar amorphous carbon foam nanostructured by SiC nanowires for tunable electromagnetic wave absorption. *Carbon* **122** 718–725.
[134] Liu X., Fan K., Shadrivov I.V. and Padilla W.J. 2017. Experimental realization of a terahertz all-dielectric metasurface absorber. *Opt. Express* **25** 281296.

Chapter 7

# TOPOLOGICAL ORIGIN OF GATE-INDUCED LOCALIZED STATES IN GRAPHENE

Lu-Yao Wang, Ching-Hong Ho, Hsien-Ching Chung, Chiun-Yan Lin, Chih-Wei Chiu, and Ming-Fa Lin

This chapter explores how the gate potential of $N_c$, the number of chains along the armchair direction (along the $y$-axis), in two-dimensional graphene can produce topological localized states (LSs). The wave functions of the LSs show strongly localized behavior around the gate and decay away from the gate. The number of LSs is shown to depend on the number of chains, $N_c$. It is remarkable that there is one branch and two branches of LSs launching from the Dirac cone for single-chain gate ($N_c = 1$) and multi-chain gate ($N_c \geq 2$), respectively. Such topological LSs are formed as long as the gate-induced potential $V_0$ is non-zero. The topological origin of the formation of LSs can be interpreted by the mechanism of pseudospin rotations, that is, the rotation upon a valley-dependent pseudospin is equivalent to another valley-dependent pseudospin. It is notable that both pseudospins are on the same side of the gate potential. The picture of pseudospin rotations can be extracted from the LS secular equation, which is derived based on the scattering process upon applying the gate potential to a graphene sheet. As the energy $E$ varies across the entire energy gap for a given $k_y$, the rotation angle of the pseudospin-rotation operator fits within the variation range of the relative angle $\Delta\theta_p$ between two pseudospins. These LS branches reveal a

Dirac-point signature, with dispersion relations arising out of the Dirac point (at $k_y = 0$). For general multiple ($N_c > 1$) chain potential cases, the number of LS branches, $N_{\text{LS}}$, increases with the potential strength $V_0$ up to a maximum of $N_{\text{LS,max}} = N_c$. It is important that there are only two LS branches coupling to the Dirac-point signature in the case of $N_c > 1$.

## 7.1 Topological-Based Energy Dispersions

The most intriguing physics in condensed matter is topology-based energy dispersions [1, 2], which consist of topologically protected localized states on either the boundaries of two-dimensional (2D) materials [3–5] or the surfaces of three-dimensional (3D) materials [6–8]. Such topological localized states within specific materials which provide strong spin–orbit coupling (SOC) are called topological insulators (TIs) or quantum spin-Hall effect due to the insulating bulk carrying a robust propagating mode against disorder [1, 2]. The topological localized states occurring near the edges of TI materials have recently attracted a lot of attention, not only for their fundamental physical importance but also because of their potential application in spintronics [9].

TIs are made of semiconductor- and alloy-based materials and have been realized experimentally [10, 11]. The edge states of TIs reveal helical characteristics arising from a spin-momentum locking mechanism, where the edge states with opposite spins propagate in opposite directions [1, 2]. The topological nature of such edge states existing in gapped 2D TI materials can be characterized by a conventional topological index called the Chern number. Fortunately, one of the best candidates for 2D materials is graphene due to its unique electron properties and wide applications [12, 13]. The novel topological edge states in a 2D graphene incorporated with an intrinsic SOC have been theoretically studied by Kane and Mele [14, 15]. The SOC mixing electron spins on different valleys (K, K$'$-valleys) and induces Dirac cones to open to form a small global gap. The topological edge states launch from an upper K (K$'$) valley to another lower K$'$ (K) valley [14, 15] within the gap. Ideally, the topological edge states in 2D TI materials are immune to backscattering by local non-magnetic disorders due to the spin-momentum locking feature [16]. However, the topological edge states in graphene with intrinsic

SOC are too insignificant to be observed because of the very small gap opening. Thus, it becomes an interesting issue to generate the topological localized states by purely electric means in a graphene sheet without SOC [17].

Several topological perspectives are available to verify the topological LSs in pristine graphene because Dirac-like dispersion without a global gap cannot provide the Chern number. The generation of topologically localized states was proposed in pristine graphene, imposed by the different types of edges. The zigzag-edge graphene supports the non-dispersive zero-mode edge state [18–26]. The topological nature of the zero-mode localized state in zigzag-edge graphene can be characterized by the $\mathbf{Z}_2$ topological class [18] and is perfectly conducting against disorder [19]. Hatsugai has proposed an unconventional topological origin in graphene with different types of edges by projecting the bulk graphene Hamiltonian in the form of $h(\boldsymbol{k}) = \boldsymbol{R}(\boldsymbol{k}) \cdot \sigma$, which gives us a pictorial view in the $\boldsymbol{k}$-parameter space and its pseudospins for the $A(B)$ site in a graphene unit cell. The verification of the topological edge state depends on the evolving path of $\boldsymbol{R}(\boldsymbol{k})$, i.e., whether the path forms a loop enclosing its origin (singular point) for a given longitudinal $k_y$ along a specific type of edge [20]. It is found that zigzag- and bearded-edge graphenes satisfy the criterion to support the topological edge states, but the path of $\boldsymbol{R}(\boldsymbol{k})$ becomes a straight line passing through the origin in armchair-edge graphene for $k_y = 0$ [20]. Alternatively, the chiral edge states can be induced due to inversion symmetry breaking, which is attributed to the asymmetric on-site potential on the $A$ and $B$ sites in zigzag-edge graphene [21]. The anisotropic electron hopping can also generate the edge state in zigzag-edge graphene, and its topological nature is determined using the winding number [22]. However, armchair-edge graphene cannot support the edge states for the isotropic electron-hopping tight-binding model [23]. For the modified armchair-edge graphene, the localized states can be induced near the reconstructed armchair-edge [27] graphene, and the topological localized state can be induced in bearded armchair-edge graphene [28]. A time-dependent periodic strain field can induce the topological edge states in armchair-edge graphene ribbon [29]. Recently, the generation of edge states by mechanical modulation [30–33] and bond modulation [34, 35] has been studied in armchair-edge graphene but their topological nature is not clear. The uniaxial strain effect induces the pseudo-magnetic field, giving rise to zero-mode edge states in armchair-edge graphene [30–33]. The tight-binding Hamiltonian with anisotropic electron hopping in the boundary [34] and inside the bulk [35, 36] of the armchair-edge graphene can also induce the edge states. For quasi-1D armchair-edge graphene ribbons, anisotropic hopping gives rise to edge states, and their topological nature is demonstrated by the Zak phase [36]. These topological edge states can be cast as the bulk Hamiltonian $H^{\mathrm{bulk}}(\boldsymbol{k})$ in armchair-edge graphene without the local tunable parameter.

The topological LSs induced by a gate potential in 2D bulk graphene are potentially more applicable in devices than the LSs produced at the edges due to the complicated modifications at the edges. The tunable LSs induced in the bulk graphene have been studied in the following cases [37–41]. The line-type impurity (parallel to nanoribbons) in graphene ribbons gives rise to another hopping constant that is different from the bulk values [37]. The grain boundary along the armchair-direction with a medium opening angle [38] is within the scheme of tight-binding model. The topological 1D LSs are induced by a gate potential on the surface state of a three-dimensional topological system, and the effective Hamiltonian is employed in the system [39]. The LSs can be induced by a tensile strain locally applied along the direction of the zigzag carbon chain within the armchair-type graphene ribbons [40]. For the quasi-1D armchair-edge graphene ribbons, where the lateral direction is quantized with different types of terminations described by $H^{\mathrm{bulk}}(\boldsymbol{k})$, the localized state is generated along the junction (the direction of the zigzag carbon chain) between two topological types of ribbons whose topological nature is characterized by the Zak phase [41]. All the aforementioned works considered the bulk Block Hamiltonian without breaking the inversion symmetry. When the inversion symmetry is broken by a local gate potential, the system cannot be treated as a bulk Hamiltonian. Chu *et al.* proposed that the localized state emerging from the Dirac point can be generated near the edges of the armchair graphene nanoribbons by applying the gate potential along the armchair carbon chain boundary [42]. Therefore, the scattering approach is appropriately employed to deal with the scattering region in a pristine graphene sheet and the local potential.

In general, the $N_c$-chain gate potential along the armchair direction is also studied in this chapter. The features of LS generation and the topological origin of the topological LS are explored in this chapter. The key finding is that the energy evolution of the Bloch state pseudospins within the energy gap for a given $k_y$ (parallel to the gate) reveals the topological origin. Also, it is shown analytically that in both of the cases of $N_c = 1$ and $N_c = 2$, in particular, the LS-secular equation can be transformed into a pictorial mechanism of pseudospin rotation in the gap region. It is different from the gated 2D Dirac Hamiltonian case [44]. Our analytical results show one (two) topological LS branches for $N_c = 1$ ($N_c = 2$). Furthermore, the dispersion of the topological LS is embedded in the Dirac-point characteristics. It is remarkable that the line gate potential along the armchair carbon chain provides more interesting physics due to the mixing of two valleys (K and K′).

## 7.2 Generation of the Localized State in a One-Chain Gate ($N_c = 1$)

This section presents the theoretical model and the coupling between bulk graphene and the one-chain gate potential. By applying the electric gate atop pristine graphene, as schematically illustrated in Fig. 7.1, the $N_c$-chain gate is biased with the potential $V(M)$, indicated by shaded areas. The branch of LS is caused by the pseudospin-associated Bloch wave in

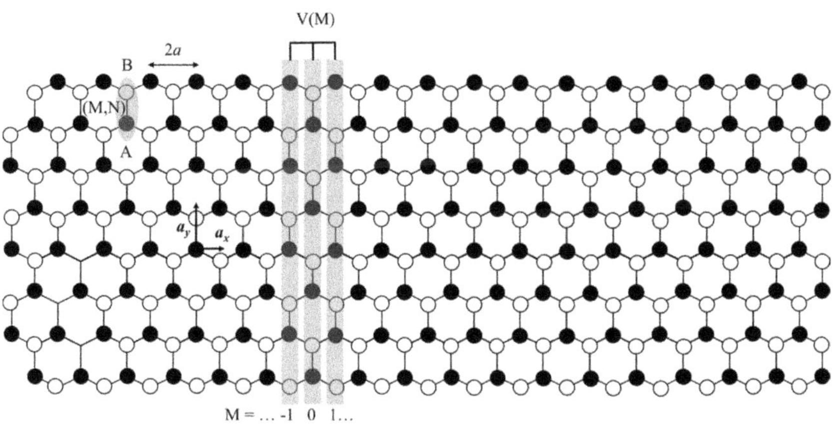


Fig. 7.1 The shaded areas represent the three-chain gate potential ($N_c = 3$) along the armchair chain with potential strength $V(M)$ in pristine graphene, where $M$ is the indices along the $x$-axis. The unit cell is labeled by $(M, N)$, and the $A$- ($B$-) sites are indicated by solid- (open-) circles in the unit cell. The primitive vector is $\boldsymbol{R}_{MN} = Ma\hat{x} + N\sqrt{3}a\hat{y}$. The unit lengths are $(a_x, a_y) = (a, \sqrt{3}a)$, and the periodicity along the $y$-axis is $2\sqrt{3}a$.

graphene scattered upon the gate potential, and the LS is generated from the Dirac point within the energy gap.

The Hamiltonian of the 2D bulk graphene given by the effective tight-binding Hamiltonian is

$$\begin{bmatrix} 0 & -t_0 h_{\boldsymbol{k}} \\ -t_0 \tilde{h}_{\boldsymbol{k}} & 0 \end{bmatrix} \begin{bmatrix} C_A(\boldsymbol{k}) \\ C_B(\boldsymbol{k}) \end{bmatrix} = E(\boldsymbol{k}) \begin{bmatrix} C_A(\boldsymbol{k}) \\ C_B(\boldsymbol{k}) \end{bmatrix}, \tag{1}$$

where $C_{A(B)}(\boldsymbol{k})$ denotes the wave function in the $A$ $(B)$ site.

Here, we define $h_{\boldsymbol{k}} = (1 + 2e^{-ik_y\sqrt{3}a} \cos k_x a)$ and $\tilde{h}_{\boldsymbol{k}} = (1 + 2e^{ik_y\sqrt{3}a} \cos k_x a)$. In this chapter, the energy unit is $t_0 = 2.66\,\text{eV}$ [23] and the nearest-neighbor hopping constant is $-t_0$ for bulk graphene. The energy can be obtained in the form of [23, 43]

$$E(\boldsymbol{k}) = \nu\sqrt{1 + 4\cos(k_x a)\cos(k_y\sqrt{3}a) + 4\cos^2(k_x a)}. \tag{2}$$

The notation $\nu = \pm 1$ denotes the positive and negative energy branches, and we introduce the dimensionless energy $\varepsilon(\boldsymbol{k})$ as $E(\boldsymbol{k}) = \nu\varepsilon(\boldsymbol{k})$. The pseudospin can be expressed by

$$\chi(\boldsymbol{k}) = \frac{1}{\sqrt{N_{\boldsymbol{k}}}} \begin{bmatrix} 1 \\ -\nu\frac{\varepsilon(\boldsymbol{k})}{h(\boldsymbol{k})} \end{bmatrix}, \tag{3}$$

where $N_{\boldsymbol{k}}$ is the normalized constant.

In order to investigate the topological gate-induced LS, the energy range is focused on the gap region. Figure 7.2 shows the energy dispersion and the energy gap for a given $k_y$. The left-going and right-going wave vectors, $p$ and $q$ are determined by the given $k_y$ and the incident energy $\varepsilon$. For convenience, the arguments $k_x a$ and $k_y\sqrt{3}a$ are replaced by $k_x$ and $k_y$ (parallel to the gate), respectively. It is noted that the wave vectors $k_x$ are obtained from the relation

$$k_x = \cos^{-1}\left(-\frac{1}{2}\cos k_y \pm \frac{1}{2}\sqrt{\varepsilon^2 - \sin^2 k_y}\right), \tag{4}$$

where the sign inside the square root determines the real or complex values of $k_x$. The right-going and left-going $k_x$ are denoted by $p$ and $q$, respectively. When the incident energy $\nu\varepsilon$ is within the gap ($\varepsilon^2 < \sin^2 k_y$), $p$ and $q$ become complex values. We can define the energy gap as $\varepsilon_g(k_y) = |\sin k_y|$. This condition considers the electron energy within a gap where the topological LS is generated by a biased gate due to a quasiparticle picture. The

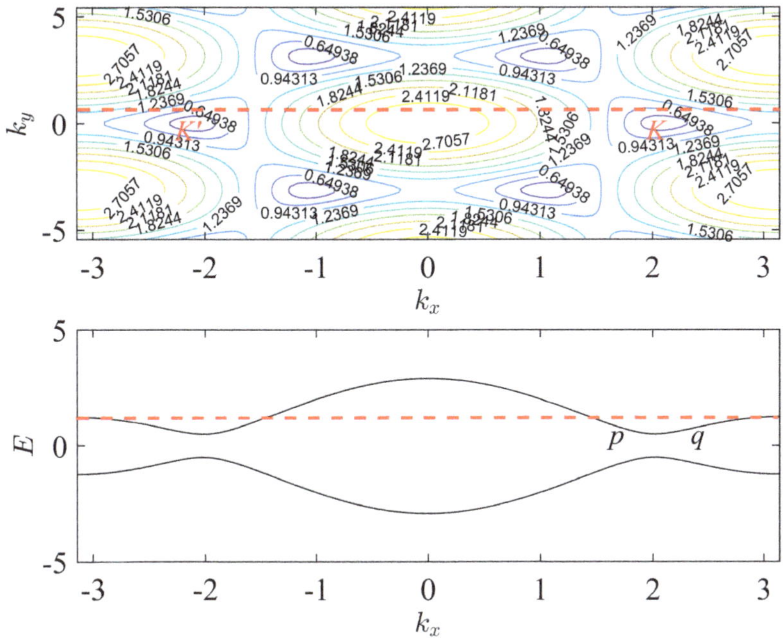


Fig. 7.2 The upper panel shows the contour of the dispersion in bulk graphene, and the K (K′) valleys are indicated. The lower panel shows the dispersion for a given $k_y = 0.3$ with a given energy gap. The left-going and right-going wavevectors are $p$ and $q$, respectively. The dashed line presents the specific energy for a given $k_y$.

orientation of the Bloch state pseudospins within the energy gap is crucial to the formation of the topological LS around the gate.

### 7.2.1 *Pseudospins rotation within the energy gap*

We focus on the electron energy within the energy gap, namely $|E| \leq \varepsilon_g(k_y)$, and the pseudospins $\chi(k_{\xi\alpha})$ expression for the Bloch state $|k_{\xi\alpha}\rangle$ is discussed as follows. For a given $(E, k_y)$, there are four Bloch states, as shown in Fig. 7.2: $|k_{\xi\alpha}\rangle$ with $\xi = \pm$ standing for left-going ($p$) or right-going ($q$) and $\alpha = \pm$ standing for $K$ ($K'$) valley. Within the energy gap, the complex $k_x$ is given by Eq. (4), where the wavevectors $k_{p\alpha} = (p_\alpha, k_y)$ and $k_{q\alpha} = (q_\alpha, k_y)$. The real part of $k_x$ is $k_{xR}$, which is positive (negative) for $\alpha = \pm$, and the imaginary part $k_{xI}$ is positive (negative) for $\xi = \pm$. From quantum mechanics, it is known that a conventional spinor with the magnitudes of its two components must to equal to the in-plane spinor. However, the pseudospin of Eq. (3) no longer lies on the $xy$-plane (the magnitude of the two-component pseudospin is not equal) because the complex $p_\alpha$ leads to $\varepsilon \neq |h(k_{p\alpha})|$. Here, $h(\boldsymbol{k}_{p\alpha}) = 1 + 2\cos p_\alpha e^{-ik_y}$. Thus, the non-equal pseudospin components result in the out-of-plane pseudospin. When Eq. (4) is plugged into $h(\boldsymbol{k}_{p\alpha})$, one can obtain

$$h(\boldsymbol{k}_{p\alpha}) = ie^{-ik_y}(\sin k_y + \alpha\sqrt{\sin^2 k_y - \varepsilon^2}). \tag{5}$$

By employing Eq. (3) and $p_\alpha = -p^*_{-\alpha}$, one can define the angle $\theta_{p\alpha}$ for $\sin\theta_{p\alpha} = \varepsilon/\sin k_y$ which gives rise to a more meaningful pseudospin form as

$$\chi(\boldsymbol{k}_{p1}) = \begin{bmatrix} \cos(\theta_{p1}/2) \\ \nu\sin(\theta_{p1}/2)e^{i\phi_y} \end{bmatrix}, \tag{6}$$

where $\phi_y = k_y + \pi/2$.

It is most important that the orientation of the pseudospin rotating as $\varepsilon$ varies can be visualized in the transparent form of Eq. (6). The novel topological origin can be interpreted as a ration operator of the pseudospin, and this analytical analysis is described in the following section. The other pseudospin state $\chi(k_{p\bar{1}})$ can be directly obtained from Eq. (6) by replacing $\theta_{p1} \leftrightarrow \pi - \theta_{p\bar{1}}$. Also, the right-going pseudospin state connected to the left-going pseudospin state is given by $\chi(\boldsymbol{k}_{q\alpha}) = \chi(\boldsymbol{k}_{p\bar{\alpha}})$.

The highlights of the topological origin of gate-induced LS in graphene are simply illustrated in Fig. 7.3. This figure demonstrates the LS-associated pseudospin state $\chi(k_{p\alpha})$ evolving in $x'z$-plane by scanning the energy $E$ from the bottom to the top of the energy gap. The relative angle $\Delta\theta_p$ between the two pseudospins can vary from 0 to $2\pi$ (from $-\varepsilon_g$ to $\varepsilon_g$), and $\Delta\theta_p$ is counted from $\chi_{p\bar{1}}(E)$ to $\chi_{p1}(E)$. The angle between $\hat{z}$ and pseudospins $\chi_{p,1}(E)$, or $\chi_{p,\bar{1}}(E)$ is $\theta_{p,1}$, or $\pi - \theta_{p,\bar{1}}$, respectively. For the cases of positive energy ($\nu > 0$), both pseudospins lie in the first and forth quadrants of the $x'z$-plane. On the other hand, the cases of negative energy ($\nu < 0$) give rise to both pseudospins lying in the second and third quadrants of the $x'z$-plane. The reason for this can be easily determined to be due to the factor $\nu$ in Eq. (6).

The rotation of the corresponding pseudospin $\chi(\boldsymbol{k}_{p\alpha}, E)$ by tracking energy $E$ from $-\varepsilon_g$ to $\varepsilon_g$ is shown in Fig. 7.3. Both pseudospins start at $-\hat{x}'$ and stop at $\hat{x}'$ with the opposite revolution. In the point of view along the direction $\hat{z} \times \hat{x}'$, $\chi(\boldsymbol{k}_{p1})$ and $\chi(\boldsymbol{k}_{p\bar{1}})$ rotate clockwise and counterclockwise,

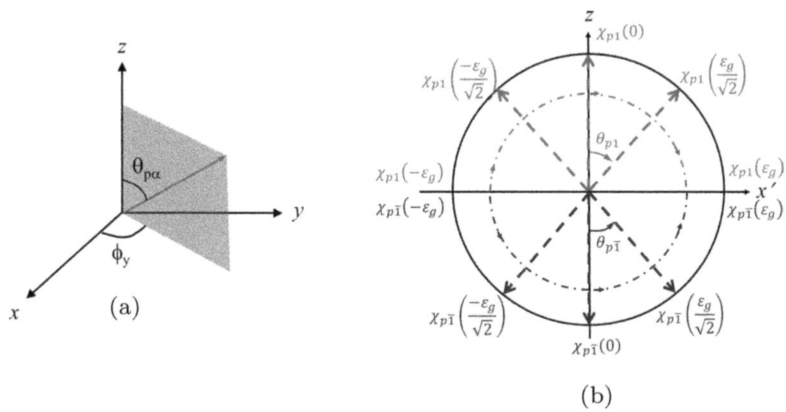


Fig. 7.3 The orientations of the pseudospin state $\chi(\boldsymbol{k}_{p\alpha}, E)$ within the energy gap, $|E| \leq \varepsilon_g(k_y)$. Illustration of the rotation of the pseudospin by letting $\chi_{p1}(E) = \chi(\boldsymbol{k}_{p\alpha}, E)$. (a) The red arrow denotes the pseudospin $\chi_{p1}(E)$ with an angle $\theta_{p\alpha}$ with respect to the $z$-axis and the set of $(\varepsilon, k_y)$ exhibits the $xz$-plane rotating to the $x'z$-plane by the angle $\phi_y$. (b) The K (K′)-valley pseudospin, indicated by the solid (dashed) arrow, rotates clockwise (counterclockwise) as $E$ varies from the bottom to the top of the energy gap.

respectively. The relative angle $\Delta\theta_p$ monotonically changing from 0 to $2\pi$ is important to establishing the topological origin of gate-induced LS.

### 7.2.2 *Secular equation of gate-induced localized state*

In this section, the LS secular equation is obtained for one-chain gate potential, and the LS dispersion is analytically obtained for the long-wavelength regime from the LS-secular equation. The one-chain gate potential provides a platform to study the pseudospins of Bloch state scattered upon the gate potential, and the symmetric environment across the one-chain gate (at $M = 0$) allows us to construct the scattering states on one side ($M < 0$) of the gate.

The total Hamiltonian with a gated one-chain potential in graphene is given by $H = h_0 + h_{\text{gate}}$, where the detailed expressions are given by

$$\begin{cases} h_0 = \sum\limits_{\langle M,M',N,N',s,s' \rangle} |M,N,s\rangle \langle M',N',s'| \\ h_{\text{gate}} = V_0 \sum\limits_{N,s} |M=0,N,s\rangle \langle M=0,N,s|. \end{cases} \tag{7}$$

The first term denotes summation over the nearest-neighbor site, and the second term denotes the one-chain gate potential at $M = 0$ with a strength of $V_0$. The LS wave function $|\Psi^L_{ky}\rangle$ on the left-hand side ($M \leq 0$) of gate potential for a given $k_y$ has the form

$$|\Psi^L_{ky}\rangle = \sum_{M\leq 0,N}\sum_{\alpha=\pm} r_{p_\alpha} e^{i\boldsymbol{k}_{p_\alpha}\cdot\boldsymbol{R}_{MN}}\chi(\boldsymbol{k}_{p_\alpha})|M,N\rangle$$
$$= {\sum_N}' e^{ik_yN}\sum_{M\leq 0} g_M|M,N\rangle, \tag{8}$$

where the primed summation indicates the convention that $M + N$ must be even integers, and the auxiliary function for $M \leq 0$ is given by

$$g_M = \sum_\alpha r_{p\alpha} e^{ip_\alpha M}\chi(\boldsymbol{k}_{p\alpha}). \tag{9}$$

On the right-hand side ($M \geq 0$) of the gate, the LS wave function $|\Psi^R_{ky}\rangle$ is similar to Eq. (8) and obtained from it by replacing the left-going wavevector $p$ in $g_M$ by the right-going wavevector $q$ for the corresponding $M$. From the Schrödinger equation, matching both sides of the wave function $|\Psi^{L(R)}_{ky}\rangle$ at the gate site ($M = 0$) gives

$$Eg_0 = -[\sigma_x g_0 + Tg_1 + Tg_{-1}] + V_0 g_0, \tag{10}$$

where

$$T = \begin{bmatrix} 0 & e^{-ik_y} \\ e^{ik_y} & 0 \end{bmatrix}. \tag{11}$$

The symmetry leads to $g_M = g_{-M}$. It is known that $g_M$ has to satisfy the bulk-graphene Schrödinger equation at $M = 0$, resulting in

$$(E + \sigma_x)g_0 + Tg_{-1} + T\tilde{g}_1 = 0. \tag{12}$$

The notation $\tilde{g}_1$ is the extension of Eq. (9) for $M = 1$. Then, the LS-secular equation is obtained in the form of

$$-V_0 g_0 + Tg_{-1} + T\tilde{g}_1 = 0. \tag{13}$$

This LS-secular equation can be rewritten using the corresponding reflection coefficients $(r_{p1}, r_{p\bar{1}})$ as

$$\begin{bmatrix} V_0 C^A_{p1} + 2i\beta^* \sin p_1 C^B_{p1} & V_0 C^A_{p\bar{1}} + 2i\beta^* \sin p_{\bar{1}} C^B_{p\bar{1}} \\ V_0 C^B_{p1} + 2i\beta \sin p_1 C^A_{p1} & V_0 C^B_{p\bar{1}} + 2i\beta \sin p_{\bar{1}} C^A_{p\bar{1}} \end{bmatrix} \begin{bmatrix} r_{p1} \\ r_{p\bar{1}} \end{bmatrix} = 0, \tag{14}$$

with $\beta = e^{ik_y}$. After some lengthy algebra, one can simplify Eq. (14) into the elegant form of

$$\begin{bmatrix} D_1 & D_2^* \\ D_2 & D_1^* \end{bmatrix} \begin{bmatrix} r_{p1} \\ r_{p\bar{1}} \end{bmatrix} = 0. \tag{15}$$

Here, $D_1 = V_0 \cos(\theta/2) - 2\nu \sin(p_1) \sin(\theta/2)$ and $D_2 = V_0 \sin(\theta/2) - 2\nu \sin(p_1) \cos(\theta/2)$. It is noted that the matrix equation in Eq. (15) leads to the real LS-secular equation, expressed by

$$V_0^2 \cos\theta - 4\cos\theta |\sin p_1|^2 - 4\nu V_0 \sin\theta \,\mathrm{Re}[p_1] = 0, \tag{16}$$

where $\sin\theta = \varepsilon / \sin k_y$ is related to the dimensionless LS-dispersion $\varepsilon$. The value of $\sin p_1$ is in the first quadrant of Eq. (4), and the LS-dispersion $E(k_y)$ is numerically obtained from the LS-secular equation in Eq. (16).

Figure 7.4 presents the LS dispersions as a function of $k_y$ (parallel to the gate direction) for one-chain gate potential for various $V_0$. The

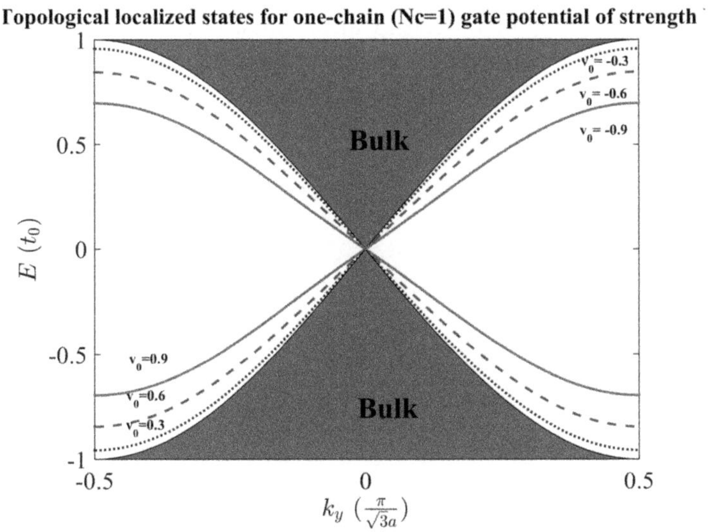


Fig. 7.4 The LS dispersion of a one-chain gate in pristine graphene as a function of $k_y$ for various $V_0$. The Dirac-point associated LSs launch from the Dirac point and extend to the zone boundary along $k_y$. The gray areas present the bulk dispersion from the $E$–$k_y$ point of view. The linear behavior of LS dispersion is observed in the small $k_y$ regime (long-wavelength limit). The solid, dashed and dotted curves show the LS dispersions for $V_0 = \pm 0.3$, $\pm 0.6$ and $\pm 0.9$, respectively. The large magnitude of $V_0$ leads to the LS dispersion being further away from the bulk dispersion.

general feature of LS dispersion is that each LS dispersion launches from the Dirac point and extends up to the Brillouin zone (BZ) boundary at $k_y = \pm 0.5$ in units of $(\frac{\pi}{\sqrt{3}a})$ within the entire gap of $-\varepsilon_g < \varepsilon < \varepsilon_g$. Each branch of the LS dispersion is fully occupied within the entire gap, corresponding to the evanescent wavevectors of $p_\alpha$ or $q_\alpha$. It is clearly shown that the LS dispersion is further away from the bulk dispersion, leading to a more significant localized state. Moreover, the same magnitude of $V_0$ with positive and negative signs shows the two LS dispersions $(\pm|V_0|)$ in mirror symmetry with respect to $E = 0$. It is important that the linear dispersion behavior of LS dispersion is near the Dirac point for the long-wavelength regime and that it is possible to identify the branch of LS dispersion in gated graphene with a bias $\pm|V_0|$. It is a key finding that the LS dispersion is generated by an arbitrary non-zero potential ($V_0 \neq 0$).

The long-wavelength approach (small $k_y$) of the LS dispersion can be calculated from Eq. (16). For the long-wavelength regime, the significant feature of the LS dispersion presents a linear behavior, as shown in Fig. 7.5. Such a linear behavior of the LS dispersion in the long-wavelength regime can be obtained using the LS energy approximation form of $E = \Gamma(V_0)|k_y|$, where $\Gamma(V_0)$ is the proportional constant depending on $V_0$. It is known that $p_1 = (\frac{2\pi}{3} + \Delta p_{1R}) - i\Delta p_I$ with positive $\Delta p_I$. One can substitute $p_1$ into

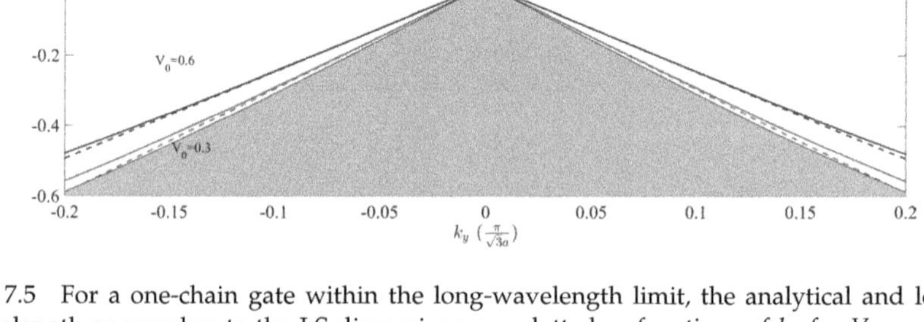


Fig. 7.5 For a one-chain gate within the long-wavelength limit, the analytical and long-wavelength approaches to the LS dispersions are plotted as functions of $k_y$ for $V_0 = \pm 0.3$ and $V_0 = \pm 0.6$. The solid and dashed curves present the analytical and long-wavelength approaches to the LS dispersions, respectively.

Eq. (4), resulting in

$$\begin{cases} \Delta p^I = |k_y|\sqrt{(1-\Gamma^2)/3} \\ \Delta p_1^R = k_y^2(\Gamma^2-4)/(6\sqrt{3}). \end{cases} \tag{17}$$

It is important to dig out the proportional constant by substituting Eq. (17) into the LS secular equation, and it gives a $V_0$-dependent expression as

$$\Gamma(V_0) = \left(\frac{V_0^2-3}{V_0^2+3}\right)\mathrm{sgn}(V_0), \tag{18}$$

where sgn($V_0$) determines the sign of $V_0$. Equation (18) is applicable for all orders of $V_0$. It reveals the positive and negative slopes corresponding to the sign of $V_0$ with $|V_0| \ll \sqrt{3}$. The LS energy is positive or negative, according to Eq. (18), within the energy gap and close to the continuum, can be explained by the LS formed from continuum states being dragged into the gap, as shown in Fig. 7.5. This figure shows that the long-wavelength LS dispersion agrees well with the fully analytical LS dispersion in the small $k_y$ region. The discrepancy between the long-wavelength and analytical approaches becomes more apparent for a large $k_y$ region.

By solving the LS energy $\varepsilon_{\mathrm{LS}}$ for a given $V_0$ and $k_y$ in Eq. (16), we can substitute $\varepsilon_{\mathrm{LS}}$ into Eq. (14) to determine the reflection coefficients $(r_{p1}, r_{p\bar{1}})$ and plug them into Eq. (8) to obtain the LS wave function $\Psi^L_{ky}$ for $M \leq 0$. Similarly, one can also obtain the wave function $\Psi^R_{ky}$ for $M > 0$. For the case of one-chain gate potential, Fig. 7.6 presents the LS wave function $|\Psi_M|^2$ varying with site $M$ along the transverse zigzag carbon chain.

The intuitive picture to observe the property of the LS can be explored in the spatial distribution of their LS wave function in Fig. 7.6 for $V_0 = 0.3$, 0.6 and 0.9. With increasing $V_0$, the LS wave function becomes more localized. The one-chain gate is located at $M = 0$ and along the armchair carbon-chain direction in bulk graphene. It is remarkable that the LS wave function decays from $M = 0$, and the decay length is decreasing as $V_0$ increases. The decaying length of the LS wave function depends on the magnitude of the complex wavevectors $p_\alpha$ $(q_\alpha)$ due to $\varepsilon_{\mathrm{LS}}$ within the energy gap. The LS wave function does not contribute to the conventionally propagating Bloch wave, but to the evanescent wave around the applied gate. The topological LS wave function based on the

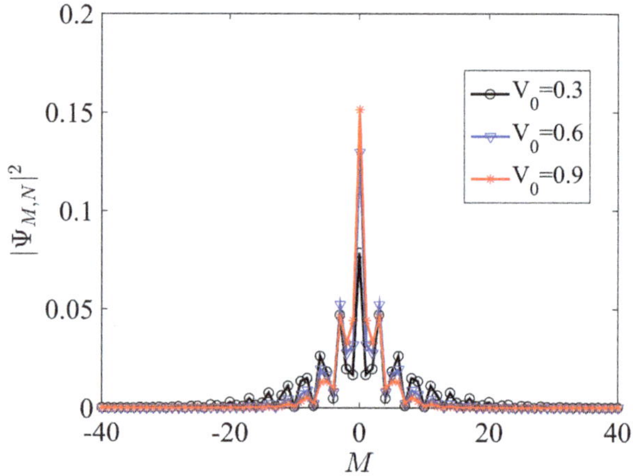


Fig. 7.6 The LS wave function $|\Psi_M|^2$ is plotted as a function of site $M$ for one-chain gate in pristine graphene for $V_0 = 0.3$, 0.6 and 0.9. The one-chain gate is located at $M = 0$. The magnitude of the LS wave function has its maximum value at the gate center, and the decay length decreases as $V_0$ increases. The open-circle, triangle and star indicate $V_0 = 0.3$, 0.6 and 0.9, respectively.

corresponding pseudospin rotating mechanism has been pictorially illustrated in Fig. 7.3.

The LS wave function $|\Psi_{M,N}|^2$ can be extended to the continuum space from discrete sites through the Gaussian-smoothing profile given by

$$|\Psi_{M,N}|^2 = \int_{-\infty}^{\infty}\int_{-\infty}^{\infty} e^{\frac{-1}{2\sigma^2}[(x-x_M)^2+(y-y_N)^2]}dxdy \tag{19}$$

where $\sigma$ is the smoothing scale. The original wave function is a one-site value with respect to $(x_M, y_N)$, which extends to the continuum space $(x, y)$. The 2D spatial wave function is plotted in Fig. 7.7 in the cases of (a) $V_0 = 0.3$ and 0.9. Figure 7.7 provides the localized wave function around the gate for a given $k_y = 0.3$. The spatial distribution is periodic along the $y$-direction, with a periodicity of $2\sqrt{3}a$. The maximum values of the wave function are located at the gate ($M = 0$).

### 7.2.3 *Topological origin: Pseudospin rotations*

The topological origin of the LS formation, given by the LS-secular equation in Eq. (13) for a non-zero $V_0$, can be transformed into a pseudospin-rotation form within the energy gap. Generally, it is important to point out

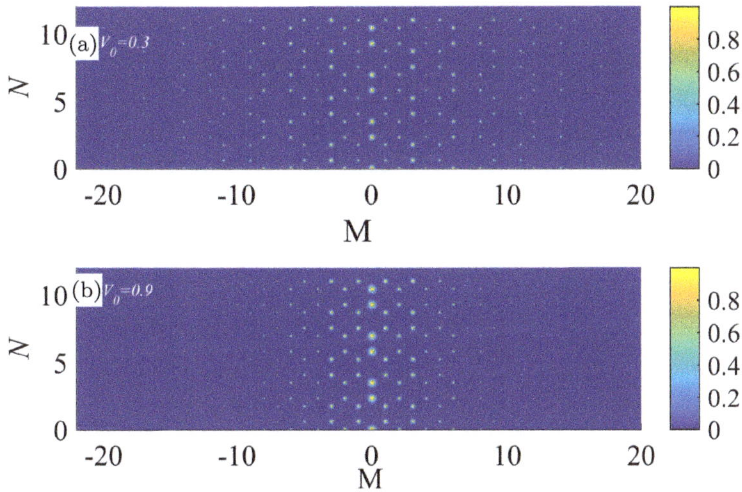


Fig. 7.7 The 2D spatial distribution of wave function $|\Psi_{M,N}|^2$ is shown in the $xy$-plane for cases: (a) $V_0 = 0.3$ and (b) $V_0 = 0.9$, where the cell indices $M$ and $N$ are along the $\hat{x}$ and $\hat{y}$ directions, respectively. The one-chain gate is located at $M = 0$.

that a two-component secular equation may not be expressed in terms of a pseudospin-rotating form. The crucial point of formation of the pseudospin rotation must be equipped with a geometric basis on which the LS formation is established.

Starting from Eq. (13), one can expand $g_M$ and $\tilde{g}_M$ in the form of

$$-V_0 \sum_\alpha r_{p\alpha} \chi(\boldsymbol{k}_{p\alpha}) + T \sum_\alpha r_{p\alpha} e^{-ip_a} \chi(\boldsymbol{k}_{p\alpha}) - T \sum_\alpha r_{p\alpha} e^{ip_a} \chi(\boldsymbol{k}_{p\alpha}) = 0. \quad (20)$$

The connection between the two pseudospin states $\chi(\boldsymbol{k}_{p1})$ and $\chi(\boldsymbol{k}_{p\bar{1}})$ is given by

$$\chi(\boldsymbol{k}_{p1}) = -\left[\frac{r_{p\bar{1}}}{r_{p1}} M(p_1)^{-1} M(p_{\bar{1}})\right] \chi(\boldsymbol{k}_{p\bar{1}}), \quad (21)$$

where the $2 \times 2$ matrix $M(p_\alpha)$ is explicitly expressed as

$$M(p_a) = \begin{bmatrix} V_0 & -2\sin p_a e^{-i\phi_y} \\ 2\sin p_a e^{i\phi_y} & V_0 \end{bmatrix}. \quad (22)$$

The successful requirement of the topological LS formation for the given $E$ and $k_y$ can be understood in Eq. (21), where the K-valley pseudospin

$\chi(\boldsymbol{k}_{p1})$ needs to be expressed by the K′-valley pseudospin $\chi(\boldsymbol{k}_{p\bar{1}})$ via the proper matrix operator $[-(r_{p\bar{1}}/r_{p1})M(p_1)^{-1}M(p_{\bar{1}})]$.

When a non-zero $V_0$ is turned on in a gated graphene sheet, the topological LS dispersion can be found using Eq. (21) for the entire gap $|E| < \varepsilon_g(k_y)$ by a rotation matrix with an appropriate rotation axis (along $\hat{z} \times \hat{x}'$, as shown in Fig. 7.3(a)). Furthermore, the determinant of $M(p_1)$ is the complex given in a simple form as

$$\det(M(p_1)) = V_0^2 + 4\sin^2 p_1 \equiv \Lambda e^{i\delta_M}. \tag{23}$$

Then, Eq. (21) turns out to be

$$\chi(\boldsymbol{k}_{p1}) = C_0 \Im(k_y, E)\chi(\boldsymbol{k}_{p\bar{1}}), \tag{24}$$

where $C_0 = -(r_{p\bar{1}}/r_{p1})e^{-i\delta_M}$ and

$$\Im(k_y, E) = \frac{Adj[M(p_1)]}{\Lambda} M(p_{\bar{1}}). \tag{25}$$

Adj$[M(p_1)]$ denotes the adjugate of the matrix $M(p_1)$.

Equation (24) can be simplified by rotating the pseudospin back onto the $xz$-plane, namely

$$\tilde{\chi}(\boldsymbol{k}_{p1}) = \Re(z, -\phi_y)\chi(\boldsymbol{k}_{p\bar{1}}), \tag{26}$$

where the rotation operator $\Re(\hat{n}, \phi) = e^{-i\hat{n}\cdot\sigma\phi/2}$ is rotated about the axis $\hat{n}$ by an angle $\phi$.

The LS-secular equation in Eq. (24) becomes

$$\tilde{\chi}(\boldsymbol{k}_{p1}) = C_0 \Re(\hat{z}, -\phi_y)\Im(k_y, E)\Re^{\dagger}(\hat{z}, -\phi_y)\tilde{\chi}(\boldsymbol{k}_{p\bar{1}}) \equiv \tilde{\Im}(k_y, E)\tilde{\chi}(\boldsymbol{k}_{p\bar{1}}). \tag{27}$$

The operator $\tilde{\Im}(k_y, E)$ is unimodular because of the relation $\Lambda^2 = (V_0 - 4|\sin p_1|^2)^2 + (4V_0 \operatorname{Re}[\sin p_1])^2$. It is more convenient to rewrite the operator $\tilde{\Im}(k_y, E)$ in the form of the rational operator denoted by $\tilde{\Im}(k_y, E) = \Re(\hat{y}, \phi)$ when the LS energy $\varepsilon_{\text{LS}}(k_y)$ is within the whole gap. Here, the rational angle $\phi$ about the rotational axis $\hat{y}$ can be obtained using the relations

$$\begin{cases} \sin(\phi/2) = \dfrac{-4V_0 \operatorname{Re}[\sin p_1]}{\Lambda} \\[2ex] \cos(\phi/2) = \dfrac{V_0^2 - 4|\sin p_1|^2}{\Lambda}. \end{cases} \tag{28}$$

The new rotational axis returns to the $\hat{y}$-axis instead of the previously tilted $\hat{y}'$-axis, as shown in Fig. 7.3(b). This reflects the fact that the applied rotational operator $\Re(\hat{z}, -\phi_y)$ changes the plane of the pseudospin to the $xz$-plane (in Eq. (27)).

It is noted that the value of $\phi$ in Eq. (28) would never cross the integer multiple of $2\pi$ in the energy gap range because $\mathrm{Re}[\sin p_1] \neq 0$ within the same range. From the relation $p_1 = p_{1R} - ip_I$ with $p_I \geq 0$, we can obtain $\mathrm{Re}[\sin p_1] = \sin p_{1R} \cosh p_I$, and the zeros of $\mathrm{Re}[\sin p_1]$ can be obtained from $p_{1R}$ due to the non-zero $\cosh p_I$. From Eq. (4), the relation $2 \sin p_{1R} \sinh p_I = \sqrt{\sin^2 k_y - \varepsilon^2}$ reveals that $\sin p_{1R}$ does not vanish in the energy gap range. On the other hand, for a given $V_0$, the rotational angle $\phi$ of the pseudospin rotation operator $\tilde{\Im}(k_y, E)$ would remain in the same $2\pi$ interval, and $\phi$ would be negative or positive in the whole gap range.

From Eq. (28), it is clear that the ungated graphene case of $V_0 = 0$ gives rise to $\tilde{\Im}(k_y, E) = -1$, which is not a pseudospin rotation operator about the rotational $\hat{y}$-axis. When the energy $E$ meets either the top or bottom of the energy gap, $E$ is coupled with the bulk dispersion, and the topological LS cannot be found.

It is also important about the additional sign factors from pseudospin rotation and about its relation to the signs of the energy and of the rotational angle $\phi$ in $\Re(\hat{y}, \phi)$. Starting from Eq. (27), one can get the equation given by

$$\tilde{\chi}(\boldsymbol{k}_{p1}) = -\nu\nu_R \Re(k_y, E) \tilde{\chi}(\boldsymbol{k}_{p\bar{1}}), \tag{29}$$

where $\nu_R$ denotes the sign of $\phi_p$. We know that $\phi_p = \Delta\theta_p$ for positive $\phi_p$ and $\phi_p = -(2\pi - \Delta\theta_p)$ for negative $\phi_p$. The relative angle is counted from $\chi_{p\bar{1}}(E)$ to $\chi_{p\bar{1}}(E)$ along the clockwise direction (see Fig. 7.3(b)). Combining the sign factor, the LS-secular equation in Eq. (27) becomes

$$\tilde{\chi}(\boldsymbol{k}_{p1}) = -\nu\nu_R \tilde{C}_0 \tilde{\Im}(k_y, E) \tilde{\chi}(\boldsymbol{k}_{p\bar{1}}), \tag{30}$$

where $\tilde{C}_0 = \nu\nu_R (r_{p\bar{1}}/r_{p1}) e^{-i\delta_M}$ and $\nu_R$ denotes the sign of $\phi$ in $\tilde{\Im}(k_y, E)$.

By choosing $\tilde{C}_0 = 1$, Eq. (30) becomes an elegant pseudospin-rotation form that makes a successful rotation from $\tilde{\chi}(\boldsymbol{k}_{p\bar{1}})$ to $\tilde{\chi}(\boldsymbol{k}_{p1})$. The choice of $\tilde{C}_0 = 1$ is always acceptable, and this choice determines the relative phase between two reflection coefficients $r_{p\alpha}$. The unit choice of $\tilde{C}_0 = 1$ also provides a physically sensible result for $r_{p\alpha}$, which implies that the LS wave function is established by equal weighting the (K, K$'$) valley

Bloch waves. The valley symmetry is due to the AB site symmetry in the gated graphene.

The geometric perspective of the pseudospin rotation demonstrates the consequence of topological LS formation, and the pseudospin rotation operator also rules out accidental possibility in such LS formation. This consequent qualitative change, with or without an LS, depends on an infinitesimal change, non-zero or zero $V_0$, so that this mechanism supports the significant topological point of view in LS. In addition, the LS branch has Dirac-point characteristics, which means that the LS dispersion launches from the Dirac point (at $k_y = 0$) and spreads out to its BZ boundary.

## 7.3 Localized State in the Two-Chain Gate ($N_c = 2$)

In this section, we use the theoretical framework developed in Section 7.2 to the case of a two-chain gate potential with homogeneous or inhomogeneous potential strength ($V(M = 0)$ or $V(M = 1)$), where we consider the two-chain gates located at sites $M = 0$ and $M = 1$, as shown in Fig. 7.7(a). There are two LS branches: the topological nature of each is shown in the pseudospin-rotation form of the LS-secular equation. Also, each LS branch carries the Dirac-point characteristics.

### 7.3.1 *LS-secular equation for the $N_c = 2$ case*

The LS-secular equation is derived from the Schrödinger equation at $M = 0$ and $M = 1$ gated chains along the direction of the armchair carbon chain. We choose the origin of the $x$-coordinate at the midpoint of the two chains to reduce the two equations into one equation. The notation $\eta = \pm$ is defined for the two LS branches. The LS-secular equation can be real by using the homogeneous equations of $r_{p\alpha}$, and the real LS-secular equation is convenient to estimate the LS dispersions.

The choice of $x$ origin gives us, for $|\Psi^L_{ky}\rangle$,

$$g_M = \sum_{\alpha} r_{p\alpha} e^{ip_\alpha(M-1/2)} \chi(\boldsymbol{k}_{p\alpha}) \tag{31}$$

in the $M \leq 0$ region. Similarly, $g_M$ for $|\Psi^R_{ky}\rangle$ in the $M \geq 1$ region is obtained from Eq. (31) by replacing $p$ by $q$. At $M = 0$, we subtract the equation obtained from the Schrödinger equation and the equation obtained from that for pristine graphene to get

$$V_0 g_0 + T(\tilde{g}_1 - g_1) = 0. \tag{32}$$

Here, $\tilde{g}_1$ is the extension to the $M = 1$ chain defined in $M \leq 0$ region. Multiply $\Re(\hat{z}, -\phi_y)T$ with the left-hand side of Eq. (32) and use the relations $T^2 = 1$, $p_{\alpha}^* = -p_{\bar{\alpha}}$ and $\Re(\hat{z}, -\phi_y)T\Re^+(\hat{z}, -\phi_y) = -\sigma_y$. We can obtain the $M = 0$ LS-secular equation

$$\begin{aligned} &r_{p1}[e^{ip_1/2} - V_0 e^{-ip_1/2}\sigma_y]\tilde{\chi}(\boldsymbol{k}_{p1}) + r_{p\bar{1}}[e^{ip_{\bar{1}}/2} - V_0 e^{-ip_{\bar{1}}/2}\sigma_y]\tilde{\chi}(\boldsymbol{k}_{p\bar{1}}) \\ &\quad - r_{q\bar{1}} e^{-ip_1/2}\tilde{\chi}(\boldsymbol{k}_{p1}) - r_{q1} e^{-ip_{\bar{1}}/2}\tilde{\chi}(\boldsymbol{k}_{p\bar{1}}) = 0. \end{aligned} \tag{33}$$

Here, $\sigma_y$ is the Pauli matrix. Similarly, the LS-secular equation at $M = 1$ is given by

$$\begin{aligned} &r_{q\bar{1}}[e^{ip_1/2} - V_0 e^{-ip_1/2}\sigma_y]\tilde{\chi}(\boldsymbol{k}_{p1}) + r_{q1}[e^{ip_{\bar{1}}/2} - V_0 e^{-ip_{\bar{1}}/2}\sigma_y]\tilde{\chi}(\boldsymbol{k}_{p\bar{1}}) \\ &\quad - r_{p1} e^{-ip_1/2}\tilde{\chi}(\boldsymbol{k}_{p1}) - r_{p\bar{1}} e^{-ip_{\bar{1}}/2}\tilde{\chi}(\boldsymbol{k}_{p\bar{1}}) = 0. \end{aligned} \tag{34}$$

Comparing Eqs. (33) and (34), the two equations have a similar form, except for the reflection coefficients, and one can get an important relation for these reflection coefficients, which are given by $r_{q\bar{\alpha}} = \eta r_{p\alpha}$ (with $\eta = \pm$). This relation indicates the existence of two LS modes with mode indices $\eta = \pm$.

Therefore, Eqs. (33) and (34) can be reduced into one LS-secular equation given by

$$\begin{aligned} &r_{p1}[e^{ip_1/2} - \eta e^{-ip_1/2} - V_0 e^{-ip_1/2}\sigma_y]\tilde{\chi}(\boldsymbol{k}_{p1}) \\ &\quad + r_{p\bar{1}}[e^{ip_{\bar{1}}/2} - \eta e^{-ip_{\bar{1}}/2} - V_0 e^{-ip_{\bar{1}}/2}\sigma_y]\tilde{\chi}(\boldsymbol{k}_{p1}) = 0. \end{aligned} \tag{35}$$

For the even mode ($\eta = +1$), Eq. (35) can be further simplified to the form of

$$V_0^2 e^{i(p_{\bar{1}}^* - p_1)/2}\cos\theta - 4|\sin(p_1/2)|^2\cos\theta - 2\nu V_0 \sin\theta \sin[(p_1 + p_{\bar{1}}^*)/2] = 0, \tag{36}$$

where $\theta = \theta_{p1}$. The equation can be used to calculate LS dispersion for the even mode ($\eta = +1$). The analytical long-wavelength dispersion is defined by $E_+(k_y) = \Gamma_+(V_0)|k_y|$ from Eq. (36), and the proportional factor is given by

$$\Gamma_+(V_0) = \frac{V_0^2 - 3}{\sqrt{V_0^4 - 3V_0^2 + 9}}\,\mathrm{sgn}(V_0). \tag{37}$$

Following a similar procedure, we can also get the odd-mode ($\eta = -1$) LS dispersion by replacing the factor $\sin(p_1/2)$ by $\cos(p_1/2)$ in Eq. (36). The long-wavelength analytical dispersion $E_+(k_y) = \Gamma_+(V_0)|k_y|$ for the odd-mode LS is straightforwardly obtained, and the proportional factor is given by

$$\Gamma_-(V_0) = \frac{V_0^2 - 1}{\sqrt{V_0^4 + V_0^2 + 1}} \operatorname{sgn}(V_0). \tag{38}$$

It is noted that the proportional factor related to the sign of $V_0$ for $N_c = 2$ homogeneous gated graphene is similar to the result for $N_c = 1$ gated graphene.

Figure 7.8 presents the LS dispersion for the case of the homogeneous two-chain gate ($V(M = 0) = V(M = 1) = V_0$). Figure 7.8(a) shows the two-chain ($N_c = 2$) gate potential applied at $M = 0$ and $M = 1$ for bulk graphene, and both gates have the homogeneous potential $V_0$. Figure 7.8(b) shows the two branches of LS dispersion for the case $N_c = 2$ at $V_0 = 0.6$ (thin blue curves) and 0.9 (thick red curves). We can observe that the two LS branches have higher energy in the odd mode ($\eta = -1$) and lower energy in the even mode ($\eta = +1$). Two branches of LS dispersion launch from the Dirac point and extend to the zone boundary. The branches of LS dispersion are further away from bulk dispersion for larger $|V_0|$. It is noted that both branches of LS dispersion degenerate at the zone boundary due to the homogeneous gate potential.

For the case of the inhomogeneous two-chain gate potential ($V(M = 0) \neq V(M = 1)$), as shown in Fig. 7.9(a), we can perform a similar analysis to obtain a $4 \times 4$ matrix for the LS-secular equation form. Though there is no simpler form in the case of the inhomogeneous two-chain gate, we can use a numerical method to find the LS dispersion. Figure 7.9(b) shows the LS dispersion for the case of two-chain gate potential ($V(M = 0) = 0.6$, $V(M = 1) = 0.8$). The degeneracy occurs at the zone boundary for the two LS branches due to the breaking of the two-chain gate symmetry. But the Dirac-point characteristics remain, as seen in Fig. 7.9(b).

### 7.3.2 *Topological origin for the $N_c = 2$ case*

For the case of the homogeneous two-chain gate, from Eq. (35), we keep the pseudospin states intact to arrive at a form where the connecting

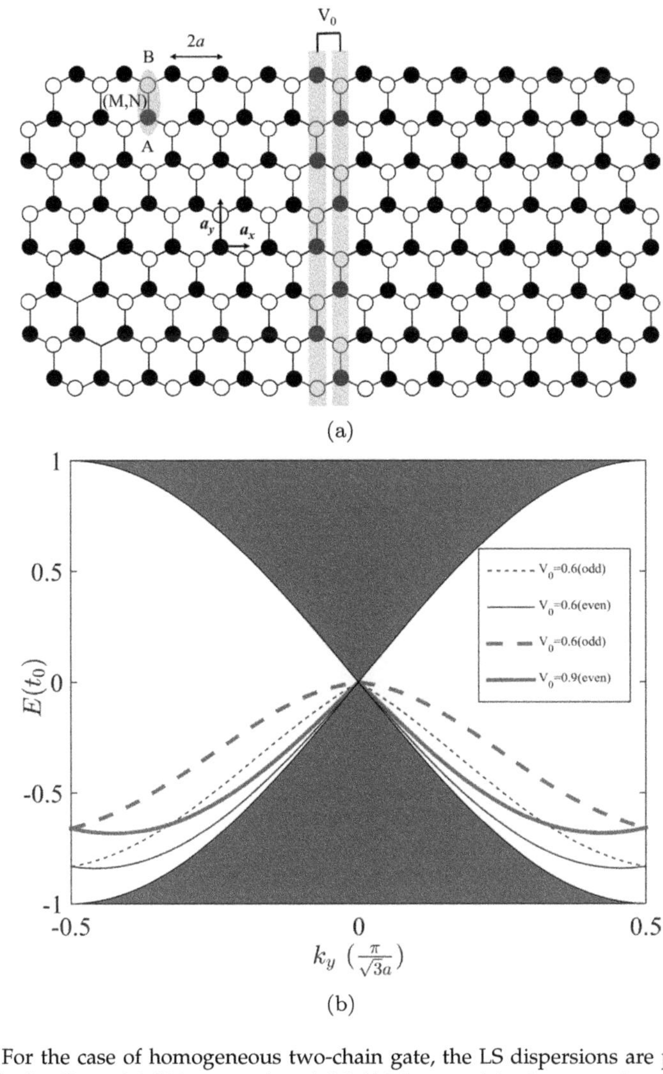


Fig. 7.8 For the case of homogeneous two-chain gate, the LS dispersions are plotted as a function $k_y$ for $V_0 = 0.6$ (thin curves) and 0.9 (thick curves) in the case of two-chain gate potential ($N_c = 2$). The solid and dashed curves correspond to the even and odd parities of their wave functions.

operator is shown to be a pseudospin-rotation operator. For the even-mode LS ($\eta = +1$), Eq. (35) becomes a matrix form:

$$\begin{bmatrix} m_+ & n_+ \\ -n_+ & m_+ \end{bmatrix} \tilde{\chi}(\boldsymbol{k}_{p1}) = C_2 \begin{bmatrix} -m_+^* & n_+^* \\ -n_+^* & -m_+^* \end{bmatrix} \tilde{\chi}(\boldsymbol{k}_{p\bar{1}}), \tag{39}$$

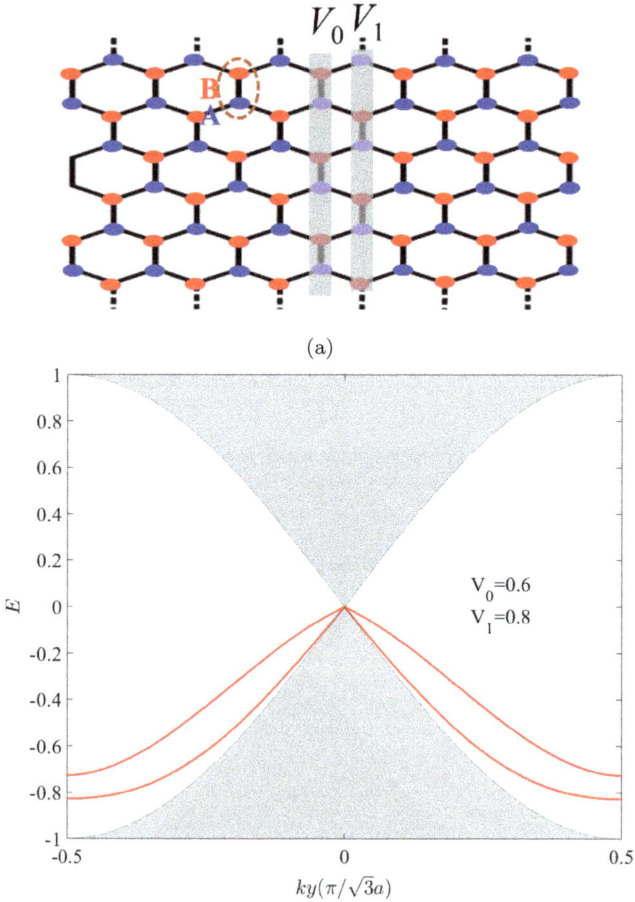


(b)

Fig. 7.9 (a) Two inhomogeneous two-gate potentials $(V_0, V_1)$ are applied to the graphene sheet. (b)The LS dispersions are plotted as a function $k_y$ for different gate potentials $(V_0, V_1) = (0.6, 0.8)$ in the case of two-chain gate potential ($N_c = 2$).

where $m_+ = 2\sin(p_1/2)$, $n_+ = V_0 e^{-ip_1/2}$ and $C_2 = r_{p\bar{1}}/r_{p1}$. Equation (39) can be rewritten as

$$\chi(\boldsymbol{k}_{p1}) = C_2 e^{i\delta_+} \Im_+(k_y, E)\chi(\boldsymbol{k}_{p\bar{1}}). \tag{40}$$

Here, $m_+^2 + n_+^2 = \Lambda_+ e^{i\delta_+}$, and the connecting operator is given by

$$\Im_+(k_y, E) = \begin{bmatrix} |n_+|^2 - |m_+|^2 & m_+ n_+^* + n_+ m_+^* \\ -(m_+ n_+^* + n_+ m_+^*) & |n_+|^2 - |m_+|^2 \end{bmatrix}. \tag{41}$$

The operator $\Im_+(k_y, E)$ is unimodular, so we define

$$\begin{cases} \sin\left(\frac{\phi_+}{2}\right) = -\frac{m_+n_+^*+n_+m_+^*}{\Lambda_+} \\ \cos\left(\frac{\phi_+}{2}\right) = \frac{|n_+|^2-|m_+|^2}{\Lambda_+}. \end{cases} \tag{42}$$

Equation (40) can be re-expressed as

$$\tilde{\chi}(\boldsymbol{k}_{p1}) = -\nu\nu_R\tilde{C}_2^+\Im_+(k_y, E)\tilde{\chi}(\boldsymbol{k}_{p\bar{1}}), \tag{43}$$

where $\tilde{C}_2^+ = -\nu\nu_R C_2 e^{i\delta_+}$ and $\nu_R$ refers to the sign of $\phi_+$. By choosing $\tilde{C}_2^+ = 1$, Eq. (43) becomes an appropriate pseudospin-rotation form. It can also be shown that $m_+n_+^* + n_+m_+^* = 2V_0\sin(p_{1R})$ so that $\phi_+$ is never zero within the energy gap region.

In the case of the two-chain gate, it is important that the LS formation for $\eta = +1$ is the same as that for the case of the one-chain gate.

Figure 7 presents the two branches of the LS dispersion for the LS modes $\eta = \pm 1$.

For the odd-mode LS ($\eta = -1$), Eqs. (40) and (41) can give the LS-secular equation.

By replacing the subscript "+" by "−" everywhere, $C_2$ becomes $-C_2$. Similarly, one can get $m_- = 2\sin(p_1/2)$, $n_- = iV_0e^{-ip_1/2}$ and $m_-^2 + n_-^2 = \Lambda_-e^{i\delta_-}$. Thus, $\tilde{C}_2^- = \nu\nu_R C_2 e^{i\delta_-}$, and $\nu_R$ refers to the sign of $\phi_-$ in $\Im_+(k_y, E) = \Re(\hat{y}, \phi_-)$ as $V_0 \neq 0$. Also, we can conclude that the odd-mode LS ($\eta = -1$) of the two-chain gate has a similar pseudospin rotation feature as in the case of the one-chain gate.

The localized wave functions for even- or odd-mode LS are shown in Figs. 7.10 and 7.11, respectively. The spatial-dependent parity of LS wave functions is exhibited by the real and imaginary parts of the wave function shown for $\eta = \pm 1$ modes in Figs. 7.10(a) and 7.11(a), respectively.

The total LS wave function $|\Psi_M|^2$ presents the even- and odd-mode LSs in Figs. 7.10(b) and 7.11(b), respectively. The odd-mode LS reveals a higher LS energy than the even-mode LS, as shown in Fig. 7.8(b). The odd-mode LS wave function also reveals more oscillating decay behavior than the even-mode LS.

### 7.3.3 *The localized state in the $N_c$-chain gate*

In this section, we develop the generalized LS-secular equation for the case of $N_c$ chain gates ($N_c > 1$). In general, the gate potential is given by

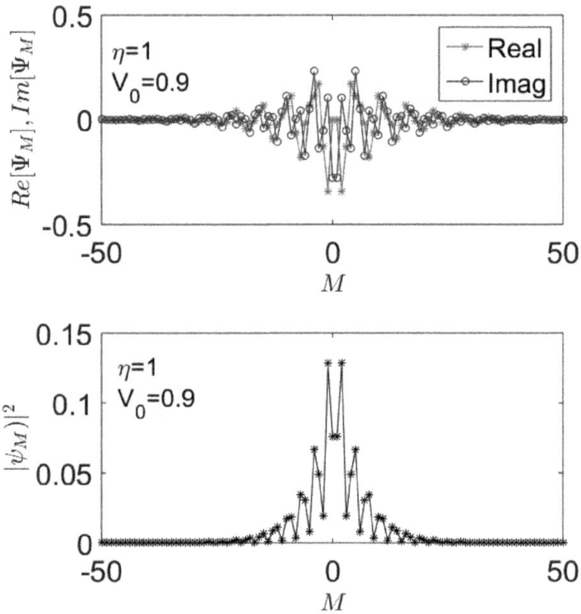


Fig. 7.10 The LS wave function for the case of the homogeneous two-chain gate with $V(M=0)=V(M=1)=V_0=0.9$. The LS wave function is plotted for the even-mode $\eta=1$ and $k_y=0.2\left(\frac{\pi}{\sqrt{3}a}\right)$. (a) The real and imaginary parts of the wave function show even-parity symmetry. (b) The LS wave function also shows oscillatory decay from the center of the two gates.

$\sum'_{M,N} V_M I_{2\times 2}|M,N\rangle\langle M,N|$, where the primed summation over the $M_L \leq M \leq M_R$ chains and $I_{2\times 2}$ denote the pseudospin space. It is convenient to focus on uniform $N_c$-gate potential cases because $V_M$ remains the same. However, our derivation still remains a flexible way to applying on the non-uniform $N_c$-gate cases. The left- and right-hand side wave functions $|\Psi^L_{ky}\rangle$ and $|\Psi^R_{ky}\rangle$ refer to the regions $M \leq M_L$ and $M \geq M_R$, respectively. The amplitude of the wave function at the $A$ and $B$ sites in the $M$-cell is given by

$$F^0_M = \begin{pmatrix} C^{(1)}_M \\ C^{(2)}_M \end{pmatrix}. \tag{44}$$

At the boundaries of gate potential denoted by $M = M_L$ or $M_R$, the wave function can be substituted by $|\Psi^L_{ky}\rangle$ and $|\Psi^R_{ky}\rangle$, respectively. As such,

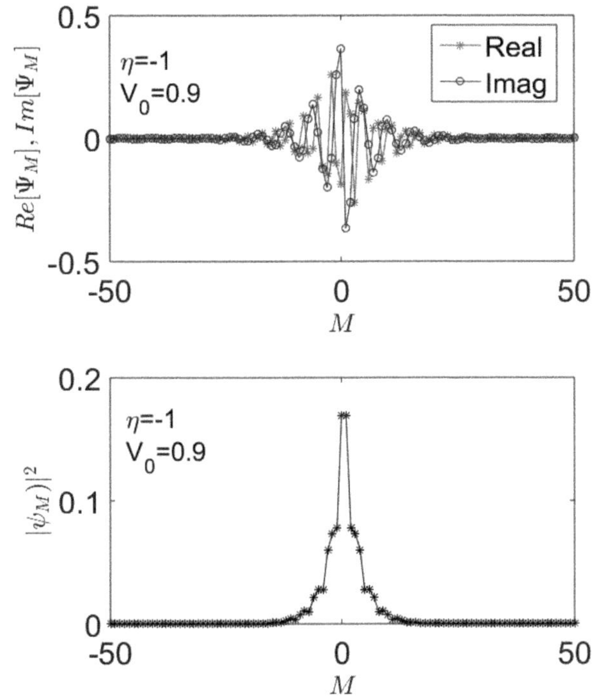


Fig. 7.11 The LS wave function for the case of the homogeneous two-chain gate with $V(M=0) = V(M=1) = V_0 = 0.9$. The LS wave function is plotted for the odd-mode $\eta = -1$ and $k_y = 0.2\left(\frac{\pi}{\sqrt{3}a}\right)$. (a) The real and imaginary parts of the wave function show odd-parity symmetry. (b) The LS wave function also shows oscillatory decay from the center of the two gates.

it can be expressed in terms of the reflection coefficients $r_{p\alpha}$ and $r_{q\alpha}$. We can define that

$$F^0_{M_L} = \begin{pmatrix} r_{p1} \\ r_{p\bar{1}} \end{pmatrix}, \quad F^0_{M_R} = \begin{pmatrix} r_{q1} \\ r_{q\bar{1}} \end{pmatrix}. \tag{45}$$

Multiplying the rotational operator $\Re(\hat{z}, -\phi_y)$ with the left-hand side of the Schrödinger equation at $M$, we find the equation at $M = M_R$:

$$Q_R(M_R, V_{M_R})F^0_{M_R} - F^0_{M_R-1} = 0 \tag{46}$$

and at $M = M_R - 1$:

$$-Q_R(M_R+1, 0)F^0_{M_R} + \varepsilon_{M_R-1}F^0_{M_R-1} - F^0_{M_R-2} = 0. \tag{47}$$

Similarly, we can deal with the matrix $Q_L$ at $M = M_L$, and the two forms of matrices are given by

$$Q_R(M,V) = \begin{bmatrix} \begin{pmatrix} e^{ip_1^*(M-1)}\sin\theta/2 \\ +ivVe^{ip_1^*M}\cos\theta/2 \end{pmatrix} & \begin{pmatrix} e^{-ip_1(M-1)}\cos\theta/2 \\ +ivVe^{-ip_1M}\sin\theta/2 \end{pmatrix} \\ \begin{pmatrix} ve^{ip_1^*(M-1)}\cos\theta/2 \\ -iVe^{ip_1^*M}\sin\theta/2 \end{pmatrix} & \begin{pmatrix} ve^{-ip_1(M-1)}\sin\theta/2 \\ -iVe^{-ip_1M}\cos\theta/2 \end{pmatrix} \end{bmatrix} \tag{48}$$

and

$$Q_L(M,V) = \begin{bmatrix} \begin{pmatrix} e^{ip_1(M+1)}\cos\theta/2 \\ +ivVe^{ip_1M}\sin\theta/2 \end{pmatrix} & \begin{pmatrix} e^{-ip_1^*(M+1)}\sin\theta/2 \\ +ivVe^{-ip_1^*M}\cos\theta/2 \end{pmatrix} \\ \begin{pmatrix} ve^{ip_1(M+1)}\sin\theta/2 \\ -iVe^{ip_1M}\cos\theta/2 \end{pmatrix} & \begin{pmatrix} ve^{-ip_1^*(M+1)}\cos\theta/2 \\ -iVe^{-ip_1^*M}\sin\theta/2 \end{pmatrix} \end{bmatrix}. \tag{49}$$

The energy-dependent matrix is defined by

$$\varepsilon_M = \begin{bmatrix} -e^{-ik_y} & -i(E-V_M) \\ i(E-V_M) & -e^{ik_y} \end{bmatrix}. \tag{50}$$

The Schrödinger equation at the gate-acting region ($M_L + 1 < M < M_R - 1$) after multiplying it by $\Re(\hat{z}, -\phi_y)$ from the left gives

$$-F^0_{M+1} + \varepsilon_M F^0_M - F^0_{M-1} = 0. \tag{51}$$

At $M = M_L + 1$, we get

$$-F^0_{M_L+2} + \varepsilon_{M_L+1}F^0_{M_L+1} - Q_L(M_L-1,0)F^0_{M_L} = 0, \tag{52}$$

and at $M = M_L$, we get

$$-F^0_{M_L+1} + Q_L(M_L, V_{M_L})F^0_{M_L} = 0. \tag{53}$$

Combining all the above equations, one can get a matrix equation as

$$\begin{bmatrix} Q_R(M_R,V_{M_R}) & -1 & & & & & \\ -Q_R(M_R+1,0) & \varepsilon_{M_R-1} & -1 & & 0 & & \\ & -1 & \varepsilon_{M_R-2} & -1 & & & \\ & & & \cdots & & & \\ & & & -1 & \varepsilon_{M_L+2} & -1 & \\ & 0 & & & -1 & \varepsilon_{M_L+1} & -Q_L(M_L-1,0) \\ & & & & & -1 & Q_L(M_L,V_{M_L}) \end{bmatrix} \begin{bmatrix} F^0_{M_R} \\ F^0_{M_R-1} \\ F^0_{M_R-2} \\ \\ F^0_{M_L+2} \\ F^0_{M_L+1} \\ F^0_{M_L} \end{bmatrix} = 0. \tag{54}$$

Equation (54) is a triblock diagonal matrix, in which each block represents a $2 \times 2$ pseudospin space. Other blocks not belonging to tridiagonal blocks are zero. The determinant of Eq. (54) can be shown to be real.

The number of branches of the LS dispersion, $N_{\text{LS}}$, is crucially dependent on $N_c$, the number of gates. It is remarkable that the minimum of $N_{\text{LS}}$ is two and the maximum of $N_{\text{LS}}$ is $N_c$. Thus, the range of $N_{\text{LS}}$ is given by $2 \leq N_{\text{LS}} \leq N_c$ for a given uniform potential $V_0$. The key finding is that there are always two Dirac-point associated branches of the LS dispersion, and they belong to a topological origin. More branches arise from the bulk dispersion, while the uniform potential $V_0$ varies for a given $N_c$. For the case of $N_c = 3$, Fig. 7.12 presents the features of the branch of LS dispersion when the potential gradually increases at $V_0 = 0.4$, 0.8 and 1.2.

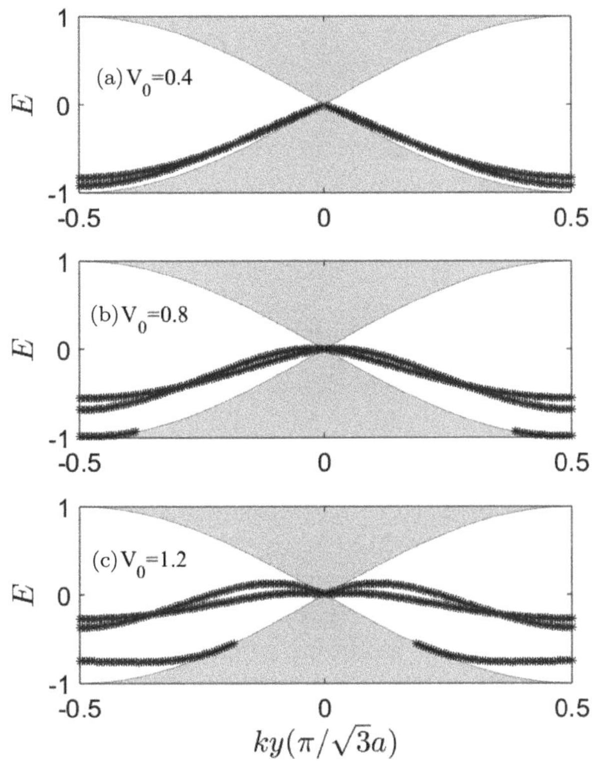


Fig. 7.12 For the $N_c = 3$ case of uniform gate potential, the branches of LS dispersion are plotted as a function of $k_y$ with $V_0 =$ (a) 0.4, (b) 0.8 and (c) 1.2. There are two branches associated with Dirac-point characteristics.

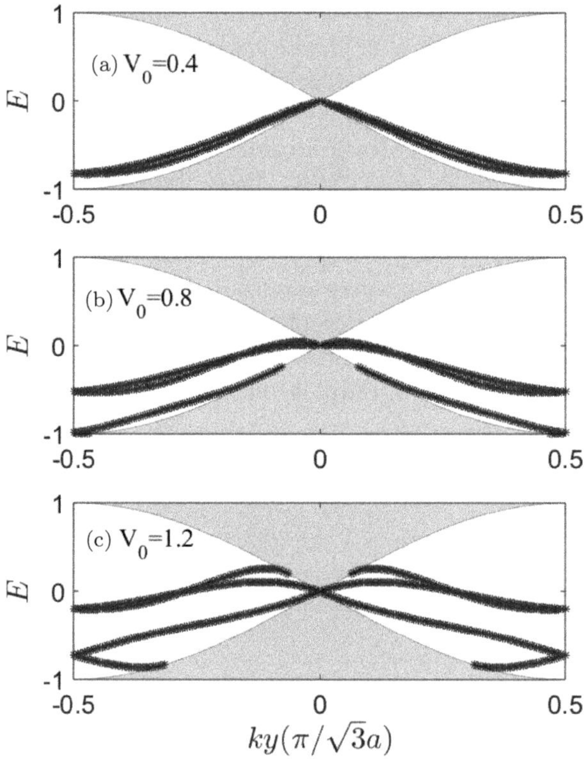


Fig. 7.13 For the $N_c = 4$ case of uniform gate potential, the branches of LS dispersion are plotted as a function of $k_y$ with $V_0$ = (a) 0.4, (b) 0.8 and (c) 1.2. There are two branches associated with Dirac-point characteristics.

The two Dirac-point associated LS branches remain for all $V_0$. For the smaller value case of $V_0 = 0.4$, $N_{\text{LS}} = 2$, as shown in Fig. 7.12(a). Increasing beyond $V_0 = 0.4$ in Fig. 7.12(b), one more side branch arises from the lower bulk dispersion, providing $N_{\text{LS,max}} = N_c = 3$. In Fig. 7.12(c), a third side branch is leveled up into the gap for $V_0 = 1.2$.

For the case of the $N_c = 4$ uniform gate, Fig. 7.13 presents the features of the number of branches of LS dispersion, $N_{\text{LS}}$, for $V_0$ = 0.4, 0.8 and 1.2. The number of branches of LS dispersion $N_{\text{LS}}$ is equal to 2 (associated with the Dirac point), 3 and 4, corresponding to $V_0$ = 0.4, 0.8 and 1.2, respectively.

## 7.4 Conclusions

In conclusion, we have obtained the LS branches on graphene for a one-chain gate that is along the armchair-carbon chain. The gate-induced LS is contributed by the superposition of two pseudospins related to K and K′ valleys when the LS energy lies in the entire gap range $|E| \leq \varepsilon_g$. As such, both pseudospins exhibit out-of-plane orientations with a relative angle of $\Delta\theta$, which varies from $2\pi$ to 0, corresponding to $E = -\varepsilon_g$ to $E = \varepsilon_g$. The topological origin comes from the geometric variation of pseudospin rotation. The appropriate pseudospin rotational operator was analytically demonstrated by the successful connection between the two pseudospins. Such a topological origin of the LS strongly depends on the LS-secular equation obtained from the pseudospin rotation mechanism. The LS can always be found within the gap range due to the topological pseudospin rotation. The topological LS branch induced by the gated graphene also reveals the Dirac-point characteristics. The branch of LS dispersion launches from the Dirac point and extends to the zone boundary through the whole gap. The LS wave function directly shows the localized length scale, which depends on the potential $V_0$ for the one-chain case. Unlike conventional topological edge states near the sample boundary, our topological LS occurs around the gated-graphene region. It is a key finding that such a topological LS can be generated by an arbitrary finite $V_0$ without a threshold $V_0$.

We also studied the cases of $N_c \geq 1$ and two LS branches remains Dirac-point characteristics. For the cases of $N_c \geq 3$, certain number of side LSs, $N_{\text{LS}}$, arising from the bulk dispersion are not directly related to Dirac-point physics. The out-of-plane pseudospin rotation is also demonstrated by the case of $N_C = 2$. The LS-secular equation reveals the same topological origin as the one-chain case. For the two-chain case, we also found the LS to contain even and odd modes due to the homogeneous gate potential.

Finally, we can add that the Dirac-point characteristics of the topological LS branches are the manifestation of both the absence of a finite threshold $V_0$ for their formation and the zero gap in the bulk graphene. Our $N_C$-chain gate potential results are expected to motivate studies on the effects of a line of adsorbed atoms or gates with different lateral configurations, oriented along the armchair chain, on the generation of topological LSs in graphene.

## References


[1] Hasan M.Z. and Kane C.L. 2010. *Rev. Mod. Phys.* **82** 3045.
[2] Xiao-Liang Qi and Shou-Cheng Zhang. 2011. *Rev. Mod. Phys.* **83** 1057.
[3] Haldane F.D.M. 1988. *Phys. Rev. Lett.* **61** 2015.
[4] Bernevig B.A., Hughes T.L. and Zhang S.C. 2006. *Science* **314** 1757.
[5] Bernevig B.A. and Zhang S.C. 2006. *Phys. Rev. Lett.* **96** 106802.
[6] Fu L., Kane C.L. and Mele E.J. 2007. *Phys. Rev. Lett.* **98** 106803.
[7] Hsieh D., Qian D., Wray L., Xia Y., Hor Y.S., Cava R.J. and Hasan M.Z. 2008. *Nature* **452** 970.
[8] Hsieh D., Xia Y., Qian D., Wray L., Dil J.H., Meier F., Osterwalder J., Patthey L., Checkelsky J.G., Ong N.P., Fedorov A.V., Lin H., Bansil A., Grauer D., Hor Y.S., Cava R.J. and Hasan M.Z. 2009. *Nature* **460** 1101.
[9] Fan Y. and Wang K.L. 2016. *Spin* **6** 1640001.
[10] König M., Wiedmann S., Brüne C., Roth A., Buhmann H., Molenkamp L., Qi X.-L. and Zhang S.-C. 2007. *Science* **318** 766.
[11] Chen Y.L., Analytis J.G., Chu J.-H., Liu Z.K., Mo S.-K., Qi X.L., Zhang H.J., Lu D.H., Dai X., Fang Z., Zhang S.C., Fisher I.R., Hussain Z. and Shen Z.-X. 2009. *Science* **325** 178.
[12] Novoselov K.S., Geim A.K., Morozov S.V., Jiang D., Zhang Y., Dubonos S.V., Grigorieva I.V. and Firsov A.A. 2004. *Science* (Supporting material) **306** 666.
[13] Castro Neto A.H., Guinea F., Peres N.M.R., Novoselov K.S. and Geim A.K. 2009. *Rev. Mod. Phys.* **81** 109.
[14] Kane C.L. and Mele E.J. 2005. *Phys. Rev. Lett.* **95** 226801.
[15] Kane C.L. and Mele E.J. 2005. *Phys. Rev. Lett.* **95** 146802.
[16] Quansheng Wu, Liang Du, and Sacksteder IV V.E. 2013. *Phys. Rev. B* **88** 045429.
[17] Wang L.Y., Chang C.Y. and Chu C.S. 2018. *Phys. Rev. B* **98** 235427.
[18] Imura J.I., Mao S., Yamakage A. and Kuramoto Y. 2011. *Nanoscale Res. Lett.* **6** 358.
[19] Wakabayashi K., Takane Y. and Sigrist M. 2007. *Phys. Rev. Lett.* **99** 036601.
[20] Ryu S. and Hatsugai Y. 2002. *Phys. Rev. Lett.* **89** 077002.
[21] Wang Yao, Young S.A. and Qian Niu. 2009. *Phys. Rev. Lett.* **102** 096801.
[22] Kariyado T. and Xia Hu. 2017. *Sci. Rep.* **7** 16515.
[23] Castro Neto A.H., Guinea F., Peres N.M.R., Novoselov K.S. and Geim A.K. 2009. *Rev. Mod. Phys.* **81** 109.
[24] Nakada K., Fujita M., Dresselhaus G. and Dresselhaus M.S. 1996. *Phys. Rev. B* **54** 17954.
[25] Barone V., Hod O. and Scuseria G.E. 2006. *Nano Lett.* **6** 2748.
[26] Son Y.W., Cohen M.L. and Louie S.G. 2006. *Phys. Rev. Lett.* **97** 216803.
[27] Park C., Ihm J. and Kim G. 2013. *Phys. Rev. B* **88** 045403.
[28] Winkler R. and Deshpande H. 2017. *Phys. Rev. B* **95** 235312.
[29] Roman-Taboada P. and Naumis G.G. 2017. *Phys. Rev. B* **96** 155435.
[30] Guinea F., Katsnelson M.I. and Geim A.K. 2010. *Nat. Phys.* **6** 30.
[31] Liu Z.F., Wu Q.P., Chen A.X., Xiao X.B., Liu N.H. and Miao G.X. 2017. *Sci. Rep.* **7** 8854.

[32] Ghaemi P., Gopalakrishnan S. and Shinsei Ryu. 2013. *Phys. Rev. B* **87** 155422.
[33] Salerno G., Ozawa T., Price H.M. and Carusotto I. 2017. *Phys. Rev. B.* **95** 245418.
[34] Maksimov P.A., Rozhkov A.V. and Sboychakov A.O. 2013. *Phys. Rev. B* **88** 245421.
[35] Kohmoto M. and Hasegawa Y. 2007. *Phys. Rev. B* **76** 205402.
[36] Delplace P., Ullmo D. and Montambaux G. 2011. *Phys. Rev. B* **84** 195452.
[37] Costa Filho R.N., Farias G.A. and Peeters F.M. 2007. *Phys. Rev. B* **76** 193409.
[38] Mesaros A., Papanikolaou S., Flipse C.F.J., Sadri D. and Zaanen J. 2010. *Phys. Rev. B* **82** 205119.
[39] Yokoyama T., Balasky A.V. and Nagaosa N. 2010. *Phys. Rev. Lett.* **104** 246806.
[40] Jeong Y.H., Kim S.C. and Eric Yang S.-R. 2015. *Phys. Rev. B.* **91** 205441.
[41] Ting Cao, Fangzhou Zhao and Louie S.G. 2017. *Phys. Rev. Lett.* **119** 076401.
[42] Chiu C.H. and Chu C.S. 2012. *Phys. Rev. B* **85** 155444.
[43] Hsieh S.H. and Chu C.S. 2016. *Appl. Phys. Lett.* **108** 033113.
[44] Yokoyama T., Balatsky A.V. and Nagaosa N. 2010. *Phys. Rev. Lett.* **104** 246806.

Chapter 8

# TWO-DIMENSIONAL DIRAC FERMIONS IN $Z_2$ TOPOLOGICAL PHASES

Ching-Hong Ho, Lu-Yao Wang, Hsien-Ching Chung, Chiun-Yan Lin, Chih-Wei Chiu, and Ming-Fa Lin

In quantum vacuum, a free electron is governed by the Dirac equation [1] as is a wave equation differing from the Schrödinger equation. The equation is given by the Dirac Hamiltonian, a construction based on the compatibility made by Paul Dirac, between the single-particle quantum mechanics and the theory of special relativity. Consequently, the intrinsic properties of spin-1/2 and antiparticle (positron) of the electron are revealed in the solution. In the more involved quantum field theory, which combines the Dirac equation and the classical field theory, Dirac fermions can emerge as the quanta of a field [2]. In addition, there are also Weyl fermions [3] and Majorana fermions [4] proposed, from the Dirac equation or its variant, despite the fact that they have not been found in quantum vacuum yet. A door to modern particle physics has thus been opened along this line. On the other hand, the quantum field theory is also useful in the description of quasiparticles in condensed matter [5]. When quasiparticles near the single-particle ground state acquire an effective Hamiltonian mimicking the Dirac Hamiltonian, they behave as Dirac fermions and obey the Dirac equation. In such a connection, it is of importance and interest to see how one can get a convenient condensed matter playground for the tabletop test and the realization of the quantum field theory. Recently, both Weyl fermions [26] and Majorana fermions [7] have been discovered from condensed materials, which can be deemed as the greatest achievements in science.

## 8.1 Dirac Fermions

The Dirac equation for a free electron in $(d+1)$-dimensional spacetime is written as follows:

$$i\partial_t\psi = H_D\psi, \tag{1}$$

with the Dirac Hamiltonian

$$H_D = -i\hbar c\alpha_i\partial_i + \beta m^2, \tag{2}$$

or in momentum representation

$$H_D = c\alpha_i p_i + \beta m^2, \tag{3}$$

where the summation over $i = 1, \ldots, d$ is implied in Einstein's convention from now on, and $m$ is the mass of an electron. The Dirac matrices $\alpha_i$ and $\beta$ are required to satisfy $\alpha_i^2 = \beta^2 = \mathbb{I}$, $\{\alpha_i, \beta\} = 0$ and $\{\alpha_i, \alpha_j\} = 2\delta_{ij}$, leading to the energy–momentum relation $E^2 = p^2c^2 + m^2c^4$, as a result of Lorentz invariance. For this, a verification can be made by using a plane-wave function $\psi$ and applying $H_D$ twice on it. Alternatively, a covariant form of the Dirac equation is given by [8]

$$(i\gamma^\mu\partial_\mu - m)\psi = 0, \tag{4}$$

where the Dirac operator $D = i\gamma^\mu\partial_\mu - m$ runs over $\mu = 0, 1, \ldots, d$, with the gamma matrices given by $\gamma^\mu = (\beta, \beta\alpha_i)$. The gamma matrices obey the Clifford algebra: $\{\gamma^\mu, \gamma^\nu\} = 2\eta^{\mu\nu}$, with $\eta^{\mu\nu}$ the Minkowski metric tensor $\eta^{00} = -\eta^{ii} = 1$, and otherwise, $\eta^{\mu\nu} = 0$. For 2D, the Dirac equation needs three $(d+1 = 3)$ irreducible matrices. The Pauli matrices $\sigma_i$ happen to meet this requirement. For 3D, four $(d+1 = 4)$ matrices are needed, and therefore, higher $4 \times 4$ matrices come into play, as given by some direct product of two sets of Pauli matrices, $\sigma_i$ and $\tau_i$. Here, those $4 \times 4$ matrices are not used up and cannot be represented uniquely, but only up to a unitary transformation. For instance, in the standard representation, the Dirac matrices read $\alpha_i = \sigma_i\tau_1$ and $\beta = \sigma_0\tau_3$. Thus, the wave functions have four components in 3D, called 4-spinors, two components for particle and antiparticle states and two for 1/2-spin orientations [1]. By contrast, in 2D, the two-component spinors consist of particle and antiparticle states while the spin degree of freedom is polarized [9–11]. Another important contrast to be stated is about the chiral symmetry, which exists in $(3+1)$ dimensions but does not in $(2+1)$ dimensions; see the following section. In a class of condensed matter dubbed Dirac matter [12–16], quasiparticles appears as Dirac fermions near the single-particle ground states, where the effective Hamiltonians mimic the Dirac Hamiltonian $H_D$ (Eqs. (2) and (3)). Therefore, the Dirac equation (Eq. (1)) holds with Lorentz invariance. Materials belonging to this class were commonly found out to be semimetals or narrow-gap semiconductors. Such semimetal turns out to have some band touch, where two bands linearly crosses each other. In the vicinity of the band touch, quasiparticles disperse linearly like massless Dirac fermions governed by the Dirac equation in quantum vacuum. Even if a band gap is opened thereby, there could be an analogous Dirac Hamiltonian with a mass term, where massive Dirac fermions can be found in the valley about the precedent band touch. Besides, in analog to positrons in the Dirac sea, there could be holes in an unbounded valence band (Fermi sea). Owing to the zero or little gap, pair production and annihilation could also occur analogously. Furthermore, Klein tunneling can also occur for massless Dirac fermions because in the massless limit, Dirac fermions acquire chirality as revealed in the effective Hamiltonian; see graphene, for example, [17]. As such, the frequently used two-band model suffices it.

Until now, everything seems fine. Nonetheless, there exist certain disparities being significant to be noted. For a spatial lattice in condensed matter, space is discretized while time is left continuous so that Lorentz invariance is absent. Here, discrete and periodic crystal momentum take place in the entire Brillouin zone (BZ). It is the band structures that are relevant for the lattice Hamiltonian. Massless Dirac fermions are quasiparticles hosted around band touch determined *a priori* accidentally [18, 19] or under symmetry constraints [18, 20]. Now, the most fundamental problem lies in the distinction that in quantum vacuum, masslessness is the nature of particles themselves. Moreover, the analogous Dirac Hamiltonians of quasiparticles might be constructed with some types of pseudospin due to the lattice symmetries [21] rather than electron spin. Certain intrinsic properties of electron spin would be lost thereby. In some circumstances, the effective Hamiltonians obtained would lack parity or isotropy [22] and the Dirac equation is not strictly obeyed. Condensed matter physicists, indeed, usually loosely define the Dirac fermions on their own. In this approach, new kinds of fermionic quasiparticles, such as tilted massless fermions, have also been investigated [23–27], which do not even have an analogy to Dirac fermions.

Given the effective Dirac Hamiltonian, massless Dirac fermions do not exist globally in the BZ, while they appear only locally where the continuum approximation is valid. But, how about the correctness of this locality? Mathematically, a continuous differential equation cannot have monotonic solutions if it is modulated by a discrete frame. That is, the solutions must have pairs of zeros, if any, and be regularized to get rid of divergence in some way. With respect to this issue, one has to be concerned with topology. It is relevant not only to the quantum field theory but also to condensed matter physics.

In the quantum field theory, the ultraviolet divergence of a single Dirac field is cut off by some regularization schemes. One of the regularization schemes appeals to the quantum lattice field theory [28]. Unfortunately, by mathematical reasoning, extra solutions would be obtained. The outstanding problem of fermion doubling would be inevitably involved and should be forbidden, since the imposed lattice regularization is an artifact. In a no-go theorem, it has been proven that in even-dimensional spacetime lattice, a local, real and free fermion action suffers from fermion doubling in case that chiral symmetry is present [29–31]. To sum up, for an even $(d+1)$-dimensional lattice, $2^{d+1}-1$ obtained species of Dirac fermions are spurious. The doubly many Dirac fermions obtained carry opposite chiralities, which are equal in number to be cancelled out according to the presence of chiral symmetry. Why is this forbidden or a no-go? Recall that in classical field theory, global symmetries dictate the conservation laws [32]. At the quantum level, however, some classical symmetries would not survive the field quantization. This is a quantum anomaly [33, 34], a distinct type of symmetry breaking arising neither explicitly nor spontaneously. Similarly, another quantum anomaly, i.e., parity anomaly, exists in odd $(d+1)$-dimensional lattice as a consequence of the absence of chiral symmetry in odd-dimensional spacetime, where the field has real $SO(d+1)$ representation. It is quantum anomaly that leads to topology in Dirac fields in quantum vacuum [34]. It will be present at the level of quantization for classical fields, while fermion doubling smears it.

In contrast to the quantum field theory, where the lattice is artificially imposed on the fields for regularization, the condensed matter lattices are naturally present such that the band structures should be free from divergence. Thus, in condensed matter, fermion doubling is not problematic as opposed to the case of quantum lattice fields theory. As compensation, the doubling of Dirac fermions signifies the possible loss of topology in condensed matter, where the anomaly would be cancelled out. It is noted that quantum anomaly can show up in the presence of a gauge field, for example, a chiral anomaly occurs in Weyl or Dirac semimetals when both electric and magnetic fields are applied [16]. Yet, as well known, single species of Dirac fermions take place at the topological phase transitions or the surfaces of certain topological insulators. The most famous example is that in the Haldane model, where parity anomaly exists in certain 2D lattices. Generally speaking, to what extent and how Dirac fermions are topologically nontrivial in condensed matter depend on the symmetry protection in relevant spatial dimensions for the possibilities of certain quantum anomalies. In this way, various topological phases can be classified and characterized with their own topological invariants.

## 8.2 Berry Curvature and Topological Invariants

Michael Berry has proposed a generalized concept about the geometrical phase of a system, which has undergone a cyclic process adiabatically [35]. The geometrical phase is the phase difference acquired as a result of the geometry of the Hamiltonian parameter space. The original expression was written in real (**R**) space for an atom moving over a closed contour $\Gamma$ with the wave function $\psi$ as follows:

$$\phi = \oint_{\Gamma} d\mathbf{l} \cdot \langle \psi | i\nabla_{\mathbf{R}} \psi \rangle. \tag{5}$$

The geometrical phase, commonly called the Berry phase, is profound. Nonzero values of it can be connected to certain types of quantum anomaly, which are anomalous by the measure of a path integral in spacetime [33, 34]. Indeed, parameter space where the integration takes place can be generalized to any appropriate ones, such as the BZs for the band structures in condensed matter lattices. As being translated with the crystal wave vector **k** and the Bloch functions $u_{n\mathbf{k}}$ ($n$ being the band index), the following expression of the Berry phase makes sense [52]:

$$\phi_n = \oint_{\Gamma} d\mathbf{k} \cdot \mathbf{A}_n(\mathbf{k}), \tag{6}$$

with the Berry connection

$$\mathbf{A}_n(\mathbf{k}) = \langle u_{n\mathbf{k}} | i\nabla_{\mathbf{k}} u_{n\mathbf{k}} \rangle. \tag{7}$$

One might get insight from Eq. (7) with respect to topological concern about the Berry connection. At first, it is the expectation value of the position operator $i\nabla_{\mathbf{k}}$ for a fixed **k**, although it is gauge-dependent because the Bloch functions are given only up to a **k**-dependent phase. The situation is just analogous to the vector potential for a magnetic field. Hence, the

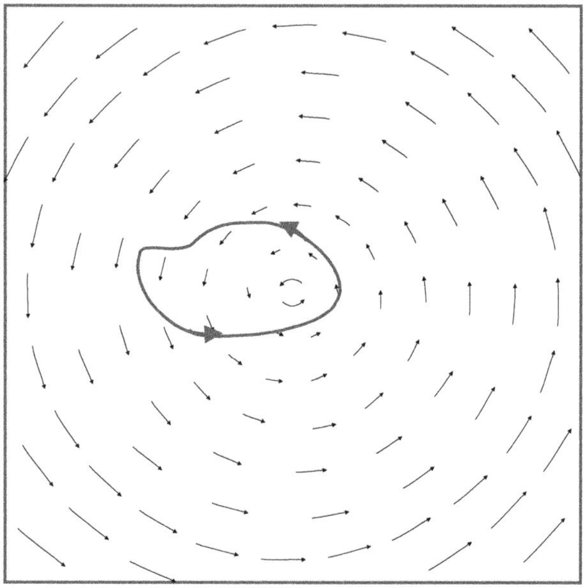

Fig. 8.1 Schematic diagram of the Berry phase. Berry connection (black arrows) is a vector field in the BZ. The system undergoes a process along a contour (red arrowed loop) or the border of the BZ (red square).

circulation of the Berry connection in the BZ can be observed by a curl on it:

$$\mathbf{\Omega}_n(\mathbf{k}) = \nabla_{\mathbf{k}} \wedge \mathbf{A}_n(\mathbf{k}). \tag{8}$$

Figure 8.1 depicts such a picture. It is also noted that integration over the surface enclosed by the Berry phase contour can be performed if no singular nodes exist therein. This integral is identical to the Berry phase according to the Stoke theorem.

Since topology is a global concept, its characterization must be, at first, performed over the relevant space, which should be compact and nonsingular. The BZ can be a satisfying one if the bands are gapped. In general, the space must be a compact manifold, in which the Berry curvature is not singular. Indeed, for $d$-dimensional condensed matter lattice, the BZ is topological equivalent to a torus $T^d$, as shown in Fig. 8.2. The manifold chosen can be generalized to the cases of gapless bands (referring to Figs. 8.3(a) and 8.3(b) for a 3D BZ), where spheres $S^2$ and $S^1$ are taken for a nodal point and a nodal line in the bulk BZ, respectively. The simplest case might be the Chern number. It is acquired as an integer over a 2D gapped

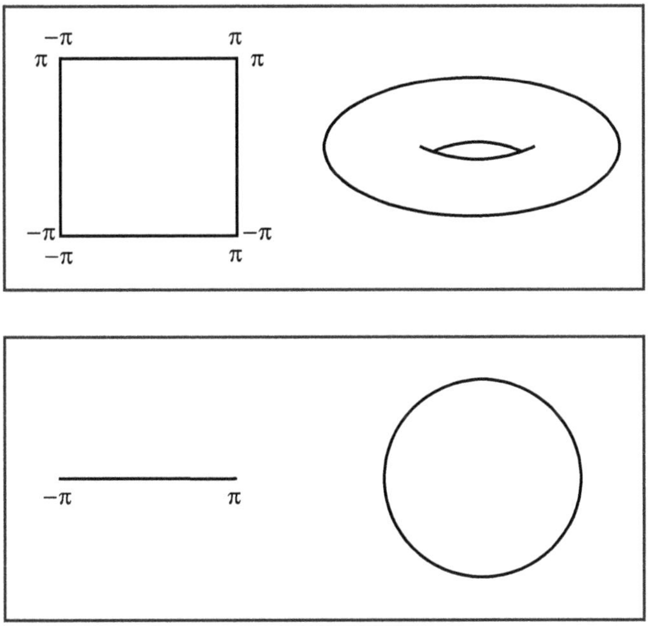


Fig. 8.2 Topological equivalence between (a) 2D BZ and a 2-torus and $T^2$ and (b) 1D BZ and a 1-torus (circle) and $T^1$.

and symmetry-free manifold, defined as

$$C = \frac{1}{2\pi}\Sigma_{n'} \int_{BZ} d\mathbf{k} \cdot \mathbf{\Omega}_{n'}(\mathbf{k}), \tag{9}$$

where $n'$ denotes the occupied bands. The integer quantum Hall effect (QHE) [37], as the first identified topological phase, has been prominently characterized by the nonzero Chern numbers, which are equivalent to the Hall conductivities at different Hall plateaus [38–40]. It is reasonable now that the topological phase with the integer IQHE can be extended to a 2D system without an external magnetic field and hence without Landau levels (LLs) as long as the Chern number is nonzero [41]. Systems with such phases are called Chern insulators. Moreover, Weyl semimetals are also characterized by the Chern numbers over $S^2$ manifolds enclosing the Weyl points [16].

In distinction to the Chern number, which is defined over a 2D manifold without any symmetries including time reversal symmetry (TRS), of course, another topological invariant defined over a 2D manifold with TRS is relevant to the subject, 2D Dirac fermions, of this chapter. It is a binary

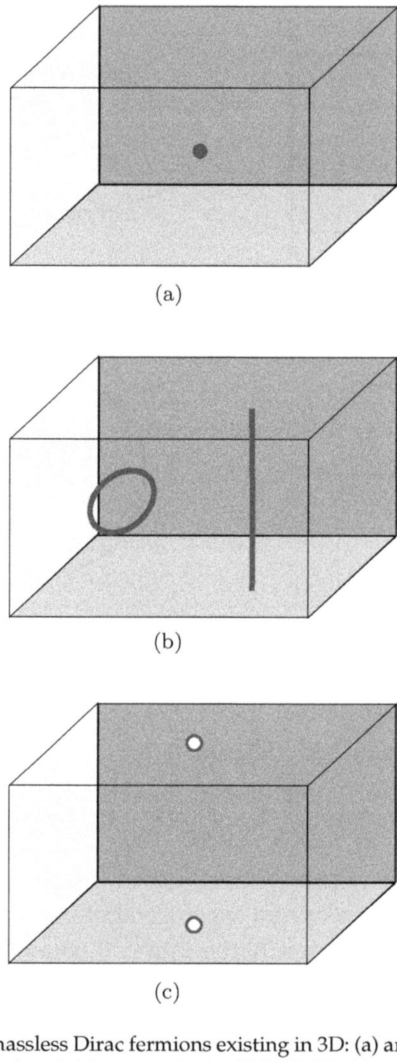

Fig. 8.3 Classes of 2D massless Dirac fermions existing in 3D: (a) around nodal points inside the 3D bulk BZ, (b) around nodal lines inside the 3D bulk BZ, (c) around nodal points on the 2D surface BZ.

($Z_2$) number because of TRS. An extension to 3D systems is possible. Under the protection of TRS, 2D topological insulators with the quantum spin Hall effect (QSHE) and 3D topological insulators are characterized by the $Z_2$ invariant [42, 43]. In the presence of TRS, the occupied eigenstates of the $Z_2$ systems must be even in number and can be divided into two TRS-related classes, called Kramers pairs, each carrying one spin orientation.

Observe Eqs. (8) and (9), where the time reversal transformation makes both $\mathbf{k}$ to $-\mathbf{k}$ and spin $\uparrow$ to $\downarrow$. Hence, the Chern numbers for the two classes are respectively given by $Z_\uparrow$ and $Z_\uparrow = -Z_\uparrow$, adding to a zero value of the total Chern number. One might think of $Z = |Z_\uparrow|$ as the topological invariant for use here. However, this possibility has to be denied. This is because in the case of topological semimetals where nodes exist in the band structure under TRS, similar consideration happens if each of the manifolds for the doubling nodes is denoted a pseudospin. Examples can be found in spinless graphene [16], inversion-symmetric gapped bilayer graphene [44] and rhombohedral graphite [45]. The presence of nodes makes an indefinite division of occupied eigenstates. A different division would lead to different $|Z|$ numbers, which differ from one another by an even integer. It is $|Z|$ mod 2 that is well defined here. This is the reason for the $Z_2$ invariant [42, 43, 46–48].

## 8.3 $Z_2$ Dirac Matter

In a noninteracting fermionic 2D lattice with SLS, there can be 2D massless Dirac fermions hosted around linearly crossing nodes, i.e., Dirac points (DPs), that could be present somewhere in the BZ. Due to the fermion doubling, the Dirac Hamiltonian should be based on two spinors, one of which is owing to the valleys about the DPs and the other is the sublattice. Take $SU(2)$ graphene as an example. Without intervalley mixing, the two-spinor effective Dirac Hamiltonian is given by $H(\mathbf{k}) = v_0\hbar\tau_3\sigma\mathbf{k}$, with $\hbar\mathbf{k}$ the crystal momentum measured from each DPs and $\sigma$ and $\tau$ the Pauli matrices acting on the sublattice and the valley spinor, respectively. With $\lambda$ and $\xi$ denoting the band and valley indices, respectively, the energy of Dirac fermions is given by $E_{\lambda\xi}(\mathbf{k}) = \lambda v_0\hbar\mathbf{k}$. The wave function around $\mathbf{k} = 0$ is given by $\psi_{\lambda\xi}(k) = (1/\sqrt{2})\left(\exp(-i\xi\theta/2), \lambda\xi\exp(i\xi\theta/2)\right)^T$, where $\mathbf{k} = (k, \theta)$ has a phase defined conventionally [17]. Thus, the winding number, $w_1 = \lambda\xi$, is derived with respect to the chirality. It is clear now that the effective CS arises from the global SLS in the whole lattice. In the presence of a perpendicular, uniform magnetic field $B\hat{\mathbf{z}}$, the effective CS is still preserved via the minimal coupling. This fact is simply reflected in clean graphene (without the disorder).

The first demonstration of 2D massless Dirac fermions in condensed matter was addressed on the 2D honeycomb lattice [49], where electron spin degeneracy is assumed ($SU(2)$ symmetry) and the two sublattices come into play as pseudospin. Hence, 2D massless Dirac fermions have been recognized in the vicinity of DPs in an analogous $(2+1)$-dimensional gauge field. Such 2D massless Dirac fermions have also been shown, in the 2D limit of graphite, to be stable against disorder due to impurity without intervalley scattering between the two DPs [50]. These results are good for 2D graphene, which was realized subsequently [17]. In a 3D system, intuitively speaking, 2D massless Dirac fermions can exist in the vicinity of DPs in 2D momentum space or in a 2D subspace of 3D momentum space. It is easy to figure out that there are only three classes of topological phases with massless Dirac fermions in a 3D noninteracting lattice system [51], where the bulk BZ is associated with projection-mapped 2D BZs. In the first class, as shown in Fig. 8.3(a), there are 3D massless Dirac fermions around isolated DPs in the bulk, where the Dirac cones exhibit linear dispersion in all three dimensions. The first class is irrelevant to 2D massless Dirac fermions [16]. In the third class (Fig. 8.3(c)), 2D massless Dirac fermions are hosted on the surfaces, also irrelevant to the definition of the topological invariant. The second class shown in Fig. 8.3(b) includes phases with the so-called Dirac nodal lines (DNLs) in the bulk. In the vicinity of a DNL, there are 2D massless Dirac fermions lying in each perpendicular plane. The intersection points could be deemed 2D DPs and they are continuously lined up. There are infinitely many Dirac cones along the DNL, each dispersing linearly in the two dimensions of its own plane. The presence of DNLs generally indicates a nontrivial nodal-line TSM. On each surface layer, the surface state manifests itself by a drumhead flat band corresponding to the bulk DNL. In the still-growing research on this class, the simplest model is, among others, the rhombohedral lattice, which results from the layer stack of honeycomb lattices in the rhombohedral configuration. The physical realization has been found from rhombohedral graphite (RG) [45, 52–57].

When a 2D layer of lattice host massless Dirac fermions, whether being in an SPT phase or not, one may ask what if the layers are stacked to 3D in some ways. For systems such as d-wave cuprate superconductors with negligible interlayer coupling [13], a quasi-2D model is employed for granted so that vertical DNLs are naively grown from the DPs existing in each layer. When the interlayer coupling is significant, take graphene, for example, nexuses are present in Bernal graphite [56]. By contrast, although there exist DPs in the $\pi$-flux square lattice, an SPT insulating phase of chiral TI in the $\pi$-flux cubic lattice had been proposed [21]. All these are examples of the dimensional crossover of topological phases from 2D to

3D [51]. The (co)-dimensional-periodic table in the 10-fold way symmetry classification dictates that the topological characterization might or might not alter after the dimensional crossover. This subchapter introduces a well-analyzed case in the following. From graphene to RG, there is a dimensional crossover of topological phases from a 2D TSM with nodal points (DPs) to 3D nodal-line TSM, where the co-dimension $p = 2$ holds [53, 57]. This case belongs to the second class of 2D massless Dirac fermions existing in 3D (Fig. 11.1(b)). Since these two TSMs belong to the same symmetry class and have the same co-dimension, they must have the same topological invariant. Indeed, both contain 2D massless Dirac fermions.

After all, the characteristic of RG is completely identical to graphene at the minimal model level. The reason is topological. It lies in the fact that both systems belong to the same topological class, but not the small value of interlayer coupling for the 2D limit. It can be proved that it is the chiral operator $\tau_0\sigma 3$ that protects the zero-mode LL. As a whole, the LL spectra and wave functions both are the same as graphene. Of course, the basis for the wave functions (sublattice pseudospinor) is different. Although the nearest hopping within the same sublattice will break the SLS, there is a generalized chiral operator that provides the protection [58]. The chiral half-filled zero mode is obviously present, showing the persistence of the effective CS in the presence of a perpendicular magnetic field. Finally, a 3D layered system has the easiest way to satisfy the Diophantine equation, which is required for the 3D integer QH effect [59]. The 3D integer QH effect in RG has been discovered [60, 61], which shows the scaling invariance in comparison to the 2D integer QH effect in graphene. This is the result of the same symmetry class and co-dimension between the two systems [62].

## References


[1] Dirac P.A.M. 1928. *The Principles of Quantum Mechanics* 4th ed. (Oxford: Oxford University Press).

[2] Perkin M.E. and Schroeder D.V. 1995. *An Introduction to Quantum Field Theory* (Boca Raton: CRC Press).

[3] Weyl H. 1929. *Z. Phys.* **56** 330.

[4] Majorana E. 1937. *Nuovo Cim.* **14** 171.

[5] Mahan G.D. 2000. *Many-Particle Physics* 3rd. ed. (New York: Kluwer Academic/Plenum Press).

[6] Xu Y., Belopolsk I., Alidoust N., Neupane M., Bian G., Zhang C. *et al.* 2015. *Science* **349** 613.
[7] Mourik V., Zuo K., Frolov S.M., Plissard S.R., Bakkers E.P.A.M. and Kouwenhoven L.P. 2012. *Science* **336** 1003.
[8] Sakurai J.J. 2017. *Modern Quantum Mechanics* 2nd. ed. (Cambridge: Cambridge University Press).
[9] Schonfeld J.F. 1981. *Nucl. Phys. B* **185** 157.
[10] Deser S., Jackiw S. and Templeton S. 1982. *Ann. Phys. (N.Y.)* **140** 372.
[11] Deser S., Jackiw S. and Templeton S. 1982. *Phys. Rev. Lett.* **8** 975.
[12] Cayssol J. 2013. *C. R. Phys.* **14** 760.
[13] Vafek O. and Vishwanath 2014. *Ann. Rev. Condens. Matter Phys.* **5** 83.
[14] Wehling T.O., Black-Schaffer A.M. and Balatsky A.V. 2014. *Adv. Phys.* **63** 1.
[15] Schnyder A.P. and Brydon P.M.R. 2015. *J. Phys.: Condens. Matter* **27** 243201.
[16] Armitage N.P., Mele E.J. and Vishwanath A. 2018. *Rev. Mod. Phys.* **90** 015001.
[17] Castro Neto A.H., Guinea F., Peres N.M.R., Novoselov K.S. and Geim A.K. 2009. *Rev. Mod. Phys.* **81** 109.
[18] von Neumann J. and Wigner E.P. 1929. *Phys. Z.* **30** 467, translated in Knox R.S. and Gold A. 1964. *Symmetry in the Solid State* (New York: W. A. Benjamin).
[19] Herring C. 1937. *Phys. Rev.* **52** 365.
[20] Bouckaert L.P., Smoluchowski R. and Wigner E.P. 1936. *Phys. Rev.* **50** 58.
[21] Hosur P., Ryu S. and Vishwanath A. 2010. *Phys. Rev. B* **81** 045120.
[22] Mañes J.L. *Phys. Rev. B* **85** 155118.
[23] Soluyanov A.A., Gresch D., Wang Z., Ponomarenko L.A., Xin N., Holwill M. *et al.* 2015. *Nature* **527** 495.
[24] Xu Y., Zhang F. and Zhang C. 2015. *Phys. Rev. Lett.* **115** 265304.
[25] Bradlyn B., Cano J., Wang Z., Vergniory M.G., Felser C., Cava R.J. *et al.* 2016. *Science* **353** aaf5037.
[26] Wieder B.J., Kim Y., Rappe A.M. and Kane C.l. 2016. *Phys. Rev. Lett.* **116** 186402.
[27] Barrier J., Kumaravadivel P., Krishna Kumar R., Ponomarenko L.A., Xin N., Holwill M. *et al.* 2020. *Nat. Commun.* **11** 5756.
[28] Montvay I. and Münster G. 1997. *Quantum Fields on a Lattice* (Cambridge: Cambridge University Press).
[29] Nielsen H.B. and Ninomiya M. 1981. *Nucl. Phys. B* **185** 20.
[30] Nielsen H.B. and Ninomiya M. 1981. *Phys. Lett. B* **105** 219.
[31] Nielsen H.B. and Ninomiya M. 1981. *Nucl. Phys. B* **193** 173.
[32] Goldstein H. 1980. *Classical Mechanics* 2nd. ed. (Massachusetts: Addison-Wesley).
[33] Bertlmann R.A. 2000. *Anomalies in Quantum Field Theory* 4th ed. (Oxford: Oxford University Press).
[34] Nakahara M. 2003. *Geometry, Topology and Physics* (New York: Taylor and Francis).
[35] Berry M.V. 1984. *Proc. R. Soc. London Ser. A* **392** 45.
[36] Xiao D., Chang M.C. and Niu Q. 2009. *Rev. Mod. Phys.* **82** 1959.
[37] von Klitzing K., Dorda G. and Pepper M. 1980. *Phys. Rev. Lett.* **45** 494.

[38] Thouless D.J., Kohmoto M., Nightingale M.P. and Nijs den M. 1982. *Phys. Rev. Lett.* **49** 405.
[39] Avron J.E., Seiler R. and Simon B. 1983. *Phys. Rev. Lett.* **51** 51.
[40] Simon B. 1983. *Phys. Rev. Lett.* **51** 2167.
[41] Haldane F.D.M. 1988. *Phys. Rev. Lett.* **61** 2015.
[42] Schnyder A.P., Ryu S., Furusaki A. and Ludwig A.W.W. 2008. *Phys. Rev. B* **78** 195125.
[43] Kitaev A. 2009. *AIP Conf. Proc.* **1134** 22.
[44] Lin K.S. and Chou M.Y. 2018. *Nano Lett.* **18** 7254.
[45] Hyart T., Ojajärvi R. and Heikkilä T.T. 2018. *J. Low Temp. Phys.* **191** 35.
[46] Hořava P. 2005. *Phys. Rev. Lett.* **95** 016405.
[47] Zhao Y.X. and Wang Z.D. 2013 *Phys. Rev. Lett.* **110** 240404.
[48] Zhao Y.X., Schnyder A.P. and Wang Z.D. 2016. *Phys. Rev. Lett.* **116** 156402.
[49] Semenoff G.W. 1984. *Phys. Rev. Lett.* **53** 2449.
[50] DiVincenzo D.P. and Mele E.J. 1984. *Phys. Rev. B* **29** 1685.
[51] Volovik G.E. 2003. *The Universe in a Helium Droplet* (Oxford: Oxford University Press).
[52] Xiao R., Tasnádi F., Koepernik K., Venderbos J.W.F., Richter M. and Taut M. 2011. *Phys. Rev. B* **84** 165404.
[53] Heikkilä T.T. and Volovik G.E. 2011. *JETP Lett.* **93** 59.
[54] Ho C.H., Chang C.P., Su W.P. and Lin M.F. 2013. *New J. Phys.* **15** 053032.
[55] Ho C.H., Chang C.P. and Lin M.F. 2014. *Solid State Commun.* **197** 11.
[56] Heikkilä T.T. and Volovik G.E. 2015. *New J. Phys.* **17** 093019.
[57] Ho C.H., Chang C.P. and Lin M.F. 2016. *Phys. Rev. B* **93** 075437.
[58] Kawarabayashi T., Hatsugai Y., Morimoto T. and Aoki H. 2011. *Phys. Rev. B* **83** 153414.
[59] Kohmoto M., Halperin B.I. and Wu Y.S. 2006. *Phys. Rev. B* **45** 13488.
[60] Kopelevich Y., Torres J.H.S., da Silva R.R., Mrowka F., Kempa H. and Esquinazi P. 2003. *Phys. Rev. Lett.* **90** 156402.
[61] Kempa H., Esquinazi P. and Kopelevich Y. 2006. *Solid State Commun.* **138** 118.
[62] Balents L. and Fisher M.P.A 1996. *Phys. Rev. Lett.* **76** 2782.
[63] Winkler R. 2015. *ANZIAM. J.* **57** 3.
[64] Jackiw R. 1984. *Topological Investigations of Quantized Ggauge Theories* in *Relativity, Groups and Topology II* Les Houches 1983 eds. DeWitt B.S. and Stora R. (Amsterdam: North-Holland).
[65] Fujikawa K. and Suzuki H. 2004. *Path Integrals and Quantum Anomalies* (Oxford: Oxford University Press).

Chapter 9

# ZERO-MODE LANDAU LEVEL OF 2D MASSLESS DIRAC FERMIONS

Ching-Hong Ho, Hsien-Ching Chung, Chiun-Yan Lin, Lu-Yao Wang, Chih-Wei Chiu, and Ming-Fa Lin

The investigation of quasiparticles in condensed matter has yielded remarkable insights into the behavior of massless Dirac fermions in two-dimensional systems. These fermions exhibit intriguing properties, including the existence of zero-mode Landau levels and the half-integer quantum Hall effect, which can be explained through the relativistic Dirac Hamiltonian. In contrast to Dirac fermions in a vacuum, the fermions in these systems possess spins that are inherent to the system. The Dirac Hamiltonian is dependent on the symmetries present in the system, with both spin and particle-hole symmetries resulting in half-filled ground states. Although the effective Hamiltonian for massless Dirac fermions is unbound at large momenta, its eigenstates are chiral, which results in an effective presence in $d$-dimensional space. In two-dimensional lattices without interaction and with spin–orbit coupling, Dirac points can be present, leading to a Dirac Hamiltonian based on two spinors. The graphene system serves as a prime example of such a system.

## 9.1 Background

One of the fascinating properties of quasiparticles, such as 2D massless Dirac fermions in condensed matter, is the robustly existing zero-mode Landau level and the associated half-integer quantum Hall effect in appropriately high magnetic fields $B \lesssim 9$ T in experiments [1, 2]. Theoretically, this can be solved using the relativistic Dirac Hamiltonian $H_D$. Since there are generally two Dirac points due to fermion doubling under the continuous chiral symmetry, as described in Chapter 8, they are not coupled by the applied magnetic fields in the present scope. In contrast to Dirac fermions in quantum vacuum, spins are taken *a priori* as opposed to the electron spin, which is derived from the genuine Dirac equation. The reason for acquiring a Dirac Hamiltonian depends on the symmetries present in the system. Of the three nonspatial symmetries, TRS is irrelevant, while each of SLS and PHS can lead to a Dirac Hamiltonian. The former gives rise to Dirac fermions about the Dirac nodes in TSMs. The latter is responsible for the TSCs, where Majorana fermions (particles being their own antiparticles) appear about the Fermi energy. It is noted that both SLS and PHS lead to half-filled ground states of their own Dirac Hamiltonian. For massless Dirac fermions that have, for example, a single spinor, the Dirac Hamiltonian is cast as $H(\mathbf{p}) = v_0\sigma\mathbf{p}$, with $v_0$ the Fermi velocity and $\sigma$ the Pauli matrices for the spinor [11]. Of course, $H(\mathbf{p})$ is a low-energy effective Hamiltonian since it is unbound at large momenta. The eigenstates of $H(\mathbf{p})$ are chiral, for they are simultaneous eigenstates of the helicity operator $h = \sigma \cdot \mathbf{p}/|\mathbf{p}|$. Therefore, $H(\mathbf{p})$ has an effective CS. This CS is effectively present in the $d$-dimensional space, differing from that defined in the $(d+1)$-dimensional Minkowski spacetime. Here, a crucial issue should be addressed: How can an effective Dirac Hamiltonian be cast? In a noninteracting fermionic 2D lattice with SLS, there can be 2D massless Dirac fermions hosted around linearly crossing nodes, i.e., Dirac points (DPs), that could be present somewhere in the BZ. Due to fermion doubling, the Dirac Hamiltonian should be based on two spinors, one of which is due to the valleys about the DPs and the other is the sublattice. Take $SU(2)$ graphene as an example.

A reducible representation $(c_{A\mathbf{K}^+}c_{B\mathbf{K}^+}c_{A\mathbf{K}^-}c_{B\mathbf{K}^-})^T$ of the Dirac matrices is taken here. That is,

$$H_D = v_F\alpha_i p_i, \tag{1}$$

with $v_F$ the Fermi velocity, where $\alpha_i = (\sigma_x\tau_z, \sigma_y\tau_0)$ with $\sigma_i$ for the two sublattices A and B and $\tau_i$ for the two valleys around the two Dirac points $K^+$ and $K^-$. Graphene has been described in this way and, as such, can be used generally. Applying a magnetic field by Peierls substitution, $\mathbf{p} \to \mathbf{p} + e\mathbf{A}$, with $A$ a generic vector potential. Without intervalley mixing, the two-spinor effective Dirac Hamiltonian is given by $H(\mathbf{k}) = v_0\hbar\tau_3\sigma\mathbf{k}$, with $\hbar\mathbf{k}$ the crystal momentum measured from each DP, and $\sigma$ and $\tau$ the Pauli matrices acting on the sublattice and valley spinor, respectively. With $\lambda$ and $\xi$ denoting the band and valley indices, respectively, the energy of Dirac fermions is given by $E_{\lambda\xi}(\mathbf{k}) = \lambda v_0 \hbar \mathbf{k}$. The wave function around $\mathbf{k} = 0$ is given by

$$\psi_{\lambda\xi}(k) = (1/\sqrt{2})\left(\exp\left(-i\xi\theta/2\right), \lambda\xi \exp\left(i\xi\theta/2\right)\right)^T, \tag{2}$$

where $\mathbf{k} = (k, \theta)$ has a phase that is defined conventionally [18]. Thus, the winding number, $w_1 = \lambda\xi$, is derived with respect to the chirality. It is now clear that the effective CS arises from the global SLS in the whole lattice. In the presence of a perpendicular, uniform magnetic field $B\hat{\mathbf{z}}$, the effective CS is still preserved via the minimal coupling. This fact is simply reflected in clean graphene (without disorder). That is, the LL energies read

$$\epsilon_n = s\hbar\omega_c\sqrt{|n|}, \tag{3}$$

with $s = \pm$ and $\omega_c = v_0\sqrt{2eB/\hbar c}$, which is characterized by a zero mode that is constant at zero energy in a wide range of magnetic fields, as shown in the Hofstadter butterfly [19]. Moreover, the LL spectrum given by a $\sqrt{B}$-dependent function reflects the linear dispersion of the Dirac fermions. The LL wave functions are given by

$$\Psi(\mathbf{r}) \propto \left(\phi_{|n|}, s\phi_{|n|-1}\right)^T, \tag{4}$$

with $\phi_{|n|}$ being the $n$th simple harmonic oscillator wave function, so that the zero mode is pseudospin polarized (to a phase of gauge). Therefore, the zero-mode LL is still chiral and half-filled with respect to the Dirac point. It is robust under the protection of the effective CS of 2D massless Dirac fermions. In other words, the chiral zero-mode LL is a signature of the existence of 2D massless Dirac fermions in a noninteracting fermionic lattice with SLS.

In the QHE limit, the half-integer quantized Hall conductivity $\sigma_{xy} = (2n+1)2e^2/h$ has also been inferred for graphene [22, 23] as a further consequence of the effective CS. Moreover, it was realized soon after [1, 2]. To date, it has been fully understood that the half-integer QH effect, if shown, reflects the half-filled ground state, which can be ascribed to either SLS for 2D massless Dirac fermions or PHS for Majorana fermions. This chapter is concerned with the former in 3D layered systems. Nevertheless, life is not so easy. It must be noted that the transition between Hall plateaus is a topological QPT, and thus, the criticality must be ascribed to localization induced by disorder [25, 26]. Even if the disorder respects CS, like an external magnetic field, the characteristic of 2D massless Dirac fermions can be washed away. This point excludes clean systems from the consideration of QHE. There is controversy among the existing explanations for the half-integer QHE of graphene [20, 24, 29]. Anyway, the characteristic of 2D massless Dirac fermions must be a necessary condition for the half-integer QHE. Intuitively, intervalley scattering between the DPs due to disorder washes away the characteristic of 2D massless Dirac fermions so that the half-integer QHE is impossible.

## 9.2 Tilted Dirac Fermions

When a 2D layer of lattice hosts massless Dirac fermions, whether in an SPT phase or not, one may ask what if the layers are stacked in 3D in some ways. For systems such as d-wave cuprate superconductors with negligible interlayer coupling, a quasi-2D model is taken for granted so that vertical DNLs are naively grown from the DPs existing in each layer. When interlayer coupling is significant, take graphene for example, nexuses are present in Bernal graphite [34, 46]. By contrast, although there exist DPs in the $\pi$-flux square lattice, an SPT insulating phase of chiral TI in the $\pi$-flux cubic lattice has been proposed. All these are examples of the dimensional crossover of topological phases from 2D to 3D [31]. The (co-)dimensional-periodic table in the 10-fold-way symmetry classification dictates that the topological characterization might or might not alter after the dimensional crossover. This section introduces a well-analyzed case in the following. From graphene to RG, there is a dimensional crossover of topological phases from a 2D TSM with nodal points (DPs) to a 3D nodal-line TSM, where the co-dimension $p = 2$ holds [31, 35]. This case belongs to the second class of 2D massless Dirac fermions existing in 3D (Fig. 9.1(b)). Since these two TSMs belong to the same symmetry class and have the same co-dimension, they must have the same topological invariant. Indeed, they both contain 2D massless Dirac fermions.

The current model is described by the 3D lattice shown in Figs. 9.1(a) and 9.1(b), which is the rhombohedral stack (along the $z$ direction) of honeycomb-lattice layers in the $(x, y)$ plane. The bulk lattice consists of two sublattices as well as the 2D honeycomb lattice. In fact, it can be taken as a 3D extension of the SSH model [31, 35], which describes Dirac fermions about tremendously coupled chains of atoms [43]. Such systems would render a nontrivial topology [30, 31]. A realization of this model is

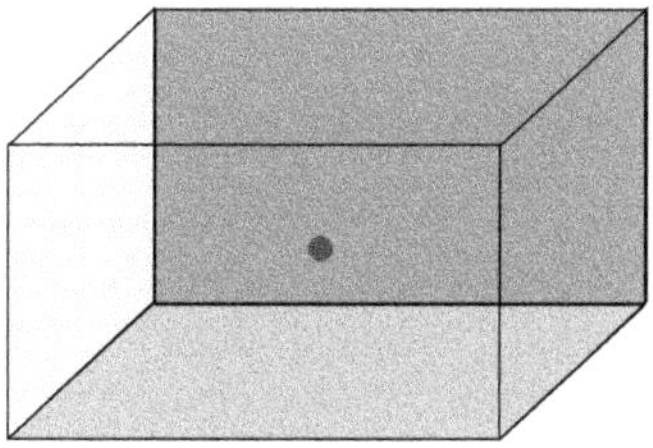

(a)

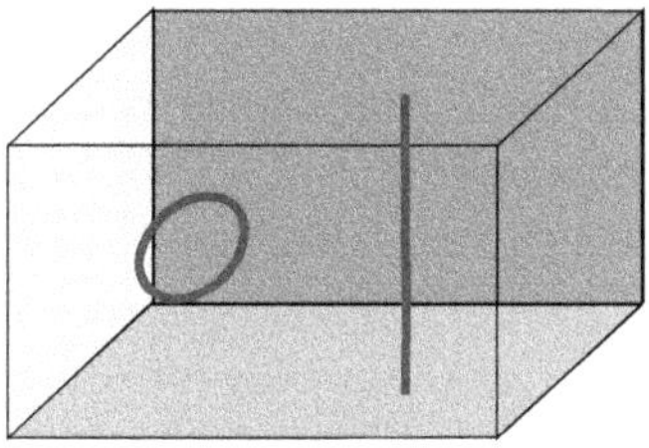

(b)

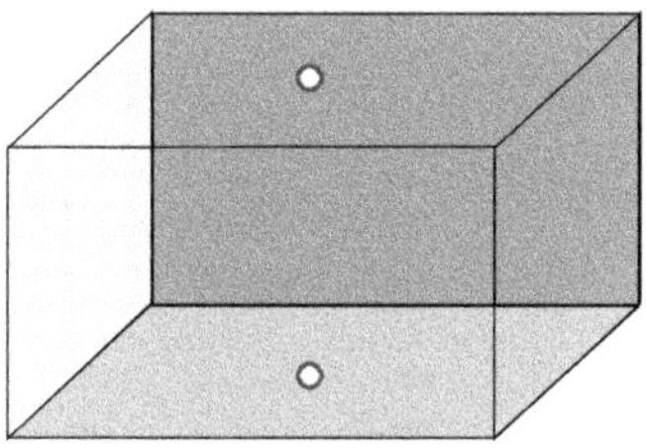

(c)

Fig. 9.1 Classes of 2D massless Dirac fermions existing in 3D: (a) around nodal points inside the 3D bulk BZ, (b) around nodal lines inside the 3D bulk BZ, (c) around nodal points on the 2D surface BZ.

given by RG with graphene layers stacked in an ABC configuration. The bulk Hamiltonian is written as

$$\mathcal{H}_0 = -t\sum_l \sum_{\langle i,j\rangle} [a^\dagger_{l,i} b_{l,j} + \text{h.c.}] + t' \sum_l \sum_i [b^\dagger_{l+1,i} a_{l,i} + \text{h.c.}], \tag{5}$$

where $l$ labels the layers and $\langle i,j\rangle$ denotes nearest-neighbor sites on each layer, and the operators $a^\dagger_{l,i}$ ($b^\dagger_{l,i}$) create fermions in the sublattice $A$ ($B$) at the $i$ site on the $l$ layer. Owing to the absence of inter-sublattice coupling, there exists the chiral symmetry operator $\mathcal{S} = \sum_{l,\langle i,j\rangle} s^\dagger_{l,\langle i,j\rangle} \sigma_z s_{l,\langle i,j\rangle}$, with $s^\dagger_{l,\langle i,j\rangle} = (a^\dagger_{l,i}, b^\dagger_{l,j})$, for $\mathcal{S}\mathcal{H}_0\mathcal{S}^{-1} = -\mathcal{H}_0$. The Fourier transformation by translation symmetry reads $a_{l,i} = N^{-\frac{1}{2}} \sum_{\mathbf{k}} e^{-i\mathbf{k}\cdot\mathbf{r}_{i,l}} a_{\mathbf{k}}$, and similarly for $b_{\mathbf{k}}$, so that the Bloch Hamiltonian is given by

$$H(\mathbf{k}) = d_1(\mathbf{k})\sigma_1 + d_2(\mathbf{k})\sigma_2, \tag{6}$$

where the Pauli matrices $(\sigma_1, \sigma_2, \sigma_3)$ act in the space of pseudospin $(a_{\mathbf{k}}, b_{\mathbf{k}})$, and $d_1(\mathbf{k}) = -t\sum_m^3 \cos(\mathbf{k}_{||}\cdot\delta_m) + t'\cos(k_z c)$ and $d_2(\mathbf{k}) = t\sum_m^3 \sin(\mathbf{k}_{||}\cdot\delta_m) - t'\sin(k_z c)$ are obtained, with $\delta_m$ being the three vectors connecting the nearest-neighbor sites and $c$ is the layer distance (Fig. 9.1(b)). Thus, the chiral operator is given by $\mathcal{S} = \sigma_z$, and SLS is written as $\mathcal{S}^{-1}H(\mathbf{k})\mathcal{S} = -H(\mathbf{k})$ in the momentum representation. It is easy to verify TRS by the operator $\mathcal{T} = \sigma_0\mathcal{K}$: $\mathcal{T}^{-1}H(-\mathbf{k})\mathcal{T} = H(\mathbf{k})$, and PHS by the operator $\mathcal{C} = \sigma_z\mathcal{K}$: $\mathcal{C}^{-1}H(-\mathbf{k})\mathcal{C} = -H(\mathbf{k})$, where $\sigma_0$ is the $2 \times 2$ identity matrix and $\mathcal{K}$ is the complex conjugation operator. Therefore, the Hamiltonian in Eq. (2) belongs to the chiral orthogonal BDI class.

The Hamiltonian described in Eqs. (1) and (2) is free from disorders and perturbations. There are three degrees of freedom $(k_x, k_y, k_z)$ and two equations $(d_1 = d_2 = 0)$ for finding the zeros, which determine the existence of nodes. In RG, a pair of DNLs are realized, as shown in Fig. 9.2 [32, 42]. The two DNLs spiral in opposite directions across the BZ boundaries from $k_z = -\pi$ to $k_z = \pi$. With polar coordinates $\mathbf{k} = (k, \phi)$, the two DNLs parameterized by $k_z$ are obtained as

$$\begin{aligned} k_{DL} &= \frac{t'}{\hbar v}, \\ \phi_{DL} &= \mp\left(k_z c - \frac{\pi}{2}\right) + \frac{\pi}{6}, \end{aligned} \tag{7}$$

where $v = \sqrt{3}at/(4\hbar)$ is the Fermi velocity. Equation (5) is in the same form as that not perturbed by $w_1$ [32], except for the spiraling axes now being shifted to the $(\mathbf{k}_{DNP}^{(\pm)}, k_z)$ lines.

The energy is derived as follows. In terms of the polar coordinates $(q, \theta)$ measured from the DNPs for fixed $kz$, the coordinate transformation runs as $q^2 = k^2 + k_D^2 - 2k_D k\cos(\phi - \phi_D)$ and $\tan\theta = (k\sin\phi - k_D\sin\phi_D)(k\cos\phi - k_D\cos\phi_D)^{-1}$. Obviously, $\theta \to -\theta$ as $\phi \to -\phi$ when changing between the two valleys (denoted by $\xi = \pm 1$). The Dirac cone is thus derived as

$$\varepsilon(q, \theta) = \lambda\xi v\hbar q, \tag{8}$$

where $\lambda$ indexes the bands. Note that it is independent of $k_z$, as an evidence of the same co-dimension $p = 2$ as that for graphene.

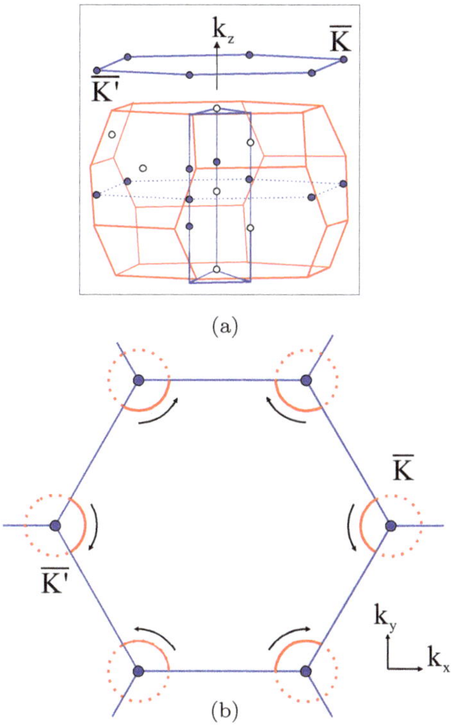


Fig. 9.2 (a) Projections (red circles) of the spiraling DNLs on the 2D projected BZ (blue hexagon), where the portions of DNLs inside (solid) and outside (dotted) are shown. To sum up, there are two inequivalent DNLs in the 3D rhombohedral BZ. The arrows indicate the spiraling senses in the increase of $k_z$. (b) 3D rhombohedral BZ (red), with the unfilled dots on the high-symmetry points, along with the 2D projected BZ.

The winding number can be obtained anywhere around the nodal lines, given by

$$w_1 = \frac{1}{2\pi i} \nabla_{\mathbf{q}} \log \tau_3 H(\mathbf{k}), \tag{9}$$

where the Pauli matrix $\tau_3$ acts as the valleys around the two nodal lines. As a result, $w_1 = \pm 1$, identically to that for graphene. The verification of the bulk correspondence as the key has been shown earlier [31]. The epitaxy for RG performed recently has shown that the surface flat bands do exist [57].

After all, the characteristic of RG is completely identical to that of graphene at the minimal model level (Eq. (1)). The reason is topological.

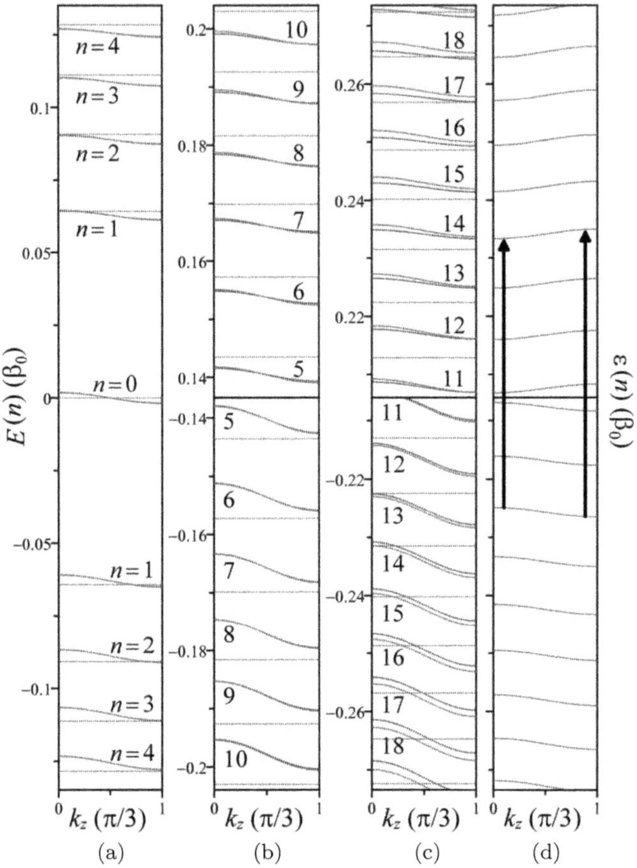


Fig. 9.3 Magnified LL spectra of RG at $B_0 = 30$ T for (a) $|E(n, B_0, k_z)| \leq 0.12$, (b) $0.195 \leq |E(n, B_0, k_z)| \leq 0.255$ and (c) $0.335 \leq |E(n, B_0, k_z)| \leq 0.365$ (in the unit of $t$): minimal model (black), Onsager quantization (red), and numerical result (blue).

It lies in the fact that both systems belong to the same topological class, but not the small value of interlayer coupling for the 2D limit. It can be proved that it is the chiral operator $\tau_0\sigma3$ that protects the zero-mode LL. As a whole, the LL spectra and wave functions are both the same as those of graphene. Of course, the basis for the wave functions (sublattice pseudospinor) is different. Although, the nearest hopping within the same sublattice will break the SLS, there is a generalized chiral operator that provides protection [52]. Numerical results for the LL energy spectrum and wave functions are shown in Figs. 11.5 and 11.6. The chiral half-filled

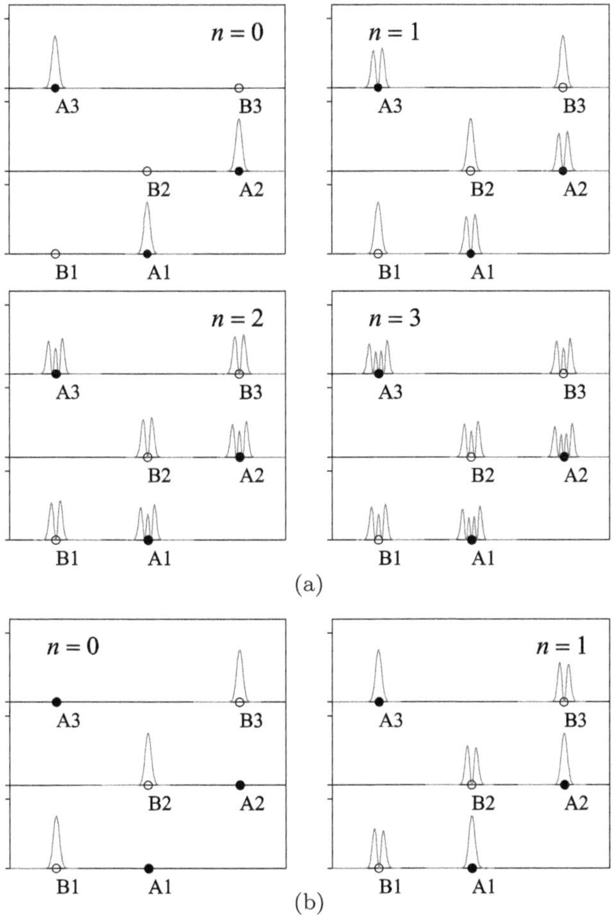


Fig. 9.4 LL wave functions for $k_z = \pi/6d$ in RG at $B_0 = 20$ T: (a) one of the two degenerate sets of LSs, plotted for $n = 0$ and unoccupied $n = 1$, 2 and 3; (b) the other set plotted for $n = 0$ and unoccupied $n = 1$. All the ordinate tick marks are labeled at 0 and 0.005.

zero mode is obviously present, showing the persistence of the effective CS in the presence of a perpendicular magnetic field. Finally, a 3D layered system is the easiest way to satisfy the Diophantine equation, which is required for the 3D integer QH effect [47], as shown in Fig. 9.2. The 3D integer QH effect in RG has been discovered [49, 50], which shows the scaling invariance in comparison to the 2D integer QH effect in graphene. This is a result of the same symmetry class and co-dimension between the two systems [48].

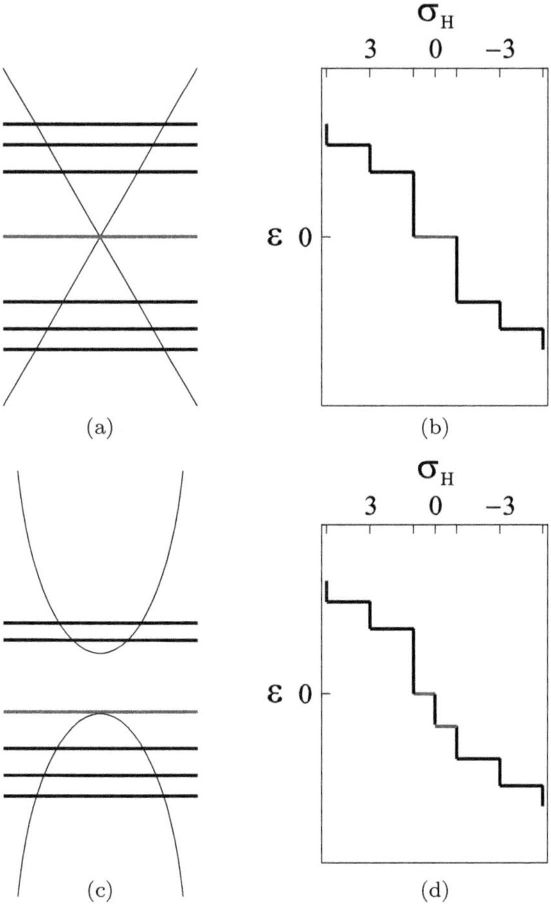


Fig. 9.5 Schematic diagrams of the LL energy spectra for: (a) LL energy spectrum as the effective CS holds in $B\hat{\mathbf{z}}$, (b) Hall plateaus arising from (a), (c) LL energy spectrum as the effective CS is broken due to magnetic doping, (d) Hall plateaus with magnetic doping in $B\hat{\mathbf{z}}$.

## References

[1] Novoselov K.S., Geim A.K., Morozov S.V., Jiang D., Katsnelson M.I., Grigorieva I.V. *et al.* 2005. Room-temperature quantum Hall effect in graphene. *Nature* **438** 197.

[2] Zhang Y., Tan Y.W., Stormer H.L. and Geim A.K. 2005. Room-temperature quantum Hall effect in graphene. *Nature* **438** 201.

[3] Hatsugai Y., Morimoto T., Kawarabayashi T., Hamamoto Y. and Aoki H. 2013. *New J. Phys.* **15** 035023.

[4] Klitzing K.V., Dorda G. and Pepper M. 1980. New method for high-accuracy determination of the fine-structure constant based on quantized Hall resistance. *Phys. Rev. Lett.* **45** 494.
[5] Thouless D.J., Kohmoto M., Nightingale M.P. and Nijs M.D. 1982. Quantized Hall conductance in a two-dimensional periodic potential. *Phys. Rev. Lett.* **49** 405.
[6] Avron J.E., Seiler R. and Simon B. 1983. Holonomy, the quantum adiabatic theorem, and Berry's phase. *Phys. Rev. Lett.* **51** 51.
[7] Simon B. 1983. Holonomy, the quantum adiabatic theorem, and Berry's phase. *Phys. Rev. Lett.* **51** 2167.
[8] Berry M.V. 1984. Quantum phase factors accompanying adiabatic changes. *Proc. R. Soc. London, Ser. A* **392** 45.
[9] Xiao D., Chang M.C. and Niu Q. 2010. Berry phase effects on electronic properties. *Rev. Mod. Phys.* **82** 1959.
[10] Hatsugai Y. 1993. Chen mumber and edge states in the integer quantum Hall effect. *Phys. Rev. Lett.* **71** 3697.
[11] Ryu S. and Hatsugai Y. 2002. Topological origin of zero-energy edge states in particle-hole symmetric systems. *Phys. Rev. Lett.* **89** 077002.
[12] Qi X.L., Wu Y.S. and Zhang S.C. 2006. General theorem relating the bulk topological number to edge states in two-dimensional insulators. *Phys. Rev. B* **74** 045125.
[13] Hatsugai Y. 2009. Bulk-surface correspondence in graphene with/without magnetic field: Chiral symmetry, Dirac fermions and edge states. *Solid State Commun.* **149** 1061.
[14] Mong R.S.K. and Shivamoggi V. 2011. Edge states and the bulk-boundary correspondence in Dirac Hamiltonians. *Phys. Rev. B* **83** 125109.
[15] Demkov Y.N. and Kurasov P.B. 2007. von Neumann-Wigner Theorem: Level's repulsion and degenerate eigenvalues. *Theor. Math. Phys.* **153** 1407.
[16] Young S.M., Zaheer S., Teo J.C.Y., Kane C.L., Mele E.J. and Rappe A.M. 2012. Dirac semimetal in three dimensions. *Phys. Rev. Lett.* **108** 140405.
[17] DiVincenzo D.P. and Mele E.J. 1984. Self-consistent effective-mass theory for intralayer screening in graphite intercalation compounds. *Phys. Rev. B* **29** 1685.
[18] Castro Neto A.H., Guinea F., Peres N.M.R., Novoselov K.S. and Geim A.K. 2009. The electronic properties of graphene. *Rev. Mod. Phys.* **81** 109.
[19] Hofstadter D.R. 1976. Energy levels and wave functions of Bloch electrons in rational and irrational magnetic fields. *Phys. Rev. B* **14** 2239.
[20] Kawarabayashi T., Morimoto T., Hatsugai Y. and Aoki H. 2010. Anomalous criticality at the n=0 quantum Hall transition in graphene: The role of disorder preserving chiral symmetry. *Phys. Rev. B* **82** 195426.
[21] Kawarabayashi T., Hatsugai Y. and Aoki H. 2010. Landau level broadening in graphene with long-range disorder — Robustness of the $n = 0$ level. *Physica E* **42** 759.
[22] Zheng Y. and Ando T. 2002. Hall conductivity of a two-dimensional graphite system. *Phys. Rev. B* **65** 245420.

[23] Gusynin V.P. and Sharapov S.G. 2005. Unconventional integer quantum Hall effect in graphene. *Phys. Rev. Lett.* **95** 146801.
[24] Kawarabayashi T., Hatsugai Y. and Aoki H. 2009. Quantum Hall plateau transition in graphene with spatially correlated random hopping. *Phys. Rev. Lett.* **103** 156804.
[25] Laughlin R.B. 1981. Quantized Hall conductivity in two dimensions. *Phys. Rev. B* **23** 5632.
[26] Halperin B.I. 1982. Quantized Hall conductance, current-carrying edge states, and the existence of extended states in a two-dimensional disordered potential. *Phys. Rev. B* **25** 2185.
[27] Ludwig A.W.W., Fisher M.P.A., Shankar R. and Grinstein G. 1994. Integer quantum Hall transition: An alternative approach and exact results. *Phys. Rev. B* **50** 7526.
[28] Hatsugai Y., Wen X.G. and Kohmoto M. 1997. Disordered critical wave functions in random-bond models in two dimensions: Random-lattice fermions at $E = 0$ without doubling. *Phys. Rev. B* **56** 1061.
[29] Ostrovsky P.M., Gornyi I.V. and Mirlin A.D. 2008. Theory of anomalous quantum Hall effects in graphene. *Phys. Rev. B* **77** 195430.
[30] Xiao R., Tasnádi F., Koepernik K., Venderbos J.W.F., Richter M. and Taut M. 2011. Density functional investigation of rhombohedral stacks of graphene: Topological surface states, nonlinear dielectric response, and bulk limit. *Phys. Rev. B* **84** 165404.
[31] Heikkilä T.T. and Volovik G.E. 2011. Dimensional crossover in topological matter: Evolution of the multiple Dirac point in the layered system to the flat band on the surface. *JETP Lett.* **93** 59.
[32] Ho C.H., Chang C.P., Su W.P. and Lin M.F. 2013. Precessing anisotropic Dirac cone and Landau subbands along a nodal spiral. *New J. Phys.* **15** 053032.
[33] Ho C.H., Chang C.P. and Lin M.F. 2014. Landau subband wavefunctions and chirality manifestation in rhombohedral graphite. *Solid State Commun.* **197** 11.
[34] Heikkilä T.T. and Volovik G.E. 2015. Nexus and Dirac lines in topological materials. *New J. Phys.* **17** 093019.
[35] Ho C.H., Chang C.P. and Lin M.F. 2016. Evolution and dimensional crossover from the bulk subbands in ABC-stacked graphene to a three-dimensional Dirac cone structure in rhombohedral graphite. *Phys. Rev. B* **93** 075437.
[36] Burkov A.A., Hook M.D. and Balents L. 2011. Topological nodal semimetals. *Phys. Rev. B* **84** 235126.
[37] Yang B.J. and Nagaosa N. 2014. Classification of stable three-dimensional Dirac semimetals with nontrivial topology. *Nat. Commun.* **5** 4898.
[38] Armitage N.P., Mele E.J. and Vishwanath A. 2018. Weyl and Dirac semimetals in three-dimensional solids. *Rev. Mod. Phys.* **90** 015001.
[39] Fu L. and Kane C.L. 2007. Topological Insulators in Three Dimensions. *Phys. Rev. Lett.* **98** 106803.
[40] Fu L. and Kane C.L. 2007. Topological insulators with inversion symmetry. *Phys. Rev. B* **76** 045302.
[41] Wallace P.R. 1947. The band theory for graphite. *Phys. Rev.* **71** 622.

[42] McClure J.W. 1969. Electron energy band structure and electronic properties of rhombohedral graphite. *Carbon* **7** 425.

[43] Guinea F., Castro Neto A.H. and Peres N.M.R. 2006. Electronic states and Landau levels in graphene stacks. *Phys. Rev. B* **73** 245426.

[44] Ho C.H., Chang C.P. and Lin M.F. 2015. Optical magnetoplasmons in rhombohedral graphite with a three-dimensional Dirac cone structure. *J. Phys. Condens. Matter* **27** 125602.

[45] Goerbig M.O., Fuch J.N., Montambaux G. and Piéchon F. 2008. Tilted anisotropic Dirac cones in quinoid-type graphene and $\alpha$-(BEDT-TTF)$_2$I$_3$. *Phys. Rev. B* **78** 045415.

[46] Hyart T. and Heikkilä T.T. 2016. Momentum-space structure of surface states in a TSM with a nexus point of Dirac lines. *Phys. Rev. B* **93** 235147.

[47] Kohmoto M., Halperin B.I. and Wu Y.S. 2006. Diophantine equation for the three-dimensional quantum Hall effect. *Phys. Rev. B* **45** 13488.

[48] Balents L. and Fisher M.P.A. 1996. Chiral surface states in the bulk quantum Hall effect. *Phys. Rev. Lett.* **76** 2782.

[49] Kopelevich Y., Torres J.H.S., da Silva R.R., Mrowka F., Kempa H. and Esquinazi P. 2003. Reentrant metallic behavior of graphite in the quantum limit. *Phys. Rev. Lett.* **90** 156402.

[50] Kempa H., Esquinazi P. and Kopelevich Y. 2006. Integer quantum Hall effect in graphite. *Solid State Commun.* **138** 118.

[51] Raghu S. and Haldane F.D.M. 2008. Analogs of quantum-Hall-effect edge states in photonic crystals. *Phys. Rev. A* **78** 033834.

[52] Kawarabayashi T., Hatsugai Y., Morimoto T. and Aoki H. 2011. Generalized chiral symmetry and stability of zero modes for tilted Dirac cones. *Phys. Rev. B* **83** 153414.

[53] Kawarabayashi T., Aoki H. and Hatsugai Y. 2016. Lattice realization of the generalized chiral symmetry in two dimensions. *Phys. Rev. B* **94** 235307.

[54] Kim Y., Wieder B.J., Kane C.L. and Rappe A.M. 2015. Dirac line nodes in inversion-symmetric crystals. *Phys. Rev. Lett.* **115** 036806.

[55] Fang C., Chen Y., Kee H.Y. and Fu L. 2015. Topological nodal line semimetals with and without spin-orbital coupling. *Phys. Rev. B* **92** 081201.

[56] Chan Y.H., Chiu C.K., Chou M.Y. and Schnyder A.P. 2016. $Ca_3P_2$ and other topological semimetals with line nodes and drumhead surface states. *Phys. Rev. B* **93** 205132.

[57] Pierucci D., Sediri H., Hajlaoui M., Girard J.C., Brumme T. and Calandra M. 2015. Evidence for flat bands near the Fermi level in epitaxial rhombohedral multilayer graphene. *ACS Nano* **9** 5432.

Chapter 10

# SPIN-POLARIZED MAGNETO-ELECTRONIC PROPERTIES OF BUCKLED MONOLAYER GaAs

Hsien-Ching Chung, Lu-Yao Wang, Chih-Wei Chiu, Ching-Hong Ho, Chiun-Yan Lin, and Ming-Fa Lin

Over the past decade, graphene [1, 2] has successfully led scientists into the new world of two-dimensional (2D) materials owing to its incredible intrinsic properties, such as high carrier mobility at room temperature (>200000 cm$^2$/Vs) [3–5], superior thermoconductivity (3000–5000 W/mK) [3, 6], high transparency for incident light over a wide range of wavelength (97.7%) [7, 8], extremely large Young's modulus (∼1 TPa) and tensile strength (∼100 GPa) [9], half-integer and integer quantum hall effect [10–12] and Klein tunneling [13]. Few-layer graphenes are observed to possess rich magnetic quantizations, e.g., the Landau levels (LLs) with the $\sqrt{B_z}$-dependent energy spectrum in monolayer graphene featuring massless Dirac fermions [14, 15] and the linear $B_z$ dependence in AB-stacked bilayer graphene featuring massive Dirac fermions [16, 17], as well as the coexistence of square-root and linear $B_z$-dependent LLs in trilayer ABA stacking [18], where $B_z$ is the strength of magnetic field. Although interest in graphene materials is still high, it is also conspicuous that such systems have their limitations. For instance, in contrast to conventional semiconductors, the lack of significant bandgaps limits their potential applications in electronic devices, where high transistor ON/OFF ratios are important [19]. This obstacle urges researches to study emergent 2D materials [20], including Group-IV [21], Group-V [22] and Group III–V compounds [23] and transition-metal dichalcogenides

(TMDs) [24–26]. (TMDs have more than 40 different types [27, 28], depending on the combination of chalcogen (S, Se, Te) and transition-metal atoms (Mo, W, Nb, Ta, Ti).) These 2D layered materials are expected to exhibit various electronic properties, being sensitive to the lattice symmetry, stacking configuration, layer number, orbital hybridization, spin–orbit interaction (SOI) and external electric and magnetic fields.

Group-IV mono-elemental 2D honeycomb materials beyond graphene, such as silicene [29–34], germanene [33, 35–38] and stanene [39–41], have been predicted to exhibit bandgaps, depending on the strength of SOI [42]. Recently, few-layer silicene, germanene and stanene have been synthesized on distinct substrates: silicene on Ag(111) [31, 43], Ir(111) [44] and $ZrB_2$(0001) [45]; germanene on Pt(111) [46], Al(111) [47] and Au(111) [35]; stanene on $Bi_2Te_3(111)$ [40]. Silicene, germanene and stanene have buckled structures with significant SOIs, which grow as the atomic number increases. They are very different from the planar hexagonal graphenes without SOI. Their low-lying electronic structures are governed by SOIs and multi-orbital hybridizations. In general, Group-IV materials with heavy atomic masses have large buckling angles and rather strong SOIs, resulting in their energy gaps being higher than the thermal energy at room temperature (25 meV) [42, 48]. Furthermore, the greatly diversified quantizations are identified from the various magnetic- and electric-field-created LLs with crossing, anticrossing and non-crossing behaviors [49]. However, the strong interactions between Group-IV 2D materials and substrates might distort the buckled structure and hybridize the electronic states near the Fermi level, creating non-negligible modifications of low-energy electronic properties. Recent experiments on tunneling spectra of silicene have evidenced the disappearance of LL sequences based on the instability of the dangling bonds of the $sp^3$-hybridized atoms [50].

Apart from the 2D materials of Group-IV elements, the binary compounds of Group III–V elements have also been predicted to have honeycomb lattices with large energy gaps [23, 34]. Although the Group III–V elemental 2D materials with a buckled structure having mixed $sp^3$–$sp^2$ bonding are more stable compared to those with a planar structure having $sp^2$ bonding [34], the dangling-bond-induced instability remains. A promising route is to saturate the dangling bonds with halogen atoms, as successfully revealed in graphene [51]. Bulk GaAs is one of the famous Group III–V elemental binary compounds, which is widely used in the manufacture of electronic and optical devices due to its direct bandgap (an indirect gap for silicon) and high mobility (higher than silicon) [52, 53].

According to first-principles calculations [54], monolayer GaAs possesses a buckled hexagonal structure, multi-orbital chemical bonding and significant SOI, which are expected to induce diverse electronic properties and rich magnetic quantization in the existence/absence of electric fields.

The tight-binding model proposed by Zhao *et al.* [54] is extended to study the electronic properties under external electric and magnetic fields. The quantized energy spectra and wave functions are calculated using the exact diagonalization method. A similar generalized tight-binding model has been widely adopted to systematically study multi-dimensional carbon-based materials and hybrid systems, ranging from three-dimensional (3D) graphites [55, 56], 2D graphenes [56, 57], 1D graphene nanoribbons (GNRs) [58–62], carbon nanotubes (CNTs) [63, 64], graphene nanoflake [65] and graphene-related hybrids [66]. It is also suitable for studying mainstream layered materials, such as Group-IV [42, 67], Group-V [68] and TMD [69–72] 2D materials.

In this work, buckled monolayer GaAs with each atom being passivated by one F adatom is chosen for a model study (Fig. 10.1(a)), in which the saturated dangling bonds greatly weaken the substrate instability. The generalized tight-binding model, simultaneously considering geometric structure, multi-orbital hybridizations, SOI and external fields, is employed to explore the essential properties. The low-energy electronic properties are thoroughly investigated under the strong effects of external magnetic and electric fields. A "semimetal–semiconductor" phase diagram is obtained for the bandgap as a function of both magnetic and electric fields. The current study sheds light on the diversified magnetic quantizations of the electronic energy spectra of GaAs and other Group III–V 2D materials. The predicted magneto-electronic properties of the monolayer GaAs could be identified by scanning tunneling spectroscopy (STS) measurements.

## 10.1 Method

### 10.1.1 *Geometric structure of monolayer buckled GaAs*

Monolayer GaAs has a buckled honeycomb lattice, with each atom being passivated by an F adatom, as clearly shown in Fig. 10.1. There exists $sp^3$ chemical bonding among (Ga, As, F) atoms (three for As or Ga and one for F), in which the Ga–As bond length is about 2.521 Å. A unit cell containing two different Ga and As sublattices is indicated by a rhombus with

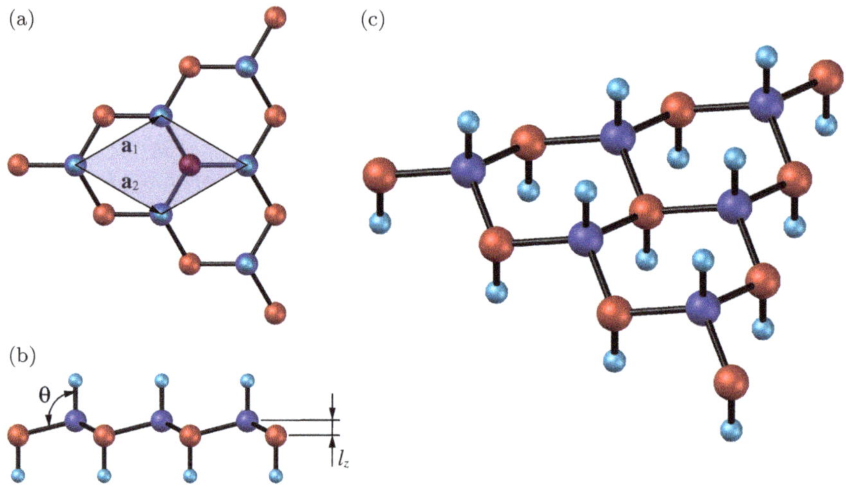


Fig. 10.1 Geometric structure of monolayer GaAs: (a) Top view of the low-buckled monolayer GaAs. The unit cell is indicated by the transparent blue rhombus, where $\mathbf{a}_1$ and $\mathbf{a}_2$ are translation vectors. The Ga, As and F atoms are indicated by the blue, red and cyan balls, respectively. (b) Side view of the low-buckled monolayer GaAs. $\theta$ and $l_z$ are the buckling angle and the distance between the Ga and As planes, respectively. (c) Schematic representation of the monolayer GaAs decorated by F adatoms.

primitive unit vectors, $\mathbf{a}_1$ and $\mathbf{a}_2$, with a lattice constant of $a = 4.226\,\text{Å}$. The altitude of the buckled structure measured from the distance between the Ga and As planes is $l_z = 0.633\,\text{Å}$ (Fig. 10.1(b)). The buckling angle $\theta$ between the $z$-axis and the Ga–As bond is about 104.54°. This configuration is free from dangling bonds and thus chemically stable. F atoms are right above/below the Ga/As atoms, with the Ga–F and As–F distances of 1.776 and 1.781 Å, respectively [54].

### 10.1.2 *Generalized tight-binding model*

A generalized tight-binding model, which is very suitable for exploring the essential properties of monolayer GaAs, is developed, where many critical factors, namely the buckled structure, multi-orbital hybridizations, SOI, electric field and magnetic field, are included in the calculation simultaneously [58, 73]. The current model provides a theoretical framework for investigating the strong cooperations/competitions among various key factors and affords systematic studies of a range of materials, from bulk materials to low-dimensional materials. The theoretical models, with both single-particle and many-body schemes, can also be combined to comprehend the essential physical properties, e.g., frequency-dependent and static Kubo formulas for exploring the optical absorption spectra [58, 59, 74] and quantum Hall effect, respectively.

To illustrate the electronic properties of monolayer GaAs explicitly, the Hamiltonian built from the tight-binding functions of $4s$, $4p_x$ and $4p_y$ orbitals is expressed as

$$\mathcal{H} = \sum_{m,\alpha} \epsilon_m^\alpha c_m^{\alpha\dagger} c_m^\alpha + \sum_{\langle m,n\rangle,\alpha,\beta} \gamma_{mn}^{\alpha\beta} (c_m^{\alpha\dagger} c_m^\beta + h.c.), \tag{1}$$

where $\epsilon_m^\alpha$, $c_m^{\alpha\dagger}$ and $c_m^\alpha$, respectively, represent the on-site energy, creation and annihilation operators of an electron at the $\alpha$ orbital of the $m$th atom. $\gamma_{mn}^{\alpha\beta}$ is the nearest-neighbor hopping integral between an $\alpha$ orbital of the $m$th atom and a $\beta$ orbital of the $n$th atom. The multi-orbital hopping integrals are

$$\gamma_{mn}^{ss} = V_{ss\sigma}, \tag{2}$$

$$\gamma_{mn}^{sp_x} = V_{sp\sigma} \cos\theta_x, \tag{3}$$

$$\gamma_{mn}^{sp_y} = V_{sp\sigma} \cos\theta_y, \tag{4}$$

$$\gamma_{mn}^{p_xp_x} = V_{pp\sigma} \cos^2\theta_x + V_{pp\pi}(1 - \cos^2\theta_x), \tag{5}$$

$$\gamma_{mn}^{p_yp_y} = V_{pp\sigma} \cos^2\theta_y + V_{pp\pi}(1 - \cos^2\theta_y), \tag{6}$$

$$\gamma_{mn}^{p_xp_y} = (V_{pp\sigma} - V_{pp\pi}) \cos\theta_x \cos\theta_y, \tag{7}$$

where $\theta_x$ and $\theta_y$ are, respectively, the angles of the vector initiated from the $m$th atom to the $n$th atom with respect to the $x$- and $y$-axes [75], and the Slater–Koster hopping parameters in the $sp^3$ bonding optimized at the equilibrium state are $V_{ss\sigma} = -1.707$ eV, $V_{sp\sigma} = 2.056$ eV, $V_{pp\sigma} = 2.650$ eV and $V_{pp\pi} = -0.827$ eV [54]. The on-site energies of $s$ and $p$ orbitals are set to the values $-12.00$ and $-5.67$ eV for Ga and $-17.68$ and $-8.30$ eV for As, which are taken from those of bulk GaAs [76]. It should be noted that the $4p_z$ orbitals are ignored in the Hamiltonian in Eq. (1). A strong fluorination could fully suppress the dangling bands of monolayer GaAs; that is, there exist very pronounced between $2p_z$ and $4p_z$ orbitals due to the top-site adsorptions. The rather stable $2p_z$–$4p_z$ chemical bonds create the deeper valence bands far away from the Fermi level [77, 78]. This is responsible for the absence of $2p_z$ and $4p_z$ orbitals in the current model. The $\sigma$ orbitals $(p_x, p_y)$ with the stronger SOI dominate low-lying electronic states, instead of the $\pi$ orbitals ($p_z$) with the weaker SOI. Similar phenomena are also revealed in fluorinated graphene theoretically [79, 80] and experimentally [51], in which the $(2p_x, 2p_y)$ orbitals of carbon atoms make important contributions to the low-lying band structures, but not the $2p_z$ orbitals. As a result, the $\pi$ bonding-related linear Dirac cone structures are suppressed thoroughly.

### 10.1.3 *Spin–orbit interaction*

When an electron of momentum $\mathbf{p}$ moves close to the atomic nuclei in a crystal with an electric potential $V$, it experiences an effective magnetic field $B_{\text{eff}} \sim \nabla V \times \mathbf{p}/m_0 c^2$ in its rest frame, where $m_0$ is the mass of a free electron and $c$ is the speed of light. Such a field induces a momentum-dependent Zeeman energy called the SO interaction, which is expressed as

$$H^{\text{SO}} = \frac{\hbar}{4m_0^2 c^2}(\nabla V \times \mathbf{p}) \cdot \sigma, \tag{8}$$

where $\hbar$ is the reduced Planck constant and $\sigma$ is the vector of Pauli matrices. In the central-field approximation, the crystal potential $V(\mathbf{r})$ is considered the spherical atomic potential. The SOI term for the same atom is taken into account, and it can be obtained by calculating the mean value:

$$H^{\text{SO}}_{i,\alpha\beta} = \lambda_i \langle \mathbf{L} \cdot \sigma \rangle_{\alpha\beta}, \tag{9}$$

where $\lambda_i$ is the SOI strength of the $i$th atom and $\mathbf{L}$ is the orbital angular momentum operator. The matrix element $\langle \mathbf{L} \cdot \sigma \rangle_{\alpha\beta}$ is given on the basis of atomic orbitals $(\alpha, \beta)$, and the dimensionless SOI operator $\mathbf{L} \cdot \sigma$ for the relevant orbitals (4$s$, 4$p_x$ and 4$p_y$) in the 2D system is given by

$$\mathbf{L} \cdot \sigma = \begin{pmatrix} 0 & 0 & 0 \\ 0 & 0 & -is_z \\ 0 & is_z & 0 \end{pmatrix}, \tag{10}$$

where $s_z = \left(\begin{smallmatrix} 1 & 0 \\ 0 & -1 \end{smallmatrix}\right)$. The SOI strengths of Ga and As atoms are chosen to be 0.058 and 0.140 eV, respectively [81].

### 10.1.4 *Peierls substitution with composite electric and magnetic fields*

When a uniform perpendicular magnetic field, $\mathbf{B} = B_z\hat{z}$, is applied to monolayer GaAs, the effective Hamiltonian could be regarded as the Peierls substitution Hamiltonian [82]. Each Hamiltonian matrix element turns into a product of the zero-field element and the extra Peierls phase, $\exp(i2\pi\theta_{mn})$, where $\theta_{mn} = (1/\phi_0)\int_m^n \mathbf{A}\cdot d\mathbf{l}$ is a line integral of the vector potential $\mathbf{A}$ from the $m$th to $n$th site, $\mathbf{A}$ is chosen as $(0, B_z x, 0)$ in the Landau gauge and $\phi_0 = h/e$ $(4.1357\times 10^{-15}\ \mathrm{T}\cdot\mathrm{m}^2)$ is the magnetic flux quantum. The unit cell becomes an enlarged rectangle with $2R_B$ Ga and $2R_B$ As atoms to satisfy the periodicity of the Peierls phase, where $R_B = \phi_0/\phi = \phi_0/(B_z\sqrt{3}/2a^2) \sim 26739\ \mathrm{T}/B_z$ is the ratio of flux quantum to magnetic flux through a hexagon $\phi$ (Fig. 10.2).

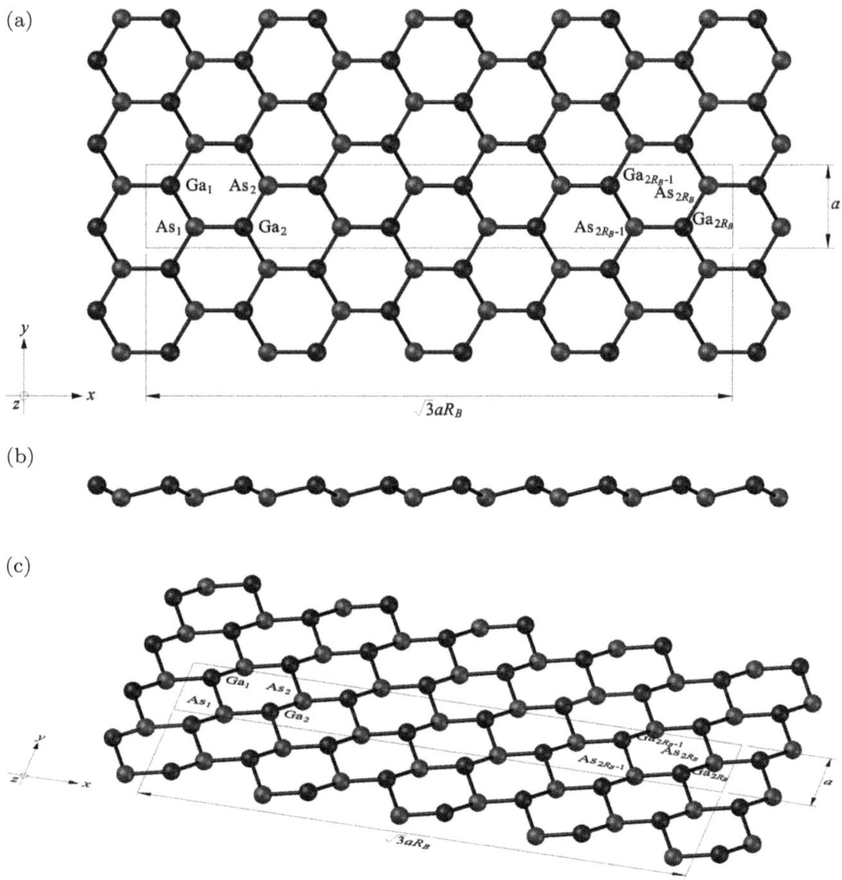


Fig. 10.2 Geometric structure of the low-buckled monolayer GaAs in a uniform perpendicular magnetic field: (a) Top view: The transparent-blue enlarged unit cell, with $2R_B$ Ga and $2R_B$ As atoms satisfies the periodicity of the Peierls phase, where $R_B$ is the ratio of flux quantum to magnetic flux through a hexagon. The Ga and As atoms are indicated by the blue and red balls, respectively; the F adatoms are omitted for a clearer presentation. (b) Side view. (c) Schematic representation.

The reduced Brillouin zone has an area of $4\pi^2/\sqrt{3}a^2R_B$. The magnetic Hamiltonian is built in the space spanned by the $24R_B$ tight-binding functions $\{|Ga_m^{\mathrm{orb}}\rangle, |As_m^{\mathrm{orb}}\rangle;\ m = 1,2,3,\ldots,2R_B;\ \mathrm{orb} = 4s, 4p_x, 4p_y\} \bigotimes \{\uparrow,\downarrow\}$. An electric field, $\mathbf{E} = E_z\hat{z}$, along the $z$-axis introduces a Coulomb potential energy, $-eE_zl_z/2$ ($eE_zl_z/2$), to the site energy of the Ga (As) sublattice. The exact diagonalization method of the giant Hamiltonian matrix $\mathcal{H}$ yields the energy spectrum $E^{c,v}$ and the wave functions $|\Psi^{c,v}\rangle$, where the superscripts $c$ and $v$ denote the conduction and valence subbands, respectively.

## 10.2 Results and Discussions

### 10.2.1 *SOI-induced spin-polarized band structure and state probabilities*

Monolayer GaAs exhibits feature-rich energy bands, mainly owing to the significant buckled structure, $sp^3$ bonding and SOIs. As shown in Fig. 10.3, there exist three low-lying energy subbands with strong wavevector dependences in the monotonous form. The unoccupied conduction subband ($n_1$) and two occupied valence subbands ($n_2$ and $n_3$), with their distinct curvatures, are initiated from the Γ point. That is, the low-lying electronic states could form the Γ valley being responsible for the rich magnetic quantization. Without the SOIs, each subband is twofold degenerate for the spin degree of freedom except for the fourfold degeneracy at the intersection of $n_2$ and $n_3$ subbands (dashed curves in Fig. 10.3). The conduction and valence subbands near the Γ point are, respectively, dominated by the 4s and ($4p_x, 4p_y$) orbitals [54]. More importantly, a direct bandgap ($E_g = 0.742$ eV) is determined by the band-edge states of $n_1$ and $n_2/n_3$ in the absence of SOIs. The significant SOI further induces a variation in the bandgap and spin-splitting (solid curves in Fig. 10.3). The bandgap shrinks to $E_g^{\mathrm{SO}} = 0.623$ eV, while the $n_2$ and $n_3$ valence subbands are separated by $\Delta_{\mathrm{SO}} = 0.237$ eV, lifting the state degeneracy at the Γ point from fourfold to twofold. The spin degeneracy is removed except for the zone from the Γ to M points. As a result, the spin-degenerate subbands become the spin-polarized ones. As for energy subbands of the same group, the splitting energies gradually grow when the electronic states deviate from the Γ point and their splittings reach the maximum values at the K (K′) points, such as 0.196 eV between the $n_1^{\uparrow}$ and $n_1^{\downarrow}$ subbands and 0.133 eV between the $n_2^{\uparrow}$ and $n_2^{\downarrow}$ ones). Such spin splittings also appear in the GaAs quantum wells verified by photocurrent measurements [83],

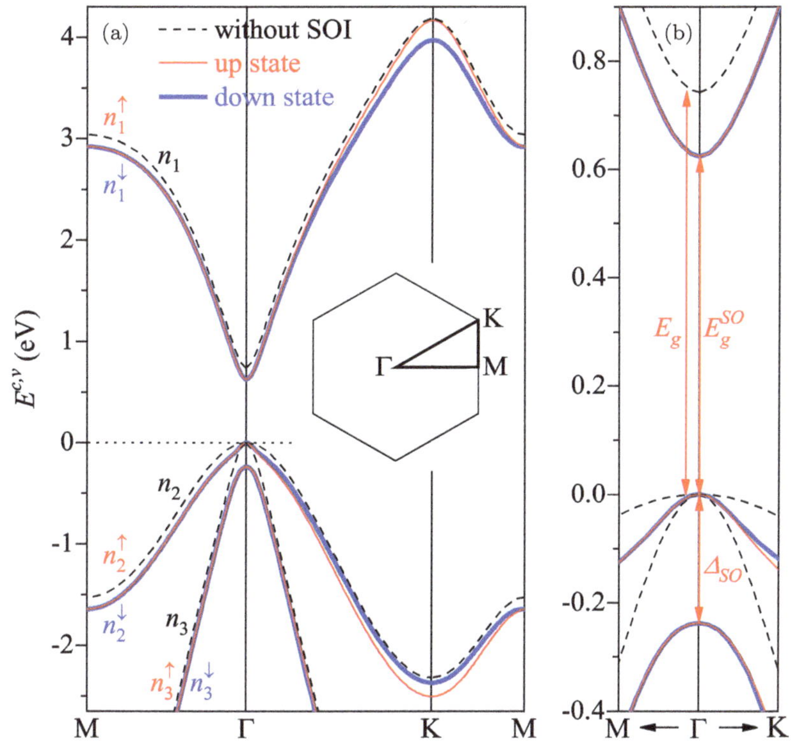


Fig. 10.3 Low-lying subbands of monolayer GaAs: (a) Spin-degenerate energy subbands without SOI ($n_1$, $n_2$ and $n_3$) and SOI-induced spin-polarized subbands ($n_1^{\uparrow}$, $n_1^{\downarrow}$, $n_2^{\uparrow}$, $n_2^{\downarrow}$, $n_3^{\uparrow}$ and $n_3^{\downarrow}$) along the high-symmetry points (Γ, M and K); the dotted line indicates the Fermi level, $E_F = 0$. (b) The energy spectra near $E_F = 0$.

where SOI results in interaction terms linear in wavevector **k** in the effective Hamiltonian [84].

The electronic state probabilities ($|\Psi^{c,v}|^2$) correspond to the spatial distributions of different orbitals, which in turn correspond to energy subbands and determine the variations of major/minor orbitals along the high-symmetric points. A whole range of orbital variations on different sublattices for the $n_1^{\uparrow\downarrow}$, $n_2^{\uparrow\downarrow}$ and $n_3^{\uparrow\downarrow}$ subbands is shown in Fig. 10.4. It is sufficient to discuss one of the polarized states (e.g., spin-up states) since the state configurations of spin-up (white zones) and spin-down (gray zones) states are almost similar. The state probabilities for different orbitals are very sensitive to the sublattices and wavevectors. In the conduction $n_1^{\uparrow\downarrow}$, valence $n_2^{\uparrow\downarrow}$ and valence $n_3^{\uparrow\downarrow}$ subbands, the $s$ orbitals (red curves),

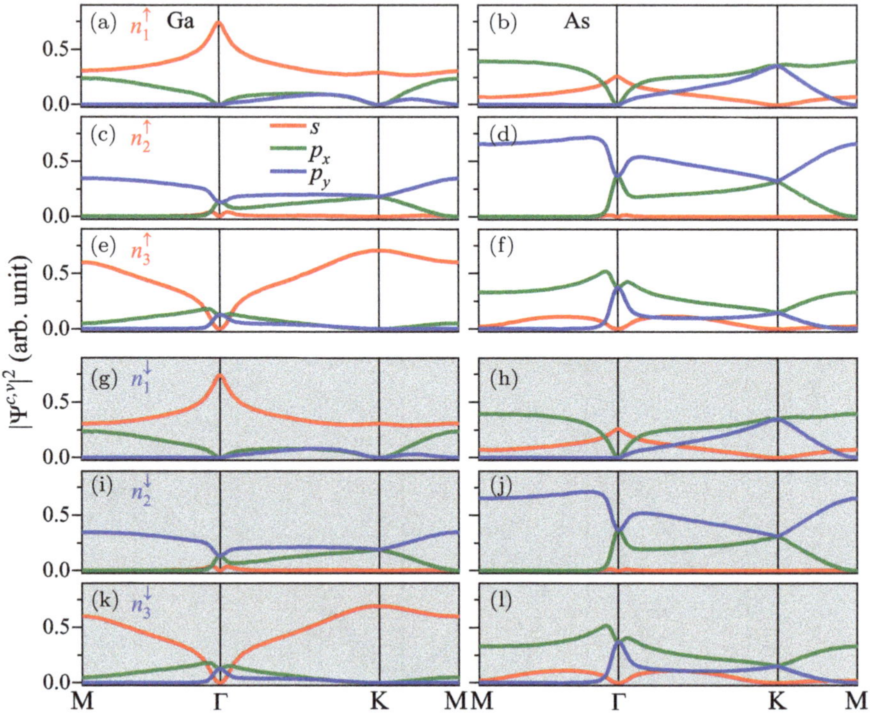


Fig. 10.4 State probabilities of monolayer GaAs: The state probabilities of various orbitals located at two sublattices Ga and As with spin-up (white zones) and spin-down (gray zones) arrangements. The $s$, $p_x$ and $p_y$ orbitals are represented by red, green and blue curves, respectively.

$p_y$ orbitals (blue curves) and $p_x$ orbitals (green curves), respectively, make the most important contributions to a wide range of wavevectors. Remarkably, the $p_x$ and $p_y$ orbitals are of identical intensity at the high-symmetric Γ and K points. The state probabilities near the Γ point, which are much different from those far away from it, apparently reveal a drastic orbital variation for the low-lying states. The conduction subbands are dominated by the $s$ orbitals, whose state probabilities on the Ga sublattice are larger than those on the As sublattice (Figs. 10.4(a), (b), (g) and (h)). The increase in the $p_x$- and $p_y$-orbital strengths and the decrease in the $s$-orbital strength arise as $\mathbf{k}$ deviates from the Γ point. The valence $n_2$ ($n_3$) subbands are dominated by $p_y$ orbitals ($p_x$ orbitals) (Figs. 10.4(c)–(f) and (i)–(l)). Instead of the Ga sublattices, the dominating orbitals on the As sublattices possess greater strength. It should be noted that the relative strength of the orbital probabilities for the low-lying states will reflect on the quantized magneto-electronic states. In other words, the low-energy Landau states feature those accumulated zero-field states near the Γ valley (discussed later).

### 10.2.2 *Spin-polarized LL spectra and state probabilities*

Magnetic fields constrain carrier motions in real space and result in the quantization of the cyclotron orbitals. They create highly degenerate dispersionless states, which are called the LLs. The initial energies of LLs might be related to the zero-field electronic structures. The energy spacing between the Landau states depends on the external fields, massless/massive characteristics of the Dirac fermions and SOIs. Near the Fermi energy, there are three groups of spin-polarized dispersionless LLs, i.e., one group of unoccupied conduction LLs ($n_1^{\uparrow}$ and $n_1^{\downarrow}$ in Fig. 10.5(a)) and two groups of occupied valence LLs ($n_2^{\uparrow}$, $n_2^{\downarrow}$, $n_3^{\uparrow}$ and $n_3^{\downarrow}$ in Figs. 10.5(d) and (e)). The distinct spin polarization in each group of LLs results from the SOI between the $4p_x$ and $4p_y$ orbitals. Such groups are, respectively, initiated near 0.62, 0 and −0.24 eV, reflecting the electronic state energies at the Γ point in the absence of magnetic fields. For each $(k_x, k_y)$, all LLs are twofold degenerate, being attributed to the one Γ-valley degree of freedom and two $B_z$-field-direction degrees of freedom. When the state energy grows, the energy spacing between LLs of the same spin-up/spin-down subgroup gradually shrinks.

Wave functions, which present the spatial information of electronic states, are very important in understanding fundamental physical properties, such as charge densities [85–88], state mixing [89, 90] and optical selection rules [58, 59, 74, 91]. Under the influence of a magnetic field, wave functions in monolayer GaAs exhibit peculiar spatial distributions, where the localization center, the dominating orbital, the oscillational form and the number of zero points are very sensitive to the wavevector, state energy and spin configurations. Each spin-polarized LL wave function can be decomposed into subenvelope functions with the ($s$, $p_x$, $p_y$) orbitals on the Ga and As sublattices at the odd and even sites. For the sake of simplicity, only the distribution probabilities of subenvelope functions at the odd sites ($Ga_o$ and $As_o$) are considered because the even-site probabilities have the same behavior as the odd-site ones. The localization centers of the LL wave functions are strongly dependent on the wavevectors. At $(k_x, k_y) = (0, 0)$, the doubly degenerate spin-polarized LL states are

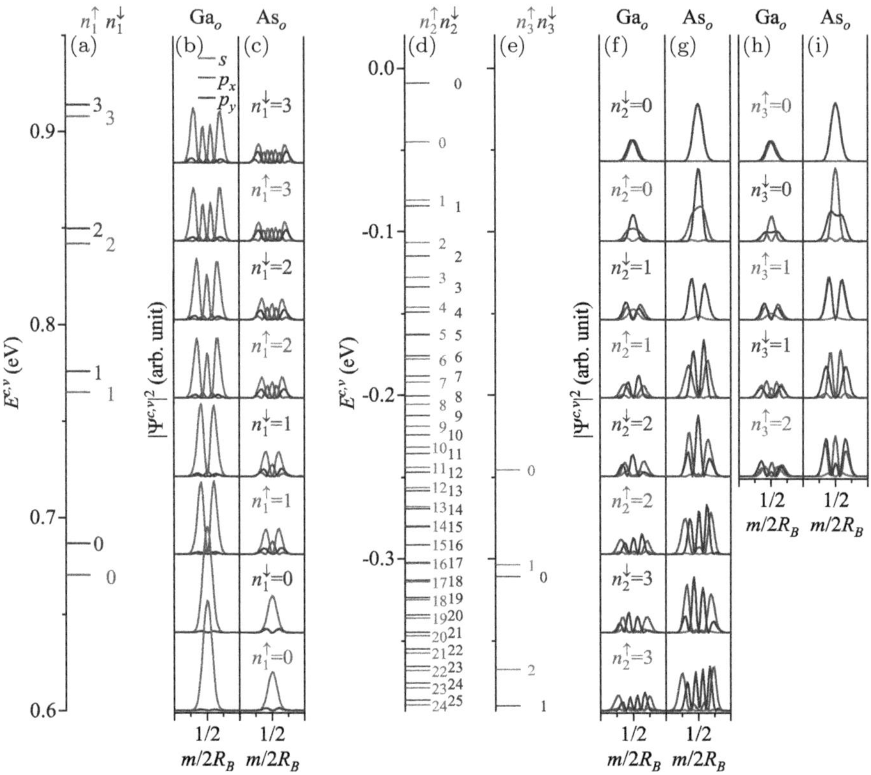


Fig. 10.5 Spin-polarized LLs and state probabilities: (a, d and e) three groups of spin-polarized dispersionless LLs are shown under $B_z = 60$ T: One group of unoccupied conduction LLs ($n_1^\uparrow$ and $n_1^\downarrow$) and two groups of occupied valence LLs ($n_2^\uparrow$, $n_2^\downarrow$, $n_3^\uparrow$ and $n_3^\downarrow$). (b, c and f–i) The corresponding probabilities of the subenvelope functions near the localization center. The $s$, $p_x$ and $p_y$ orbitals are indicated by red, green and blue curves, respectively. The major properties, such as the dominating orbitals, node regularities and localization centers, are the same for various strengths of the magnetic field, indicating that the main features of state probabilities can be discussed for a specific $B_z$.

localized at the 1/2 and 0 positions of the enlarged unit cell ($m/2R_B = 1/2$ and 0). The former is chosen for a model study since both of them have the state probabilities.

The probabilities of subenvelope functions are well behaved in their spatial distributions. Their patterns of oscillation around the localization center are similar to those of harmonic oscillators, clearly revealing regular node (zero-point) numbers. For any particular LL, the node number of the $s$-/$p_x$-/$p_y$-decomposed orbital subenvelope function is identical for the Ga and As sublattices. The subenvelope function of the dominating orbital is significant for characterizing the Landau state. Its node number gradually grows as the state energy increases, indicating that it could serve as the quantum number of the LL. In the $n$th conduction LLs ($n_1^{\uparrow} = n$ and $n_1^{\downarrow} = n$), the state probabilities of dominating $s$-orbital subenvelope functions, with $n$ zero points, are much stronger than those of $p_x$- and $p_y$-decomposed ones, with $n+1$ nodes (Figs. 10.5(b) and (c)). In the $n_2^{\uparrow} = n$ ($n_2^{\downarrow} = n$) valence LLs, there are $n$, $n$ and $n+1$ ($n-1$) nodes in the major $p_y$ orbitals and minor $p_x$ and $s$ orbitals, respectively (Figs. 10.5(f) and (g)). In the $n_3^{\uparrow} = n$ ($n_3^{\downarrow} = n$) valence LLs, $n$, $n$ and $n-1$ ($n+1$) nodes, respectively, correspond to the dominant $p_x$ orbitals and minor $p_y$ and $s$ orbitals (Figs. 10.5(h) and (i)).

The Landau state reflects the average of the accumulated neighboring zero-field electronic states with very close energies. In other words, the relative strength among the LL subenvelope function probabilities of distinct orbitals and sublattices directly arises from that of the zero-field wave functions. In each LL group, the energy-dependent relative orbital strength is closely related to the **k**-dependent one at $B_z = 0$, mainly owing to the monotonous band structure near the $\Gamma$ point. The dominant $s$-orbitals on the Ga sublattice have strengths stronger than those on the As sublattice in the conduction LLs. Moreover, the enhancement of $p_x$- and $p_y$-orbital strengths, accompanied by the decline of $s$-orbital strength, occurs as $n_1^{\uparrow}$ and $n_1^{\downarrow}$ increase (comparison between Figs. 10.5(b) and (c) and Figs. 10.4(a), (b), (g) and (h)). Instead of the Ga sublattice, the dominating orbitals on the As sublattice in the valence LLs exhibit higher strengths. The $p_x$ and $p_y$ orbitals on a specific sublattice are of the same strength in the $n_2^{\downarrow} = 0$ and $n_3^{\uparrow} = 0$ valence LLs (Figs. 10.5(f)–(i)), reflecting the fact that the zero-field $p_x$ and $p_y$ orbitals possess the equivalent strength at the $\Gamma$ point (Figs. 10.4(c)–(f) and (i)–(l)). As the subband index increases, the $n_2^{\uparrow\downarrow}$ ($n_3^{\uparrow\downarrow}$) valence LLs become $p_y$-orbital- ($p_x$-orbital-) dominated, which resembles the **k** dependence of the dominated orbitals near the $\Gamma$ point. The aforementioned LL node regularities and energy-dependent orbital variation provide fundamental information for further research in optical and transport properties, such as magneto-optical absorption selection rules, including major/minor optical transitions and the available/forbidden transport channels.

### 10.2.3 *Magnetic-field-dependent LL spectra and DOS*

The energies of low-lying LLs exhibit a monotonic variation with the strength of a magnetic field, directly reflecting the feature of the monotonic band structure near the Γ point at zero field (Fig. 10.6(a)). A simple relation between the energies of LLs and $B_z$ is absent, meaning that there are no linear or square-root $B_z$ dependences, as revealed in graphene systems. Between the $n$th spin-up and spin-down LLs of the same group ($n_i^{\uparrow} = n_i^{\downarrow} = n; i \in \{1,2,3\}$), their energy spacing grows with the increase in $B_z$, arising from the enhanced SOI by the more localized LL wave functions. For instance, the energy spacing is 24 meV between $n_1^{\uparrow} = 0$ and $n_1^{\downarrow} = 0$ LLs at $B_z = 100$ T (comparable to the thermal energy at room temperature). Under a very small magnetic field ($B_z \to 0$), the energy spacing between the lowest conduction LL and the highest valence LL approaches (or converges to) the zero-field energy gap. For an increasing magnetic field, these two LLs deviate from the Fermi level, thus leading to the growth of the energy gap. Most importantly, the non-crossing and crossing behaviors appear in the $B_z$-dependent energy spectra. The first spin-up and spin-down groups always have the lower and higher energies, respectively, so that these two subgroups do not cross each other during the variation of $B_z$. However, the distinct spin LLs might exhibit the crossing phenomenon, especially for the second and third groups. The significant features of the LL energy spectra further demonstrate the strong competitions/cooperations among the lattice symmetry, multi-orbital chemical bondings, SOIs and magnetic field.

The density of state (DOS), defined as [58, 59, 92] $\sum_{\mathbf{k}} \sum_{n_i^{\uparrow\downarrow}; i\in\{1,2,3\}} \delta[\omega - E^{c,v}(\mathbf{k}, n_i^{\uparrow\downarrow})]$, directly reflects the main features of the LL energy spectra, as depicted in Fig. 10.6(b). The dispersionless and highly degenerate LLs can create the delta-function-like prominent peaks known as the van Hove singularities in the DOS. Three groups of delta-function-like symmetric peaks, respectively, appear at ∼0.62, 0 and −0.24 eV. Their peak heights are the same, implying the identical degeneracy of LLs. In each spin-polarized LL subgroup, the peak spacing is shrunk for a larger subband index. The aforementioned characteristics of LL peaks, including peak structure, height and spacing, could be verified through experimental measurements using STS [14, 15]. Moreover, it is expected that the optical absorption peaks are contributed by the specific inter-LL transitions with regular spatial probability distributions.

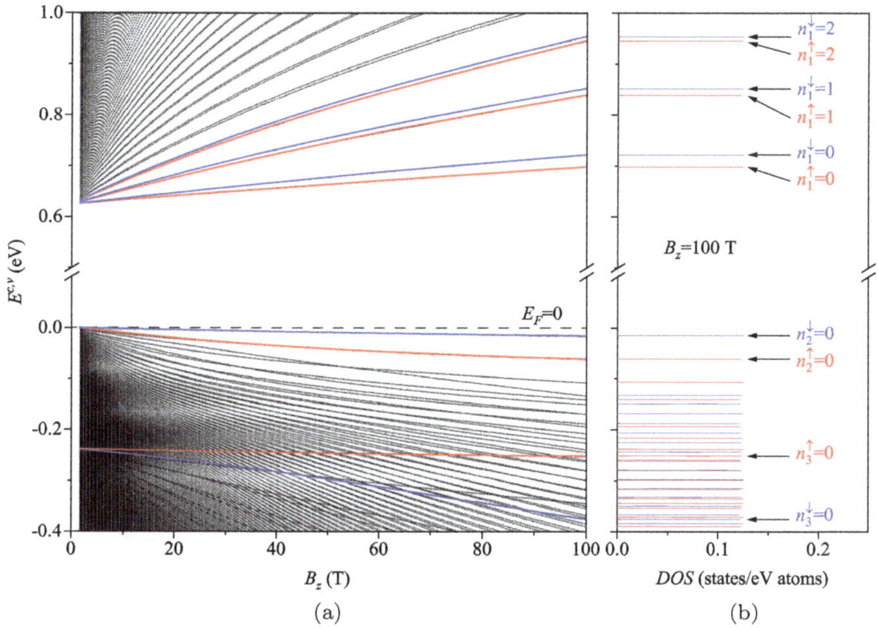


Fig. 10.6 Magnetic-field-dependent LLs and DOS: (a) The LL energies have no simple dependence on the strength of the magnetic field. With the increase in $B_z$, the spin splitting between the $n_i^{\uparrow} = n$ and $n_i^{\downarrow} = n$ LLs is enhanced, and the gap between the lowest conduction $n_1^{\uparrow} = 0$ LL and the highest valence $n_2^{\downarrow} = 0$ LL grows. Parts of the spin-up (red lines) and spin-down (blue lines) LLs are indicated for guidance. The Fermi level is marked by the dash line. (b) Spin-polarized DOS at $B_z = 100$ T. The delta-function-like peaks correspond to dispersionless and highly degenerate features. The peaks relating to the spin-up and spin-down states are depicted in red and blue, respectively.

### 10.2.4 *Electric-field-dependent LLs*

The magneto-electronic properties of monolayer GaAs with a buckled structure, which are completely different from those of monolayer graphene with a planar structure, can be diversified by a perpendicular electric field, $E_z$. The Coulomb potential difference $V_z = E_z l_z$ between the planes of Ga and As sublattices clearly results in monotonous/non-monotonous dispersion relations, crossing of LL spectra, enhancement of spin splitting and modulation of the energy gap. For a moderate magnetic field (Fig. 10.7(a)), the $n_1^{\uparrow\downarrow}$ and $n_2^{\uparrow\downarrow}/n_3^{\uparrow\downarrow}$ LL energy spectra, respectively, exhibit monotonous decreases and increases as the electric field grows. An intergroup LL crossing occurs between $n_1^{\uparrow} = 0$ and $n_2^{\downarrow} = 0$ LLs at the

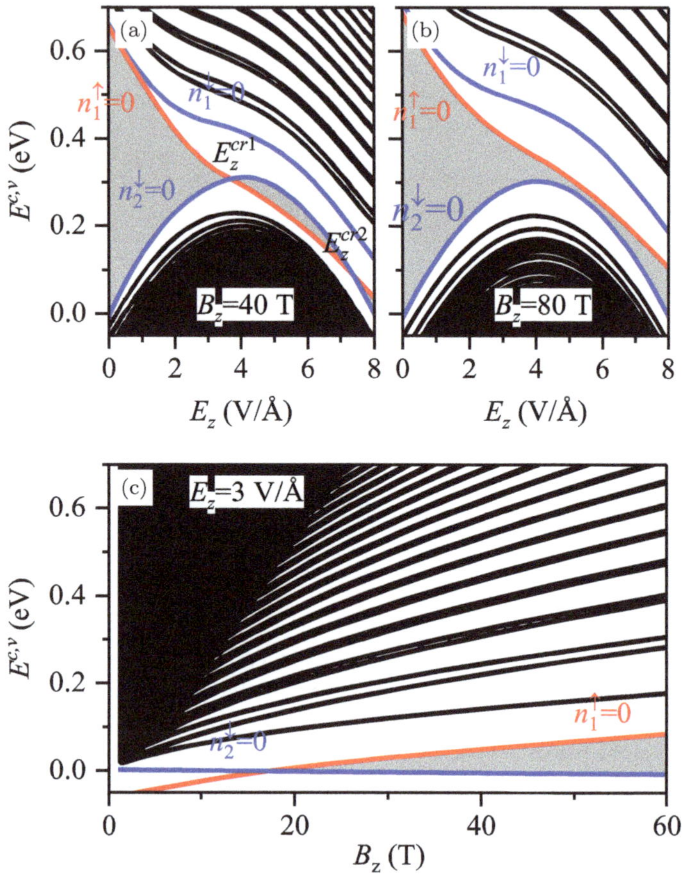


Fig. 10.7 External-field-dependent LLs: (a and b) Electric-field-dependent LL energy spectra at $B_z = 40$ and 80 T. The monotonous/non-monotonous dispersion relations, intergroup LL crossing, enhancement of spin splitting and gap modulation are obviously shown. (c) Magnetic-field-dependent LL energies at $E_z = 3$ V/Å.

critical electric fields ($E_z^{\text{cr1}} = 3.6$ V/Å and $E_z^{\text{cr2}} = 7.3$ V/Å), where the variation in the energy gap is indicated by the gray zone. The former and latter become the occupied and unoccupied states after $E_z^{\text{cr1}}$, respectively, and return to their original states after $E_z^{\text{cr2}}$, so that monolayer GaAs exhibits gapless behavior at the specific critical fields of $E_z^{\text{cr1}}$ and $E_z^{\text{cr2}}$. Meanwhile, the splitting of spin is enhanced; furthermore, the energy spacing between the $n_1^{\uparrow} = 0$ and $n_1^{\downarrow} = 0$ LLs is about 100 meV (much higher than the room-temperature thermal energy). For a large magnetic field (Fig. 10.7(b)), the

$n_1^{\uparrow\downarrow}$ LLs present a monotonous decline, while the $n_2^{\uparrow\downarrow}/n_3^{\uparrow\downarrow}$ LLs vary non-monotonously, with the turning points at $E_z \sim 4$ V/Å. The energy gap gradually shrinks and then reaches a minimum finite value ($E_g = 26$ meV), without the intergroup crossing between $n_1^{\uparrow\downarrow}$ and $n_2^{\uparrow\downarrow}$ LLs. There is no semiconductor–metal transition during the variation of $E_z$ under a very high $B_z$. Moreover, the $B_z$-dependent LL energy spectrum clearly shows a similar intergroup crossing at the critical magnetic field under a sufficiently high electric field, e.g., $(B_z, E_z^{cr1}) = (18\ \text{T}, 3\ \text{V/Å})$ in Fig. 10.7(c). Briefly, the electric and magnetic fields induce complex variations of the energy gap; therefore, their strong competitions create the semiconductor–semimetal transitions only under the critical fields.

The gap modulation due to the competition between the magnetic and electric fields is shown in detail using a color map shown in Fig. 10.8(a). For a magnetic field lower than the critical one ($B_z < B_z^{cr} \sim 64$ T), the gap gradually shrinks to zero, becomes finite, reduces to zero again and then grows as $E_z$ increases. In this variation, the intergroup LL crossings between $n_1^{\uparrow} = 0$ and $n_2^{\downarrow} = 0$ LLs occur at $E_z^{cr1}$ and $E_z^{cr2}$, which corresponds to the occurrence of the semiconductor–semimetal transitions at these critical electric fields. For $B_z > B_z^{cr}$, the gap gradually reduces to a finite size and then increases as $E_z$ grows. In addition, $E_0^{cr1} \sim 2.59$ V/Å and $E_0^{cr2} \sim 8.75$ V/Å at $B_z \to 0$ can be straightforwardly determined using the energy spectrum at $B_z = 0$, owing to the direct correspondence between the LL initial energies at small $B_z$ and the energies of the electronic states at the $\Gamma$ point at $B_z = 0$ (Figs. 10.8(b) and (c)). The aforementioned external-field-controlled gap modulation and phase transitions are helpful in developing top-gated electronic/optical devices and enable potential applications in phase-change electronic devices [93].

### 10.2.5 *Experimental verification of theoretical predictions*

The main feature and the external field-induced modulations of the prominent symmetric Landau peaks in the DOS could be verified by experimental measurements through STS. This experiment is an extension of scanning tunneling microscopy (STM) [94] and provides detailed information about the DOSs on a sample surface, such as silicon [95] and CNTs [96]. The tunneling differential conductance ($dI/dV$), which is roughly proportional to the DOS [97], directly reveals the van Hove singularities of the electronic energy spectra, such as the form, number, position

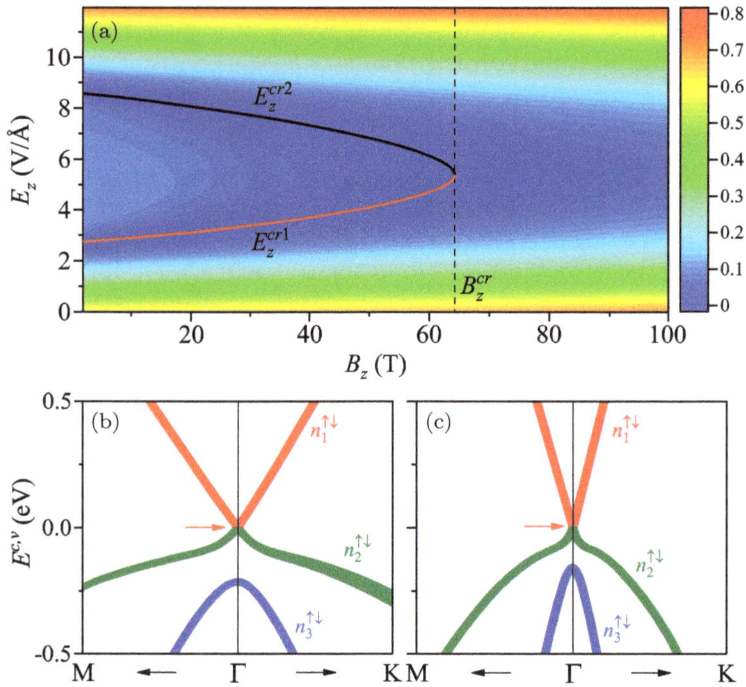


Fig. 10.8 External-field-dependent gap modulation and phase transition: (a) The $E_z$–$B_z$ color contour diagram clearly illustrates the complex gap modulation and phase transition. The red and black curves indicate $E_z^{cr1}$ and $E_z^{cr2}$ at a specific magnetic field $B_z$, respectively. (b and c) In the absence of a magnetic field, the zero gap occurs at $E_z = 2.59$ V/Å and $E_z = 8.75$ V/Å (arrows), indicating the strength of the critical electric fields $E_0^{cr1}$ and $E_0^{cr2}$ at $B_z \to 0$. The $n_1^{\uparrow\downarrow}$, $n_2^{\uparrow\downarrow}$ and $n_3^{\uparrow\downarrow}$ subbands are, respectively, colored in red, green and blue.

and intensities of the special structures. Some theoretical predictions of the LL energy spectra of few-layered graphene have been identified from the STS measurements, e.g., the $\sqrt{B_z}$-dependent LL energies in monolayer graphene [14, 15], the linear $B_z$-dependent LL spectra in AB-stacked bilayer graphene [16, 17] and the concurrence of square-root and linear $B_z$ dependences in trilayer ABA stacking [18]. The predicted magneto-electronic properties of the monolayer GaAs, including three groups of LLs without a simple $B_z$ dependence, the external-field-controlled gap modulation and the SOI-induced spin splitting, could be directly examined through the STS experiments. Such verifications are very useful in fully understanding the very strong competitions/cooperations among the critical factors.

The aforementioned main features of wave functions could be confirmed by the spectroscopic-imaging STM [98–103], which is available for resolving the spatial distribution of charge from the local DOS. This experimental technique is very suitable for identifying standing waves and Landau wave functions on the surfaces of various condensed matter systems. Standing waves have been directly observed at the surface steps of Au(111) and Cu(111) [103–106] as well as in finite-length metallic CNTs [107–109]. Also, the spatial mapping of the electronic states in the troughs between self-organized Pt nanowires on Ge(001) is presented [110, 111]. Recently, the Landau orbits without zero points have been observed [112, 113], and then, observations of the concentric-ring-like nodal structures were also obtained [114, 115]. In monolayer GaAs, the predicted orbital dominance for various groups of LLs and the relative strength of various orbitals (or different sublattices) for a specific LL could be examined through spectroscopic-imaging STM measurements of nodal structures.

### 10.2.6 *Comparisons between the planar graphene and buckled GaAs*

Monolayer planar graphene and buckled GaAs possess very different electronic properties and responses to external fields. For example, the low-energy electronic structure of the former exhibits a pair of single-orbital-dominated ($p_z$) conduction and valence bands, in which they linearly intersect at the K (or K′) points and are doubly degenerate in the spin degree of freedom. This zero-gap system sharply contrasts the middle-gap monolayer GaAs. The latter exhibit a direct gap of 0.623 eV at the Γ point, which is related to the multi-orbital-induced parabolic bands ($s$, $p_x$ and $p_y$ orbitals). Moreover, the significant SOIs and $sp^3$ bondings create the **k**-dependent spin-polarized energy subbands. The distinct electronic properties are also revealed in magnetic quantizations, such as the magnetic field dependence of LL energies, spin splitting/non-splitting, localization centers of Landau wave functions, non-crossing/crossing behaviors and quantum mode regularities (i.e., number and ordering of nodes in the LL wave functions) and the electric-field effects. The LLs directly reflect the main features of zero-field energy dispersions, revealing $\sqrt{B_z}$-dependent spin-degenerate LLs and non-specific $B_z$-dependent spin-split LLs in monolayer graphene and GaAs, respectively. The localization centers of the spin-degenerate states (spin-polarized states) are situated at the 1/6, 2/6, 4/6 and 5/6 (0 and 1/2) positions of the enlarged unit cell. Only the valence LLs of GaAs display the crossing phenomenon during the variation of $B_z$. For a specific LL of graphene (GaAs), the major node numbers of subenvelope functions in different sublattices differ by one (are identical). Electric fields further induce an on-site energy difference between two buckled sublattices of GaAs, leading to gap modulation, phase transition and enhancement of spin splitting.

## 10.3 Conclusion

We developed a generalized tight-binding model from the basic one proposed by Zhao *et al.* [54] to explore the magneto-electronic properties of monolayer GaAs. This theoretical framework could be further used to investigate the diverse magnetic quantizations of the emergent 2D layered materials. A number of critical factors, including the buckled structure, multi-orbital hybridizations, SOIs, electric field and magnetic field, are considered in the calculations simultaneously. Three groups of SOI-induced spin-polarized subbands ($n_1^{\uparrow\downarrow}$, $n_2^{\uparrow\downarrow}$ and $n_3^{\uparrow\downarrow}$) initiated from the $\Gamma$ point exhibit monotonous energy dispersions and strong **k**-dependent spin splitting. There is a direct bandgap ($E_g^{SO} = 0.623\,\mathrm{eV}$) between $n_1^{\uparrow\downarrow}$ and $n_2^{\uparrow\downarrow}$ subbands and an SOI-induced energy splitting ($\Delta_{SO} = 0.237\,\mathrm{eV}$) between $n_2^{\uparrow\downarrow}$ and $n_3^{\uparrow\downarrow}$ subbands at the $\Gamma$ point. The electronic state probabilities demonstrate that the conduction $n_1^{\uparrow\downarrow}$, valence $n_2^{\uparrow\downarrow}$ and valence $n_3^{\uparrow\downarrow}$ energy subbands are $s$-, $p_y$- and $p_x$-orbital dominated, respectively. The calculated results apparently demonstrate that monolayer GaAs exhibits great contrast with Group-IV and Group-V layered systems in the essential magneto-electronic properties, which include the localization centers, state degeneracy, orbital-dependent subenvelope functions, spin splittings, $B_z$ and $E_z$ dependences, and non-crossing/crossing/anticrossing phenomena. These important differences are attributed to the distinct multi-orbital hybridizations and SOIs.

Magnetic quantization induces three groups of spin-polarized LLs with initial energies near 0.62, 0 and 0.24 eV, reflecting the zero-field electronic state energies at the initial $\Gamma$ point. Each LL exhibits twofold degeneracy based on one $\Gamma$-valley and two $z$-axis degrees of freedom. The magnetic state probabilities are well behaved in their spatial distributions, possessing oscillation patterns with regular nodes at the

**k**-dependent localization centers (e.g., the 0 and 1/2 positions of the enlarged unit cell for $(k_x, k_y) = (0,0)$), as revealed by a harmonic oscillator. In each LL, the node numbers of various orbital subenvelope functions on the Ga and As sublattices are identical, and the $s$- and $(p_x, p_y)$-orbital node numbers differ from each other by one. Such predicted characteristics of magnetic wave functions could be examined through spectroscopic-imaging STM measurements on nodal structures. Moreover, the non-specific $B_z$ dependence of the LL energy spectrum obviously shows significant multi-orbital bondings and SOIs. Both the energy gap and energy spacing of spin splitting are gradually enhanced with the growth of the magnetic field. The conduction LLs only show the non-crossing behavior during the variation of $B_z$. However, the inter-subgroup LL crossings appear in the first/second valence group, and they frequently occur between these two valence groups. There are six subgroups of spin-polarized LL DOS peaks featuring rich van Hove singularities. The delta-function-like symmetric structure, the initial frequencies for each subgroup, the degeneracy-induced peak height and the reduced energy spacing of the higher quantum numbers could be examined through the STS measurements.

The electric field leads to drastic changes in energy dispersions, LL crossings, enhancement of spin splitting, gap modulation and phase transition. For a magnetic field lower than the critical one, $B_z < B_z^{cr}$, the intergroup LL crossings occur between the conduction and valence LLs at the critical electric fields, resulting in semiconductor–semimetal transitions. On the other hand, energy remains finite in the absence of the intergroup LL crossing near the Fermi level. Moreover, the splitting of spin is enhanced with an energy spacing larger than the room-temperature thermal energy. The complex gap modulations and phase transitions based on the rather strong competitions/cooperations between magnetic and electric fields are investigated in detail. The $E_z$–$B_z$ phase diagram clearly demonstrates the critical electric fields under a specific magnetic field. Specifically, the $E_z$-induced LL anticrossings are absent, which is completely different from what is observed in few-layer Group-IV and Group-V layered systems. The predicted magneto-electronic properties could be verified by scanning tunneling spectroscopy. These studies can help design top-gated phase-change electronic devices and future energy materials [116–120].

## Acknowledgments

The author (H. C. Chung) thanks Prof. Ming-Fa Lin for the book chapter invitation and for inspiring him to study this topic. The author (H. C. Chung) thanks Pei-Ju Chien for discussions on English language usage and corrections, as well as Ming-Hui Chung, Su-Ming Chen, Lien-Kuei Chien, and Mi-Lee Kao for financial support. This work was supported in part by Super Double Power Technology Co., Ltd., Taiwan, under the project "Development of Cloud-native Energy Management Systems for Medium-scale Energy Storage Systems (https://osf.io/7fr9z/)" (Grand number: SDP-RD-PROJ-001-2020).

## References

[1] Novoselov K.S., Jiang D., Schedin F., Booth T.J., Khotkevich V.V., Morozov S.V. and Geim A.K. 2005. Two-dimensional atomic crystals. *Proc. Natl. Acad. Sci. USA* **102** 10451. DOI: 10.1073/pnas.0502848102.

[2] Novoselov K.S., Geim A.K., Morozov S.V., Jiang D., Zhang Y., Dubonos S.V., Grigorieva I.V. and Firsov A.A. 2004. Electric field effect in atomically thin carbon films. *Science* **306** 666. DOI: 10.1126/science.1102896.

[3] Morozov S.V., Novoselov K.S., Katsnelson M.I., Schedin F., Elias D.C., Jaszczak J.A. and Geim A.K. 2008. Giant intrinsic carrier mobilities in graphene and its bilayer. *Phys. Rev. Lett.* **100** 016602. DOI: 10.1103/PhysRevLett.100.016602.

[4] Bolotin K.I., Sikes K.J., Jiang Z., Klima M., Fudenberg G., Hone J., Kim P. and Stormer H.L. 2008. Ultrahigh electron mobility in suspended graphene. *Solid State Commun.* **146** 351. DOI: 10.1016/j.ssc.2008.02.024.

[5] Berger C., Song Z.M., Li X.B., Wu X.S., Brown N., Naud C., Mayou D., Li T.B., Hass J., Marchenkov A.N., Conrad E.H., First P.N. and de Heer W.A. 2006. Electronic confinement and coherence in patterned epitaxial graphene. *Science* **312** 1191. DOI: 10.1126/science.1125925.

[6] Balandin A.A., Ghosh S., Bao W.Z., Calizo I., Teweldebrhan D., Miao F. and Lau C.N. 2008. Superior thermal conductivity of single-layer graphene. *Nano Lett.* **8** 902. DOI: 10.1021/nl0731872.

[7] Bae S., Kim H., Lee Y., Xu X.F., Park J.S., Zheng Y., Balakrishnan J., Lei T., Kim H.R., Song Y.I., Kim Y.J., Kim K.S., Ozyilmaz B., Ahn J.H., Hong B.H. and Iijima S. 2010. Roll-to-roll production of 30-inch graphene films for transparent electrodes. *Nat. Nanotechnol.* **5** 574. DOI: 10.1038/nnano.2010.132.

[8] Nair R.R., Blake P., Grigorenko A.N., Novoselov K.S., Booth T.J., Stauber T., Peres N.M.R. and Geim A.K. 2008. Fine structure constant defines visual transparency of graphene. *Science* **320** 1308. DOI: 10.1126/science.1156965.

[9] Lee C., Wei X.D., Kysar J.W. and Hone J. 2008. Measurement of the elastic properties and intrinsic strength of monolayer graphene. *Science* **321** 385. DOI: 10.1126/science.1157996.
[10] Rickhaus P., Weiss M., Marot L. and Schoenenberger C. 2012. Quantum Hall effect in graphene with superconducting electrodes. *Nano Lett.* **12** 1942–1945. DOI: 10.1021/nl204415s.
[11] Novoselov K.S., Geim A.K., Morozov S.V., Jiang D., Katsnelson M.I., Grigorieva I.V., Dubonos S.V. and Firsov A.A. 2005. Two-dimensional gas of massless Dirac fermions in graphene. *Nature* **438** 197. DOI: 10.1038/nature04233.
[12] Zhang Y.B., Tan Y.W., Stormer H.L. and Kim P. 2005. Experimental observation of the quantum Hall effect and Berry's phase in graphene. *Nature* **438** 201. DOI: 10.1038/nature04235.
[13] Katsnelson M.I., Novoselov K.S. and Geim A.K. 2006. Chiral tunnelling and the Klein paradox in graphene. *Nat. Phys.* **2** 620–625. DOI: 10.1038/nphys384.
[14] Song Y.J., Otte A.F., Kuk Y., Hu Y., Torrance D.B., First P.N., de Heer W.A., Min H., Adam S., Stiles M.D., MacDonald A.H. and Stroscio J.A. 2010. High-resolution tunnelling spectroscopy of a graphene quartet. *Nature* **467** 185. DOI: 10.1038/nature09330.
[15] Li G., Luican A. and Andrei E.Y. 2009. Scanning tunneling spectroscopy of graphene on graphite. *Phys. Rev. Lett.* **102** 176804. DOI: 10.1103/PhysRevLett.102.176804.
[16] Yin L.J., Zhang Y., Qiao J.B., Li S.Y. and He L. 2016. Experimental observation of surface states and Landau levels bending in bilayer graphene. *Phys. Rev. B* **93** 125422. DOI: 10.1103/PhysRevB.93.125422.
[17] Henriksen E.A., Jiang Z., Tung L.C., Schwartz M.E., Takita M., Wang Y.J., Kim P. and Stormer H.L. 2008. Cyclotron resonance in bilayer graphene. *Phys. Rev. Lett.* **100** 087403. DOI: 10.1103/PhysRevLett.100.087403.
[18] Yin L.J., Li S.Y., Qiao J.B., Nie J.C. and He L. 2015. Landau quantization in graphene monolayer, Bernal bilayer, and Bernal trilayer on graphite surface. *Phys. Rev. B* **91** 115405. DOI: 10.1103/PhysRevB.91.115405.
[19] Xia F., Farmer D.B., Lin Y.M. and Avouris P. 2010. Graphene field-effect transistors with high on/off current ratio and large transport band gap at room temperature. *Nano Lett.* **10** 715. DOI: 10.1021/nl9039636.
[20] Bhimanapati G.R., Lin Z., Meunier V., Jung Y., Cha J., Das S., Xiao D., Son Y., Strano M.S., Cooper V.R., Liang L.B., Louie S.G., Ringe E., Zhou W., Kim S.S., Naik R.R., Sumpter B.G., Terrones H., Xia F.N., Wang Y.L., Zhu J., Akinwande D., Alem N., Schuller J.A., Schaak R.E., Terrones M. and Robinson J.A. 2015. Recent advances in two-dimensional materials beyond graphene. *ACS Nano* **9** 11509. DOI: 10.1021/acsnano.5b05556.
[21] Matthes L., Gori P., Pulci O. and Bechstedt F. 2013. Universal infrared absorbance of two-dimensional honeycomb group-IV crystals. *Phys. Rev. B* **87** 035438. DOI: 10.1103/PhysRevB.87.035438.

[22] Liu H., Neal A.T., Zhu Z., Luo Z., Xu X.F., Tomanek D. and Ye P.D. 2014. Phosphorene: An unexplored 2D semiconductor with a high hole mobility. *ACS Nano* **8** 4033. DOI: 10.1021/nn501226z.

[23] Zhuang H.L.L., Singh A.K. and Hennig R.G. 2013. Computational discovery of single-layer III-V materials. *Phys. Rev. B* **87** 165415. DOI: 10.1103/PhysRevB.87.165415.

[24] Wang Q.H., Kalantar-Zadeh K., Kis A., Coleman J.N. and Strano M.S. 2012. Electronics and optoelectronics of two-dimensional transition metal dichalcogenides. *Nat. Nanotechnol.* **7** 699. DOI: 10.1038/nnano.2012.193.

[25] Coleman J.N., Lotya M., O'Neill A., Bergin S.D., King P.J., Khan U., Young K., Gaucher A., De S., Smith R.J., Shvets I.V., Arora S.K., Stanton G., Kim H.Y., Lee K., Kim G.T., Duesberg G.S., Hallam T., Boland J.J., Wang J.J., Donegan J.F., Grunlan J.C., Moriarty G., Shmeliov A., Nicholls R.J., Perkins J.M., Grieveson E.M., Theuwissen K., McComb D.W., Nellist P.D. and Nicolosi V. 2011. Two-dimensional nanosheets produced by liquid exfoliation of layered materials. *Science* **331** 568. DOI: 10.1126/science.1194975.

[26] Benameur M.M., Radisavljevic B., Heron J.S., Sahoo S., Berger H. and Kis A. 2011. Visibility of dichalcogenide nanolayers. *Nanotechnology* **22** 125706. DOI: 10.1088/0957-4484/22/12/125706.

[27] Wilson J.A. and Yoffe A.D. 1969. The transition metal dichalcogenides discussion and interpretation of the observed optical, electrical and structural properties. *Adv. Phys.* **18** 193. DOI: 10.1080/00018736900101307.

[28] Marseglia E.A. 1983. Transition metal dichalcogenides and their intercalates. *Int. Rev. Phys. Chem.* **3** 177. DOI: 10.1080/01442358309353343.

[29] Resta A., Leoni T., Barth C., Ranguis A., Becker C., Bruhn T., Vogt P. and Le Lay G. 2013. Atomic structures of silicene layers grown on Ag(111): Scanning tunneling microscopy and noncontact atomic force microscopy observations. *Sci. Rep.* **3** 2399. DOI: 10.1038/srep02399.

[30] De Padova P., Vogt P., Resta A., Avila J., Razado-Colambo I., Quaresima C., Ottaviani C., Olivieri B., Bruhn T., Hirahara T., Shirai T., Hasegawa S., Asensio M.C. and Le Lay G. 2013. Evidence of Dirac fermions in multilayer silicene. *Appl. Phys. Lett.* **102** 163106. DOI: 10.1063/1.4802782.

[31] Vogt P., De Padova P., Quaresima C., Avila J., Frantzeskakis E., Asensio M.C., Resta A., Ealet B. and Le Lay G. 2012. Silicene: Compelling experimental evidence for graphenelike two-dimensional silicon. *Phys. Rev. Lett.* **108** 155501. DOI: 10.1103/PhysRevLett.108.155501.

[32] Lalmi B., Oughaddou H., Enriquez H., Kara A., Vizzini S., Ealet B. and Aufray B. 2010. Epitaxial growth of a silicene sheet. *Appl. Phys. Lett.* **97** 223109. DOI: 10.1063/1.3524215.

[33] Cahangirov S., Topsakal M., Akturk E., Sahin H. and Ciraci S. 2009. Two- and one-dimensional honeycomb structures of silicon and germanium. *Phys. Rev. Lett.* **102** 236804. DOI: 10.1103/PhysRevLett.102.236804.

[34] Sahin H., Cahangirov S., Topsakal M., Bekaroglu E., Akturk E., Senger R.T. and Ciraci S. 2009. Monolayer honeycomb structures of group-IV elements and III-V binary compounds: First-principles calculations. *Phys. Rev. B* **80** 155453. DOI: 10.1103/PhysRevB.80.155453.

[35] Davila M.E., Xian L., Cahangirov S., Rubio A. and Le Lay G. 2014. Germanene: A novel two-dimensional germanium allotrope akin to graphene and silicene. *New J. Phys.* **16** 095002. DOI: 10.1088/1367-2630/16/9/095002.

[36] Bampoulis P., Zhang L., Safaei A., van Gastel R., Poelsema B. and Zandvliet H.J.W. 2014. Germanene termination of $Ge_2Pt$ crystals on Ge(110). *J. Phys. Condens. Matter* **26** 442001. DOI: 10.1088/0953-8984/26/44/442001.

[37] Li L., Lu S.Z., Pan J., Qin Z., Wang Y.Q., Wang Y., Cao G.Y., Du S. and Gao H.J. 2014. Buckled germanene formation on Pt(111). *Adv. Mater.* **26** 4820. DOI: 10.1002/adma.201400909.

[38] Bianco E., Butler S., Jiang S., Restrepo O.D., Windl W. and Goldberger J.E. 2013. Stability and exfoliation of germanane: A germanium graphane analogue. *ACS Nano* **7** 4414. DOI: 10.1021/nn4009406.

[39] Xu Y., Tang P. and Zhang S.C. 2015. Large-gap quantum spin Hall states in decorated stanene grown on a substrate. *Phys. Rev. B* **92** 081112. DOI: 10.1103/PhysRevB.92.081112.

[40] Zhu F.F., Chen W.J., Xu Y., Gao C.L., Guan D.D., Liu C.H., Qian D., Zhang S.C. and Jia J.F. 2015. Epitaxial growth of two-dimensional stanene. *Nat. Mater.* **14** 1020. DOI: 10.1038/nmat4384.

[41] Garcia J.C., de Lima D.B., Assali L.V.C. and Justo J.F. 2011. Group IV graphene- and graphane-like nanosheets. *J. Phys. Chem. C* **115** 13242. DOI: 10.1021/jp203657w.

[42] Liu C.C., Jiang H. and Yao Y.G. 2011. Low-energy effective hamiltonian involving spin-orbit coupling in silicene and two-dimensional germanium and tin. *Phys. Rev. B* **84** 195430. DOI: 10.1103/PhysRevB.84.195430.

[43] Tao L., Cinquanta E., Chiappe D., Grazianetti C., Fanciulli M., Dubey M., Molle A. and Akinwande D. 2015. Silicene field-effect transistors operating at room temperature. *Nat. Nanotechnol.* **10** 227. DOI: 10.1038/nnano.2014.325.

[44] Meng L., Wang Y.L., Zhang L.Z., Du S.X., Wu R.T., Li L.F., Zhang Y., Li G., Zhou H.T., Hofer W.A. and Gao H.J. 2013. Buckled silicene formation on Ir(111). *Nano Lett.* **13** 685. DOI: 10.1021/nl304347w.

[45] Fleurence A., Friedlein R., Ozaki T., Kawai H., Wang Y. and Yamada-Takamura Y. 2012. Experimental evidence for epitaxial silicene on diboride thin films. *Phys. Rev. Lett.* **108** 245501. DOI: 10.1103/PhysRevLett.108.245501.

[46] Li L.F., Lu S.Z., Pan J.B., Qin Z.H., Wang Y.Q., Wang Y.L., Cao G.Y., Du S.X. and Gao H.J. 2014. Buckled germanene formation on Pt(111). *Adv. Mater.* **26** 4820. DOI: 10.1002/adma.201400909.

[47] Derivaz M., Dentel D., Stephan R., Hanf M.C., Mehdaoui A., Sonnet P. and Pirri C. 2015. Continuous germanene layer on Al(111). *Nano Lett.* **15** 2510. DOI: 10.1021/acs.nanolett.5b00085.

[48] Acun A., Zhang L., Bampoulis P., Farmanbar M., van Houselt A., Rudenko A.N., Lingenfelder M., Brocks G., Poelsema B., Katsnelson M.I. and Zandvliet H.J.W. 2015. Germanene: The germanium analogue of graphene. *J. Phys. Condens. Matter* **27** 443002. DOI: 10.1088/0953-8984/27/44/443002.

[49] Tabert C.J. and Nicol E.J. 2013. Magneto-optical conductivity of silicene and other buckled honeycomb lattices. *Phys. Rev. B* **88** 085434. DOI: 10.1103/PhysRevB.88.085434.

[50] Lin C.L., Arafune R., Kawahara K., Kanno M., Tsukahara N., Minamitani E., Kim Y., Kawai M. and Takagi N. 2013. Substrate-induced symmetry breaking in silicene. *Phys. Rev. Lett.* **110** 076801. DOI: 10.1103/Phys RevLett.110.076801.

[51] Nair R.R., Ren W.C., Jalil R., Riaz I., Kravets V.G., Britnell L., Blake P., Schedin F., Mayorov A.S., Yuan S.J., Katsnelson M.I., Cheng H.M., Strupinski W., Bulusheva L.G., Okotrub A.V., Grigorieva I.V., Grigorenko A.N., Novoselov K.S. and Geim A.K. 2010. Fluorographene: A two-dimensional counterpart of Teflon. *Small* **6** 2877. DOI: 10.1002/smll.201001555.

[52] Arora N.D., Hauser J.R. and Roulston D.J. 1982. Electron and hole mobilities in silicon as a function of concentration and temperature. *IEEE Trans. Electron Devices* **29** 292. DOI: 10.1109/t-ed.1982.20698.

[53] Sze S.M. and Irvin J.C. 1968. Resistivity mobility and impurity levels in GaAs, Ge, and Si at 300 degrees K. *Solid-State Electron.* **11** 599. DOI: 10.1016/0038-1101(68)90012-9.

[54] Zhao M.W., Chen X., Li L.Y. and Zhang X.M. 2015. Driving a GaAs film to a large-gap topological insulator by tensile strain. *Sci. Rep.* **5** 8441. DOI: 10.1038/srep08441.

[55] Chen R.B., Chiu C.W. and Lin M.F. 2015. Magnetoplasmons in simple hexagonal graphite. *RSC Adv.* **5** 53736. DOI: 10.1039/c5ra08171h.

[56] Wallace P.R. 1947. The band theory of graphite. *Phys. Rev.* **71** 622. DOI: 10.1103/PhysRev.71.622.

[57] Wu H.C., Chaika A.N., Hsu M.C., Huang T.W., Abid M., Aristov V.Y., Molodtsova O.V., Babenkov S.V., Niu Y.R., Murphy B.E., Krasnikov S.A., Lubben O., Liu H.J., Chun B.S., Janabi Y.T., Molotkov S.N., Shvets I.V., Lichtenstein A.I., Katsnelson M.I. and Chang C.R. 2017. Large positive in-plane magnetoresistance induced by localized states at nanodomain boundaries in graphene. *Nat. Commun.* **8** 14453. DOI: 10.1038/ncomms14453.

[58] Chung H.C., Chang C.P., Lin C.Y. and Lin M.F. 2016. Electronic and optical properties of graphene nanoribbons in external fields. *Phys. Chem. Chem. Phys.* **18** 7573. DOI: 10.1039/c5cp06533j.

[59] Chung H.C., Lin Y.T., Lin S.Y., Ho C.H., Chang C.P. and Lin M.F. 2016. Magnetoelectronic and optical properties of nonuniform graphene nanoribbons. *Carbon* **109** 883. DOI: 10.1016/j.carbon.2016.08.091.

[60] Chung H.C., Lee M.H., Chang C.P., Huang Y.C. and Lin M.F. 2011. Effects of transverse electric fields on quasi-Landau levels in zigzag graphene nanoribbons. *J. Phys. Soc. Jpn.* **80** 044602. DOI: 10.1143/jpsj.80.044602.

[61] Chung H.C., Huang Y.C., Lee M.H., Chang C.C. and Lin M.F. 2010. Quasi-Landau levels in bilayer zigzag graphene nanoribbons. *Physica E* **42** 711. DOI: 10.1016/j.physe.2009.11.090.

[62] Fujita M., Wakabayashi K., Nakada K. and Kusakabe K. 1996. Peculiar localized state at zigzag graphite edge. *J. Phys. Soc. Jpn.* **65** 1920. DOI: 10.1143/JPSJ.65.1920.

[63] Kane C.L. and Mele E.J. 1997. Size, shape, and low energy electronic structure of carbon nanotubes. *Phys. Rev. Lett.* **78** 1932. DOI: 10.1103/PhysRevLett.78.1932.

[64] Saito R., Fujita M., Dresselhaus G. and Dresselhaus M.S. 1992. Electronic-structure of graphene tubules based on C-60. *Phys. Rev. B* **46** 1804. DOI: 10.1103/PhysRevB.46.1804.

[65] Szalowski K. 2017. Ferrimagnetic and antiferromagnetic phase in bilayer graphene nanoflake controlled with external electric fields. *Carbon* **118** 78. DOI: 10.1016/j.carbon.2017.03.019.

[66] Zhang L., Lin B.C., Wu Y.F., Wu H.C., Huang T.W., Chang C.R., Ke X., Kurttepeli M., Van Tendeloo G., Xu J., Yu D. and Liao Z.M. 2017. Electronic coupling between graphene and topological insulator induced anomalous magnetotransport properties. *ACS Nano* **11** 6277. DOI: 10.1021/acsnano.7b02494.

[67] Gert A.V., Nestoklon M.O. and Yassievich I.N. 2015. Band structure of silicene in the tight binding approximation. *J. Exp. Theor. Phys.* **121** 115–121. DOI: 10.1134/s1063776115060072.

[68] Takao Y., Asahina H. and Morita A. 1981. Electronic structure of black phosphorus in tight binding approach. *J. Phys. Soc. Jpn.* **50** 3362. DOI: 10.1143/jpsj.50.3362.

[69] Ridolfi E., Le D., Rahman T.S., Mucciolo E.R. and Lewenkopf C.H. 2015. A tight-binding model for $MOS_2$ monolayers. *J. Phys. Condens. Matter* **27** 365501. DOI: 10.1088/0953-8984/27/36/365501.

[70] Liu G.B., Shan W.Y., Yao Y.G., Yao W. and Xiao D. 2013. Three-band tight-binding model for monolayers of group-vib transition metal dichalcogenides. *Phys. Rev. B* **88** 085433. DOI: 10.1103/PhysRevB.88.085433.

[71] Cappelluti E., Roldan R., Silva-Guillen J.A., Ordejon P. and Guinea F. 2013. Tight-binding model and direct-gap/indirect-gap transition in single-layer and multilayer $MOS_2$. *Phys. Rev. B* **88** 075409. DOI: 10.1103/PhysRevB.88.075409.

[72] Zahid F., Liu L., Zhu Y., Wang J. and Guo H. 2013. A generic tight-binding model for monolayer, bilayer and bulk $MOS_2$. *AIP Adv.* **3** 052111. DOI: 10.1063/1.4804936.

[73] Chung H.C., Chiu C.W. and Lin M.F. 2019. Spin-polarized magneto-electronic properties in buckled monolayer GaAs. *Sci. Rep.* **9** 2332. DOI: 10.1038/s41598-018-36516-8.

[74] Chung H.C., Lee M.H., Chang C.P. and Lin M.F. 2011. Exploration of edge-dependent optical selection rules for graphene nanoribbons. *Opt. Express* **19** 23350–23363. DOI: 10.1364/OE.19.023350.

[75] Slater J.C. and Koster G.F. 1954. Simplified LCAO method for the periodic potential problem. *Phys. Rev.* **94** 1498–1524. DOI: 10.1103/PhysRev.94.1498.

[76] Vogl P., Hjalmarson H.P. and Dow J.D. 1983. A semi-empirical tight-binding theory of the electronic structure of semiconductors. *J. Phys. Chem. Solids* **44** 365–378. DOI: 10.1016/0022-3697(83)90064-1.

[77] Si C., Liu J.W., Xu Y., Wu J., Gu B.L. and Duan W.H. 2014. Functionalized germanene as a prototype of large-gap two-dimensional topological insulators. *Phys. Rev. B* **89** 115429. DOI: 10.1103/PhysRevB.89.115429.

[78] Xu Y., Yan B.H., Zhang H.J., Wang J., Xu G., Tang P.Z., Duan W.H. and Zhang S.C. 2013. Large-gap quantum spin hall insulators in tin films. *Phys. Rev. Lett.* **111** 136804. DOI: 10.1103/PhysRevLett.111.136804.

[79] Robinson J.T., Burgess J.S., Junkermeier C.E., Badescu S.C., Reinecke T.L., Perkins F.K., Zalalutdniov M.K., Baldwin J.W., Culbertson J.C., Sheehan P.E. and Snow E.S. 2010. Properties of fluorinated graphene films. *Nano Lett.* **10** 3001. DOI: 10.1021/nl101437p.

[80] Medeiros P.V.C., Mascarenhas A.J.S., Mota F.D. and de Castilho C.M.C. 2010. A DFT study of halogen atoms adsorbed on graphene layers. *Nanotechnology* **21** 485701. DOI: 10.1088/0957-4484/21/48/485701.

[81] Zutic I., Fabian J. and Das Sarma S. 2004. Spintronics: Fundamentals and applications. *Rev. Mod. Phys.* **76** 323. DOI: 10.1103/RevModPhys.76.323.

[82] Peierls R. 1933. Zur theorie des diamagnetismus von leitungselektronen. *Z. Phys.* **80** 763.

[83] Giglberger S., Golub L.E., Bel'kov V.V., Danilov S.N., Schuh D., Gerl C., Rohlfing F., Stahl J., Wegscheider W., Weiss D., Prettl W. and Ganichev S.D. 2007. Rashba and Dresselhaus spin splittings in semiconductor quantum wells measured by spin photocurrents. *Phys. Rev. B* **75** 035327. DOI: 10.1103/PhysRevB.75.035327.

[84] Bychkov Y.A. and Rashba E.I. 1984. Oscillatory effects and the magnetic susceptibility of carriers in inversion layers. *J. Phys. C-Solid State Phys.* **17** 6039. DOI: 10.1088/0022-3719/17/33/015.

[85] Perez-Piskunow P.M., Usaj G., Balseiro C.A. and Foà Torres L.E.F. 2014. Floquet chiral edge states in graphene. *Phys. Rev. B* **89** 121401. DOI: 10.1103/PhysRevB.89.121401.

[86] Usaj G., Perez-Piskunow P.M., Foà Torres L.E.F. and Balseiro C.A. 2014. Irradiated graphene as a tunable floquet topological insulator. *Phys. Rev. B* **90** 115423. DOI: 10.1103/PhysRevB.90.115423.

[87] Tu M.W.Y. and Zhang W.M. 2008. Non-Markovian decoherence theory for a double-dot charge qubit. *Phys. Rev. B* **78** 235311. DOI: 10.1103/Phys RevB.78.235311.

[88] Schrodinger E. 1926. An undulatory theory of the mechanics of atoms and molecules. *Phys. Rev.* **28** 1049. DOI: 10.1103/PhysRev.28.1049.

[89] Chung H.C., Su W.P. and Lin M.F. 2013. Electric-field-induced destruction of quasi-Landau levels in bilayer graphene nanoribbons. *Phys. Chem. Chem. Phys.* **15** 868. DOI: 10.1039/c2cp43631k.

[90] Chung H.C., Yang P.H., Li T.S. and Lin M.F. 2014. Effects of transverse electric fields on Landau subbands in bilayer zigzag graphene nanoribbons. *Philos. Mag.* **94** 1859. DOI: 10.1080/14786435.2014.897009.

[91] Saroka V.A., Shuba M.V. and Portnoi M.E. 2017. Optical selection rules of zigzag graphene nanoribbons. *Phys. Rev. B* **95** 155438. DOI: 10.1103/Phys RevB.95.155438.

[92] Lo P.Y., Xiong H.N. and Zhang W.M. 2015. Breakdown of Bose-Einstein distribution in photonic crystals. *Sci. Rep.* **5** 9423. DOI: 10.1038/srep09423.
[93] Li Y., Duerloo K.A.N., Wauson K. and Reed E.J. 2016. Structural semiconductor-to-semimetal phase transition in two-dimensional materials induced by electrostatic gating. *Nat. Commun.* **7** 10671. DOI: 10.1038/ncomms10671.
[94] Binnig G., Rohrer H., Gerber C. and Weibel E. 1982. Surface studies by scanning tunneling microscopy. *Phys. Rev. Lett.* **49** 57. DOI: 10.1103/PhysRevLett.49.57.
[95] Hamers R.J., Tromp R.M. and Demuth J.E. 1986. Surface electronic structure of Si(111)-(7x7) resolved in real space. *Phys. Rev. Lett.* **56** 1972. DOI: 10.1103/PhysRevLett.56.1972.
[96] Wildoer J.W.G., Venema L.C., Rinzler A.G., Smalley R.E. and Dekker C. 1998. Electronic structure of atomically resolved carbon nanotubes. *Nature* **391** 59. DOI: 10.1038/34139.
[97] Tersoff J. and Hamann D.R. 1985. Theory of the scanning tunneling microscope. *Phys. Rev. B* **31** 805. DOI: 10.1103/PhysRevB.31.805.
[98] Richardella A., Roushan P., Mack S., Zhou B., Huse D.A., Awschalom D.D. and Yazdani A. 2010. Visualizing critical correlations near the metal-insulator transition in $Ga_{1-x}Mn_xAs$. *Science* **327** 665. DOI: 10.1126/science.1183640.
[99] Moon C.R., Mattos L.S., Foster B.K., Zeltzer G., Ko W. and Manoharan H.C. 2008. Quantum phase extraction in isospectral electronic nanostructures. *Science* **319** 782. DOI: 10.1126/science.1151490.
[100] Moon C.R., Lutz C.P. and Manoharan H.C. 2008. Single-atom gating of quantum-state superpositions. *Nat. Phys.* **4** 454. DOI: 10.1038/nphys930.
[101] Ilani S., Martin J., Teitelbaum E., Smet J.H., Mahalu D., Umansky V. and Yacoby A. 2004. The microscopic nature of localization in the quantum Hall effect. *Nature* **427** 328. DOI: 10.1038/nature02230.
[102] Maltezopoulos T., Bolz A., Meyer C., Heyn C., Hansen W., Morgenstern M. and Wiesendanger R. 2003. Wave-function mapping of InAs quantum dots by scanning tunneling spectroscopy. *Phys. Rev. Lett.* **91** 196804. DOI: 10.1103/PhysRevLett.91.196804.
[103] Crommie M.F., Lutz C.P. and Eigler D.M. 1993. Confinement of electrons to quantum corrals on a metal surface. *Science* **262** 218. DOI: 10.1126/science.262.5131.218.
[104] Hormandinger G. 1994. Comment on "direct observation of standing wave formation at surface steps using scanning tunneling spectroscopy". *Phys. Rev. Lett.* **73** 910. DOI: 10.1103/PhysRevLett.73.910.
[105] Hasegawa Y. and Avouris P. 1993. Direct observation of standing wave formation at surface steps using scanning tunneling spectroscopy. *Phys. Rev. Lett.* **71** 1071. DOI: 10.1103/PhysRevLett.71.1071.
[106] Crommie M.F., Lutz C.P. and Eigler D.M. 1993. Imaging standing waves in a two-dimensional electron gas. *Nature* **363** 524. DOI: 10.1038/363524a0.

[107] Lee J., Eggert S., Kim H., Kahng S.J., Shinohara H. and Kuk Y. 2004. Real space imaging of one-dimensional standing waves: Direct evidence for a Luttinger liquid. *Phys. Rev. Lett.* **93** 166403. DOI: 10.1103/PhysRevLett.93.166403.

[108] Lemay S.G., Janssen J.W., van den Hout M., Mooij M., Bronikowski M.J., Willis P.A., Smalley R.E., Kouwenhoven L.P. and Dekker C. 2001. Two-dimensional imaging of electronic wavefunctions in carbon nanotubes. *Nature* **412** 617. DOI: 10.1038/35088013.

[109] Venema L.C., Wildoer J.W.G., Janssen J.W., Tans S.J., Tuinstra H., Kouwenhoven L.P. and Dekker C. 1999. Imaging electron wave functions of quantized energy levels in carbon nanotubes. *Science* **283** 52. DOI: 10.1126/science.283.5398.52.

[110] Heimbuch R., van Houselt A., Farmanbar M., Brocks G. and Zandvliet H.J.W. 2013. Interfering Bloch waves in a 1D electron system. *J. Phys. Condens. Matter* **25** 014014. DOI: 10.1088/0953-8984/25/1/014014.

[111] van Houselt A., Oncel N., Poelsema B. and Zandvliet H.J.W. 2006. Spatial mapping of the electronic states of a one-dimensional system. *Nano Lett.* **6** 1439. DOI: 10.1021/nl060593x.

[112] Miller D.L., Kubista K.D., Rutter G.M., Ruan M., de Heer W.A., Kindermann M., First P.N. and Stroscio J.A. 2010. Real-space mapping of magnetically quantized graphene states. *Nat. Phys.* **6** 811. DOI: 10.1038/nphys1736.

[113] Hashimoto K., Sohrmann C., Wiebe J., Inaoka T., Meier F., Hirayama Y., Roemer R.A., Wiesendanger R. and Morgenstern M. 2008. Quantum Hall transition in real space: From localized to extended states. *Phys. Rev. Lett.* **101** 256802. DOI: 10.1103/PhysRevLett.101.256802.

[114] Fu Y.S., Kawamura M., Igarashi K., Takagi H., Hanaguri T. and Sasagawa T. 2014. Imaging the two-component nature of Dirac-Landau levels in the topological surface state of Bi2Se3. *Nat. Phys.* **10** 815. DOI: 10.1038/nphys3084.

[115] Hashimoto K., Champel T., Florens S., Sohrmann C., Wiebe J., Hirayama Y., Roemer R.A., Wiesendanger R. and Morgenstern M. 2012. Robust nodal structure of Landau level wave functions revealed by Fourier transform scanning tunneling spectroscopy. *Phys. Rev. Lett.* **109** 116805. DOI: 10.1103/PhysRevLett.109.116805.

[116] Chung H.C. and Cheng Y.C. 2020. Summary of safety standards for repurposing batteries. *Monthly J. Taipower's Eng.* **860** 35. DOI: 10.31224/osf.io/d4n3s.

[117] Chung H.C. 2020. Engineering integrations, potential applications, and outlooks of Li-ion batteries. *engrXiv*. DOI: 10.31224/osf.io/swcyg.

[118] Chung H.C. 2021. Charge and discharge profiles of repurposed $LiFePO_4$ batteries based on the UL 1974 standard. *Sci. Data.* **8** 165. DOI: 10.1038/s41597-021-00954-3.

[119] Chung H.C. and Cheng Y.C. 2019. Action planning and situation analysis of repurposing battery recovery and application in China. *J. Taiwan Energy* **6** 425. DOI: 10.31224/osf.io/nxv7f.

[120] Chung H.C. 2018. Failure mode and effects analysis of LFP battery module. *engrXiv*. DOI: 10.31224/osf.io/acxsp.

Chapter 11

# CONCLUDING REMARKS

Chiun-Yan Lin, Lu-Yao Wang, Ching-Hong Ho, Hsien-Ching Chung, Chih-Wei Chiu, and Ming-Fa Lin

This book aims to establish a comprehensive framework for understanding the intricate relationships between phenomenological models, first-principles simulations, experimental measurements and potential applications in the study of quasiparticles. Chapters 1–11 provide a thorough analysis of the physical properties of materials, and Chapter 12 introduces the application of 2D materials. The modified quasiparticles have revealed intrinsic interactions between orbitals and spins, as well as many-body interactions, which are verified through single- to many-particle properties in various 3D, 2D and 1D emergent systems. However, their behavior is drastically altered under different interactions, lattice structures, atomic interactions and spin–orbital coupling strengths. Detailed calculation results are combined with physical pictures to clearly illustrate the key factors in exploring quasiparticles in these systems. While most theoretical predictions require further testing, some have been verified by the latest experimental measurements. It is worth mentioning that Chapter 2 discussed the development of a unified theoretical framework for single- to many-particle physical pictures, including intrinsic interactions of charges and spins, crystal symmetries of chemical bonds, quantum confinement of edge structure and external fields. By analyzing the diverse Hamiltonians resulting from these mechanisms, this book explores the theory behind quasiparticle modes and topological properties in condensed matter systems.

Several well-known concepts in condensed matter physics that emerge under electron–electron interactions, such as the Coulomb excitations and decays, dynamic plasmon-assisted charge density waves and Friedel oscillations, are investigated and explained in detail in Chapters 3–5. These quasiparticle properties have been qualitatively and quantitatively elucidated through the development of the modified random-phase approximation (RPA) and modified self-energy within the linear response scheme. In Chapter 3, the concept of quasiparticles is used to describe the Coulomb decay rate in condensed matter physics. The state energy and lifetime of quasiparticles are calculated using the self-consistent Dyson equation, reflecting the elastic and inelastic scatterings of quasiparticles in the crystal potential. In the generalized tight-binding model based on layer and sublattice structures, all important mechanisms of the Hamiltonian are considered to be consistent with the modified RPA and self-energy methods. The band-structure effects play an important role in the excitation and de-excitation processes. However, only the case of electron doping is considered because the asymmetric property of the conduction and valence bands is not significant.

In monolayer graphene, the excited states decay through electron–hole (e–h) or plasmon de-excitation channels. In detail, the decay processes can be categorized into four kinds for valence holes, conduction holes and conduction electrons, as described in Section 3.2.1. The decay rate of the excited conduction electrons near $E^c \lesssim 1.5\, E_F$ is roughly proportional to $(E^c - E_F)^2 ln|E^c - E_F|$, i.e., a characteristic of 2D electron gas, because the intraband process is the only de-excitation channel. Such a characteristic is also applicable to multilayer graphenes. Moreover, the non-zero decay rate at the Dirac point under electron doping indicates its robustness. It should be noted that plasmons provide efficient channels for excited holes to decay, especially in the region of $|E^{c,v}| < 0.5\, E_F$, where the quasiparticles exhibit fast Coulomb decay. On the other hand, the decay rate spectra of AA-, AB- and ABC-stacked graphenes are in sharp contrast with each other due to their unique quasiparticle excitation behavior. For any stacking configuration within the range close to and far from the Fermi level, the anisotropy of the interpair Coulomb decay rate is observed in the $k$ space. The symmetrical properties of the quasiparticle behavior about the K point are changed strongly for the ABC-stacked graphene, which is mainly due to the surface-localized states. Regarding the plasmon resonance played by the optical plasma mode, this influence is revealed by the Coulomb scattering of a deeper excited state but may be suppressed by

Landau damping. However, in these three systems, the prominent acoustic plasmon mode causes rapid attenuation of quasiparticles, as observed in conventional 2D systems. In AA-stacked graphene, the Coulomb decay rates indicate the hybrid nature of the multi-Dirac cones. Even though the AA-stacked graphene has n-type doping, the plasmon-resonance peak can appear in electron decay or hole decay. Perhaps under suitable conditions, ARPES observation of the valence state can verify the properties of the quasiparticles in the multi-Dirac cones. In AB-stacked graphene, the quasiparticle decay rate cannot be deduced as a composite effect of monolayer and bilayer graphenes, mainly due to the collective oscillations and Landau dampings of energy bands. Stronger experimental evidence is still needed to distinguish the characteristics of AB stacking. Moreover, the additional de-excitation channels due to the band hybridization around the K point enrich and diversify the Coulomb decay rate in the ABC-stacked graphene. The physical pictures of e–h and electron–plasma coupling are necessary to realize the dynamics of Dirac quasiparticles in related materials, such as layered Group-IV materials [6, 7], the semimetal ZrSiSe [8], $Bi_2Te_3$ [9] and 3D HgTe [10, 11].

In metallic systems, the Friedel oscillation is one of the most typical static screening phenomena caused by impurities or defects and is the primary mechanism of long-range interactions, dominating carrier transport and residual resistivity at low temperatures. This quantum phenomenon manifests itself on the Fermi surface structure by the induced charge density oscillating at a distance far away from the impurity with the Yukawa decaying potential. Scanning tunneling microscopy (STM) results have shown standing-wave patterns due to the Friedel oscillations with a 2D electron-like Fermi surface on surfaces of Cu(111) [14], Au(111) [13] and Be(0001) [14]. The observation of anisotropic screening charge density oscillations on the Be($10\bar{1}0$) surface using STM clearly shows a consequence of the anisotropic Fermi surface [15]. The unique information of the oscillation of the charge density (Friedel oscillation) in graphene provides an insight into the quasiparticle, chirality [16] and Berry phase [17, 18]. In Chapter 4, the charge screening phenomena for monolayer and multilayer graphenes are sharply contrasted to those from carbon nanotubes in the static limit $\omega \to 0$. The intralayer and interlayer orbital hybridizations and the Coulomb interactions between charged particles are unified in a theoretical framework to calculate the electric polarization of cylindrical or planar honeycomb lattices. Using the Fourier transform, the static screening ability is discussed in detail through the long-range oscillation of the effective potential and the charge density to show the real-space Friedel oscillation. Furthermore, multi-mode oscillations in graphene-related layered systems are induced because the Coulomb scattering process involves the intra-valley and inter-valley scatterings on equal footing in order to calculate the electric polarization. This chapter explains the phenomenon of static screening of layered graphene systems, 1D carbon nanotubes and electron gas, especially for the effective screening length and long-range charge oscillation in the layered system, as well as their relationship with the number of layers and stacking configurations. This shows the important roles played by the symmetry of the lattice and crystal structure in the electron–electron interactions.

In the past year, with the development of ultrafast spectroscopy, the dynamic behavior of quasiparticles has emerged as one of the mainstream topics in solid-state physics. Ultrafast laser pulses are used to generate and monitor the dynamic oscillation of electrons. Efforts to integrate ultrashort laser pulse technology with STM have yielded a new microscopy technique that can realize the real-space imaging of transient carrier dynamics in nanostructures. Recently, ultrafast coherent dynamics and related nanoscale phenomena have been observed through coherent nanoimaging and nanospectroscopy with nanometer and femtosecond temporal resolutions. In 2001, an analysis using femtosecond spectroscopy found that screening requires a finite time ($\sim$10 fs) to be established in GaAs in order to cause the collective behavior of electrons in the system [19]. Therefore, there has been increased interest in plasmon propagation in real space and time, as near-field optical microscopes have achieved nanoscale imaging. The large optical field confinement and enhancement of plasma waves can be used to design active plasmonic devices. Up to now, apart from the electron-gas systems, there have been few theoretical studies on the dynamic charge oscillation phenomenon focusing on the frequency response of an external perturbation of a step potential. The screening of electrons in a medium is based on the concept of a quasiparticle in quantum mechanics. Graphene systems are outstanding candidates for clearly illustrating the diverse carrier dynamics due to the diversity of stacked configurations. Previous studies have shown that electron concentration, temperature, gate voltage, tube radius, chirality and stacking configuration are part of a critical mechanism for the screening effect in 1D carbon nanotubes and 2D few-layer graphene systems. It should be noted that all the aforementioned effects on the screened response functions are unified and simultaneously combined with the development of the theoretical framework (as detailed in Chapter 2). This approach allows for the charge response to be decomposed into two parts: one caused by the plasmon mode and the other related to the e–h pair based on a clear classification of collective and single-particle excitation. The division is one of the most important aspects of elementary excitations. In Chapter 5, we discussed the plasmon wave by investigating the Fourier components of the energy loss function or the screened charge density, demonstrating its propagation in the real space and time domains. At a sufficiently high Fermi level, plasmon oscillations on different layers can reflect the resonance mode with the acoustic plasmon described in Chapter 3. We further consider the dynamic frequency response on a finite time scale, showing the importance of the external perturbation to the excitation spectrum and plasmon waves. As the density charge propagates in the time and position domains, it reflects the dynamic screening produced by the single-particle and collective excitations, especially for those excitations near the Fermi surface, which can be realized from the characteristics of the energy loss spectrum and can be pumped by high-resolution femtosecond spectroscopy to verify it [20, 21].

Moreover, the findings presented in Chapter 6 demonstrate that the optical scattering behavior of layered materials, which has been extensively explored using the theoretical framework of quasiparticles [22], can be elucidated through the physical phenomenon of photon–quasiparticle couplings. This approach is particularly suitable for capturing the rich and unique internal interactions and external fields involved. By combining the fundamental physical properties, these interactions can be comprehensively understood. Notably, the layered graphene system with an AA-, AB-, ABC- or AAB-stacking configuration exhibits distinct reflectance and transmittance spectra due to the quantum-size effect. The pronounced frequency and thickness dependences of these spectra are indicative of the significant coupling between $\pi$ electron excitations and electromagnetic waves.

On the other hand, this book delves into the effects of gate potential, spin–orbit coupling, edge structure, geometric symmetry and dimensionality on the topological properties of massless 2D Dirac fermions. In most condensed matter systems, fermionic quasiparticles in the vicinity of their single-particle ground state possess continuous effective Hamiltonians that emulate the Dirac Hamiltonian of Dirac fermions in quantum vacuum. The difference between the two lies in the fact that quantum vacuum is continuous, whereas naturally occurring condensed matter lattices are discrete. In the language of quantum field theory, both scenarios entail a single continuous Dirac field, which displays some forms of quantum anomaly that give rise to topology. However, as both Dirac fields suffer from ultraviolet divergence, they must be regularized by some means. The discrete nature of condensed matter lattices offers a natural regularization method; however, this can lead to diminished topology due to the reduced quantum anomaly. Nonetheless, various symmetries or other mechanisms can restore topology in condensed matter lattices, resulting in a plethora of topological phases being discovered in condensed matter.

Chapter 7 presented noteworthy conclusions on the emergence of topological localized states (LSs) in graphene, specifically for a one-chain gate along the armchair-carbon chain. These LSs originate from the superposition of two pseudospins linked to the K and K$'$ valleys, whereby both pseudospins display out-of-plane orientations with a relative angle ranging from $2\pi$ to 0. This angle corresponds to the energy range $|E| \leq \epsilon_g$, where $\epsilon_g$ denotes the entire gap range. The topological origin of these LSs arises from the geometric variation of pseudospin rotation, which is demonstrated analytically using the appropriate pseudospin rotational operator that connects the two pseudospins. The LS-secular equation obtained from this mechanism confirms the topological origin of the LSs, which can be found within the gap range due to the pseudospin rotation. These topological LS branches induced by the gated graphene exhibit the Dirac-point characteristics, with their dispersion launching from the Dirac point and extending to the zone boundary throughout the gap. The LS wave function reveals the localized length scale, which depends on the potential $V_0$ for the one-chain case. Notably, these topological LSs differ from conventional topological edge states near the sample boundary, as they occur around the gated graphene region. Additionally, these LSs can be generated by an arbitrary finite $V_0$ without requiring a threshold $V_0$.

We also investigated the cases of $N_c \geq 1$, where two LS branches exhibit Dirac-point characteristics. For $N_c \geq 3$, a certain number, $N_{LS}$, of side LSs arises from the bulk dispersion but is not directly related to Dirac-point physics. The out-of-plane pseudospin rotation is also demonstrated in the case of $N_C = 2$, where the LS-secular equation confirms the same topological origin as the one-chain case. For the two-chain case, even and odd modes are found in the LS due to the homogeneous gate potential. In conclusion, the Dirac-point characteristics of the topological LS branches manifest in both the absence of a finite threshold $V_0$ for their formation and the zero gap in bulk graphene. These results are expected to motivate further research on the effects of a line of adsorbed atoms or gates with different lateral configurations oriented along the armchair chain on the generation of topological LSs in graphene.

Chapters 8 and 9 reveal that rhombohedral graphite shares the same topological characteristics as graphene in the minimal model. The reason for this similarity is that both systems belong to the same topological class, despite having different interlayer couplings for the 2D limit. The protection of the zero-mode Landau level (LL) is guaranteed by the chiral operator $\tau_0\sigma 3$. The LL spectra and wave functions in rhombohedral graphite are identical to those in graphene, although the basis for the wave functions (sublattice pseudospinor) is different. Even though nearest hopping within the same sublattice breaks the sublattice symmetry, there exists a generalized chiral operator that provides protection [23]. The chiral half-filled zero mode is evident, indicating the persistence of effective chiral symmetry in the presence of a perpendicular magnetic field. Furthermore, a 3D layered system offers the easiest way to satisfy the Diophantine equation required for the 3D integer quantum Hall effect [24]. The discovery of the 3D integer quantum Hall effect in rhombohedral graphite [25, 26] shows scaling invariance compared to the 2D integer quantum Hall effect in graphene. This scaling invariance is a result of the same symmetry class and co-dimension between the two systems [27].

In Chapter 10, the tight-binding model from the basic model established by Zhao *et al.* [28] was generalized to explore the magneto-electronic properties of monolayer GaAs. This theoretical framework could be further used to investigate the diverse magnetic quantizations of the emergent 2D layered materials. A number of critical factors, including the buckled structure, multi-orbital hybridizations, SOIs, electric field and magnetic field, are considered in the calculations simultaneously. Three groups of SOI-induced spin-polarized subbands ($n_1^{\uparrow\downarrow}$, $n_2^{\uparrow\downarrow}$, $n_3^{\uparrow\downarrow}$) initiated from the $\Gamma$ point exhibit monotonous energy dispersions and strong **k**-dependent spin splitting. There is a direct bandgap ($E_g^{SO} = 0.623$ eV) between $n_1^{\uparrow\downarrow}$ and $n_2^{\uparrow\downarrow}$ subbands and an SOI-induced energy splitting ($\Delta_{SO} = 0.237$ eV) between $n_2^{\uparrow\downarrow}$ and $n_3^{\uparrow\downarrow}$ subbands at the $\Gamma$ point. The electronic state probabilities demonstrate that the conduction $n_1^{\uparrow\downarrow}$, valence $n_2^{\uparrow\downarrow}$ and valence $n_3^{\uparrow\downarrow}$ energy subbands are $s$-, $p_y$- and $p_x$-orbital-dominated, respectively. The calculated results apparently demonstrate that monolayer GaAs is in great contrast with Group-IV and Group-V layered systems in the essential magneto-electronic properties, including the localization centers, state degeneracy, orbital-dependent subenvelope functions, spin splittings, $B_z$ and $E_z$ dependences, and non-crossing/crossing/anticrossing phenomena. These important differences are attributed to the distinct multi-orbital hybridizations and SOIs.

Magnetic quantization induces three groups of spin-polarized LLs with initial energies, respectively, near 0.62, 0 and 0.24 eV, reflecting the zero-field electronic state energies at the initial $\Gamma$ point. Each LL exhibits two-fold degeneracy based on one $\Gamma$-valley and two $z$-axis degrees of freedom. The magnetic state probabilities are well behaved in their spatial distributions, possessing oscillation patterns with regular nodes at the **k**-dependent localization centers (e.g., the 0 and 1/2 positions of the enlarged unit cell for $(k_x, k_y) = (0, 0)$), as revealed by a harmonic oscillator. In each LL, the node numbers of various orbital subenvelope functions on the Ga and As sublattices are identical, and the $s$- and $(p_x, p_y)$-orbital node numbers differ from each other by one. Such predicted characteristics of magnetic wave functions could be examined through spectroscopic-imaging STM measurements on nodal structures. Moreover, the non-specific $B_z$ dependence of the LL energy spectrum obviously shows the significant multi-orbital bondings and SOIs. Both the energy gap and energy spacing of spin splitting are gradually enhanced with the growth of the magnetic field. The conduction LLs only show the non-crossing behavior during the variation of $B_z$. However, the inter-subgroup LL crossings appear in the first/second valence group, and they frequently occur between these two valence groups. There are six subgroups of spin-polarized LL DOS peaks featuring rich van Hove singularities. The delta-function-like symmetric structure, the initial frequencies for each subgroup, the degeneracy-induced peak height and the reduced energy spacing of the higher quantum numbers could be examined through scanning tunneling spectroscopy (STS) measurements.

The electric field leads to drastic changes in energy dispersions, LL crossings, enhancement of spin splitting, gap modulation and phase transition. For a magnetic field lower than the critical one, $B_z < B_z^{\mathrm{cr}}$, the intergroup LL crossings occur between the conduction and valence LLs at the critical electric fields, resulting in semiconductor–semimetal transitions. On the other hand, energy remains finite in the absence of the intergroup LL crossing near the Fermi level. Moreover, the splitting of spin is enhanced with an energy spacing larger than the room-temperature thermal energy. The complex gap modulations and phase transitions based on the rather strong competitions/cooperations between magnetic and electric fields are investigated in detail. The $E_z$–$B_z$ phase diagram clearly demonstrates the critical electric fields under a specific magnetic field. Specifically, the $E_z$-induced LL anticrossings are absent, which is completely different from what is observed in few-layer Group-IV and Group-V layered systems. The predicted magneto-electronic properties could be verified by STS. These studies can help in the design of top-gated phase-change electronic devices and future energy materials [29–33].

## References


[1] Foulkes W.M.C. and Haydock R. 1989. Tight-binding models and density-functional theory. *Phys. Rev. B* **39** 12520–12526.

[2] Ahuja R., Auluck S., Wills J.M., Alouani M., Johansson B. and Eriksson O. 1997. Optical properties of graphite from frst-principles calculations. *Phys. Rev. B* **55** 4999.

[3] Cheng C.M., Hsu C.J., Peng J.L., Chen C.H., Yuh J.Y. and Tsuei K.D. 2015. Tightbinding parameters of graphite determined with angle-resolved photoemission spectra. *Appl. Surf. Sci.* **354** 229–234.

[4] Zhang Y., Tan Y.W., Stormer H.L. and Kim P. 2005. Experimental observation of the quantum Hall effect and Berry's phase in graphene. *Nature* **438** 201.

[5] Zhang X., Rajaraman B.R.S., Liu H. and Ramakrishna S. 2014. Graphene's potential in materials science and engineering. *RSC Adv.* **4** 28987.

[6] Li Z.Q., Henriksen E.A., Jiang Z., Hao Z. and Martin M.C. 2008. Dirac charge dynamics in graphene by infrared spectroscopy. *Nat. Phys.* **4** 532.

[7] Matthes L., Pulci O. and Bechstedt F. 2013. Massive Dirac quasiparticles in the optical absorbance of graphene, silicene, germanene, and tinene. *J. Condens. Matter Phys.* **25** 395305.

[8] Gatti G., Crepaldi A., Puppin M., Tancogne-Dejean N. and Xian L. 2020. Light-induced renormalization of the Dirac quasiparticles in the nodal-line semimetal ZrSiSe. *Phys. Rev. Lett.* **125** 076401.

[9] Miao L., Wang Z.F., Ming W.M., Yao M.Y. and Wang M.X. 2013. Quasiparticle dynamics in reshaped helical Dirac cone of topological insulators. *PNAS* **110** 2758.

[10] Liu C., Bian G., Chang T.R., Wang K. and Xu S.Y. 2020. Tunable spin helical Dirac quasiparticles on the surface of three-dimensional HgTe. *Phys. Rev. B* **92** 115436.

[11] Jost A., Bendias M., Bottcher J., Hankiewicz E. and Brune C. 2017. Electron-hole asymmetry of the topological surface states in strained HgTe. *PNAS* **114** 3381.

[12] Crommie M.F., Lutz C.P. and Eigler D.M. 1993. Imaging standing waves in a two-dimensional electron gas. *Nature* **363** 524.

[13] Hasegawa Y. and Avouris P. 1993. Direct observation of standing wave formation at surface steps using scanning tunneling spectroscopy. *Phys. Rev. Lett.* **71** 1071.

[14] Crommie M.F., Lutz C.P. and Eigler D.M. 1993. Imaging standing waves in a two-dimensional electron gas. *Nature* **363** 524.

[15] Hofmann P., Briner B.G., Doering M., Rust H.-P., Plummer E.W. *et al.* 1997. Anisotropic two-dimensional Friedel oscillations. *Phys. Rev. Lett.* **79** 265.

[16] Brihuega I., Mallet P., Bena C., Bose S., Michaelis C. *et al.* 2021. Quasiparticle chirality in epitaxial graphene probed at the nanometer scale. *Phys. Rev. Lett.* **101** 206802.

[17] Dutreix C., Gonzalez-Herrero H., Brihuega I., Katsnelson M.I., Chapelier C. *et al.* 2019. Measuring the Berry phase of graphene from wavefront dislocations in Friedel oscillations. *Nature* **574** 219.

[18] Zhang Y., Su Y. and He L. 2020. Local Berry phase signatures of bilayer graphene in intervalley quantum interference. *Phys. Rev. Lett.* **125** 116804.

[19] Huber R., Tauser F., Brodschelm A., Bichler M., Abstreiter G. *et al.* 2001. How many-particle interactions develop after ultrafast excitation of an electron-hole plasma. *Nature* **414** 286.

[20] Roskos H.G., Thomson M.D., Kreß M. and Loffler T. 2007. Broadband THz emission from gas plasmas induced by femtosecond optical pulses: From fundamentals to applications. *Laser & Photon. Rev.* **4** 349.

[21] Margetic V., Pakulev A., Stockhaus A., Bolshov M., Niemax K. *et al.* 2000. A comparison of nanosecond and femtosecond laser-induced plasma spectroscopy of brass samples. *Spectrochim. Acta B* **55** 31771.

[22] Lin C.Y., Wu J.Y., Chiu C.W. and Lin M.F. 2019. *Coulomb Excitations and Decays in Graphene-Related Systems* (CRC Press).

[23] Kawarabayashi T., Hatsugai Y., Morimoto T. and Aoki H. 2011. *Phys. Rev. B* **83** 153414.

[24] Kohmoto M., Halperin B.I. and Wu Y.S. 2006. *Phys. Rev. B* **45** 13488.

[25] Kopelevich Y., Torres J.H.S., da Silva R.R., Mrowka F., Kempa H. and Esquinazi P. 2003. *Phys. Rev. Lett.* **90** 156402.

[26] Kempa H., Esquinazi P. and Kopelevich Y. 2006. *Solid State Commun.* **138** 118.

[27] Balents L. and Fisher M.P.A. 1996. *Phys. Rev. Lett.* **76** 2782.

[28] Zhao M.W., Chen X., Li L.Y. and Zhang X.M. 2015. Driving a GaAs film to a large-gap topological insulator by tensile strain. *Sci. Rep.* **5** 8441.

[29] Chung H.C. and Cheng Y.C. 2020. Summary of safety standards for repurposing batteries. *Monthly J. Taipower's Eng.* **860** 35.

[30] Chung H.C. 2020. Engineering integrations, potential applications, and outlooks of Li-ion batteries. *engrXiv*.

[31] Chung H.C. 2020. Charge/discharge profiles of LiFePO4 repurposing batteries based on UL 1974. *engrXiv*.

[32] Chung H.C. and Cheng Y.C. 2019. Action planning and situation analysis of repurposing battery recovery and application in China. *J. Taiwan Energy* **6** 425.

[33] Chung H.C. 2018. Failure mode and effects analysis of LFP battery module. *engrXiv*.

Chapter 12

# PERSPECTIVES AND OPEN ISSUES

Hsien-Ching Chung, Lu-Yao Wang, Chiun-Yan Lin, Ching-Hong Ho,
Chih-Wei Chiu, and Ming-Fa Lin

To establish a comprehensive quasiparticle framework, it is important to closely examine the relationships between phenomenological models [1], first-principles applications [2, 3], experimental measurements [4], and potential applications [5], which were explored in this book considering previous and future proposals [6]. Through analysis of consistent physical, chemical and material properties, both basic and applied science research can be conducted. This book successfully identified concise mechanisms associated with the orbital- and spin-dependent intrinsic interactions [7, 8] and the essential properties of many 3D, 2D and 1D emergent systems [9–11]. The dynamic excitation processes of modified quasiparticles were also evaluated, such as the transformation from electron–photon into orbital–photon couplings under electromagnetic-wave perturbation and the coexistence of stable and intermediate chemical bondings. Additionally, some experimental examinations cannot be accomplished with purely physical limits, such as the non-conserved perpendicular momentum transfer from angle-resolved photoemission spectroscopy (ARPES) observations on 3D lithium oxides with numerous valence energy subbands [12] and their van Hove singularities closely related to quantum tunneling currents [13]. The main focus of systematic investigations in near-future scientific pursuits will be on how to present theoretical and experimental progress simultaneously, which is clearly illustrated in the following sections.

## 12.1 Rich Magnetic Quantization

The theoretical developments of phenomenological models have been successfully done for the graphene-related $sp^2$-bonding materials, as clearly illustrated through thorough investigations of all consistent fundamental properties in previously published books [14] and the current one. A number of mainstream systems consist of AA-/AB-/ABC-stacked bulk graphite systems (Fig. 12.1) [15–17], layered graphene crystals with

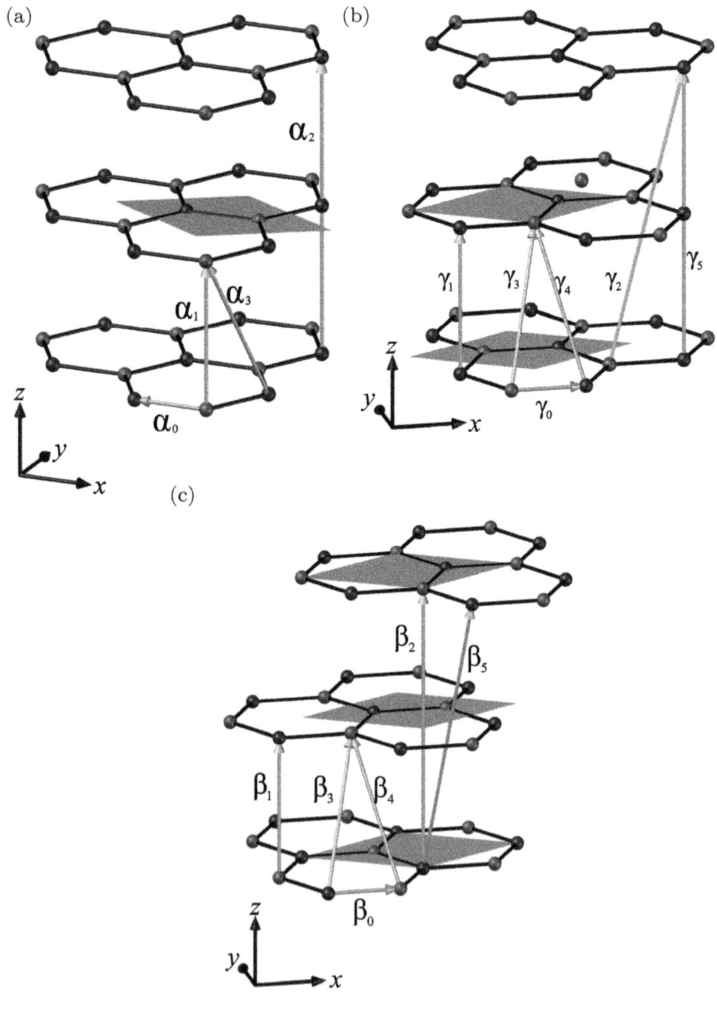


Fig. 12.1 (a) AA-, (b) AB- and (c) ABC-stacked graphite systems.

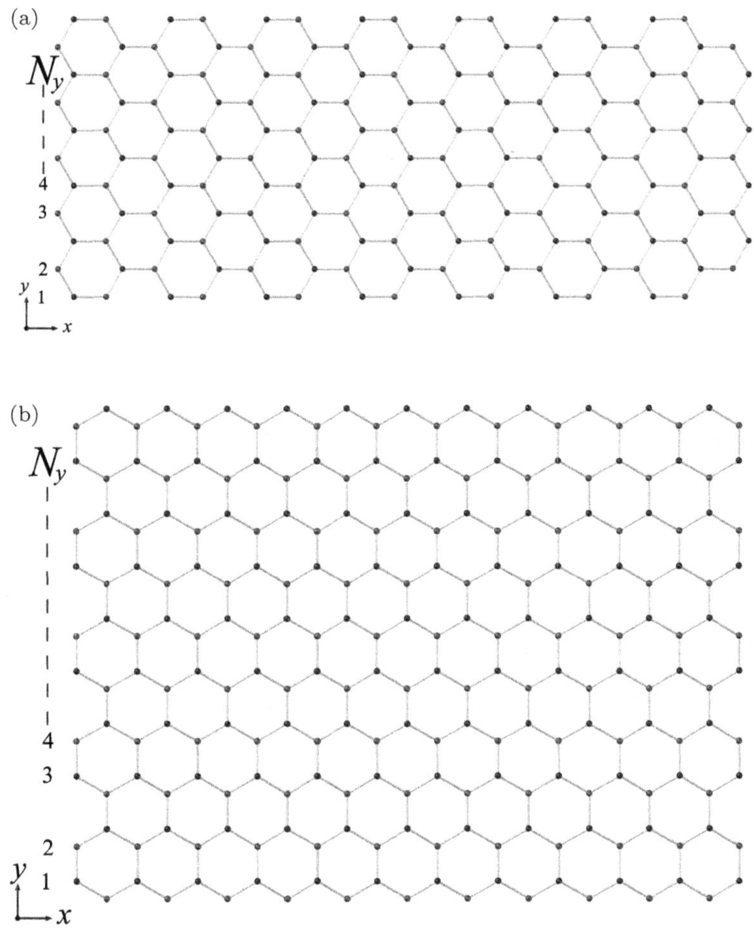


Fig. 12.2 1D (a) armchair and (b) zigzag graphene nanoribbons.

distinct stacking configurations and layer numbers [18, 19], cylindrical carbon nanotubes [20] and graphene nanoribbons (Fig. 12.2) [21], in which they obviously display diversified quasiparticle phenomena. More interestingly, the $\pi$ bondings, being closely related to the carbon-2p$_z$ orbitals, are responsible for the intralayer and interlayer atomic interactions and thus the low-energy fundamental properties. The whole hopping integrals of the tight-binding model [22] can be determined by properly fitting the first-principles results [23]. This method becomes the generalized tight-binding model in the presence of a uniform perpendicular magnetic field,

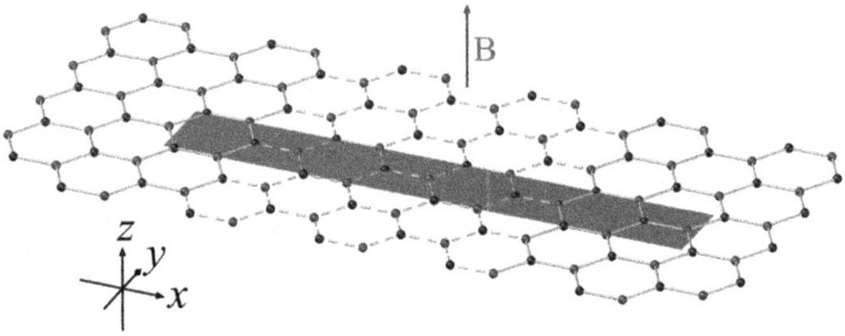


Fig. 12.3 An enlarged Moire superlattice of monolayer graphene in a uniform perpendicular magnetic field.

in which a Moire superlattice presents the vector-potential- and position-dependent Peierls phases (Fig. 12.3) [24]. Most importantly, this developed approach, which can directly link the modified RPA [25], the corrected self-energy [26] and the dynamic [27] and static Kubo formulas [27], displays rich and unique magnetic quantization. For example, the delicate calculations and analyses show that few-layer graphene systems exhibit stacking- and layer-number-dominated Landau-level (LL) energy spectra and oscillatory localization of wave functions with regular or abnormal zero points. There exist three types of LLs: (I) a normal LL with a standard oscillation for an isolated oscillator [28], (II) a perturbed one through a composite form of the dominant and minor zero points [29] and (III) an undefined mode under the field-dependent irregular probability oscillations [22]. These crystal-symmetry-based features are very convenient for further theoretical studies on the magneto-Coulomb excitations (magneto-plasmon modes in Ref. [30]), the Hall quantum conductivities [31] and the magneto-optical selective rules [32]. However, the concise physical and chemical pictures might become very complicated in buckled layered structures, as shown by monolayer silicene in Fig. 12.4. The few-layer silicene/germanene/tinene/plumbene materials [33–36] possess complex crystal symmetries and unusual intrinsic interactions, in which the former and latter show significant couplings of buckling, stacking configuration and layer number [18, 19], as well as intralayer and interlayer multi-orbital hybridizations and spin-dependent interactions. There are no reliable tight-binding models, even for bilayer structures. Systematic investigations are urgently required on this unresolved issue.

Chemical absorptions [37], decorations [38] and substitutions [39] are very powerful in creating diverse quasiparticle phenomena. Numerous

(a)

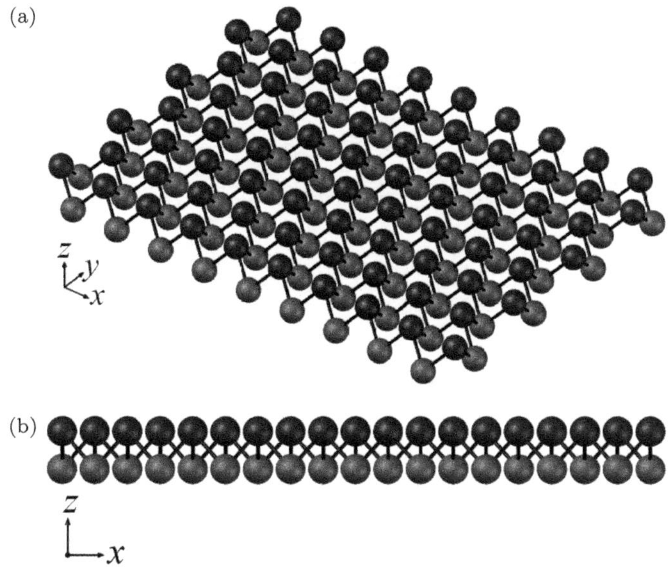


Fig. 12.4 A buckled silicene: (a) top and (b) side views.

theoretical [40] and experimental [41] studies have been successfully conducted on chemical modifications, where in the former, the main features are explored through VASP simulations [42] even for mainstream honeycomb crystals [43]. The current results clearly show that only graphite-/graphene-/carbon nanotube-/graphene-nanoribbon–alkali intercalation compounds [44–47] could be roughly characterized by the phenomenological models under a simple rigid-band approximation [48], such as a pristine $\pi$-electronic model for the stage-1 graphite alkali interaction compounds (Fig. 12.5). Apparently, the various Hamiltonians, corresponding to the different adatom concentrations and configurations, should be built from well fitting first-principles calculations. This is worthy of systematic investigations. It is well known that alkalizations [49], hydrogenations [50], oxidizations [51] and halogenations [52] play critical roles in the basic [53] and applied sciences [54].

The concise pictures of physical, chemical and material properties, as previously published in a series of books [55], are thoroughly examined, considering all consistent quantities for the diversified behaviors. After determining the multi-/single-orbital hybridizations of chemical bonds [56, 57] and spin-dependent magnetic configurations [58], the reliable

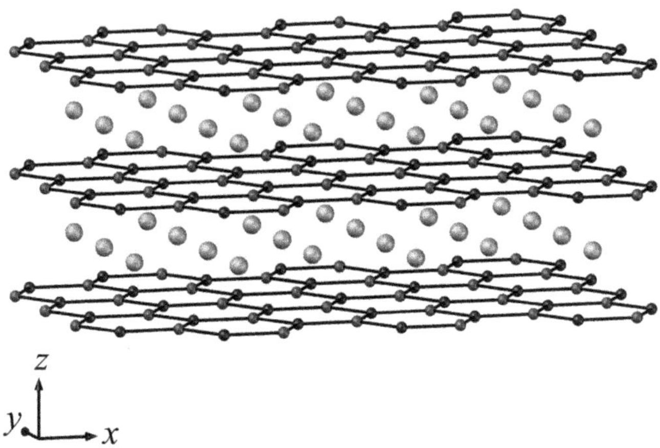


Fig. 12.5 A stage-1 graphite alkali intercalation compound.

parameters of the tight-binding model [59], which include the distinct hopping integrals of the neighboring atoms and the spin-dominated single- [60] or bi-particle [61] interactions, are urgently required in future studies on the other essential properties. For example, the significant couplings of chemical and magnetic-field environments are expected to exhibit unusual quantization phenomena [62], e.g., the sensitive dependences on field strength and free carrier density for the LLs [63], the magneto-Coulomb excitations (the diversified phase diagrams) [61], the quantized Hall conductivities (the integer and/or non-integer; the regular or abnormal behaviors) [53, 64] and the rich and unique magneto-optical transitions (several inter-LL sub-category channels) [28]. The close relations among the numerical evaluations [40], phenomenological models, and experimental observations [65, 66] should be established quickly for their further progress. This research strategy can provide an excellent platform for developing a grand quasiparticle framework, e.g., the composite quasiparticles of the same/different orbitals, the orbital-photons and the stable and intermediate chemical bondings.

## 12.2 Optical Excitations

The current work clearly illustrates that charge screening abilities and boundary conditions play critical roles in the rich and unique optical properties. Both mechanisms are very sensitive to finite-size quantum effects.

In general, any measured sample has a layered or non-planar structure, e.g., an $sp^2$ graphene or an $sp^3$ diamond with upper and lower boundaries (Fig. 6.2). The electronic properties and dielectric functions of the former were discussed using Eqs. (1) and (2) in Chapter 6. They could be calculated in the presence of perpendicular electric and magnetic fields since sublattice-based superposition is utilized in the generalized tight-binding model.

Also, the rich intrinsic interactions that arise from stacking symmetries [67], multi-orbital hybridizations [68] and spin-orbital couplings [68] are included in the various Hamiltonians. Reflectance and transmittance spectra of few-layer graphene systems are thus expected to exhibit diverse magnetic quantization phenomena under the various stacking configurations [68]. However, when a bulk system gradually reduces its thickness, electronic energy spectra and wave functions will undergo a dramatic transformation and thus resulting in dynamic charge responses. The width-created drastic changes of essential properties become an extra challenge. This issue is worthy of systematic investigations. Apparently, the current optical scattering of layered materials has been covered in the theoretical development of the quasiparticle framework [6]. That is, electronic, magnetic, optical, Coulomb-excitation and transport properties could be explored simultaneously from the modified and unified phenomenological models.

## 12.3 Many-Particle Coulomb Interactions

Many-particle effects, due to Coulomb and electron–phonon interactions, might exist in any system at low temperatures, which determine the energy and lifetime of quasiparticles. The elegant couplings of many-body scattering and intrinsic interactions in emergent 2D materials might attract research on the various properties of quasiparticles. That is, the diversified many-particle phenomena will be clearly revealed in various systems by means of the low-energy valence and conduction bands, the higher-energy electronic ones for the background screening, and the electron–phonon scatterings. The current predictions and experiments on the elementary excitations due to the honeycomb lattice indicate that only in graphene-related systems have the low-energy acoustic $\pi$ or $\sigma$ plasmons and Landau damping under different dopings been systematically studied [69–74, 76–89]. In addition, experimental measurements of EELS, ISX and various optical spectroscopies can be used to study the electronic excitation behaviors by analyzing the momentum-dependent collective plasmon modes [82–89]. EELS has successfully verified the Coulomb scattering strength related to the envelope function of two equivalent A and B sublattices in graphene [82–84]. The most important thing is that these evaluations need to be generalized to other orbitals, which should be the core issue under the current theoretical framework. Due to the multi-/single-orbital hybridizations of different chemical bonds, the Coulomb matrix is attributed to the composition of atomic orbitals and the spatial charge distributions in crystals, such as in transition metals [90–92] and binary, ternary and quaternary compounds of rare-earth metals [93–95], including s-, p-, d- and f-dominant scattering amplitudes. The energy and wave function of quasiparticles addressed according to the lattice symmetry and the tight-binding functions are capable of fully exploring the many-body effects due to the separation and recombination of quasiparticles, such as plasmons [69–75], phonons [75], polaritons [96, 97] and excitons [97–99]. However, crystal symmetry and lattice structure determine the electronic states as well as the optical and elementary excitations. There is theoretical and experimental evidence that the vertical optical transitions and many-body excitonic effects are well characterized in electromagnetic-wave excitations in low-dimensional materials [97–109]. However, how to characterize the elementary excitations during the electron–electron inelastic scatterings in condensed matter systems is still a key research challenge, especially for low-dimensional nanomaterials. High-energy IXS [110] could possibly provide an ideal means of solving this long-term issue.

## 12.4 III–VI/III–V/II–VI/II–V/IV–IV Semiconductor Compounds

The single [111], binary [112], ternary [113] and quaternary [114] semiconducting crystals, which consist of the elements of the II, III, IV, V and VI groups of the periodic table [115], present rich and unique quasiparticle phenomena [116]. More interestingly, such diversified systems dominate the previous, current and near-future scientific revolutions in electronic industries. Each element possesses an unusual atomic configuration [117], thus leading to dimensionality-enriched allotropes. For example, carbon atoms are capable of generating $sp^3$ diamond (Fig. 12.6(a)) [118], $sp^2$-related graphenes/carbonnanotubes/graphenenanoribbons

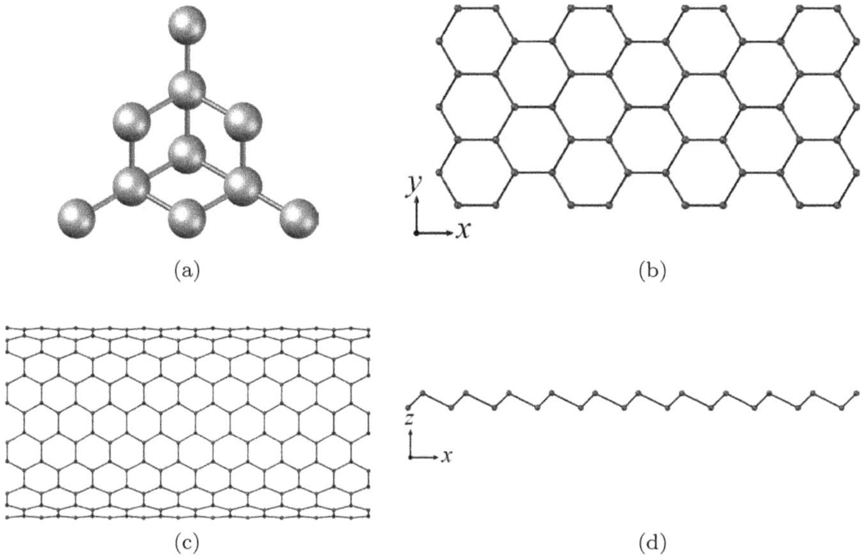


Fig. 12.6 The crystal strictures of (a) 3D diamond, (b) 2D graphene, (c) 1D carbon nanotube and (d) 0D carbon chain.

(Figs. 12.6(b)/12.6(c)/12.2), sp-based carbon chains (Fig. 12.6(d)) and other nanoscrolls/carbon onions/carbon tori/carbon fullerenes/nanoscrolls/carbon onions/carbon tori/carbon fullerenes. Moreover, the different elements could be successfully synthesized together to generate multi-compound compounds [112–114], in which the essential properties are very sensitive to their relative ratios of distinct atoms. There exist well-known semiconductors of III–VI [119], III–V [120], II–VI [121], II–V [122] and IV–IV [123]. Apparently, their basic and applied sciences have spurred extensive theoretical [124] and experimental research [125]. However, the close relations among the phenomenological models, the numerical simulations and the experimental identifications need to be well established under a unified framework, e.g., the absence of concise mechanisms to fully understand the diverse quasiparticle behaviors (discussed in the following paragraphs). This strategy is worthy of thorough investigation in the near future.

The III–VI-related materials, which have been chosen for model studies, are currently under development through VASP simulations [126] and experimental examinations [127]. These interesting condensed matter systems include 3D B/Al/Ga/In/O/S/Se/Te pure crystals, 3D and 2D

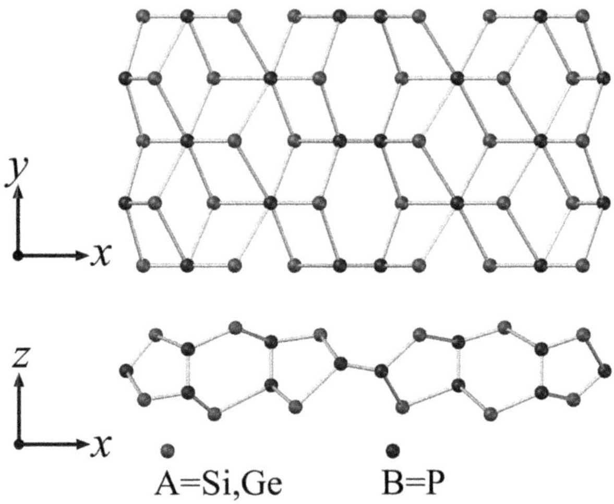


Fig. 12.7 2D gallium selenium binary compounds.

III–VI compounds (e.g., 2D Ga–Se crystal shown in Fig. 12.7) [128, 129], the unusual heterojunctions between few-layer compounds and substrate [150] and 1D III–VI nanotubes (Fig. 12.8) [130] and nanoribbons [131]. The critical factors, which are driven by the physical, chemical and material environments, are fully included in the delicate calculations and analyses, such as dimensionalities [132], layer numbers [19], stacking configurations [18], Moire superlattices [133], mechanical strains [148], boundary conditions and quantum confinements [135]. To achieve concise pictures of quasiparticles, all calculated results should be consistent with one another under a unified framework. The multi-/single-orbital hybridizations are thoroughly identified from the optimal crystal symmetries, the featured band structures and wave functions, the strongly modified orbital distributions between the neighboring atoms, the merged van Hove singularities of the atom- and orbital-decomposed density of states [136] and the orbital-hybridization-dominated optical absorption structure of the initial and final states in the presence of excitonic effects [137]. Also, the coherent atomic vibrations are thoroughly explored for the various chemical bonds to establish the atom- and orbital-dependent phonon spectra. The developed viewpoints, the composite quasiparticles of orbital–photon couplings and the intermediate orbital hybridizations during the optical excitations are clearly illustrated by the well-behaved III–VI semiconductors. The quasiparticle dynamics

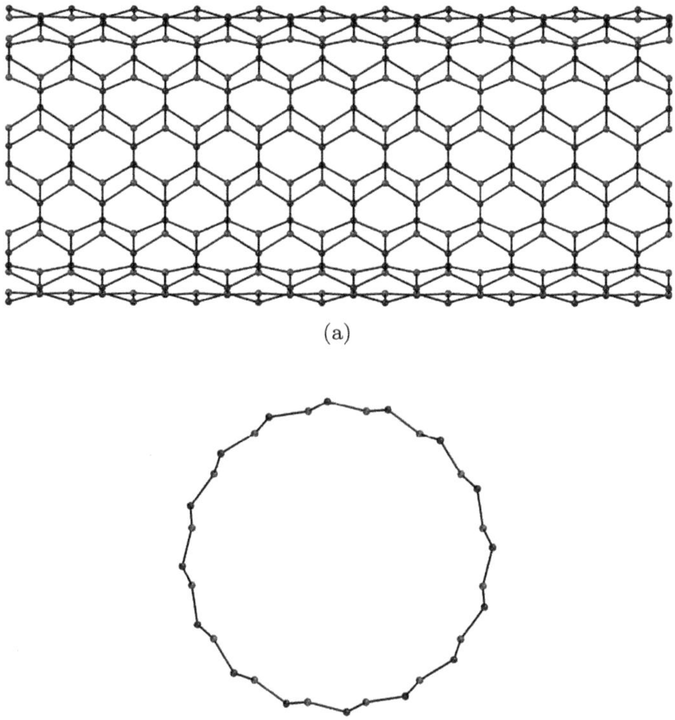

(a)

(b)

Fig. 12.8 (a) Side view and (b) top view of single-walled Ga nanotube.

are expected to be greatly diversified among the binary III–VI/III–V/II–VI/II–V/IV–IV compounds [119–123].

The next book proposal, which is jointly being prepared by the NYCU and NCKU study groups, can complete parts of experimental and theoretical research, respectively. The former is mainly focused on high-precision measurements, covering the successful syntheses of thin films by molecular beam epitaxy [138] and other growth techniques [139], *in situ* reflection high-energy electron diffraction (RHEED) [140], surface morphology measurements using atomic force/scanning electron/scanning tunneling/tunneling electron microscopies (AFM/SEM/STM/TEM) [141–144], optical reflectance/absorption/transmission/photoluminescence spectroscopies [145, 146] (where the last one is used to study excitonic effects) and Raman scatterings for the transverse and longitudinal vibration modes [147]. Also, mechanical strains [148], single- and many-particle vertical

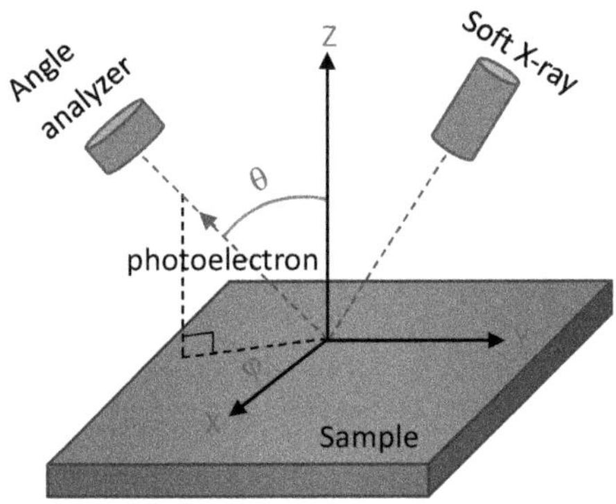


Fig. 12.9 Angle-resolved photoemission spectroscopy.

excitations [149], thickness- and crystal phase-dependent phonons (substrate effects) [150], the significant differences between 2D and 3D are included in the current experimental examinations. However, certain important detections are urgently required in the near-future studies. For example, the high-technique syntheses of few-layer binary semiconductors [112], the highly non-uniform Moire superlattices [151], the spatial orbital distributions, the featured band structures and atom dominances [152], and the atom- and orbital-projected van Hove singularities across the Fermi level [152], in which the fourth and fifth quasiparticle properties are directly tested through ARPES (Fig. 12.9) [12] and scanning tunneling spectroscopy (STS) [13], respectively. This further illustrates that the close relations among the experimental observations, the first-principle calculations, and the phenomenological models can greatly promote their simultaneous progress. More interestingly, the reliable establishment of the last one is under focus in the present study.

## 12.5 Anode/Electrolyte/Cathode Materials in Ion-Based Batteries

Ion-driven batteries are outstanding candidates for further research in the basic [53] and applied [54] sciences as well as industrial products [153]. In general, they are built from an anode [154], an electrolyte [155], a cathode [156] and a separator [157], in which such subsystems belong to the multi-component compounds. The current experimental investigations clearly show that there will be strong competition in the near future among lithium- [158], sodium- [159], aluminum- [160] and iron-based batteries [161] with sufficiently high efficiencies. Notably, a number of 3D solid/liquid materials are combined together to achieve their commercial applications [162]. For example, graphite/LiTiO [163], LiSiO/LiGeO [164, 165] and LiFeO/LiCoO/LiNiO/LiFePO [166–169] could serve as the anode, electrolyte and cathode materials, respectively, in $Li^+$-based batteries. The experimental syntheses are very successful in getting excellent samples, while high-precision examinations of the essential properties are mostly absent, except for the crystal structures obtained using X-ray techniques [170] and transport properties [171]. The STS measurements, which are associated with weak but significant quantum tunneling currents [13], cannot be used because of the too wide thickness. The ARPES observations on the occupied valence subbands are forbidden owing to the non-conservation of the perpendicular momentum transfers and the many complex energy dispersions. Moreover, a lot of absorption structures in the reflectance/energy loss function/absorption coefficient/photoluminescence [172–175], which are created by the unusual band structures of Moire crystals [176], are very difficult to thoroughly identify in the absence of a quasiparticle framework.

The first-principles method is very useful in fully understanding certain important quasiparticle properties of battery materials in the absence of magnetic fields (details in Ref. [177]). Many mainstreams compounds are chosen for a greatly enlarged quasiparticle framework [116], which is delicately achieved from all consistent quantities. For example, the ternary lithium oxide compounds are thoroughly examined to present the highly non-uniform chemical environments through the significant and observable bond-length fluctuations [178], the rich and unique valence and conduction subbands under the complicated atom dominances [55], the clear orbital formations under the multi-/single-orbital chemical bondings [179], the merged van Hove singularities of the atom- and orbital-decomposed density of states [180], the specific orbital mixings of the initial and final states for each optical absorption peak [181] and the stable or quasi-stable excitons [182]. Most importantly, the orbital–photon coupling has been proposed to generalize the electron–photon one, or the pristine quasiparticle is replaced by the modified one. The stable and intermediate chemical bondings will arise frequently during the dynamic excitation processes. Such viewpoints have been quickly developed for the other emergent materials, e.g., the Group-IV-based emergent materials considered in recent works (books by Yang and Hien [177]). The numerical simulations would be very successful only under a unified quasiparticle framework. However, this approach cannot deal with the diverse magnetic quantization [22].

The emergent batteries will be one of the focuses of studies using the VASP simulations. For example, the lithium-, phosphorus- and sulfur-based systems, which consist of numerous multi-/single-element allotrope crystals [183], are outstanding candidates for scientific research [184] and commercial products [185]. Interestingly, these three-element-enriched materials have clearly shown the diversified quasiparticle properties [186], while the complicated phenomena associated with the previous experimental [189] and theoretical studies [187] must be unified together to present a soft story through concise physical/chemical/material pictures. That is, many unusual allotrope crystals are suitable for further development in first-principles simulations [188], phenomenological models and experimental observations [189] through an enlarged framework. This strategy is currently under investigation, which was set to be completed before the end of 2021. This significant proposal includes the current applications of, at least, lithium bcc metal crystal/graphite Li-ion and adatom intercalation compounds (e.g., Li-ion battery, Fig. 12.10), 3D black and blue phosphorus crystals/2D phosphorene systems (Fig. 12.11) [190, 191], the most stable sulfur crystals/isomer

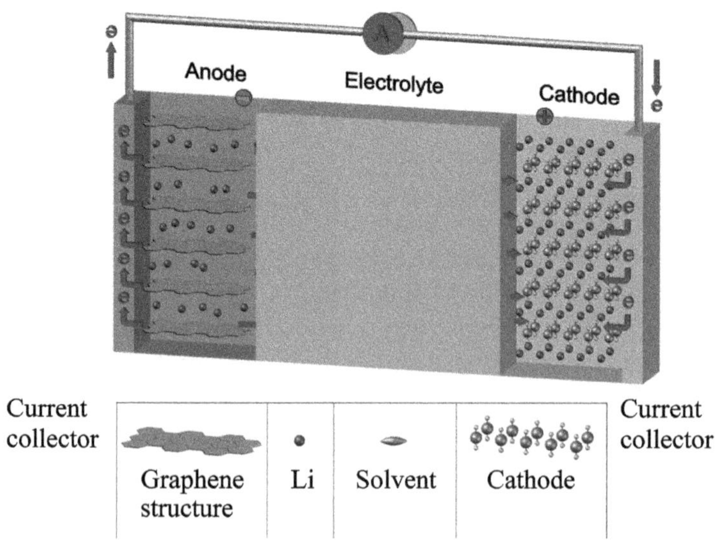


Fig. 12.10 A standard ion-based battery.

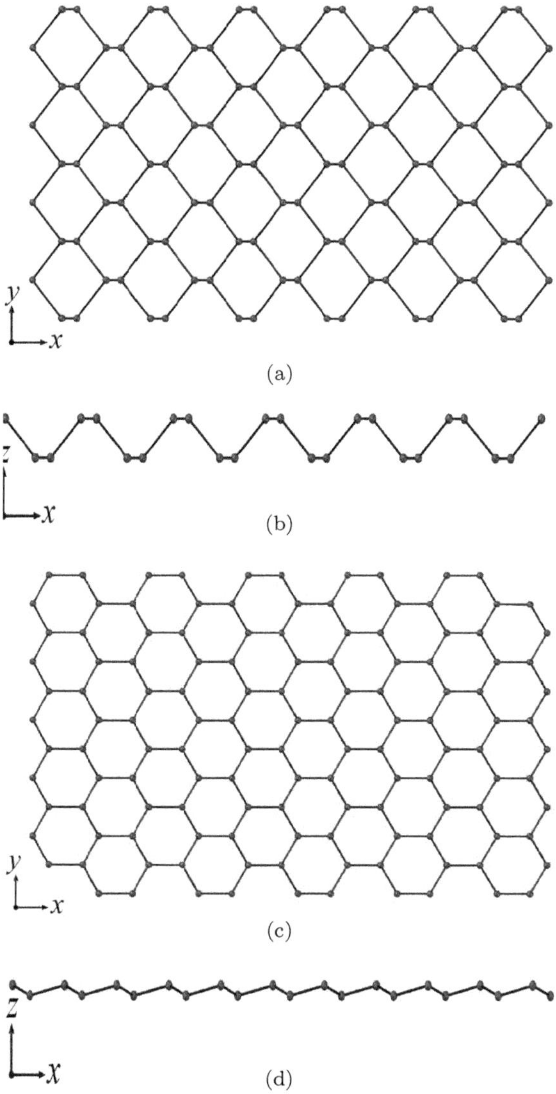


Fig. 12.11 (a) Side view and (b) top view of 2D black phosphorene. (c) Side view and (d) top view of 2D blue phosphorene.

3D structures/diversified 2D compounds with transition metals, binary lithium sulfide [192], binary lithium phosphorus materials [183], binary phosphorus sulfur of $P4S_x$ (where $x \leq 10$, Fig. 12.12) [193], ternary lithium phosphorus sulfur compounds, quaternary lithium phosphorus

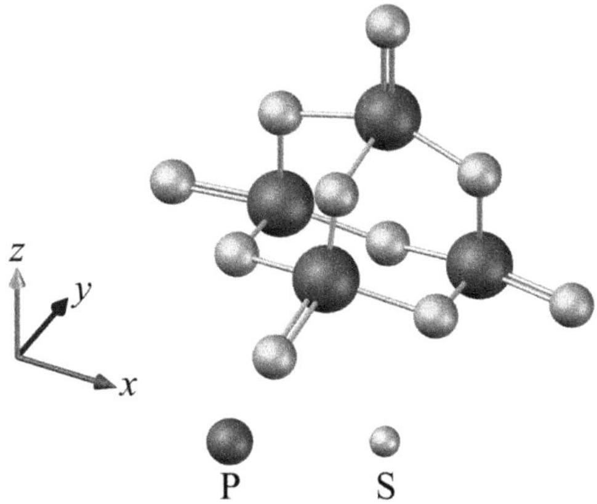


Fig. 12.12 The binary phosphorus sulfide compounds of $P_4S_{10}$.

sulfide chloride, other commensurate cathode/electrolyte/anode compounds [194], lithium sulfur batteries [195] and common lithium-, sodium-, aluminum- and iron-related batteries [158–161]. The various multi-element compounds might exhibit unusual functionalities [196] and thus generate highly potential applications [197]. This is also a focus of the study.

Only a few phenomenological models [198] are conducted on the above-mentioned cathode/separator/electrolyte/anode materials and lithium-, phosphorus- and sulfur-related systems. Obviously, certain particle phenomena could be investigated through delicate calculations and analyses, such as the rich magnetic quantization (details in Section 12.2). Most importantly, their participation can provide great simultaneous progress in the theoretical and experimental sciences, making dominant contributions to an expanded quasiparticle framework [116]. The reliable establishment of the tight-binding model/generalized tight-binding model [67, 199] lies in the atom- and orbital-dependent hopping integrals and on-site ionization energies [200], as well as their significant position-induced fluctuations. The necessary parameters become rather complicated in a highly non-uniform environment [201]; furthermore, such a case is fully supported by a Moire superlattice with many atoms [176]. Too many unusual valence and conduction energy subbands near the Fermi level, as clearly indicated by the VASP simulations [181], will lead to super-high barriers to numerical fitting. How to break through this very interesting issue is worthy of further systematic investigations, e.g., the initial studies on simple orbital-hybridization strengths in a small unit cell [202]. This strategy is also useful in comprehending the close relations among the phenomenological models [1], the first-principles simulations [203] and the experimental observations [189] under a grand quasiparticle framework [116].

## 12.6 Solar Cell

Solar cell devices have become mainstream systems in basic sciences and engineering applications [204–209], mainly due to their environmental friendliness, high power conversion efficiency [210, 211] and special means of converting the energy of light into electrical energy [212–215]. The well-characterized architecture of a commercial perovskite solar cell, as clearly shown in Figs. 12.13 and 12.14, possesses an absorber layer

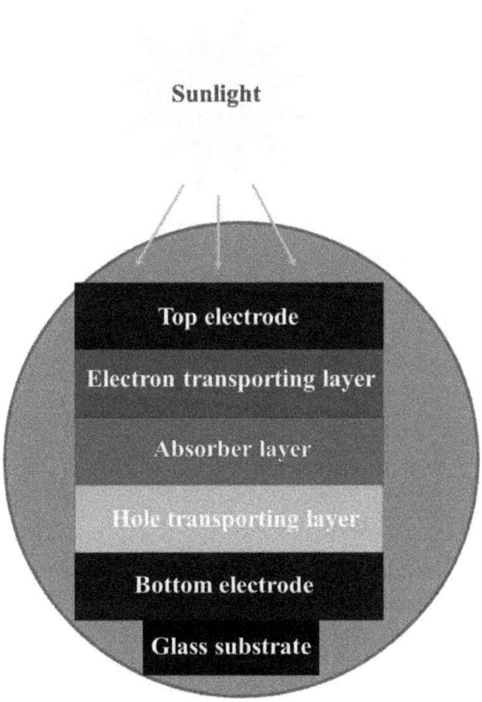


Fig. 12.13 A sketch of the standard structure of a solar cell, including the absorber layer sandwiched between the electron- and hole-transporting layers.

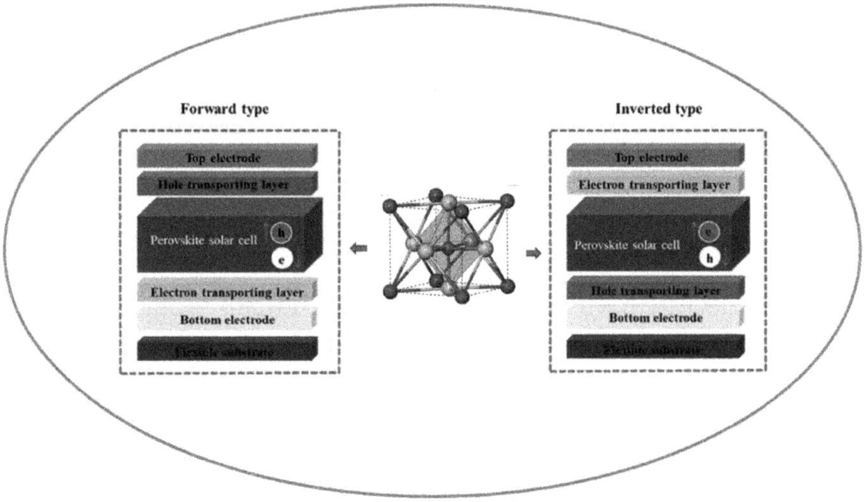


Fig. 12.14 The crystal structure of the metal halide perovskites: $AMX_3$ (A = cation, M = metal ion and X = halogen anion) depicted in the center. Schematic diagram illustrating the two types of transformation of perovskite solar cells.

sandwiched between the electron- and hole-transport layers [216, 217]. This core component efficiently absorbs the external electromagnetic waves (photons) and generates a number of electron–hole pairs due to the vertical optical transitions. Apparently, the featured optical properties are strongly dependent on the main characteristics of electronic energy spectra and wave functions, such as the threshold absorption frequency and its relation with bandgap, the prominent absorption structures associated with the joint van Hove singularities and the excitonic effects [218].

How to examine the close relations among the rich-featured properties is worthy of systematic investigations. Similarly, high-precision experimental observations are urgently required for the consistent verification of distinct physical quantities, e.g., the X-ray diffraction peaks [219, 220] due to specific crystal symmetry, the ARPES measurements [221, 222] of the occupied energy spectrum, STS to identify the van Hove singularity [223] and the reflectance/absorption/transmittance/photoluminescence spectra by the corresponding optical spectroscopies [222, 224].

Recently, the metal halide perovskites [225, 226] (Fig. 12.14), with a general formula of $AMX_3$ (A = cation, M = metal cation and X = halide anion), are expected to be outstanding candidates for illustrating the diverse quasiparticle phenomena [227–229]. Such solar-cell materials, including $CsGeCl_3$, $CsPbI_3$, $BiFeO_3$ and $BaTiO_3$ [232–238] should be included in the current development of the quasiparticle framework [230, 231].

## 12.7 Quantum Topological Insulators

Given the simple lattice models that bear various symmetries, quasiparticles have been analytically shown to have topological phases in the single-particle regime, as partly described in Chapters 8 and 9 for 2D Dirac fermions. However, there is still room for research. One direct approach is to comprehensively investigate the topological consequences of all the crystalline lattices using group theory. In this way, one could determine which crystalline 2D or 3D lattices can host chiral (massless) fermions [239]. Lately, an extension of the study to superlattices has been noted. As known from a classic symmetry consideration [240], certain super-honeycomb lattices can induce interaction between quasiparticles. A recent examination of intercalated graphene with the Kekulé commensurate pattern has shown that it exhibits spontaneously chiral symmetry breaking and electron fractionalization as a result of interaction [241, 242]. In contrast to the Kekulé structures, attention has also been drawn to the incommensurate Moire heterostructures, which can be obtained by twisting between a graphene layer and a hexagonal boron nitride substrate [243, 244]. More topological considerations are required for these complicated lattices.

Regarding the localization problem, the critical behavior at quantum phase transition is intractable [245], though non-trivial topological phases classified here [246, 247] are robust against disorder. Take the half-integer quantum Hall effect (QHE) in graphene as an example. At first glance, the zero-mode LL protected by the chiral symmetry is not broadened by the magnetic field, and thus, charge carriers cannot be accelerated across this LL. It is noted that the quantum phase transition must occur at the broadened zero-mode LL from localized to extended and then to the localized states in the presence of disorder [248].

In the exploration of realistic topological materials, great achievements have also been made with first-principles techniques [249–251]. With these techniques, the calculations for complicated band structures are feasible and reliable. This rests on two facts. The first has a topological origin, such that non-trivial topological phases are robust against disorder. For the second, a weak correlation between quasiparticles is frequently encountered in the current exploration within the single-particle approximation. One might conclude that first-principles calculation, together with a relevant group-theory analysis, succeeds in finding out all topological materials. Nevertheless, several issues are present in the first-principle calculation, bringing about various challenges. The primary one is intrinsic: Either the local density approximation (LDA) or the generalized gradient approximation (GGA) is based on the mean-field theory [252]. Consequently, the first-principles calculations do not work for strongly correlated systems with fractional QHE or high-$T_c$ superconductors, where the single-particle approximation fails [253]. Moreover, the present capability for topological superconductors is still limited, though they have been analytically characterized in various topological classes, with both the single-particle bands and the renormalized pairing terms combined in the Hamiltonians [254, 255].

Even in the single-particle regime, there are certain technical problems associated with the first-principles calculations. Usually, the calculated bandgaps are smaller than the actual values [252]. Hence, the resulting unclear distinction between a narrow-gap phase and a gapless phase ought to be overcome [250, 251]. Another technical problem comes from the treatment of the Bloch wave functions in the calculations. Here, the Bloch wave functions are taken randomly over all the gauge-dependent eigenstates, though the Berry flux derived from the Bloch wave functions is gauge independent [250, 251]. The acquired topological invariants can thus be involved and need more sophisticated interpretation [251].

Topological materials are much superior to conventional materials from an application point of view. At first, the robustness of topological phases implies the high mobility of electrons. Specifically, the edge states in the QHE of massless Dirac fermions in graphene, say, can serve as dissipation-less electric current-carrying channels [256], where the charge carriers sensitively respond to external electric and/or magnetic fields. This advantage could lead to perfect electric conduction. Moreover, the Berry phase manifests itself by handness with electron spin or some kinds of pseudospin so that the application of (pseudo)spin in contrast to spin current filtration is expected. This provides the basis for spintronics [257, 258] and quantum computation [258]. However, despite the prospect of application, the state of the art is far from successful. As for the materialization, there is still a lot of room for improvement for more desirable topological materials.

## 12.8 Electronic Applications of 2D Materials

Since the discovery of graphene [259, 260], its incredible essential properties (such as high carrier mobility at room temperature (>200000 $cm^2/Vs$) [261–263], superior thermoconductivity (3000–5000 W/mK) [262, 264], extremely high modulus (~1 TPa) [265–267] and tensile strength (~130 GPa) [266, 268], high transparency to incident light over a broad range of wavelength (97.7%) [269, 270] and anomalous QHE [271–274]) have stimulated the development of many 2D materials and studies in various fields over the past decade [275–279]. Up to now, 2D materials with various sizes of bandgap can be synthesized (e.g., gapless graphene, insulating hexagonal boron nitride (*h*-BN) and semiconducting transition-metal dichalcogenides (TMDs)), indicating that they can be building blocks similar to bulk materials. Moreover, due to their unique features (e.g., ultrathin, tunneling barrier), 2D materials are expected to be used in next-generation applications beyond bulk materials. In this section, the applications of 2D materials in electronic and optoelectronic devices as well as in friction and lubrication are reported.

Early studies on 2D materials mainly focused on the basic electrical properties, e.g., high carrier mobility [261–263] and on/off switching ratio in the fundamental transistors [280, 281]. Recently, studies on 2D devices have been broadened to the next-generation electronic applications (e.g., flexible and transparent devices) and classic applications (e.g., tunneling transistors and memories). Layered 2D materials (which are free from dangling bonds on the basal plane) can be assembled into van der Waals (vdW) heterostructures by simply stacking them without the requirement for lattice commensurability [282, 283]. A wide variety of 2D materials (e.g., metallic graphene, semiconducting TMDs and insulating *h*-BN) enable the creation of vdW heterostructures with myriad combinations. In this section, recent experimental progress on 2D-based devices in various electronic applications is described.

### 12.8.1 *Flexible and transparent devices*

Owing to their high mechanical strength and flexibility, 2D material devices are potential candidates to be used in flexible and wearable electronics. Lee *et al.* demonstrated $MoS_2$ field-effect transistors (FETs) on a polymer substrate by stacking *h*-BN/$MoS_2$/graphene as dielectric/semiconducting channel/electrode, respectively. Field-effect high mobility up to 45 $cm^2$/Vs is shown for trilayer $MoS_2$ FETs with low operating gate voltage below 10 V and unchanged performance of up to 1.5% strain [284]. Wu *et al.* reported the piezoelectric properties of 2D $MoS_2$ and showed that the cyclic stretching and releasing of thin $MoS_2$ flakes with an odd number of atomic layers produces oscillating piezoelectric voltage and current outputs (because of the broken inversion symmetry of the 2H structure), whereas no output is observed for flakes with an even number of layers (because of centrosymmetry of the structure) [285]. A single monolayer flake strained by 0.53% generates a peak output of 15 mV and 20 pA, for a power density of 2 mW/$m^2$ and a 5.08% efficiency of mechanical-to-electrical energy conversion. Transport measurements show a strong piezotronic effect in monolayer $MoS_2$, but not in bilayer and bulk $MoS_2$ [285].

$MoS_2$–graphene heterostructure curved image sensor array was reported for the soft optoelectronic device by Choi *et al.* [286]. Six-layer $MoS_2$ and four-layer graphene act as photoabsorbing layers and interconnections, respectively. $MoS_2$ channels generate photocurrent under illumination of light, except for the infrared region (due to its wide bandgap). Because of the infrared blindness of $MoS_2$ phototransistors, the infrared filter is unnecessary, leading to clearer images than from conventional silicon photodetectors. More importantly, the total thickness of the device is about 51 nm, which is more appropriate for soft optoelectronic applications. When synthesized on a suitable substrate, it is possible to make a large-scale array of 2D materials for flexible applications. Hence, the kirigami structures of 2D materials (e.g., graphene) have been proposed owing to their paper-like features [287, 288]. The 2D kirigami structures show uniform distributions of stress during stretching, resulting in extremely high stretchability [288]. Recently, the kirigami structure of $PtSe_2$ layers on a polymeric polyimide substrate exhibited a very slight current drop even at 50% tensile stretch, indicating that a 2D kirigami structure is promising for flexible and stretchable applications [289].

### 12.8.2 *2D tunneling field-effect transistors*

So far, metal-oxide-semiconductor FETs (MOSFETs) [290–293] have been broadly used for digital integrated circuits and electronic applications, and there has been a long-lasting demand for high-performance transistors with short switching times, continuous scaling ability, and low power consumption. On this subject, tunneling FETs have been considered a potential candidate owing to their high on/off ratio with a small sweep voltage and low leakage current [294, 295]. Britnell *et al.* have reported a bipolar FET that exploits the low density of states in monolayer graphene [294]. The prototype devices are graphene heterostructures with atomically thin *h*-BN or $MoS_2$ acting as a vertical transport barrier, i.e., atomically thin tunnel barriers of *h*-BN or $MoS_2$ sandwiched between the top and bottom graphene layers. Vertical tunneling devices based on vdW heterostructures demonstrate a steep transition slope of turn-on current under low powers owing to the sharpness of the band edges in atomically smooth hetero-interfaces. The devices of graphene/*h*-BN/graphene exhibited gate-tunability in on-current owing to field-effect control of the Fermi level of graphene. An increase in the Fermi level of graphene results in a lowered effective height, followed by an increased tunneling current. The room-temperature on/off ratios for *h*-BN/graphene and $MoS_2$/graphene devices are ~50 and ~10000, respectively [294].

Georgiou *et al.* have fabricated similar vertical 2D heterostructure FETs with a tunneling barrier of few-layer $WS_2$ sandwiched by graphenes, with flexible and transparent features [296]. The combination of tunneling and thermionic transport allows for unprecedented current modulation over $10^6$ at room temperature and very high on-current. One of the advantages of tunneling FETs is that the subthreshold swing has no fundamental limit of ~60 meV/decade at room temperature, as is the case for a silicon-based MOSFET [297, 298]. The subthreshold swing limit in the vertical $WS_2$ tunneling FET could be overcome by the band-to-band tunneling mechanism in graphene/$WS_2$/graphene heterostructure, leading to an expected value of ~20 mV/decade (smaller than the fundamental limit), which enables fast operation of the device [296]. Sarkar *et al.* further improved the subthreshold swing by fabricating a vertical tunneling FET composed of highly doped *p*-type Ge (as source) and bilayer $MoS_2$ (as channel) [299]. Electrons are transported through band-to-band tunneling [298, 300–305] from the valence band of Ge to the conduction band of $MoS_2$. The subthreshold swing reached a minimum of 3.9 mV/decade, which is much smaller than the fundamental thermionic limit [299]. Roy *et al.* demonstrated interlayer band-to-band tunneling in vertical $MoS_2/WSe_2$ vdW heterostructures using a dual-gate device architecture [306]. The electric potential and carrier concentration of $MoS_2$ and $WSe_2$ layers can be separately controlled by the two symmetric gates. Such a device can be gate modulated to behave as either an Esaki diode with a negative differential resistance [307–309], a forward rectifying diode with a low reverse bias current or a backward diode with a large reverse bias current.

### 12.8.3 *Memory applications*

In recent years, complementary metal-oxide-semiconductor (CMOS) size scaling has been slowing down owing to fundamental limitations associated with the physics of conventional silicon-based FETs [310]. Meanwhile, transistors applying 2D materials, such as TMDs [311] or black phosphorous [312] as semiconducting channels, are considered potential alternatives. This is due to the more achievable electrostatic gate control inherent in the extremely thin bodies of 2D materials (exhibiting mobilities comparable to those of silicon).

Traditional memory technologies consisting of static random-access memory (SRAM) [313], dynamic RAM (DRAM) [314] and flash [315, 316] are used at various levels in the memory hierarchy of today's computers [317]. In recent years, the size-scaling decrease of these charge/transistor-based technologies has presented challenges, such as power dissipation, reliability and performance degradation, similar to the CMOS technology. Therefore, alternative memory ideas are being explored to replace traditional memory technologies, such as magnetic RAM (MRAM) [318, 319], resistive RAM (RRAM) [320–322], phase-change RAM (PCRAM) [323, 324] and ferroelectric FETs (Fe-FETs) [325–327]. These ideas have emerged as potential candidates for non-volatile memory (NVM) technologies. For example, Everspin and IBM/Global-Foundries have demonstrated spin-transfer-torque (STT)-MRAM as a discrete and embedded NVM alternative to DRAM [328–330].

Recently, more and more researchers have been focusing on 2D-based memory applications [331]. For conventional memory technologies, such as SRAM [332], DRAM [333] and flash [334], 2D materials are utilized as FETs. Moreover, emerging NVM technologies would benefit from 2D materials with the ability to store information using their intrinsic material properties, such as magnetic moments or electric dipoles. Up to now, these properties have not been observed in most common 2D materials (such as TMDs or black phosphorous). Recent experiments on new 2D materials have revealed ferromagnetism [335, 336] and ferroelectricity [337].

Several recent results have demonstrated how 2D materials can be used as potential building blocks for conventional and emerging NVM elements. Zhang *et al.* have reported an electric-field-induced structural transition in vertical $MoTe_2$- and $Mo_{1-x}W_xTe_2$-based RRAM. The programming voltages of RRAM can be tuned by modifying the thickness of TMD, and the devices showed reproducible resistive switching within 10 ns between high and low resistive states. Moreover, using an $Al_2O_3/MoTe_2$ stack, the on/off current ratios of the order of $10^6$ with programming currents lower than 1 $\mu$A were reached in a selectorless RRAM architecture [338]. Song *et al.* have shown multiple-spin-filter magnetic tunnel junctions (sf-MTJs) based on 2D heterostructures, where atomically thin chromium triiodide ($CrI_3$) acts as a spin-filter tunnel barrier sandwiched between graphene contacts [339]. The tunneling magnetoresistance is drastically enhanced with the increasing $CrI_3$ layer thickness, reaching a record 19000% at low temperatures ($\sim$2 K). Jiang *et al.* have reported spin tunnel field-effect transistors (TFETs) based on dual-gated graphene/$CrI_3$/graphene tunnel junctions [340]. The devices exhibit ambipolar behavior and tunnel conductance, depending on the magnetic order in the $CrI_3$ tunnel barrier. The gate voltage switches the tunneling barrier between interlayer antiferromagnetic and ferromagnetic states under a constant magnetic bias near the spin-flip transition, effectively and reversibly altering the device between the low and high conductance states. By electrically controlling the magnetization configurations instead of the spin current, the spin TFETs achieve a high/low conductance ratio approaching 400%. Ferroelectricity has been observed not only in insulating 2D materials [337] but also in semiconducting [341] and metallic ones [342]. These diverse electronic properties in 2D ferroelectrics are essential for implementing innovative ferroelectric memory devices, such as ferroelectric semiconductor FETs (FeS-FETs) [341]. Moreover, supplementary device components with functionalities necessary for memory devices (e.g., tunneling barriers for MRAM [343], access transistors [344] and selectors for crossbar RRAM [345, 346]) can also be implemented using 2D materials. Owing to the ability of 2D materials to form 2D heterostructures without the need to match the crystal lattice, one can assemble technology-relevant memory devices by coupling 2D materials with different functionalities [344, 347].

## 12.9 Optoelectronic Applications of 2D Materials

2D materials can serve as promising candidates for conventional optoelectronics with efficient photoconversion and light emission due to their unique electronic structures and exceptional optical properties. In this section, optoelectronic properties and their applications are reported.

### 12.9.1 *2D light emitters*

Monolayer TMDs have been applied to transparent light-emitting devices with a two-terminal configuration. Lien *et al.* have shown a transient-mode electroluminescent device based on TMD monolayers ($MoS_2$, $WS_2$, $MoSe_2$ and $WSe_2$) [348]. Electroluminescence from this dopant-free two-terminal device is obtained by applying an AC voltage between the gate and the semiconductor. The electroluminescence intensity is weakly dependent on the Schottky barrier height or the polarity of the contact.

2D materials with a direct bandgap, such as monolayer TMDs and black phosphorous, can be used for light emission and harvesting owing to their high luminescence efficiency, large exciton binding energy and wide-range color tunability. Efficient 2D light emitters were realized by the engineering of the band structure in vertical vdW heterostructures. Withers *et al.* synthesized light-emitting devices by stacking metallic graphene, insulating *h*-BN, and various monolayer TMDs into carefully built sequences [349]. Especially, a quantum well (QW) composed of an *h*-BN barrier and a TMD well is introduced, enabling effective carrier confinement and efficient light emission. The external quantum efficiencies (EQEs) measured at 6 K were 1%, 6% and 8.4% for the devices with single, triple and quadruple QWs, respectively. This high value of QE is comparable to the efficiencies of the best modern-day organic light-emitting diodes (OLEDs) [350]. Recently, the same group demonstrated that monolayer $WSe_2$-based LEDs exhibited an increase in EQE with increasing temperature. The obtained EQE reached up to 5% at room temperature [351]. These results suggest that band-structure engineering in 2D heterostructures can be applied to highly efficient surface-light emission, potentially also in flexible devices. Pospischil *et al.* fabricated a *p*–*n* junction diode based on an electrostatically doped [352] $WSe_2$ monolayer [353]. The built devices can serve as a photovoltaic solar cell, a photodiode or a light-emitting diode, obtaining light-power conversion and electroluminescence efficiencies of $\sim$0.5% and $\sim$0.1%, respectively.

### 12.9.2 *2D photodetectors*

Lopez-Sanchez *et al.* realized ultrasensitive monolayer $MoS_2$ phototransistors by improving the channel mobility and lowering the contact barriers. The devices exhibit a maximum external photoresponsivity of 880 A/W at a wavelength of 561 nm and a photoresponse in the 400–680 nm range [354]. Britnell *et al.* vertically assembled 2D photodetectors composed of a photoactive TMD layer sandwiched between two graphene electrodes. In this vertical heterostructure based on $WS_2$, van Hove singularities in the electronic density of states of the TMD layer guarantee enhanced light–matter interactions, resulting in enhanced photon absorption and electron–hole creation (which are collected in transparent graphene electrodes). This led to efficient photocurrent generation over the entire junction area of the device, with a measured photoresponsivity of 0.1 A/W and an EQE of 30% [355]. Yu *et al.* reported layer-number-dependent photocurrent generation in graphene/$MoS_2$/graphene heterostructures by synthesizing a device with two distinct regions containing one-layer (1L) and seven-layer (7L) $MoS_2$ [356]. Photoresponse studies showed that the photoresponsivity of 1L $MoS_2$ is surprisingly higher than that of 7L $MoS_2$ by seven times. Spectral-dependent studies further confirmed that the internal quantum efficiency (IQE) in 1L $MoS_2$ can reach a maximum of 65%, far higher than the 7% in 7L $MoS_2$. The theoretical modeling reveals that asymmetric potential barriers in the top and bottom interfaces of the graphene/1L $MoS_2$/graphene heterojunction enable asymmetric carrier tunneling, generating unusually high photoresponsivity in 1L $MoS_2$ devices. The wavelength-dependent studies indicated the EQE of 1L $MoS_2$ increases with decreasing photon energy and can reach up to 2.5% at 633 nm (1.96 eV), whereas the EQE of 7L $MoS_2$ maintains a relatively low value of 0.3–0.4% throughout the entire measured wavelength range [356].

With the advent of atomically thin 2D materials and their vdW heterostructures, it is now possible to realize a *p–n* junction at the ultimate thickness limit [357–359]. The vdW junctions composed of *p*-type and *n*-type semiconductors (with just one unit cell thick) are predicted to possess completely different charge transport characteristics than bulk heterojunctions [360, 361]. Recently, Lee *et al.* demonstrated the electrical and optoelectronic properties of 2D semiconductor *p–n* heterojunctions $MoS_2/WSe_2$ (consisting of monolayer *n*-$MoS_2$ and *p*-$WSe_2$) [362]. The tunneling-mediated recombination of the majority carriers between layers can be tuned by adjusting the gate bias, governing the charge transport and photoresponse characteristics of atomically thin *p–n* heterojunctions. Moreover, graphene-sandwiched atomic *p–n* junctions with various thicknesses showed EQE to be a function of excitation energy. The measured EQEs at 532 nm were 2.4%, 12% and 34% for monolayer, bilayer and multilayer *p–n* junctions, respectively. These results highlighted the great potential of 2D material heterojunctions for highly efficient and rapid photoconverters.

Despite many demonstrations of 2D photodetectors with various device geometries, there are still challenges in increasing the light absorption efficiency of atomically thin 2D materials because the intrinsic light absorption of 2D materials is limited by their atomic thickness. One of the promising approaches to overcome this problem is to use plasmonic nanostructures. It was demonstrated that gold nanoparticles decorated on the surface of 2D materials enhance photocurrent by an order of magnitude owing to the enhancement of the local optical field [355]. Similarly, Liu *et al.* [363] observed strong light–matter coupling in monolayer $MoS_2$ with silver nanodisk arrays, and Wang *et al.* [364] identified strong light–matter coupling in monolayer $WS_2$ placed on metallic optical nanocavities. As an alternative approach, dual-gated graphene/$MoS_2$/graphene vertical heterostructures were proposed. The weak electrostatic screening effect of graphene allows for the integration of single/dual gates under and/or above the vertical 2D heterostructure to modulate the band slope and photocurrent generation. The amplitude and polarity of the photocurrent in the gated vertical heterostructures can be readily tuned by the electric field of an external gate to achieve a maximum EQE of 55% and an IQE of 85% [365]. To minimize the thermal loss of high-energy photons, multijunction solar cells made of stacked 2D materials of different bandgaps are proposed to absorb different wavelengths from the solar spectrum, reducing thermalization losses [366, 367]. These structures could potentially be constructed using different TMDs with varying bandgaps, ranging from the visible to the near-infrared.

## 12.10 Friction and Lubrication Applications of 2D Materials

Lubrication is an effective way to control friction and wear in mechanical tools. About 4000 years ago, according to ancient Egyptian murals, water, gypsum and animal fats were first used as lubricants to reduce friction [368]. Liquid lubricants have been widely used due to their effective anti-friction effect. However, the performance of liquid lubricants can be seriously affected by severe working environments, resulting in the failure of the tools. Hence, solid lubricants possessing many advantages (such as being more stable than liquid lubricants under the severe conditions of heavy loads, high speeds, etc.) have been proposed in the industry. Recently, traditional lubricants have begun to reveal their limitations in the case of microscale friction. On the contrary, micro- and nanoscale phenomena in tribology play an increasingly crucial role in modern technologies, e.g., micro-electro-mechanical systems. The small size and large surface features of 2D materials and nanoparticles provide excellent lubrication performance, leading to extensive studies on the friction properties and wear resistance of nanomaterials [369–371].

Before the development of nanotechnology, the understanding of friction and lubrication phenomena was mainly from a macroscopic point of view. After the development of micro- and nano-mechanical devices, macroscopic tribology theories have become obsolete, and the characteristics of friction and lubrication at the micro- and nanoscale have been extensively developed [372, 373]. Following this developmental trend, tribology has also gradually evolved into nanotribology, i.e., microscopic and atomic-scale tribology has become a new field. Since the 1990s, several scientific research teams have developed numerous microscopy technologies, such as scanning electron microscopy (SEM) [374, 375], scanning tunneling microscopy (STM) [376–380] and atomic force microscopy (AFM) [266, 381] in combination with first-principles calculations [382–384], molecular dynamics [385–387] and finite element analyses [388, 389]. It is concluded that the nanoscale friction mechanisms can be divided into interfacial and surface-sliding friction mechanisms. The interfacial friction mechanism is prominently influenced by load, size, stacking, defects, layer spacing, layer number and temperature [390–393]. Feng *et al.* found that graphene nanoflakes slide easily on a graphene surface at a temperature of $\sim$5 K. The average sliding distance of the graphene nanoflakes is larger at 5 K than at 77 K, indicating that thermal fluctuations are likely to trigger their transitions from superlubric back to commensurate ground states [394]. Li *et al.* observed that the friction between incommensurate $MoS_2$ monolayers gives a friction coefficient of $\sim 10^{-4}$ in the regime of superlubricity [395]. On the other hand, the surface friction mechanism is mainly categorized into electron–phonon coupling [396, 397], out-of-plane puckering [389, 398, 399] and deformation energy dissipation [400–403].

Filleter *et al.* observed that the dramatic difference in friction between monolayer and bilayer graphenes is attributed to the significant variation in electron–phonon coupling, which was confirmed by ARPES [396]. Cho *et al.* found that mechanically stripped graphene causes low friction on an atomically smooth substrate [404]. Deng *et al.* found that the friction is proportional to the layer number of graphene at low loads and attributed this result to the local deformation of the surface layer and vdW forces between the AFM tip and the surface layer [405].

2D materials provide interlaminar sliding and a low shear force, giving them superlubricity properties [394, 395]. Researchers have studied the mechanisms of lubricant enhancement using nanomaterial suspensions via surface analysis techniques, e.g., ball-bearing mechanisms [406–408], film formation mechanisms [409–412] and self-healing mechanisms [413, 414]. In practical applications, 2D materials can be used as lubricant additives [415, 416], nanoscale lubricating films [417–419] and vacuum space lubricating materials [420] owing to their unique anti-friction and anti-wear properties. Sanes *et al.* showed that graphene can serve as an lubricant additive to reduce more than 70% of the friction coefficient of motor oil at 150°C [416]. Kim *et al.* demonstrated graphene as the thinnest solid lubricant. Superior adhesion and frictional characteristics were observed for graphene grown on Cu and Ni metal catalysts by chemical vapor deposition and transferred onto the $SiO_2$/Si substrate [418]. Watanabe *et al.* studied the tribological characteristics of $WS_2/MoS_2$ solid lubricating multilayer films and found that the $WS_2/MoS_2$ multilayer film showed a significantly improved tribological performance in the air compared to the single-layer $MoS_2$ or $WS_2$ film, with an improvement in wear life of $\sim$7 times and a low friction coefficient of $\sim$0.05 [419]. Song *et al.* observed that graphene can overcome the shortcomings of graphite in terms of lubrication in a vacuum. Since graphene can self-assemble an ordered layer-by-layer structure and it self-forms a highly ordered lamellar structure at the sliding interface during the friction process, graphene exhibits excellent macro-tribological behavior in a vacuum at the engineering scale [420].

## 12.11 Role of 2D Materials in Developing Lithium-Ion Batteries

Electrical energy storages based on lithium-ion (Li-ion) batteries are attracting more interest nowadays, considering the expanding market for the electrification of the transportation sector, battery-powered vehicles, grid energy storages and portable electronics [421]. Recent advances in emerging 2D materials have demonstrated their role in developing better Li-ion batteries. These materials can be utilized as anode materials, cathode materials, separators, electrolytes, catalysts and current collectors. 2D materials with outstanding tunable electrical conductivity are promising candidates for overcoming the poor electrical properties of semiconductive anode/cathode materials since their added inactive mass to the batteries will not be significant. Combining with semiconductive active materials possessing high electrical conductivity, 2D materials have been shown to largely enhance the electron conductivity through the anode/cathode materials, resulting in a better electrochemical performance [422, 423]. Zhang *et al.* showed that 2D titanium carbide or carbonitride nanosheets (known as MXenes) can be used as a conductive binder for silicon electrodes, where the nanosheets form a continuous metallic network, enabling fast charge transport and providing good mechanical reinforcement for the thick electrode (up to 450 $\mu$m), and anodes with very high areal capacities (up to 23.3 mAh/cm$^2$) have been demonstrated [424]. Gullapalli *et al.* have shown that the graphene-coated 3D stainless steel current collector can be used with a $MoS_2$ electrode material to achieve high stability and rate capacity in Li-ion batteries [425]. The elemental and layer-structured 2D materials have demonstrated promising results for their application as cathodes or anodes with high specific and volumetric capacities [426–429]. 2D heterostructures are proven to possess high electrochemical activity and high electrochemical stability during charge and discharge cycling.

The enhanced mechanical properties of 2D materials can provide protection for electrode materials from disintegration during charge and discharge cycling. Their outstanding mechanical properties can accommodate the mechanical strains that occur in the high-capacity electrode particles during lithiation/delithiation processes, i.e., their enhanced ability to buffer the stress induced during the lithiation and delithiation processes. By controlling the shrinkage/expansion of high-capacity electrode materials during charge/discharge cycling with 2D protective layers, other electrode materials can be introduced, indicating more flexibility in choosing electrode materials. Furthermore, owing to their large surface area and functionalization-induced surface tunability, these 2D materials can serve as a supportive matrix for the active material particles and effectively avoid pulverization [430–432].

After the success of Li-ion batteries in the energy market, lithium–sulfur (Li–S) batteries with high energy capacity and low cost were developed for practical applications [433, 434]. The Li–S batteries are predicted to have high specific energies of 2500 Wh/kg and 2200 Wh/l, more than those exhibited by current Li-ion batteries, which have specific energies of 380 Wh/kg and 1000 Wh/l [435]. Although their average working voltage is about 2.15 V (lower than the 3.7 V of Li-ion batteries), their high theoretical specific capacity of 1672 mAh/g can compensate for this shortage [436]. The practical energy density for a Li–S battery pack may reach 500–600 Wh/kg or 500–600 Wh/l, which is sufficient for driving an electric vehicle for about 500 km [435, 437, 438]. The obstacles to realizing this high energy density of Li–S batteries mainly include high internal resistance, self-discharge and rapid capacity fading on cycling. The ability of 2D materials to block the lithium polysulfide species as a major obstacle in Li–S batteries is quite intriguing. 2D materials (e.g., graphene [259, 260], graphene-analogous [439–442] and MXenes [443]) exhibit an anchoring effect and can trap the polysulfides on their surface. Moreover, the abundant edges, defects and functional groups on these nanosheets/nanoflakes serve as catalysts for Li–S redox reactions and improve the electrochemical performance [444–446]. Sun *et al.* demonstrated a conductive porous vanadium nitride nanoribbon/graphene composite accommodating the catholyte as the cathode of a Li–S battery. The vanadium nitride/graphene composite provides strong anchoring for polysulfides and fast polysulfide conversion. Due to the high conductivity of vanadium nitride, the composite cathode shows lower polarization and faster redox reaction kinetics than a reduced graphene oxide cathode, exhibiting good rate and cycling performances [447].

Lithium–air (Li–$O_2$) batteries, based on their high theoretical specific energy ($\sim$3500 Wh/kg), are an attractive technology for future electrical energy storage that could make long-range electric vehicles possible [435]. However, the implementation of this technology still encounters several challenges. The oxygen reduction reaction (ORR) during the discharge process and the oxygen evolution reaction (OER) during the charge process of Li–$O_2$ batteries occur at a three-phase boundary involving the solid electrode, liquid electrolyte and oxygen gas, which makes the Li–$O_2$ system more complicated than the conventional Li-ion battery system. In non-aqueous Li–$O_2$ batteries, the deposition of an electrically resistive discharge product appears to limit their capacity, as do the parasitic processes occurring at the reaction product–electrolyte, product–cathode, electrolyte–cathode and electrolyte–anode interfaces [448]. In aqueous Li–$O_2$ batteries, the water-stable lithium-protection membranes are thick, heavy and highly resistive. Both types of batteries suffer from poor oxygen redox kinetics at the cathode electrode and deleterious volume and morphology changes at the anode electrode. Closed Li–$O_2$ batteries that include oxygen storage are much larger and heavier than the aforementioned open ones; however, so far, oxygen- and $OH^-$-selective membranes have not been effective in preventing the contamination of batteries [449]. Recently, 2D materials have shown their roles in electrocatalytic activities in Li–$O_2$ batteries and potential application as porous cathodes and/or catalysts for oxygen-related reactions, enabling controlled growth and facile decomposition of reaction products. In non-aqueous Li–$O_2$ batteries, Jung *et al.* demonstrated that the synthesized hybrid materials of Ru-based nanoparticles supported on reduced graphene oxide (rGO) efficiently functioned as electrocatalysts for $Li_2O_2$ oxidation reactions, maintaining cycling stability for 30 cycles without any sign of electrolyte decomposition, reducing the average charge potential to $\sim$3.7 V at a high current density of 500 mA/g and a high specific capacity of 5000 mAh/g [450]. In aqueous Li–$O_2$ batteries, Liu *et al.* used an rGO electrode, the additive LiI and the solvent dimethoxyethane to reversibly form and remove crystalline LiOH during discharge and charge, leading to high specific capacities, excellent energy efficiency (93.2%) with a voltage gap of only 0.2 V and impressive rechargeability [451].

2D materials can serve as the electrode–electrolyte interface, separator coating and electrolyte additives in uniform Li-ion electrodeposition, improving the synergy of Li-ion deposition and diminishing charge accumulation on the anode electrode [423]. These materials are also able to facilitate cationic transfer and eliminate sluggish reactions by assisting the distribution of Li-ion flux through their structure. Recently, the excellent mechanical properties of 2D materials have led to their application in impeding dendrite growth, nucleation and penetration, making the use of the Li metal as an anode approach the ultrahigh specific capacity of 3860 mAh/g and a very low reduction potential (−3.04 V vs. standard hydrogen electrode) [435, 452–455]. Shin *et al.* demonstrated that nitrogen and sulfur codoped graphene (NSG) nanosheets coated on a polyethylene separator can stabilize the lithium electrode in lithium metal batteries by effectively suppressing dendrite growth and maintaining a uniform Li-ion flux on the metal surface. Meanwhile, the ultrathin layer of NSG nanosheets improved the dimensional stability of the polymer separator at elevated temperatures [456]. Foroozan *et al.* have reported a 3D conformal GO nanosheet coating, confined within the woven structure of a glass fiber separator, permitting facile transport of Li ions through its structure, regulating the Li deposition. A remarkably enhanced cycle life and stability of the Li-metal anode were also demonstrated [457].

The safety problems of Li-ion batteries are mainly a result of thermal runaway, which causes fire and explosion. The thermal runaway of Li-ion batteries can be divided into several levels, i.e., the material level, cell level and system level. Many control and management strategies are proposed to reduce the possibility of battery failure, which can function at the material, cell or system level in practical situations [458–460]. Up to now, it is clear that no existing thermal control and management strategy or technology meets all level requirements [461]. The thermal runaway problem remains an open issue not only for current Li-ion batteries but also for next-generation batteries. The development of 2D materials has gradually shown their potential for material-level control and management strategies, such as shutdown separators, chemical shuttles, non-flammable electrolytes and coatings [462], solid electrolyte interface decomposition, anode active material-electrolyte reaction, cathode active material-electrolyte reaction, electrolyte decomposition and anode active material-binder reaction [463, 464]. Mortazavi *et al.* developed multiscale modeling techniques to evaluate the thermal conductivity of paraffin reinforced with graphene or *h*-BN nanosheet additives [465]. Luo *et al.* have developed a thermally conductive separator coated with BN nanosheets to improve the stability of the Li metal anodes, resulting in the Coulombic efficiency stabilizing at 92% over 100 cycles at a current rate of 0.5 $mA/cm^2$ and 88% at 1.0 $mA/cm^2$ [452]. Applying these 2D materials to electrode materials also exhibited improved thermal stability by impeding their high-temperature degradation [466–468].

One of the many benefits of 2D material additives is the enhancement of ionic conductivity by introducing a high surface area and changing the crystal structure of the host electrolyte. These 2D materials can increase the dissociation degree of Li ions from their anion counterparts and enhance the Li-ion transference number and ionic conductivity [469]. Yuan *et al.* showed that 2D GO fillers with high surface area and excellent mechanical properties can improve the thermo-mechanical stability of the polymer electrolyte, and they appear to significantly enhance the performance of the Li-ion battery. Nearly two orders of magnitude improvement in ion conductivity and a 260% increase in the tensile strength of the polymer electrolyte are reached with only 1 wt% GO content [470].

Owing to the popularization of electric vehicles worldwide, the number of retired power batteries has largely increased, causing environmental protection threats and a waste of resources [471]. A waste management hierarchy, ranking waste management options from the most to the least environmentally desirable, has been proposed to deal with this problem [472]. In the waste management hierarchy, reuse is considered preferable to recycling. Because considerable value is embedded in manufactured Li-ion batteries, it has been suggested that their use should be cascaded through a hierarchy of applications to optimize material use and life-cycle impacts [473, 474]. Recently, many governments have turned to the waste management hierarchy of Li-ion batteries and have made their policies on the management system [475–477] and linked the supply chain [478, 479]. Standards and codes are established [480, 481], and related research has been reported [482–485]. 2D materials are proposed to play their roles in the waste management hierarchy. In recycling the raw materials, such as Li metal, anode/cathode materials and electrolyte [486–488], 2D materials can be used as an additive to enhance the performance of re-manufactured Li-ion batteries or next-generation batteries.

Currently, 2D materials continue to suffer from several challenges, such as poor electrochemical properties [489] and poor chemical and structural stability at ambient conditions (including the expensive and often complicated processability of high-quality materials [490–492]). The high surface area feature of 2D materials may also participate in increasing parasitic reactions at the electrode–electrolyte interface [422]. Hence, we should consider developing techniques to overcome these challenges, such as developing 2D heterostructured and hybrid materials to improve electrochemical properties and mitigate undesired side reactions [493, 494]. On the other hand, the commercial availability and cost of 2D materials are very important factors for battery manufacturers in the industry. Some 2D materials, such as graphene and BN nanosheets, are commercially available at relatively low costs nowadays. However, a majority of the 2D materials, such as black phosphorene and MXene, are only available for use in small-scale experiments and require complicated synthesis processes. Hence, it is important to realize cost-effective methods in synthesis, purification and industrial-scale yield to ensure the application of 2D materials in next-generation batteries.

## 12.12 Phonon

Information about phonon properties is very useful to characterize the dynamical and electrical transport behaviors, for example, investigations of electron–phonon interactions play a critical role in normal and superconducting transport properties [495]. How to characterize the phonon properties, such as the phonon energy spectrum and the atom-decomposed phonon density of states, by using first-principles calculations and tight-binding models is the main focus of studies [495–498]. A lattice vibrates because of temperature and external fields. Even at absolute zero temperature, the zero-point motion will move the cores away from their equilibrium positions [499]. The crystal potential energy $U$ is presumed to be an analytic function of the displacements of the atoms. For most excitations, the displacements of a core are much less than the lattice constant; thus, the problems of atomic vibrations could be easily solved as the second-order term as the harmonic approximations (Fig. 12.15) and the higher-order terms to determine the anharmonic contributions will be ignored. As has been expressed in many textbooks [496, 497, 499], the phonon dispersions and polarizations could be obtained by diagonalizing the dynamical matrices given at each wavevector $k$:

$$D^{i,j}(\mathbf{k}) = \frac{1}{\sqrt{M_i M_j}} \sum_j C_{ij} \mathbf{e}^{i\mathbf{k}\Delta\mathbf{R}_{ij}}, \tag{1}$$

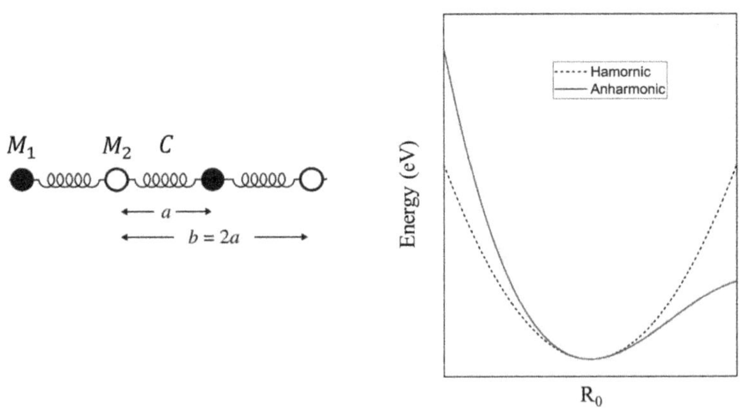


(a) (b)

Fig. 12.15

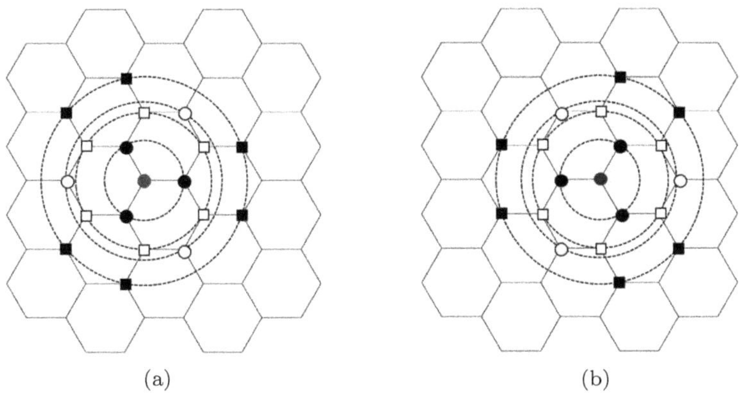


Fig. 12.16

in which, the first term $C_ij$ is the force constant tensor and the second term $\mathbf{e}^{i\mathbf{k}\Delta\mathbf{R}_{ij}}$ is related to the phase difference factors. The summation here is normally taken over only a few neighbor distances relative to the $i$th side, e.g., in the case of 2D graphene, the inclusion of four nearest-neighbor interactions is sufficient for reproducing the experimental phonon dispersion relations (Fig. 12.16). The remaining problem is how to construct the force constant tensor $C_ij$.

The tight-binding model evaluates the force constants by fitting the phonon dispersion relations over the Brillouin zone, as determined experimentally, for example, from inelastic neutron scattering [497]. In a different manner, the ability to exactly calculate the total ground-state energy and the force on atoms [496] allows for the first-principles study of the lattice dynamics of a system using the above scheme of harmonic approximations. Here, the force constant tensors could be evaluated from the second derivative of the total ground-state energy versus the atomic coordinate. Prior to the calculations of the force constants, the equilibrium geometries and their energies are obtained. Finite values for the displacement of atoms (usually less than 0.05 a.u. for each displacement) for each of the $x, y$ and $z$ directions were guessed. Then, the force constants $C_ij$ of all vibration modes could be obtained by comparing the Hellmann–Feynmann forces before and after the displacement of each atom from its equilibrium position [501].

The dynamic behavior and other thermal properties are thoroughly studied through the energy-dispersion phonon spectrum and the atom-decomposed phonon density of states. Moreover, such information is also

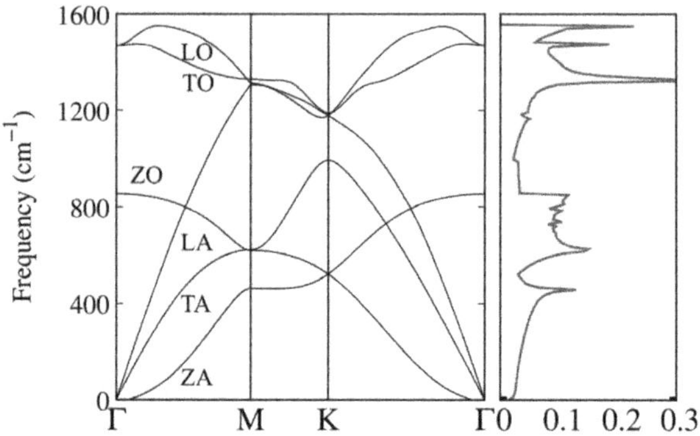


Fig. 12.17

useful for calculating the electron–phonon properties [502, 503]. The essential properties of the phonon energy dispersions could be further verified by the inelastic scattering of neutrons with the emission or absorption of a phonon [504, 505], e.g., the $k^2$ energy dependence of the out-of-plane ZA mode and the highest frequency of the degenerate LO and TO modes at the Γ point of 2D graphene (Fig. 12.17) have been determined experimentally [506, 507]. The harmonic approximation has given very accurate phonon results for a variety of materials, ranging from metals to semiconductors to complex oxides. However, for some systems or at high temperatures, anharmonic terms can modify the phonon structure significantly [508]. How to build a unified theory to deal with such problems as anharmonic oscillations could be considered an open issue.

## References


[1] Foulkes W.M.C. and Haydock R. 1989. Tight-binding models and density-functional theory. *Phys. Rev. B* **39** 12520–12526.

[2] Kresse G. and Furthmüller J. 1996. Efficient iterative schemes for ab initio total-energy calculations using a plane-wave basis set. *Phys. Rev. B* **54** 11169.

[3] Hafner J. 2008. Ab-initio simulations of materials using VASP: Density-functional theory and beyond. *J. Comput. Chem.* **29** 2044–2078.

[4] Zhang Y., Tan Y.W., Stormer H.L. and Kim P. 2005. Experimental observation of the quantum Hall effect and Berry's phase in graphene. *Nature* **438** 201–204.

[5] Notarianni M., Liu J., Vernon K. and Motta N. 2016. Synthesis and applications of carbon nanomaterials for energy generation and storage. *Beilstein J. Nanotechnol.* **7** 149–196.

[6] Lin C.Y., Wu J.Y., Chiu C.W. and Lin M.F. 2019. *Coulomb Excitations and Decays in Graphene-Related Systems* (CRC Press).
[7] Chen S., Wu J., Lin C. and Lin M. 2016. Theory of magnetoelectric properties of 2d systems. *New J. Phys.* **18** 103024.
[8] Konschuh S., Gmitra M. and Fabian J. 2010. Tight-binding theory of the spin-orbit coupling in graphene. *Phys. Rev. B* **82** 245412.
[9] Lai Y.H., Ho J.H., Chang C.P. and Lin M.F. 2008. Magnetoelectronic properties of bilayer Bernal graphene. *Phys. Rev. B* **77** 085426.
[10] Castro Neto A.H. *et al.* 2009. The electronic properties of graphene. *Rev. Mod. Phys.* **81** 109.
[11] Wilder J.W., Venema L.C., Rinzler A.G., Smalley R.E. and Dekker C. 1998. Electronic structure of atomically resolved carbon nanotubes. *Nature* **391** 59–62.
[12] Damascelli A. 2004. Probing the electronic structure of complex systems by ARPES. *Phys. Scr.* **2004** 61.
[13] Matsui T., Kambara H., Niimi Y., Tagami K., Tsukada M. and Fukuyama H. 2005. STS observations of Landau levels at graphite surfaces. *Phys. Rev. Lett.* **94** 226403.
[14] Lin C.Y., Do T.N., Wu J.Y., Shih P.H., Lin S.Y., Ho C.H. and Lin M.F. 2019. *Diverse Quantization Phenomena in Emergent Layered Materials.* (Boca Raton, FL: CRC Press).
[15] Lu C.L., Chang C.P., Huang Y.C., Ho J.H., Hwang C.C. and Lin M.F. 2007. Electronic properties of AA- and ABC-stacked few-layer graphites. *J. Phys. Soc. Jpn.* **76** 024701.
[16] Ho J.-H., Lu C., Hwang C., Chang C. and Lin M.-F. 2006. Coulomb excitations in AA-and AB-stacked bilayer graphites. *Phys. Rev. B* **74** 085406.
[17] Lin C.Y., Chen R.B., Ho Y.H. and Lin M.F. 2018. *Electronic and Optical Properties of Graphite-Related Systems* (Boca Raton, FL: CRC Press).
[18] Fu H.X., Zhang J., Ding Z.J., Li H. and Menga S. 2014. Stacking-dependent electronic structure of bilayer silicene. *Appl. Phys. Lett.* **104** 131904.
[19] Starodub E., Bostwick A., Moreschini L., Nie S., Gabaly F. E., McCarty K.F. and Rotenberg E. 2011. In-plane orientation effects on the electronic structure, stability, and Raman scattering of monolayer graphene on Ir(111). *Phys. Rev. B* **83** 125428.
[20] Lin M.F. and Shung K.W.K. 1995. Magnetoconductance of carbon nanotubes. *Phys. Rev. B* **51** 7592.
[21] Suenaga K. and Koshino M. 2010. Atom-by-atom spectroscopy at graphene edge. *Nature* **468** 1088.
[22] Huang Y.K., Chen S.C., Ho Y.H., Lin C.Y. and Lin M.F. 2014. Feature-rich magnetic quantization in sliding bilayer graphenes. *Sci. Rep.* **4** 7509.
[23] Tran N.T.T., Lin S.Y., Lin C.Y. and Lin M.F. 2017. *Geometric and Electronic Properties of Graphene-Related Systems: Chemical Bonding Schemes* (CRC Press).
[24] Lin C.Y., Wu J.Y., Ou Y.J., Chiu Y.H. and Lin M.F. 2015. Magnetoelectronic properties of multilayer graphenes. *Phys. Chem. Chem. Phys.* **17** 26008.

[25] Mak K.F., Sfeir M.Y., Misewich J.A. and Heinza T.F. 2010. The evolution of electronic structure in few-layer graphene revealed by optical spectroscopy. *Proc. Natl. Acad. Sci. USA* **107** 14999.

[26] Das Sarma S., Hwang E.H. and Tse W.K. 2007. Many-body interaction effects in doped and undoped graphene: Fermi liquid versus non-fermi liquid. *Phys. Rev. B* **75** 121406.

[27] Nakayama T. and Shima H. 1998. Computing the Kubo formula for large systems. *Phys. Rev. E* **58** 3984.

[28] Lin C.Y., Do T.N., Huang Y.K. and Lin M.F. 2017. *Optical Properties of Graphene in Magnetic and Electric Fields.* IOP Concise Physics (San Raefel, CA, USA: IOP Publishing; Morgan & Claypool Publishers).

[29] Ho C.H., Chang C.P., Su W.P. and Lin M.F. 2013. Precessing anisotropic Dirac cone and Landau subbands along a nodal spiral. *New J. Phys.* **15** 053032.

[30] Ju L., Geng B., Horng J., Girit C., Martin M., Hao Z. *et al.* 2011. Graphene plasmonics for tunable terahertz metamaterials. *Nat. Nanotechnol.* **6** 630.

[31] Chen S.C., Wu J.Y., Lin C.Y. and Lin M.F. 2017. *Theory of Magnetoelectric Properties of 2D Systems* (IOP Publishing Ltd).

[32] Ou Y.C., Chiu Y.H., Yang P.H. and Lin M.F. 2014. The selection rule of graphene in a composite magnetic field. *Opt. Express* **22** 7473.

[33] Tahir M. and Schwingenschlogl U. 2013. Valley polarized quantum Hall effect and topological insulator phase transitions in Silicene. *Sci. Rep.* **3** 1075.

[34] Tabert C.J. and Nicol E.J. 2013. AC/DC spin and valley Hall effects in silicene and germanene. *Phys. Rev. B* **87** 235426.

[35] Mu Y.S., Xue Y., Zhou T. and Yang Z.Q. 2019. Quantum anomalous Hall effects and various topological mechanisms in functionalized Sn monolayers. *New J. Phys.* **21** 023010.

[36] Ganz E., Hwang I.S., Xiong F., Theiss S.K. and Golovchenko J. 1991. Growth and morphology of Pb on Si(111). *Surf. Sci.* **257** 259.

[37] Caragiu M. and Finberg S. 2005. Alkali metal adsorption on graphite: A review. *J. Phys. Condens. Matter* **17** R995.

[38] Li J., Zhang L., Zhang L., Hao W., Wang H., Qu Q. and Zheng H. 2014. In-situ growth of graphene decorations for high-performance $LiFePO_4$ cathode through solid-state reaction. *J. Power Sources* **249** 311–319.

[39] Lv R., Li Q., Botello-Méndez A.R., Hayashi T., Wang B., Berkdemir A., Hao Q., Elías A.L., Cruz-Silva R. and Gutiérrez H.R. 2012. Nitrogen-doped graphene: Beyond single substitution and enhanced molecular sensing. *Sci. Rep.* **2** 586.

[40] Do T.N., Shih P.H., Gumbs G., Huang D., Chiu C.W. and Lin M.F. 2018. Diverse magnetic quantization in bilayer silicene. *Phys. Rev. B* **97** 125416.

[41] Briggs N., Gebeyehu Z.M., Vera A., Zhao T., Wang K., Duran A.D., Bersch B., Bowen T., Knappenberger K.L. and Robinson J.A. 2019. Epitaxial graphene/silicon carbide intercalation: A minireview on graphene modulation and unique 2D materials. *Nanoscale* **11** 15440–15447.

[42] Matthes L., Pulci O. and Bechstedt F. 2014. Optical properties of two-dimensional honeycomb crystals graphene, silicene, germanene, and tinene from first principles. *New J. Phys.* **16**.

[43] Suarez-Martinez I., El-Barbary A., Savini G. and Heggie M. 2007. First-principles simulations of boron diffusion in graphite. *Phys. Rev. Lett.* **98** 015501.
[44] Li W.-B., Lin S.-Y., Tran N.T.T., Lin K.-I. and Lin M.-F. 2020. Essential electronic properties on stage-1 Li/Li+-graphite-intercalation compounds for different concentrations. *arXiv preprint arXiv:2006.12055.*
[45] Mak K.F., Shan J. and Heinz T.F. 2010. Electronic structure of few-layer graphene: Experimental demonstration of strong dependence on stacking sequence. *Phys. Rev. Lett.* **104** 176404.
[46] Clauss W., Bergeron D. and Johnson A. 1998. Atomic resolution STM imaging of a twisted single-wall carbon nanotube. *Phys. Rev. B* **58** R4266.
[47] Lin C.Y., Chen S.C., Wu J.Y. and Lin M.F. 2012. Curvature effects on magnetoelectronic properties of nanographene ribbons. *J. Phys. Soc. Jpn.* **81** 064719.
[48] Stern E.A. 1967. Rigid-band model of alloys. *Phys. Rev.* **157** 3.
[49] Sahin H. and Peeters F.M. 2013. Adsorption of alkali, alkaline-earth, and 3D transition metal atoms on silicene. *Phys. Rev. B* **78** 085423.
[50] Elias D.C. *et al.* 2009. Control of graphene properties by reversible hydrogenation: Evidence for graphane. *Science* **323** 610–613.
[51] Compton O.C. and Nguyen S.T. 2010. Graphene oxide, highly reduced graphene oxide, and graphene: Versatile building blocks for carbon-based materials. *Small* **6** 711–723.
[52] Karlicky F., Kumara Ramanatha Datta K., Otyepka M. and Zboril R. 2013. Halogenated graphenes: Rapidly growing family of graphene derivatives. *ACS Nano* **7** 6434.
[53] Zhang Y., Tan Y.W., Stormer H.L. and Kim P. 2005. Experimental observation of the quantum Hall effect and Berry's phase in graphene. *Nature* **438** 201.
[54] Zhang X., Rajaraman B.R.S., Liu H. and Ramakrishna S. 2014. Graphene's potential in materials science and engineering. *RSC Adv.* **4** 28987.
[55] Lin S.Y., Liu H.Y., Nguyen D.K., Tran N.T.T., Pham H.D., Chang S.L., Lin C.Y. and Lin M.F. 2020. *Silicene-Based Layered Materials* (Bristol: IOP Publishing).
[56] Pham H.D., Su W.P., Nguyen T.D.H., Tran N.T.T., Lin M.F. 2020. Rich p-type-doping phenomena in boron-substituted silicene systems. *R. Soc. Open Sci.* **7** 200723.
[57] Pham H.D., Lin S.Y., Gumbs G., Khanh N.D. and Lin M.F. 2020. Diverse properties of carbon-substituted silicenes. *Front. Phys.* **8** 561350.
[58] Pham H.D. *et al.* 2020. Rich essential properties of boron, carbon, and nitrogen substituted germanenes. *Appl. Phys. Express* **13** 085502.
[59] Harrison W.A. 1981. New tight-binding parameters for covalent solids obtained using Louie peripheral states. *Phys. Rev. B* **24** 5835–5843.
[60] Liu C.C., Jiang H. and Yao Y. 2011. Low-energy effective Hamiltonian involving spin-orbit coupling in silicene and two-dimensional germanium and tin. *Phys. Rev. B* **84** 195430.
[61] Wu J.Y., Chen S.C. and Lin M.F. 2014. Temperature-dependent Coulomb excitations in silicene. *New J. Phys.* **16** 125002.
[62] Shih P.H., Do T.N., Gumbs G., Huang D., Pham H.D. and Lin M.F. 2019. Rich magnetic qantization phenomena in AA bilayer silicene. *Sci. Rep.* **9** 14799.

[63] Ho Y.H., Wu J.Y., Chen R.B., Chiu Y.H. and Lin M.F. 2010. Optical transitions between Landau levels: AA-stacked bilayer graphene. *Appl. Phys. Lett.* **97** 101905.

[64] Jiang Z., Zhang Y., Tan Y.-W., Stormer H. and Kim P. 2007. Quantum Hall effect in graphene. *Solid State Commun.* **143** 14–19.

[65] Yin L.J., Li S.Y., Qiao J.B., Nie J.C. and He L. 2015. Landau quantization in graphene monolayer, Bernal bilayer, and Bernal trilayer on graphite surface. *Phys. Rev. B* **91** 115405.

[66] Tabert C.J. and Nicol E.J. 2013. Magneto-optical conductivity of silicene and other buckled honeycomb lattices. *Phys. Rev. B* **88** 085434.

[67] Lin C.Y., Chen R.B., Ho Y.H. and Lin M.F. 2017. *Electronic and Optical Properties of Graphite-Related Systems* (Boca Raton, FL: CRC Press).

[68] Lin C.Y., Do T.N., Wu J.Y., Shih P.H., Lin S.Y., Ho C.H. and Lin M.F. 2019. *Diverse Quantization Phenomena in Emergent Layered Materials* (Boca Raton, FL: CRC Press).

[69] Wunsch B., Stauber T., Sols F. and Guinea F. 2006. Dynamical polarization of graphene at finite doping. *New J. Phys.* **8** 318.

[70] Borghi G., Polini M., Asgari R. and MacDonald A.H. 2009. Dynamical response functions and collective modes of bilayer graphene. *Phys. Rev. B* **80** 241402.

[71] Sensarma R., Hwang E.H. and Das Sarma S. 2010. Dynamic screening and low-energy collective modes in bilayer graphene. *Phys. Rev. B* **82** 195428.

[72] Hwang E.H. and Das Sarma S. 2007. Dielectric function, screening, and plasmons in two-dimensional graphene. *Phys. Rev. B* **75** 205418.

[73] Lin M.F. and Shung K.W.-K. 1996. The self-energy of electrons in graphite intercalation compounds. *Phys. Rev. B* **53** 1109.

[74] Lin M.F. and Shung K.W.-K. 1994. Plasmons and optical properties of carbon nanotubes. *Phys. Rev. B* **50** 17744.

[75] Caruso F., Novko D. and Draxl C. 2018. Phonon-assisted damping of plasmons in three- and two-dimensional metals. *Phys. Rev. B* **97** 205118.

[76] Chuang Y.C., Wu J.Y. and Lin M.F. 2013. Electric-field-induced plasmon in AA-stacked bilayer graphene. *Ann. Phys.* **339** 298.

[77] Ho J.H., Lu C.L., Hwang C.C., Chang C.P. and Lin M.F. 2006. Coulomb excitations in AA- and AB-stacked bilayer graphites. *Phys. Rev. B* **74** 085406.

[78] Chuang Y.C., Wu J.Y. and Lin M.F. 2013. Electric field dependence of excitation spectra in AB-stacked bilayer graphene. *Sci. Rep.* **3** 1368.

[79] Ho J.H., Chang C.P. and Lin M.F. 2006. Electronic excitations of the multilayered graphite. *Phys. Lett. A* **352** 446.

[80] Lin C.Y., Lee M.H. and Lin M.F. 2018. Coulomb excitations in ABC-stacked trilayer graphene. *Phys. Rev. B* **98** 041408.

[81] Ho C.H., Chang C.P. and Lin M.F. 2015. Optical magnetoplasmons in rhombohedral graphite with a three-dimensional Dirac cone structure. *J. Phys. Condens. Matter* **27** 125602.

[82] Shin S.Y., Hwang C.G., Sung S.J., Kim N.D., Kim H.S. *et al.* 2011. Observation of intrinsic intraband pi-plasmon excitation of a single-layer graphene. *Phys. Rev. B* **83** 161403.

[83] Shin S.Y., Kim N.D., Kim J.G., Kim K.S., Noh D.Y. *et al.* 2011. Control of the pi plasmon in a single layer graphene by charge doping. *App. Phys. Lett.* **99** 082110.
[84] Lu J., Loh K.P., Huang H., Chen W. and Wee A.T. 2009. Plasmon dispersion on epitaxial graphene studied using high-resolution electron energy-loss spectroscopy. *Phys. Rev. B* **80** 113410.
[85] Kramberger C., Einarsson E., Huotari S., Thurakitseree T., Maruyama S. *et al.* 2010. Interband and plasma excitations in single-walled carbon nanotubes and graphite in inelastic x-ray and electron scattering. *Phys. Rev. B* **81** 205410.
[86] Petkovi I., Williams F.I.B., Bennaceur K., Portier F., Roche P. *et al.* 2013. Carrier drift velocity and edge magnetoplasmons in graphene. *Phys. Rev. Lett.* **110** 016801.
[87] Poumirol J.M., Yu W., Chen X., Berger C., de Heer W.A. *et al.* 2013. Magnetoplasmons in quasineutral epitaxial graphene nanoribbons. *Phys. Rev. Lett.* **110** 246803.
[88] Gleeson M., Kasemo B. and Chakarov D. 2003. Thermal and adsorbate induced plasmon energy shifts in graphite. *Sur. Sci.* **524** L77.
[89] Nelson F.J., Idrobo J.-C., Fite J.D., Mišković Z.L., Pennycook S.J. *et al.* 2014. Electronic excitations in graphene in the 1–50 eV range: The $\pi$ and $\pi$+$\sigma$ peaks are not plasmons. *Nano Lett.* **14** 3827.
[90] Schaibley J.R., Yu H., Clark G., Rivera P., Jason Ross S. *et al.* 2016. Valleytronics in 2D materials. *Nat. Rev. Mater.* **1** 16055.
[91] Manzeli S., Ovchinnikov D., Pasquier D., Yazyev O.V. and Kis A. 2017. 2D transition metal dichalcogenides. *Nat. Rev. Mater.* **2** 17033.
[92] Rasmussen F.A. and Thygesen K.S. 2015. Computational 2D materials database: Electronic structure of transition-metal dichalcogenides and oxides. *J. Phys. Chem. C* **119** 13169.
[93] Chen P., Han W., Zhao M., Su J.W., Li Z.X. *et al.* 2020. Recent advances in 2D rare earth materials. *Adv. Func. Mater.* **31** 2008790.
[94] Bai G.X., Yuan S.G., Zhao Y.D., Yang Z.B., Choi S.Y. *et al.* 2016. 2D Layered materials of rare-earth Er-doped MoS2 with NIR-to-NIR down- and up-conversion photoluminescence. *Adv. Func. Mater.* **28** 7472.
[95] Mir P., Audiffred M. and Heine T. 2014. An atlas of two-dimensional materials. *Chem. Soc. Rev.* **43** 6537.
[96] Echtermeyer T.J., Milana S., Sassi U., Eiden A., Wu M. *et al.* 2016. Surface plasmon polariton graphene photodetectors. *Nano Lett.* **16** 8.
[97] Low T., Chaves A., Caldwell J.D., Kumar A., Fang N.X. *et al.* 2017. Polaritons in layered two-dimensional materials. *Nat. Mater.* **16** 182.
[98] Olsen T., Latini S., Rasmussen F. and Thygesen K.S. 2016. Simple screened hydrogen model of excitons in two-dimensional materials. *Phys. Rev. Lett.* **116** 056401.
[99] Yu H., Cui X., Xu X. and Yao W. 2015. Valley excitons in two-dimensional semiconductors. *Natl. Sci. Rev.* **2** 57.

[100] Spataru C.D., Ismail-Beigi S., Benedict L.X. and Louie S.G. 2004. Excitonic effects and optical spectra of single-walled carbon nanotubes. *Phys. Rev. Lett.* **92** 077402.
[101] Yang L., Deslippe J., Park C.-H., Cohen M.L. and Louie S.G. 2009. Excitonic effects on the optical response of graphene and bilayer graphene. *Phys. Rev. Lett.* **103** 186802.
[102] Kira M. and Koch S.W. 2006. Many-body correlations and excitonic effects in semiconductor spectroscopy. *Prog. Quantum Electron.* **30** 155.
[103] Ugeda M.M., Bradley A.J., Shi S.-F., da Jornada F.H., Zhang Y. *et al.* 2014. Giant bandgap renormalization and excitonic effects in a monolayer transition metal dichalcogenide semiconductor. *Nat. Mater.* **13** 1091.
[104] Ramasubramaniam A. 2012. Large excitonic effects in monolayers of molybdenum and tungsten dichalcogenides. *Phys. Rev. B* **86** 115409.
[105] Ajiki H. and Ando T. 1996. Energy bands of carbon nanotubes in magnetic fields. *J. Phys. Soc. Jap* **65** 505.
[106] Shyu F.L., Chang C.P., Chen R.B., Chiu C.W. and Lin M.F. 2003. Magnetoelectronic and optical properties of carbon nanotubes. *Phys. Rev. B* **67** 045405.
[107] Varsano D., Sorella S., Sangalli D., Barborini M., Corni S. *et al.* 2017. Carbon nanotubes as excitonic insulators. *Nat. Commun.* **8** 1461.
[108] Zaric S., Ostojic G.N., Kono J., Shaver J., Moore V.C. *et al.* 2004. Optical signatures of the Aharonov-Bohm phase in single-walled carbon nanotubes. *Science* **304** 1129.
[109] Lui C.H., Li Z., Mak K.F., Cappelluti E. and Heinz T.F. 2011. Observation of an electrically tunable band gap in trilayer graphene. *Nat. Phys.* **97** 205118.
[110] Gao X., Burns C., Casa D., Upton M., Gog T. *et al.* 2011. Development of a graphite polarization analyzer for resonant inelastic x-ray scattering. *Rev. Sci. Instrum.* **82** 113108.
[111] Becker P., Scyfried P. and Siegert H. 1982. The lattice parameter of highly pure silicon single crystals. *Z. Phys. B Condens. Matter* **48** 17–21.
[112] Qiu Y. and Gao L. 2003. Chemical synthesis of turbostratic carbon nitride, containing CN crystallites, at atmospheric pressure. *Chem. Commun.* **18** 2378–2379.
[113] Kim D.H., Byon E., Lee S., Kim J.-K. and Ruh H. 2004. Characterization of ternary boron carbon nitride films synthesized by RF magnetron sputtering. *Thin Solid Films* **447–448** 192–196.
[114] Yu K.M. *et al.* 2006. Multiband GaNAsP quaternary alloys. *Appl. Phys. Lett.* **88** 092110.
[115] Rath M.C. 2018. Periodic table of elements revisited for accommodating elements of future years. *Curr. Sci.* **115** 1644.
[116] Pham H.D., Dien V.K., Hanh T.T.T., Tran N.T.T., Ho C.H. and Lin M.F. *Diverse Quasiparticle Properties of Emergent Materials: First-Principles Simulations* (IOP Publishing).
[117] Villars P. and Daams J.L.C. 1993. Atomic-environment classification of the chemical elements. *J. Alloys Compd.* **197** 177.
[118] Ferro S. 2002. Synthesis of diamond. *J. Mater. Chem.* **12** 2843–2855.

[119] Cai H., Gu Y., Lin Y.-C., Yu Y., Geohegan D.B. and Xiao K. 2019. Synthesis and emerging properties of 2D layered III-VI metal chalcogenides. *Appl. Phys. Rev.* **6** 041312.

[120] Zhao J., Zeng H. and Yao G. 2021. Computational design of a polymorph for 2D III-V orthorhombic monolayers by first principles calculations: Excellent anisotropic, electronic and optical properties. *Phys. Chem. Chem. Phys.* **23** 3771.

[121] Adachi S. 2009. *Properties of Semiconductor Alloys: Group-IV, III-V and II-VI Semiconductors* (John Wiley & Sons).

[122] Joshi P.S., Molyneaux B.J., Feng L., Xie X., Macklis J.D. and Gan L. 2008. Bhlhb5 regulates the postmitotic acquisition of area identities in layers II-V of the developing neocortex. *Neuron* **60** 258.

[123] Chabi S., Chang H., Xia Y. and Zhu Y. 2016. From graphene to siliconcarbide: Ultrathin silicon carbide flakes. *Nanotechnology* **7** 075602.

[124] Sevik C. and Bulutay C. 2007. Theoretical study of the insulating oxides and nitrides: $SiO_2$, $GeO_2$, $Al_2O_3$, $Si_3N_4$, and $Ge_3N_4$. *J. Mater. Sci.* **42** 6555–6565.

[125] Bussolotti F., Chi D., Goh K.J., Huang Y.L. and Wee A.T. 2020. *2D Semiconductor Materials and Devices* (Elsevier).

[126] Wang H. *et al.* 2019. First-principles study of electronic, optical and thermal transport properties of group III-VI monolayer MX (M = Ga, In; X = S, Se). *J. Appl. Phys.* **125** 245104.

[127] Cai H., Gu Y., Lin Y.-C., Yu Y., Geohegan D.B. and Xiao K. 2019. Synthesis and emerging properties of 2D layered III-VI metal chalcogenides. *Appl. Phys. Rev.* **6** 041312.

[128] Voevodin V.G. *et al.* 2004. Large single crystals of gallium selenide: Growing, doping by In and characterization. *Opt. Mater.* **26** 495–499.

[129] Du H., Tu M., Luo S., Liu Y., Qiu X., Lu H., Li S., Yuan S., Huang W., Jie W. and Hao J. 2020. Reversible transition between bipolar resistive switching and threshold switching in 2D layered III-VI semiconductor GaSe. *Appl. Phys. Lett.* **116** 253102.

[130] Kim Y.H. *et al.* 2010. Synthesis of shape-controlled $\beta$-$In_2S_3$ nanotubes through orientedattachment of nanoparticles. *Chem. Commun.* **46** 2292–2294.

[131] Xiong X. *et al.* 2016. One-step synthesis of p-type GaSe nanoribbonsand their excellent performance inphotodetectors and phototransistors. *J. Mater. Chem. C* **4** 7817–7823.

[132] Zhou S.Y., Gweon G.-H. and Lanzara A. 2006. Low energy excitations in graphite: The role of dimensionality and lattice defects. *Ann. Phys.* **321** 1730–1746.

[133] Yang Z. and Hao J. 2019. Recent progress in 2D layered III-VI semiconductors and their heterostructures for optoelectronic device applications. *Adv. Mater. Technol.* **4** 1900108.

[134] Das D.K., Sarkar J. and Singh S.K. 2018. Effect of sample size, temperature and strain velocity on mechanical properties of plumbene by tensile loading along longitudinal direction: A molecular dynamics study. *Comput. Mater. Sci.* **151** 196–203.

[135] Cipriano L.A., Di Liberto G., Tosoni S. and Pacchioni G. 2020. Quantum confinement in group III-V semiconductor 2D nanostructures. *Nanoscale* **12** 17494–174501.
[136] Van Hove L. 1953. The occurrence of singularities in the elastic frequency distribution of a crystal. *Phys. Rev.* **89** 1189.
[137] Kira M. and Koch S.W. 2006. Many-body correlations and excitonic effects in semiconductor spectroscopy. *Prog. Quantum Electron.* **30** 155–296.
[138] Jeong S.Y., Kim J.Y., Yang H.D., Yoon B.N., Choi S.H., Kang H.K., Yang C.W. and Lee Y.H. 2003. Synthesis of silicon nanotubes on porous alumina using molecular beam epitaxy. *Adv. Mater.* **15** 1172.
[139] Lourie O.R., Jones C.R., Bartlett B.M., Gibbons P.C., Ruoff R.S. and Buhro W.E. 2000. CVD growth of boron nitride nanotubes. *Chem. Mater.* **12** 1808.
[140] Dabrowska-Szata M. 2003. Analysis of RHEED pattern from semiconductor surfaces. *Mater. Chem. Phys.* **81** 257–259.
[141] Garcéa R. and Perez R. 2002. Dynamic atomic force microscopy methods. *Surf. Sci. Rep.* **47** 197–301.
[142] Flegler S.L. and Flegler S.L. 1997. *Scanning & Transmission Electron Microscopy* (Oxford University Press).
[143] Kubby J.A. and Boland J.J. 1996. Scanning tunneling microscopy of semiconductor surfaces. *Surf. Sci. Rep.* **26** 61–204.
[144] Xu Z., Bando Y., Liu L., Wang W., Bai X. and Golberg D. 2011. Electrical conductivity, chemistry, and bonding alternations under graphene oxide to graphene transition as revealed by in situ TEM. *ACS Nano* **5** 4401.
[145] Matthes L., Pulci O. and Bechstedt F. 2014. Optical properties of two-dimensional honeycomb crystals graphene, silicene, germanene, and tinene from first principles. *New J. Phys.* **16**.
[146] Gilliland G. 1997. Photoluminescence spectroscopy of crystalline semiconductors. *Mater. Sci. Eng. R Rep.* **18** 99.
[147] Wang Y., Ni Z., Shen Z., Wang H. and Wu Y. 2008. Interference enhancement of Raman signal of graphene. *Appl. Phys. Lett.* **92** 043121.
[148] Ko W., Hus S.M., Li X., Berlijn T., Nguyen G.D., Xiao K. and Li A.-P. 2018. Tip-induced local strain on MoS2/graphite detected by inelastic electron tunneling spectroscopy. *Phys. Rev. B* **97** 125401.
[149] Fox M. 2001. *Optical Properties of Solids* (Oxford University Press).
[150] Ishigami M., Chen J., Cullen W., Fuhrer M. and Williams E. 2007. Atomic structure of graphene on $SiO_2$. *Nano Lett.* **7** 1643–1658.
[151] Crepaldi A., Pons S., Frantzeskakis E., Calleja F., Etzkorn M., Seitsonen A. *et al.* 2013. Combined ARPES and STM study of Pb/Au (111) Moiré structure: One overlayer, two symmetries. *Phys. Rev. B* **87** 115138.
[152] Khuong Dien V., Thi Han N., Nguyen T.D.H., Huynh T.M.D., Pham H.D. and Lin M.F. 2020. Geometric and electronic properties of $Li_2GeO_3$. *Front. Mater.* **7**.
[153] Nitta N., Wu F., Lee J.T. and Yushin G. 2015. Li-ion battery materials: Present and future. *Mater. Today* **18** 252–264.

[154] Liu K., Kong B., Liu W., Sun Y., Song M.-S., Chen J., Liu Y., Lin D., Pei A. and Cui Y. 2018. Stretchable lithium metal anode with improved mechanical and electrochemical cycling stability. *Joule* **2** 1857–1865.

[155] Song J.Y., Wang Y.Y. and Wan C.C. 1999. Review of gel-type polymer electrolytes for lithium-ion batteries. *J. Power Sources* **77** 183–197.

[156] Mukherjee S., Schuppert N., Bates A., Lee S.C. and Park S. 2017. Novel mesoporous microspheres of Al and Ni doped LMO spinels and their performance as cathodes in secondary lithium ion batteries. *Int. J. Green Energy* **14** 656–664.

[157] Inaguma Y. and Nakashima M. 2013. A rechargeable lithium-air battery using a lithium ion-conducting lanthanum lithium titanate ceramics as an electrolyte separator. *J. Power Sources* **228** 250–255.

[158] Scrosati B., Hassoun J. and Sun Y.-K. 2011. Lithium-ion batteries. A look into the future. *Energy Environ. Sci.* **4** 3287–3295.

[159] Liu J., Banis M.N., Xiao B., Sun Q., Lushington A., Li R., Guo J., Sham T.-K. and Sun X. 2015. Atomically precise growth of sodium titanates as anode materials for high-rate and ultralong cycle-life sodium-ion batteries. *J. Mater. Chem. A* **3** 24281–24288.

[160] Zhang K.Q., Kirlikovali K.O., Suh J.M., Choi J.W., Jang H.W., Varma R.S., Farha O.K. and Shokouhimehr M. 2020. Recent advances in rechargeable aluminum-ion batteries and considerations for their future progress. *ACS Appl. Energy Mater.* **3** 6019–6035.

[161] Saroja A., Samantaray S.S. and Sundara R. 2019. A room temperature multivalent rechargeable iron ion battery with an ether based electrolyte: A new type of post-lithium ion battery. *Chem. Commun.* **55** 10416–10419.

[162] Li G., Chen Z. and Lu J. 2018. Lithium-sulfur batteries for commercial applications. *Chem* **4** 3.

[163] Leonidov I., Leonidova O., Perelyaeva L., Samigullina R., Kovyazina S. and Patrakeev M. 2003. Structure, ionic conduction, and phase transformations in lithium titanate $Li_4Ti_5O_{12}$. *Phys. Solid State* **45** 2183–2188.

[164] Kuganathan N., Tsoukalas L. and Chroneos A. 2019. Defects, dopants and Li-ion diffusion in $Li_2SiO_3$. *Solid State Ionics* **335** 61–66.

[165] Haussühl S., Liebertz J. and Stähr S. 1982. Single crystal growth and pyroelectric, dielectric, piezoelectric, elastic, and thermoelastic properties of orthorhombic $Li_2SiO_3$, $Li_2GeO_3$, and $Na_2GeO_3$. *Cryst. Res. Technol.* **17** 521–526.

[166] Buyukyazi M. and Mathur S. 2015. 3D nanoarchitectures of a-$LiFeO_2$ and alpha-$LiFeO_2$/C nanofibers for high power lithium-ion batteries. *Nano Energy* **13** 28–35.

[167] Andriyevsky B., Doll K. and Jacob T. 2014. Electronic and transport properties of $LiCoO_2$. *Phys. Chem. Chem. Phys.* **16** 23412–23420.

[168] Laubach S. *et al.* 2009. Changes in the crystal and electronic structure of $LiCoO_2$ and $LiNiO_2$ upon Li intercalation and de-intercalation. *Phys. Chem. Chem. Phys.*

[169] Islam M.S., Driscoll D.J., Fisher C.A.J. and Slater P.R. 2005. Atomic-scale investigation of defects, dopants, and lithium transport in the $LiFePO_4$ olivine-type battery material. *Chem. Mater.*

[170] Li J. and Sun J.L. 2017. Application of X-ray diffraction and electron crystallography for solving complex structure problems. *Acc. Chem. Res.* **50** 2737–2745.

[171] Fong K.D., Self J., Diederichsen K.M., Wood B.M., McCloskey B.D. and Persson K.A. 2019. Ion transport and the true transference number in nonaqueous polyelectrolyte solutions for lithium ion batteries. *ACS Cent. Sci.* **5** 1250–1260.

[172] Karrai K. and Warburton R.J. 2003. Optical transmission and reflection spectroscopy of single quantum dots. *Superlattices Microstruct.* **33** 311–337.

[173] Nozieres P. and Pines D. 1959. Electron interaction in solids — characteristic energy loss spectrum. *Phys. Rev.* **113** 1254–1267.

[174] Gangopadhyay P., Kesavamoorthy R., Bera S., Magudapathy P., Nair K., Panigrahi B. *et al.* 2005. Optical absorption and photoluminescence spectroscopy of the growth of silver nanoparticles. *Phys. Rev. Lett.* **94** 047403.

[175] Trukhin A.N., Rogulis U. and Spingis M. 1997. Self-trapped exciton in $Li_2GeO_3$. *J. Lumin.* **72** 890–892.

[176] Reidy K., Varnavides G., Thomsen J.D., Kumar A., Pham T., Blackburn A.M., Anikeeva P., Narang P., LeBeau J.M. and Ross F.M. 2021. Direct imaging and electronic structure modulation of moiré superlattices at the 2D/3D interface. *Nat. Commun.* **12** 1290.

[177] Nguyen T.D.H., Lin S.Y., Chung H.C., Li W.B., Tran N.T.T., Nguyen T.H., Liu H.Y., Pham H.D. and Lin M.F. 2021. *First-Principles Calculations for Cathode/Electrolyte/Anode Battery Materials* (IOP Publishing).

[178] Wang L., He X., Sun W., Wang J., Li Y. and Fan S. 2012. Crystal orientation tuning of LiFePO4 nanoplates for high rate lithium battery cathode materials. *Nano Lett.* **12** 5632–5636.

[179] Shi J., Zhang P., Xiao D. and Niu Q. 2006. Proper definition of spin current in spin-orbit coupled systems. *Phys. Rev. Lett.* **96** 076604.

[180] Gao J.F., Zhang J.F., Liu H.S., Zhang Q.F. and Zhao J.J. 2013. Structures, mobilities, electronic and magnetic properties of point defects in silicene. *Nanoscale* **5** 9785–9792.

[181] Nguyen T.D.H., Pham H.D., Lin S.Y. and Lin M.F. 2020. Featured properties of Li+-based battery anode: $Li_4Ti_5O_{12}$. *RSC Adv.* **10** 14071–14079.

[182] Trukhin A.N., Rogulis U. and Spingis M. 1997. Self-trapped exciton in Li2GeO3. *J. Lumin.* **72** 890–892.

[183] Li C., He L., Luo J., Tang J., Cheng N. and Chen Z. 2020. Theoretical anchoring effect of new phosphorus allotropes for lithium-sulfur batteries. *Nanoscale* **12** 11095.

[184] Chan C.K., Peng H., Liu G., McIlwrath K., Zhang X.F., Huggins R.A. and Cui Y. 2008. High-performance lithium battery anodes using silicon nanowires. *Nat. Nanotechnol.*

[185] Xu J., Dou Y., Wei Z., Ma J., Deng Y., Li Y., Liu H. and Dou S. 2017. Recent progress in graphite intercalation compounds for rechargeable metal (Li, Na, K, Al)-ion batteries. *Adv. Sci.* **4** 1700146.
[186] Lin S.Y., Lin Y.T., Tran N.T.T., Su W.P. and Lin M.F. 2017. Feature-rich electronic properties of aluminum-adsorbed graphenes. *Carbon* **120** 209–218.
[187] Nakayama M., Yamada S., Jalem R. and Kasuga T. 2016. Density functional studies of olivine-type $LiFePO_4$ and $NaFePO_4$ as positive electrode materials for rechargeable lithium and sodium ion batteries. *Solid State Ionics.*
[188] Ong S.P., Wang L., Kang B. and Ceder G. 2008. Li-Fe-P-O2 phase diagram from first principles calculations. *Chem. Mater.*
[189] Gatta F.M., Geri A., Lamedica R., Lauria S., Maccioni M., Palone F., Rebolini M. and Ruvio A. 2016. Application of a LiFePO4 battery energy storage system to primary frequency control: Simulations and experimental results. *Energies* **9** 887.
[190] Kou L., Chen C. and Smith S.C. 2015. Phosphorene: Fabrication, properties, and applications. *J. Phys. Chem. Lett.* **6** 2794–2805.
[191] Liu H. *et al.* 2014 Phosphorene: An unexplored 2D semiconductor with a high hole mobility. *ACS Nano* **8** 4033–4041.
[192] Xu Z.L., Lin S., Onofrio N., Zhou L., Shi F., Lu W., Kang K., Zhang Q. and Lau S.P. 2018. Exceptional catalytic effects of black phosphorus quantum dots in shuttling-free lithium sulfur batteries. *Nat. Commun.* **9** 4164.
[193] Tahri Y., Chermette H. and Hollinger G. 1991. Electronic structures of phosphorus oxide $P_4O_{10}$ and phosphorus sulfides $P_4S_{10}$ and $P_4S_7$. *J. Electron Spectros. Relat. Phenomena* **56** 51–69.
[194] Fergus J.W. 2010. Recent developments in cathode materials for lithium ion batteries. *J. Power Sources.*
[195] Kang W., Deng N., Ju J., Li Q., Wu D., Ma X., Li L., Naebe M. and Cheng B. 2016. A review of recent developments in rechargeable lithium–sulfur batteries. *Nanoscale* **8** 16541–16588.
[196] Selvy P.E., Lavieri R.R., Lindsley C.W. and Brown H.A. 2011. Phospholipase D: Enzymology, functionality, and chemical modulation. *Chem. Rev.* **111** 6064–6119.
[197] Bello A., Makgopa K., Fabiane M., Dodoo-Ahrin D., Ozoemena K.I. and Manyala N. 2013. Chemical adsorption of NiO nanostructures on nickel foam-graphene for supercapacitor applications. *J. Mater. Sci.* **48** 6707.
[198] Li W.B., Lin S.Y., Tran N.T.T., Lin M.F. and Lin K.I. 2020. Essential geometric and electronic properties in stage-n graphite alkali-metal-intercalation compounds. *RSC Adv.* **10** 23573–23581.
[199] Lin C.Y., Wu J.Y., Ou Y.J., Chiu Y.H. and Lin M.F. 2015. Magnetoelectronic properties of multilayer graphenes. *Phys. Chem. Chem. Phys.* **17** 26008.
[200] Hubbard J. 1958. The description of collective motions in terms of many-body perturbation theory. II. The correlation energy of a free-electron gas. *Proc. R. Soc. Lond.* **14** 336–352.
[201] Lin S.-Y., Chang S.-L., Tran N.T.T., Yang P.-H. and Lin M.-F. 2015. H-Si bonding-induced unusual electronic properties of silicene: A method to identify hydrogen concentration. *Phys. Chem. Chem. Phys.* **17** 26443–26450.

[202] Dien V.K., Pham H.D., Tran N.T.T., Han N.T., Huynh T.M.D., Nguyen T.D.H. and Fa-Lin M. 2020. Orbital-hybridization-created optical excitations in $Li_8Ge_4O_{12}$. *arXiv preprint arXiv:2009.02160.*

[203] Ouyang C., Zhong Z. and Lei M. 2007. Ab initio studies of structural and electronic properties of $Li_4Ti_5O_{12}$ spinel. *Electrochem. Commun.* **9** 1107–1112.

[204] Khare S., Dell'Amico M., Knight C. and McGarry S. 2012. Selection of materials for high temperature latent heat energy storage. *Sol. Energy Mater. Sol. Cells* **107** 20–27.

[205] Solomin E., Kaixuan G., Yanming J. and Minghan D. 2019. Off-shore PV and wind based energy generation. In *2019 International Conference on Industrial Engineering, Applications and Manufacturing (ICIEAM)*, pp. 1–6 (IEEE).

[206] Birnbaum J., Eck M., Fichtner M., Hirsch T., Lehmann D. and Zimmermann G. 2010. A direct steam generation solar power plant with integrated thermal storage.

[207] Sharma S., Jain K.K. and Sharma A. 2015. Solar cells: In research and applications-a review. *Mater. Sci. Appl.* **6**(12) 1145.

[208] Hou W., Xiao Y., Han G. and Lin J.-Y. 2019. The applications of polymers in solar cells: A review. *Polymers* **11**(1) 143.

[209] Day J., Senthilarasu S. and Mallick T.K. 2019. Improving spectral modification for applications in solar cells: A review. *Renew. Energy* **132** 186–205.

[210] Zhao J. *et al.* 2016. Efficient organic solar cells processed from hydrocarbon solvents. *Nat. Energy* **1**(2) 1–7.

[211] Hao F., Stoumpos C.C., Cao D.H., Chang R.P. and Kanatzidis M.G. 2014. Lead-free solid-state organic-inorganic halide perovskite solar cells. *Nat. Photonics* **8**(6) 489–494.

[212] Bagher A.M., Vahid M.M.A. and Mohsen M. 2015. Types of solar cells and application. *Am. J. Opt. Photonics* **3**(5) 94–113.

[213] Pearson G. 1957. Conversion of solar to electrical energy. *Am. J. Phys.* **25**(9) 591–598.

[214] Würfel P. 2007. Photovoltaic principles and organic solar cells. *CHIMIA Int. J. Chem.* **61**(12) 770–774.

[215] Fahrenbruch A. and Bube R. 2012. *Fundamentals of Solar Cells: Photovoltaic Solar Energy Conversion* (Elsevier).

[216] Arora N. *et al.* 2017. Perovskite solar cells with CuSCN hole extraction layers yield stabilized efficiencies greater than 20. *Science* **358**(6364) 768–771.

[217] Muduli S.K. *et al.* 2017. 2D black phosphorous nanosheets as a hole transporting material in perovskite solar cells. *J. Power Sources* **371** 156–161.

[218] Albrecht S., Reining L., Del Sole R. and Onida G. 1998. Ab initio calculation of excitonic effects in the optical spectra of semiconductors. *Phys. Rev. Lett.* **80**(20) 4510.

[219] Warren B.E. 1990. *X-Ray Diffraction* (Courier Corporation).

[220] Whittig L. and Allardice W. 1986. X-ray diffraction techniques. *Methods Soil Anal. Part 1 Phys. Mineral. Methods* **5** 331–362.

[221] Palczewski A.D. 2010. *Angle-Resolved Photoemission Spectroscopy (ARPES) Studies of Cuprate Superconductors* (Iowa State University).

[222] Lv B., Qian T. and Ding H. 2019. Angle-resolved photoemission spectroscopy and its application to topological materials. *Nat. Rev. Phys.* **1**(10) 609–626.
[223] Feenstra R.M. 1994. Scanning tunneling spectroscopy. *Surf. Sci.* **299** 965–979.
[224] French R.H., Müllejans H. and Jones D.J. 1998. Optical properties of aluminum oxide: Determined from vacuum ultraviolet and electron energy-loss spectroscopies. *J. Am. Ceram. Soc.* **81**(10) 2549–2557.
[225] Zhang W., Eperon G.E. and Snaith H.J. 2016. Metal halide perovskites for energy applications. *Nat. Energy* **1**(6) 1–8.
[226] Baranowski M. and Plochocka P. 2020. Excitons in metal-halide perovskites. *Adv. Energy Mater.* **10**(26) 1903659.
[227] Han N.T., Dien V.K. and Lin M.-F. 2021. Excitonic effects in the optical spectra of $Li_3SiO_3$ compound. *Sci. Rep.* **11**(1) 1–10.
[228] Han N.T., Dien V.K., Tran N.T.T., Nguyen D.K., Su W.-P. and Lin M.-F. 2020. First-principles studies of electronic properties in lithium metasilicate (Li 2 SiO 3). *RSC Adv.* **10**(41) 24721–24729.
[229] Dien V.K. *et al.* 2021. Orbital-hybridization-created optical excitations in $Li_2GeO_3$. *Sci. Rep.* **11**(1) 1–10.
[230] Van Neck D., Verdonck S., Bonny G., Ayers P. and Waroquier M. 2006. Quasiparticle properties in a density-functional framework. *Phys. Rev. A* **74**(4) 042501.
[231] Bäppler S.A., Plasser F., Wormit M. and Dreuw A. 2014. Exciton analysis of many-body wave functions: Bridging the gap between the quasiparticle and molecular orbital pictures. *Phys. Rev. A* **90**(5) 052521.
[232] Rahaman M.Z. and Hossain A.M.A. 2018. Effect of metal doping on the visible light absorption, electronic structure and mechanical properties of non-toxic metal halide $CsGeCl_3$. *RSC Adv.* **8**(58) 33010–33018.
[233] Schwarz U., Wagner F., Syassen K. and Hillebrecht H. 1996. Effect of pressure on the optical-absorption edges of $CsGeBr_3$ and $CsGeCl_3$. *Phys. Rev. B* **53**(19) 12545.
[234] Wang Y., Zhang T., Kan M. and Zhao Y. 2018. Bifunctional stabilization of all-inorganic $\alpha$-$CsPbI_3$ perovskite for 17 efficiency photovoltaics. *J. Am. Chem. Soc.* **140**(39) 12345–12348.
[235] Palewicz A., Sosnowska I., Przenioslo R. and Hewat A. 2010. $BiFeO_3$ crystal structure at low temperatures. *Acta Phys. Pol.-Ser. A Gen. Phys.* **117**(2) 296.
[236] Acosta M. *et al.* 2017. $BaTiO_3$-based piezoelectrics: Fundamentals, current status, and perspectives. *Appl. Phys. Rev.* **4**(4) 041305.
[237] Choi K.J. *et al.* 2004. Enhancement of ferroelectricity in strained BaTiO3 thin films. *Science* **306**(5698) 1005–1009.
[238] Buttner R. and Maslen E. 1992. Structural parameters and electron difference density in $BaTiO_3$. *Acta Crystallogr. Sect. B: Struct. Sci.* **48**(6) 764–769.
[239] Mañes J.L. 2012. *Phys. Rev. B* **85** 155118.
[240] Shima N. and Aoki H. 1993. *Phys. Rev. Lett.* **71** 4389.
[241] Hou C.Y., Chamon C. and Mudry C. 2007. *Phys. Rev. Lett.* **98** 186809.
[242] Chamon C., Hou C.Y., Jackiw R., Mudry C., Pi S.Y. and Semenoff G. 2008. *Phys. Rev. B* **77** 235431.

[243] Kindermann M., Uchoa B. and Miller D.L. 2012. *Phys. Rev. B* **86** 115415.
[244] Wallbank J.R., Patel A.A., Mucha-Kruczyński M., Geim A.K. and Fal'ko V.I. 2013. *Phys. Rev. B* **87** 245408.
[245] Evers F. and Mirlin A.D. 2008. *Rev. Mod. Phys.* **80** 1355.
[246] Zirnbauer M.R. 1996. *J. Math. Phys.* **37** 4986.
[247] Altland A. and Zirnbauer M.R. 1997. *Phys. Rev. B* **55** 1142.
[248] Ostrovsky P.M., Gornyi I.V. and Mirlin A.D. 2008. *Phys. Rev. B* **77** 195430.
[249] Zhang H. and Zhang S.C. 2012. *Nat. Rev. Phys.* **3** 283.
[250] Li J., Zhang Z., Wang C., Huang H., Gu B.L. and Duan W. 2020. *J. Appl. Phys.* **128** 191101.
[251] Xiao J. and Yan B. 2021. *Nat. Rev. Phys.* **3** 283.
[252] Martin R.M. 2020. *Electronic Structure: Basic Theory and Practical Methods*, 2nd ed. (Cambridge: Cambridge University Press).
[253] Wen X.G. 2017. *Rev. Mod. Phys.* **89** 041004.
[254] Qi X.L. and Chang S.C. 2011. *Rev. Mod. Phys.* **83** 1057.
[255] Schnyder A.P. and Brydon P.M.R. 2015. *J. Phys. Condens. Matter* **27** 243201.
[256] Castro Neto A.H., Guinea F., Peres N.M.R., Novoselov K.S. and Geim A.K. 2009. *Rev. Mod. Phys.* **81** 109.
[257] Pesin D. and MacDonald A.H. 2012. *Nat. Mater.* **11** 409.
[258] He M., Sun H. and He Q.L. 2019. *Front. Phys.* **14** 43401.
[259] Novoselov K.S., Jiang D., Schedin F., Booth T.J., Khotkevich V.V., Morozov S.V. and Geim A.K. 2005. Two-dimensional atomic crystals. *Proc. Natl. Acad. Sci. USA* **102** 10451.
[260] Novoselov K.S., Geim A.K., Morozov S.V., Jiang D., Zhang Y., Dubonos S.V., Grigorieva I.V. and Firsov A.A. 2004. Electric field effect in atomically thin carbon films. *Science* **306** 666.
[261] Bolotin K.I., Sikes K.J., Jiang Z., Klima M., Fudenberg G., Hone J., Kim P. and Stormer H.L. 2008. Ultrahigh electron mobility in suspended graphene. *Solid State Commun.* **146** 351.
[262] Morozov S.V., Novoselov K.S., Katsnelson M.I., Schedin F., Elias D.C., Jaszczak J.A. and Geim A.K. 2008. Giant intrinsic carrier mobilities in graphene and its bilayer. *Phys. Rev. Lett.* **100** 016602.
[263] Berger C., Song Z.M., Li X.B., Wu X.S., Brown N., Naud C., Mayou D., Li T.B., Hass J., Marchenkov A.N., Conrad E.H., First P.N. and de Heer W.A. 2006. Electronic confinement and coherence in patterned epitaxial graphene. *Science* **312** 1191.
[264] Balandin A.A., Ghosh S., Bao W.Z., Calizo I., Teweldebrhan D., Miao F. and Lau C.N. 2008. Superior thermal conductivity of single-layer graphene. *Nano Lett.* **8** 902.
[265] Koenig S.P., Boddeti N.G., Dunn M.L. and Bunch J.S. 2011. Ultrastrong adhesion of graphene membranes. *Nat. Nanotechnol.* **6** 543–546.
[266] Lee C., Wei X.D., Kysar J.W. and Hone J. 2008. Measurement of the elastic properties and intrinsic strength of monolayer graphene. *Science* **321** 385.
[267] Bunch J.S., Verbridge S.S., Alden J.S., van der Zande A.M., Parpia J.M., Craighead H.G. and McEuen P.L. 2008. Impermeable atomic membranes from graphene sheets. *Nano Lett.* **8** 2458–2462.

[268] Papageorgiou D.G., Kinloch I.A. and Young R.J. 2017. Mechanical properties of graphene and graphene-based nanocomposites. *Prog. Mater. Sci.* **90** 75–127.
[269] Bae S., Kim H., Lee Y., Xu X.F., Park J.S., Zheng Y., Balakrishnan J., Lei T., Kim H.R., Song Y.I., Kim Y.J., Kim K.S., Ozyilmaz B., Ahn J.H., Hong B.H. and Iijima S. 2010. Roll-to-roll production of 30-inch graphene films for transparent electrodes. *Nat. Nanotechnol.* **5** 574.
[270] Nair R.R., Blake P., Grigorenko A.N., Novoselov K.S., Booth T.J., Stauber T., Peres N.M.R. and Geim A.K. 2008. Fine structure constant defines visual transparency of graphene. *Science* **320** 1308.
[271] Novoselov K.S., Jiang Z., Zhang Y., Morozov S.V., Stormer H.L., Zeitler U., Maan J.C., Boebinger G.S., Kim P. and Geim A.K. 2007. Room-temperature quantum Hall effect in graphene. *Science* **315** 1379.
[272] Novoselov K.S., McCann E., Morozov S.V., Fal'ko V.I., Katsnelson M.I., Zeitler U., Jiang D., Schedin F. and Geim A.K. 2006. Unconventional quantum Hall effect and Berry's phase of 2 pi in bilayer graphene. *Nat. Phys.* **2** 177.
[273] Novoselov K.S., Geim A.K., Morozov S.V., Jiang D., Katsnelson M.I., Grigorieva I.V., Dubonos S.V. and Firsov A.A. 2005. Two-dimensional gas of massless Dirac fermions in graphene. *Nature* **438** 197.
[274] Zhang Y.B., Tan Y.W., Stormer H.L. and Kim P. 2005. Experimental observation of the quantum Hall effect and Berry's phase in graphene. *Nature* **438** 201.
[275] Kalantar-zadeh K., Ou J.Z., Daeneke T., Strano M.S., Pumera M. and Gras S.L. 2015. Two-dimensional transition metal dichalcogenides in biosystems. *Adv. Funct. Mater.* **25** 5086–5099.
[276] Chhowalla M., Shin H.S., Eda G., Li L.J., Loh K.P. and Zhang H. 2013. The chemistry of two-dimensional layered transition metal dichalcogenide nanosheets. *Nat. Chem.* **5** 263–275.
[277] Xu M.S., Liang T., Shi M.M. and Chen H.Z. 2013. Graphene-like two-dimensional materials. *Chem. Rev.* **113** 3766–3798.
[278] Osada M. and Sasaki T. 2012. Two-dimensional dielectric nanosheets: Novel nanoelectronics from nanocrystal building blocks. *Adv. Mater.* **24** 210–228.
[279] Novoselov K.S., Fal'ko V.I., Colombo L., Gellert P.R., Schwab M.G. and Kim K. 2012. A roadmap for graphene. *Nature* **490** 192–200.
[280] Li X.L., Wang X.R., Zhang L., Lee S.W. and Dai H.J. 2008. Chemically derived, ultrasmooth graphene nanoribbon semiconductors. *Science* **319** 1229.
[281] Wang X.R., Ouyang Y.J., Li X.L., Wang H.L., Guo J. and Dai H.J. 2008. Room-temperature all-semiconducting sub-10-nm graphene nanoribbon field-effect transistors. *Phys. Rev. Lett.* **100** 206803.
[282] Kang K., Lee K.H., Han Y.M., Gao H., Xie S.E., Muller D.A. and Park J. 2017. Layer-by-layer assembly of two-dimensional materials into wafer-scale heterostructures. *Nature* **550** 229–233.
[283] Cui X., Lee G.H., Kim Y.D., Arefe G., Huang P.Y., Lee C.H., Chenet D.A., Zhang X., Wang L., Ye F., Pizzocchero F., Jessen B.S., Watanabe K., Taniguchi

T., Muller D.A., Low T., Kim P. and Hone J. 2015. Multi-terminal transport measurements of MoS2 using a van der Waals heterostructure device platform. *Nat. Nanotechnol.* **10** 534–540.

[284] Lee G.H., Yu Y.J., Cui X., Petrone N., Lee C.H., Choi M.S., Lee D.Y., Lee C., Yoo W.J., Watanabe K., Taniguchi T., Nuckolls C., Kim P. and Hone J. 2013. Flexible and transparent MoS2 field-effect transistors on hexagonal boron nitride-graphene heterostructures. *ACS Nano* **7** 7931–7936.

[285] Wu W.Z., Wang L., Li Y.L., Zhang F., Lin L., Niu S.M., Chenet D., Zhang X., Hao Y.F., Heinz T.F., Hone J. and Wang Z.L. 2014. Piezoelectricity of single-atomic-layer MoS2 for energy conversion and piezotronics. *Nature* **514** 470–474.

[286] Choi C., Choi M.K., Liu S.Y., Kim M.S., Park O.K., Im C., Kim J., Qin X.L., Lee G.J., Cho K.W., Kim M., Joh E., Lee J., Son D., Kwon S.H., Jeon N.L., Song Y.M., Lu N.S. and Kim D.H. 2017. Human eye-inspired soft optoelectronic device using high-density MoS2-graphene curved image sensor array. *Nat. Commun.* **8** 1664.

[287] Blees M.K., Barnard A.W., Rose P.A., Roberts S.P., McGill K.L., Huang P.Y., Ruyack A.R., Kevek J.W., Kobrin B., Muller D.A. and McEuen P.L. 2015. Graphene kirigami. *Nature* **524** 204–207.

[288] Shyu T.C., Damasceno P.F., Dodd P.M., Lamoureux A., Xu L.Z., Shlian M., Shtein M., Glotzer S.C. and Kotov N.A. 2015. A kirigami approach to engineering elasticity in nanocomposites through patterned defects. *Nat. Mater.* **14** 785–789.

[289] Han S.S., Kim J.H., Noh C., Ji E., Kwon J., Yu S.M., Ko T.J., Okogbue E., Oh K.H., Chung H.S., Jung Y. and Lee G.H. 2019. Horizontal-to-vertical transition of 2d layer orientation in low-temperature chemical vapor deposition-grown PtSe2 and its influences on electrical properties and device applications. *ACS Appl. Mater. Interfaces* **11** 13598–13607.

[290] Mendiratta N. and Tripathi S.L. 2020. A review on performance comparison of advanced MOSFET structures below 45 nm technology node. *J. Semicond.* **41** 061401.

[291] Hinkle C.L., Vogel E.M., Ye P.D. and Wallace R.M. 2011. Interfacial chemistry of oxides on inxga(1-x)as and implications for MOSFET applications. *Curr. Opin. Solid State Mater. Sci.* **15** 188–207.

[292] Wu M., Alivov Y.I. and Morkoc H. 2008. High-kappa dielectrics and advanced channel concepts for Si MOSFET. *J. Mater. Sci. Mater. Electron.* **19** 915–951.

[293] Yu H.N., Reisman A., Osburn C.M. and Critchlow D.L. 1979. 1 mu-m MOSFET VLSI technology: Part I-an overview. *IEEE Trans. Electron Devices* **26** 318–324.

[294] Britnell L., Gorbachev R.V., Jalil R., Belle B.D., Schedin F., Mishchenko A., Georgiou T., Katsnelson M.I., Eaves L., Morozov S.V., Peres N.M.R., Leist J., Geim A.K., Novoselov K.S. and Ponomarenko L.A. 2012. Field-effect tunneling transistor based on vertical graphene heterostructures. *Science* **335** 947–950.

[295] Britnell L., Gorbachev R.V., Jalil R., Belle B.D., Schedin F., Katsnelson M.I., Eaves L., Morozov S.V., Mayorov A.S., Peres N.M.R., Neto A.H.C., Leist J., Geim A.K., Ponomarenko L.A. and Novoselov K.S. 2012. Electron tunneling through ultrathin boron nitride crystalline barriers. *Nano Lett.* **12** 1707–1710.

[296] Georgiou T., Jalil R., Belle B.D., Britnell L., Gorbachev R.V., Morozov S.V., Kim Y.J., Gholinia A., Haigh S.J., Makarovsky O., Eaves L., Ponomarenko L.A., Geim A.K., Novoselov K.S. and Mishchenko A. 2013. Vertical field-effect transistor based on graphene-ws2 heterostructures for flexible and transparent electronics. *Nat. Nanotechnol.* **8** 100–103.

[297] Choi W.Y., Park B.G., Lee J.D. and Liu T.J.K. 2007. Tunneling field-effect transistors (TFETS) with subthreshold swing (SS) less than 60 mv/dec. *IEEE Electron Device Lett.* **28** 743–745.

[298] Zhang Q., Zhao W. and Seabaugh A. 2006. Low-subthreshold-swing tunnel transistors. *IEEE Electron Device Lett.* **27** 297–300.

[299] Sarkar D., Xie X.J., Liu W., Cao W., Kang J.H., Gong Y.J., Kraemer S., Ajayan P.M. and Banerjee K. 2015. A subthermionic tunnel field-effect transistor with an atomically thin channel. *Nature* **526** 91–95.

[300] Datta S., Liu H.C. and Narayanan V. 2014. Tunnel FET technology: A reliability perspective. *Microelectron. Reliab.* **54** 861–874.

[301] Ionescu A.M. and Riel H. 2011. Tunnel field-effect transistors as energy-efficient electronic switches. *Nature* **479** 329–337.

[302] Khatami Y. and Banerjee K. 2009. Steep subthreshold slope n- and p-type tunnel-FET devices for low-power and energy-efficient digital circuits. *IEEE Trans. Electron Devices* **56** 2752–2761.

[303] Bhuwalka K.K., Sedlmaier S., Ludsteck A.K., Tolksdorf A., Schulze J. and Eisele I. 2004. Vertical tunnel field-effect transistor. *IEEE Trans. Electron Devices* **51** 279–282.

[304] Baba T. 1992. Proposal for surface tunnel transistors. *Jpn. J. Appl. Phys.* **31** L455–L457.

[305] Quinn J.J., Kawamoto G. and McCombe B.D. 1978. Subband spectroscopy by surface channel tunneling. *Surf. Sci.* **73** 190–196.

[306] Roy T., Tosun M., Cao X., Fang H., Lien D.H., Zhao P.D., Chen Y.Z., Chueh Y.L., Guo J. and Javey A. 2015. Dual-gated MoS2/WSe2 van der Waals tunnel diodes and transistors. *ACS Nano* **9** 2071–2079.

[307] Yan R.S., Fathipour S., Han Y.M., Song B., Xiao S.D., Li M.D., Ma N., Protasenko V., Muller D.A., Jena D. and Xing H.G. 2015. Esaki diodes in van der Waals heterojunctions with broken-gap energy band alignment. *Nano Lett.* **15** 5791–5798.

[308] Agarwal S. and Yablonovitch E. 2014. Band-edge steepness obtained from esaki/backward diode current-voltage characteristics. *IEEE Trans. Electron Devices* **61** 1488–1493.

[309] Esaki L. 1958. New phenomenon in narrow germanium p-n junctions. *Phys. Rev.* **109** 603–604.

[310] Thompson S.E. and Parthasarathy S. 2006. Moore's law: The future of Si microelectronics. *Mater. Today* **9** 20–25.

[311] Radisavljevic B., Radenovic A., Brivio J., Giacometti V. and Kis A. 2011. Single-layer MoS2 transistors. *Nat. Nanotechnol.* **6** 147.
[312] Li L.K., Yu Y.J., Ye G.J., Ge Q.Q., Ou X.D., Wu H., Feng D.L., Chen X.H. and Zhang Y.B. 2014. Black phosphorus field-effect transistors. *Nat. Nanotechnol.* **9** 372.
[313] Van der Wagt J.P.A. 1999. Tunneling-based SRAM. *Proc. IEEE* **87** 571–595.
[314] Parker L.H. and Tasch A.F. 1990. Ferroelectric materials for 64 mb and 256 mb drams. *IEEE Circuits Devices Mag.* **6** 17–26.
[315] Bez R., Camerlenghi E., Modelli A. and Visconti A. 2003. Introduction to flash memory. *Proc. IEEE* **91** 489–502.
[316] Pavan P., Bez R., Olivo P. and Zanoni E. 1997. Flash memory cells — an overview. *Proc. IEEE* **85** 1248–1271.
[317] Wong H.S.P. and Salahuddin S. 2015. Memory leads the way to better computing. *Nat. Nanotechnol.* **10** 191–194.
[318] Na T., Kang S.H. and Jung S.O. 2021. STT-MRAM sensing: A review. *IEEE Trans. Circuits Syst. II — Express Briefs* **68** 12–18.
[319] Nozaki T., Yamamoto T., Miwa S., Tsujikawa M., Shirai M., Yuasa S. and Suzuki Y. 2019. Recent progress in the voltage-controlled magnetic anisotropy effect and the challenges faced in developing voltage-torque MRAM. *Micromachines* **10** 327.
[320] Zahoor F., Zulkifli T.Z.A. and Khanday F.A. 2020. Resistive random access memory (RRAM): An overview of materials, switching mechanism, performance, multilevel cell (MLC) storage, modeling, and applications. *Nanoscale Res. Lett.* **15** 90.
[321] Wang Y., Kang K.M., Kim M., Lee H.S., Waser R., Wouters D., Dittmann R., Yang J.J. and Park H.H. 2019. Mott-transition-based RRAM. *Mater. Today* **28** 63–80.
[322] Wong H.S.P., Lee H.Y., Yu S.M., Chen Y.S., Wu Y., Chen P.S., Lee B., Chen F.T. and Tsai M.J. 2012. Metal-oxide RRAM. *Proc. IEEE* **100** 1951–1970.
[323] Hatayama S., Sutou Y., Shindo S., Saito Y., Song Y.H., Ando D. and Koike J. 2018. Inverse resistance change Cr2Ge2Te6-based PCRAM enabling ultralow-energy amorphization. *ACS Appl. Mater. Interfaces* **10** 2725–2734.
[324] Wei X.Q., Shi L.P., Walia R., Chong T.C., Zhao R., Miao X.S. and Quek B.S. 2006. HSPICE macromodel of PCRAM for binary and multilevel storage. *IEEE Trans. Electron Devices* **53** 56–62.
[325] Wang C.C., Ye Y. and Cao Y. 2012. The potential of Fe-FET for robust design under variations: A compact modeling study. *Microelectron. J.* **43** 898–903.
[326] Salvatore G.A., Lattanzio L., Bouvet D. and Ionescu A.M. 2011. Modeling the temperature dependence of Fe-FET static characteristics based on Landau's theory. *IEEE Trans. Electron Devices* **58** 3162–3169.
[327] Kim W.S., Lee H.S., Park H.H. and Hwang Y.T. 2011. Properties of Pb(Zr0.52Ti0.48)O-3, SrBi2Ta2O9, and Nd2Ti2O7 in a mfis of Y2O3 insulator base structure for Fe FET. *Ferroelectrics* **413** 1–10.
[328] Moyer B. 2021. MRAM evolves in multiple directions. *Semicond. Eng.*

[329] Freund K. 2021. IBM nears breakthrough in new memory class — IBM research has been working on new non-volatile magnetic memory for over two decades. *Forbes*.
[330] Everspin Technologies, Inc. 2016. Everspin's ST-MRAM technology to be deployed in globalfoundries 22FDX eMRAM platform.
[331] Wang C.Y., Wang C., Meng F.H., Wang P.F., Wang S., Liang S.J. and Miao F. 2020. 2d layered materials for memristive and neuromorphic applications. *Adv. Electron. Mater.* **6** 1901107.
[332] Pang C.S., Thakuria N., Gupta S.K. and Chen Z. 2019. First demonstration of WSe2 based CMOS-SRAM. In *2018 IEEE International Electron Devices Meeting (IEDM)*, 22.2.1.
[333] Kshirsagar C.U., Xu W.C., Su Y., Robbins M.C., Kim C.H. and Koester S.J. 2016. Dynamic memory cells using MoS2 field-effect transistors demonstrating femtoampere leakage currents. *ACS Nano* **10** 8457–8464.
[334] Bertolazzi S., Krasnozhon D. and Kis A. 2013. Nonvolatile memory cells based on MoS2/graphene heterostructures. *ACS Nano* **7** 3246–3252.
[335] Sun X.D., Li W.Y., Wang X., Sui Q., Zhang T.Y., Wang Z., Liu L., Li D., Feng S., Zhong S.Y., Wang H.W., Bouchiat V., Regueiro M.N., Rougemaille N., Coraux J., Purbawati A., Hadj-Azzem A., Wang Z.H., Dong B.J., Wu X., Yang T., Yu G.Q., Wang B.W., Han Z., Han X.F. and Zhang Z.D. 2020. Room temperature ferromagnetism in ultra-thin van der Waals crystals of 1T-CrTe2. *Nano Res.* **13** 3358–3363.
[336] Huang B., Clark G., Navarro-Moratalla E., Klein D.R., Cheng R., Seyler K.L., Zhong D., Schmidgall E., McGuire M.A., Cobden D.H., Yao W., Xiao D., Jarillo-Herrero P. and Xu X.D. 2017. Layer-dependent ferromagnetism in a van der Waals crystal down to the monolayer limit. *Nature* **546** 270.
[337] Liu F.C., You L., Seyler K.L., Li X.B., Yu P., Lin J.H., Wang X.W., Zhou J.D., Wang H., He H.Y., Pantelides S.T., Zhou W., Sharma P., Xu X.D., Ajayan P.M., Wang J.L. and Liu Z. 2016. Room-temperature ferroelectricity in CuInP2S6 ultrathin flakes. *Nat. Commun.* **7** 12357.
[338] Zhang F., Zhang H.R., Krylyuk S., Milligan C.A., Zhu Y.Q., Zemlyanov D.Y., Bendersky L.A., Burton B.P., Davydov A.V. and Appenzeller J. 2019. Electric-field induced structural transition in vertical MoTe2- and Mo1-xWxTe2-based resistive memories. *Nat. Mater.* **18** 55.
[339] Song T.C., Cai X.H., Tu M.W.Y., Zhang X.O., Huang B.V., Wilson N.P., Seyler K.L., Zhu L., Taniguchi T., Watanabe K., McGuire M.A., Cobden D.H., Xiao D., Yao W. and Xu X.D. 2018. Giant tunneling magnetoresistance in spin-filter van der waals heterostructures. *Science* **360** 1214–1218.
[340] Jiang S.W., Li L.Z., Wang Z.F., Shan J. and Mak K.F. 2019. Spin tunnel field-effect transistors based on two-dimensional van der Waals heterostructures. *Nat. Electron.* **2** 159–163.
[341] Si M.W., Saha A.K., Gao S.J., Qiu G., Qin J.K., Duan Y.Q., Jian J., Niu C., Wang H.Y., Wu W.Z., Gupta S.K. and Ye P.D.D. 2019. A ferroelectric semiconductor field-effect transistor. *Nat. Electron.* **2** 580–586.

[342] Sharma P., Xiang F.X., Shao D.F., Zhang D.W., Tsymbal E.Y., Hamilton A.R. and Seidel J. 2019. A room-temperature ferroelectric semimetal. *Sci. Adv.* **5** eaax5080.

[343] Wang Z., Sapkota D., Taniguchi T., Watanabe K., Mandrus D. and Morpurgo A.F. 2018. Tunneling spin valves based on Fe3GeTe2/hBN/Fe3GeTe2 van der Waals heterostructures. *Nano Lett.* **18** 4303–4308.

[344] Sivan M., Li Y.D., Veluri H., Zhao Y.S., Tang B.S., Wang X.H., Zamburg E., Leong J.F., Niu J.X., Chand U. and Thean A.V.Y. 2019. All WSe2 1T1R resistive RAM cell for future monolithic 3d embedded memory integration. *Nat. Commun.* **10** 5201.

[345] Kim S., Zhou J.T. and Lu W.D. 2014. Crossbar RRAM arrays: Selector device requirements during write operation. *IEEE Trans. Electron Devices* **61** 2820–2826.

[346] Zhou J.T., Kim K.H. and Lu W. 2014. Crossbar rram arrays: Selector device requirements during read operation. *IEEE Trans. Electron Devices* **61** 1369–1376.

[347] Salahuddin S., Ni K. and Datta S. 2018. The era of hyper-scaling in electronics. *Nat. Electron.* **1** 442–450.

[348] Lien D.H., Amani M., Desai S.B., Ahn G.H., Han K., He J.H., Ager J.W., Wu M.C. and Javey A. 2018. Large-area and bright pulsed electroluminescence in monolayer semiconductors. *Nat. Commun.* **9** 1229.

[349] Withers F., Del Pozo-Zamudio O., Mishchenko A., Rooney A.P., Gholinia A., Watanabe K., Taniguchi T., Haigh S.J., Geim A.K., Tartakovskii A.I. and Novoselov K.S. 2015. Light-emitting diodes by band-structure engineering in van der Waals heterostructures. *Nat. Mater.* **14** 301–306.

[350] Reineke S., Lindner F., Schwartz G., Seidler N., Walzer K., Lussem B. and Leo K. 2009. White organic light-emitting diodes with fluorescent tube efficiency. *Nature* **459** 234–238.

[351] Withers F., Del Pozo-Zamudio O., Schwarz S., Dufferwiel S., Walker P.M., Godde T., Rooney A.P., Gholinia A., Woods C.R., Blake P., Haigh S.J., Watanabe K., Taniguchi T., Aleiner I.L., Geim A.K., Fal'ko V.I., Tartakovskii A.I. and Novoselov K.S. 2015. Wse2 light-emitting tunneling transistors with enhanced brightness at room temperature. *Nano Lett.* **15** 8223–8228.

[352] Gabor N.M., Zhong Z.H., Bosnick K., Park J. and McEuen P.L. 2009. Extremely efficient multiple electron-hole pair generation in carbon nanotube photodiodes. *Science* **325** 1367–1371.

[353] Pospischil A., Furchi M.M. and Mueller T. 2014. Solar-energy conversion and light emission in an atomic monolayer p-n diode. *Nat. Nanotechnol.* **9** 257–261.

[354] Lopez-Sanchez O., Lembke D., Kayci M., Radenovic A. and Kis A. 2013. Ultrasensitive photodetectors based on monolayer MoS2. *Nat. Nanotechnol.* **8** 497–501.

[355] Britnell L., Ribeiro R.M., Eckmann A., Jalil R., Belle B.D., Mishchenko A., Kim Y.J., Gorbachev R.V., Georgiou T., Morozov S.V., Grigorenko A.N., Geim A.K., Casiraghi C., Castro Neto A.H. and Novoselov K.S. 2013. Strong

light-matter interactions in heterostructures of atomically thin films. *Science* **340** 1311–1314.

[356] Yu W.J., Vu Q.A., Oh H., Nam H.G., Zhou H.L., Cha S., Kim J.Y., Carvalho A., Jeong M., Choi H., Neto A.H.C., Lee Y.H. and Duan X.F. 2016. Unusually efficient photocurrent extraction in monolayer van der waals heterostructure by tunnelling through discretized barriers. *Nat. Commun.* **7** 13278.

[357] Geim A.K. and Grigorieva I.V. 2013. Van der Waals heterostructures. *Nature* **499** 419.

[358] Mak K.F., Lee C., Hone J., Shan J. and Heinz T.F. 2010. Atomically thin MoS2: A new direct-gap semiconductor. *Phys. Rev. Lett.* **105** 136805.

[359] Splendiani A., Sun L., Zhang Y.B., Li T.S., Kim J., Chim C.Y., Galli G. and Wang F. 2010. Emerging photoluminescence in monolayer MoS2. *Nano Lett.* **10** 1271.

[360] Bernardi M., Palummo M. and Grossman J.C. 2013. Extraordinary sunlight absorption and one nanometer thick photovoltaics using two-dimensional monolayer materials. *Nano Lett.* **13** 3664–3670.

[361] Gregg B.A. and Hanna M.C. 2003. Comparing organic to inorganic photovoltaic cells: Theory, experiment, and simulation. *J. Appl. Phys.* **93** 3605–3614.

[362] Lee C.H., Lee G.H., van der Zande A.M., Chen W., Li Y., Han M., Cui X., Arefe G., Nuckolls C., Heinz T.F., Guo J., Hone J. and Kim P. 2014. Atomically thin p-n junctions with van der Waals heterointerfaces. *Nat. Nanotechnol.* **9** 676.

[363] Liu W.J., Lee B., Naylor C.H., Ee H.S., Park J., Johnson A.T.C. and Agarwal R. 2016. Strong exciton-plasmon coupling in MoS2 coupled with plasmonic lattice. *Nano Lett.* **16** 1262–1269.

[364] Wang S.J., Li S.L., Chervy T., Shalabney A., Azzini S., Orgiu E., Hutchison J.A., Genet C., Samori P. and Ebbesen T.W. 2016. Coherent coupling of WS2 monolayers with metallic photonic nanostructures at room temperature. *Nano Lett.* **16** 4368–4374.

[365] Yu W.J., Liu Y., Zhou H.L., Yin A.X., Li Z., Huang Y. and Duan X.F. 2013. Highly efficient gate-tunable photocurrent generation in vertical heterostructures of layered materials. *Nat. Nanotechnol.* **8** 952–958.

[366] Wang Q.H., Kalantar-Zadeh K., Kis A., Coleman J.N. and Strano M.S. 2012. Electronics and optoelectronics of two-dimensional transition metal dichalcogenides. *Nat. Nanotechnol.* **7** 699.

[367] Polman A. and Atwater H.A. 2012. Photonic design principles for ultrahigh-efficiency photovoltaics. *Nat. Mater.* **11** 174–177.

[368] Nosonovsky M. 2007. Oil as a lubricant in the ancient middle east. *Tribol. Online* **2**.

[369] Ky D.L.C., Khac B.C.T., Le C.T., Kim Y.S. and Chung K.H. 2018. Friction characteristics of mechanically exfoliated and CVD-grown single-layer MoS2. *Friction* **6** 395–406.

[370] Spear J.C., Ewers B.W. and Batteas J.D. 2015. 2d-nanomaterials for controlling friction and wear at interfaces. *Nano Today* **10** 301–314.

[371] Berman D., Erdemir A. and Sumant A.V. 2014. Graphene: A new emerging lubricant. *Mater. Today* **17** 31–42.

[372] Marchetto D., Feser T. and Dienwiebel M. 2015. Microscale study of frictional properties of graphene in ultra high vacuum. *Friction* **3** 161–169.
[373] Liu Z., Yang J.R., Grey F., Liu J.Z., Liu Y.L., Wang Y.B., Yang Y.L., Cheng Y. and Zheng Q.S. 2012. Observation of microscale superlubricity in graphite. *Phys. Rev. Lett.* **108** 205503.
[374] Ruska E. 1987. The development of the electron microscope and of electron microscopy. *Rev. Mod. Phys.* **59** 627–638.
[375] Ruska E. 1987. The development of the electron microscope and of electron microscopy (nobel lecture). *Angew. Chem. Int. Ed.* **26** 595–605.
[376] Binnig G. and Rohrer H. 1987. Scanning tunneling microscopy — from birth to adolescence. *Rev. Mod. Phys.* **59** 615–625.
[377] Binnig G. and Rohrer H. 1983. Scanning tunneling microscopy. *Surf. Sci.* **126** 236–244.
[378] Binnig G., Rohrer H., Gerber C. and Weibel E. 1983. 7 x 7 reconstruction on Si(111) resolved in real space. *Phys. Rev. Lett.* **50** 120.
[379] Binnig G., Rohrer H., Gerber C. and Weibel E. 1982. Surface studies by scanning tunneling microscopy. *Phys. Rev. Lett.* **49** 57.
[380] Binnig G. and Rohrer H. 1982. Scanning tunneling microscopy. *Helv. Phys. Acta* **55** 726.
[381] Binnig G., Quate C.F. and Gerber C. 1986. Atomic force microscope. *Phys. Rev. Lett.* **56** 930–933.
[382] Wang J.J., Li J.M., Li C., Cai X.L., Zhu W.G. and Jia Y. 2016. Tuning the nanofriction between two graphene layers by external electric fields: A density functional theory study. *Tribol. Lett.* **61** 4.
[383] Wang C.Q., Chen W.G., Zhang Y.S., Sun Q. and Jia Y. 2015. Effects of VDW interaction and electric field on friction in MoS2. *Tribol. Lett.* **59** 7.
[384] Kwon S., Ko J.H., Jeon K.J., Kim Y.H. and Park J.Y. 2012. Enhanced nanoscale friction on fluorinated graphene. *Nano Lett.* **12** 6043–6048.
[385] Yoon H.M., Jung Y., Jun S.C., Kondaraju S. and Lee J.S. 2015. Molecular dynamics simulations of nanoscale and sub-nanoscale friction behavior between graphene and a silicon tip: Analysis of tip apex motion. *Nanoscale* **7** 6295–6303.
[386] Smolyanitsky A. and Killgore J.P. 2012. Anomalous friction in suspended graphene. *Phys. Rev. B* **86** 125432.
[387] Ansari R., Ajori S. and Motevalli B. 2012. Mechanical properties of defective single-layered graphene sheets via molecular dynamics simulation. *Superlattices Microstruct.* **51** 274–289.
[388] Mohammadpour E. and Awang M. 2012. Nonlinear finite-element modeling of graphene and single- and multi-walled carbon nanotubes under axial tension. *Appl. Phys. A-Mater. Sci. Process.* **106** 581–588.
[389] Lee C., Li Q.Y., Kalb W., Liu X.Z., Berger H., Carpick R.W. and Hone J. 2010. Frictional characteristics of atomically thin sheets. *Science* **328** 76–80.
[390] Dienwiebel M., Pradeep N., Verhoeven G.S., Zandbergen H.W. and Frenken J.W.M. 2005. Model experiments of superlubricity of graphite. *Surf. Sci.* **576** 197–211.

[391] Dienwiebel M., Verhoeven G.S., Pradeep N., Frenken J.W.M., Heimberg J.A. and Zandbergen H.W. 2004. Superlubricity of graphite. *Phys. Rev. Lett.* **92** 126101.

[392] Martin J.M., Donnet C., Lemogne T. and Epicier T. 1993. Superlubricity of molybdenum-disulfide. *Phys. Rev. B* **48** 10583–10586.

[393] Hirano M. and Shinjo K. 1990. Atomistic locking and friction. *Phys. Rev. B* **41** 11837–11851.

[394] Feng X.F., Kwon S., Park J.Y. and Salmeron M. 2013. Superlubric sliding of graphene nanoflakes on graphene. *ACS Nano* **7** 1718–1724.

[395] Li H., Wang J.H., Gao S., Chen Q., Peng L.M., Liu K.H. and Wei X.L. 2017. Superlubricity between MoS2 monolayers. *Adv. Mater.* **29** 1701474.

[396] Filleter T., McChesney J.L., Bostwick A., Rotenberg E., Emtsev K.V., Seyller T., Horn K. and Bennewitz R. 2009. Friction and dissipation in epitaxial graphene films. *Phys. Rev. Lett.* **102** 086102.

[397] Cannara R.J., Brukman M.J., Cimatu K., Sumant A.V., Baldelli S. and Carpick R.W. 2007. Nanoscale friction varied by isotopic shifting of surface vibrational frequencies. *Science* **318** 780–783.

[398] Li Q.Y., Lee C., Carpick R.W. and Hone J. 2010. Substrate effect on thickness-dependent friction on graphene. *Phys. Status Solidi B-Basic Solid State Phys.* **247** 2909–2914.

[399] Lee C., Wei X.D., Li Q.Y., Carpick R., Kysar J.W. and Hone J. 2009. Elastic and frictional properties of graphene. *Phys. Status Solidi B-Basic Solid State Phys.* **246** 2562–2567.

[400] Sun X.Y., Qi Y.Z., Ouyang W.G., Feng X.Q. and Li Q.Y. 2016. Energy corrugation in atomic-scale friction on graphite revisited by molecular dynamics simulations. *Acta Mech. Sin.* **32** 604–610.

[401] Smolyanitsky A., Killgore J.P. and Tewary V.K. 2012. Effect of elastic deformation on frictional properties of few-layer graphene. *Phys. Rev. B* **85** 035412.

[402] Deng Z., Smolyanitsky A., Li Q.Y., Feng X.Q. and Cannara R.J. 2012. Adhesion-dependent negative friction coefficient on chemically modified graphite at the nanoscale. *Nat. Mater.* **11** 1032–1037.

[403] Reguzzoni M., Fasolino A., Molinari E. and Righi M.C. 2012. Friction by shear deformations in multilayer graphene. *J. Phys. Chem. C* **116** 21104–21108.

[404] Cho D.H., Wang L., Kim J.S., Lee G.H., Kim E.S., Lee S., Lee S.Y., Hone J. and Lee C. 2013. Effect of surface morphology on friction of graphene on various substrates. *Nanoscale* **5** 3063–3069.

[405] Deng Z., Klimov N.N., Solares S.D., Li T., Xu H. and Cannara R.J. 2013. Nanoscale interfacial friction and adhesion on supported versus suspended monolayer and multilayer graphene. *Langmuir* **29** 235–243.

[406] Wu Y.Y., Tsui W.C. and Liu T.C. 2007. Experimental analysis of tribological properties of lubricating oils with nanoparticle additives. *Wear* **262** 819–825.

[407] Chinas-Castillo F. and Spikes H.A. 2003. Mechanism of action of colloidal solid dispersions. *J. Tribol. Trans. ASME* **125** 552–557.

[408] Rapoport L., Leshchinsky V., Lvovsky M., Nepomnyashchy O., Volovik Y. and Tenne R. 2002. Mechanism of friction of fullerenes. *Ind. Lubr. Tribol.* **54** 171–176.

[409] Kalin M., Kogovsek J. and Remskar M. 2012. Mechanisms and improvements in the friction and wear behavior using MoS2 nanotubes as potential oil additives. *Wear* **280** 36–45.

[410] Chou R., Battez A.H., Cabello J.J., Viesca J.L., Osorio A. and Sagastume A. 2010. Tribological behavior of polyalphaolefin with the addition of nickel nanoparticles. *Tribol. Int.* **43** 2327–2332.

[411] Verma A., Jiang W.P., Abu Safe H.H., Brown W.D. and Malshe A.P. 2008. Tribological behavior of deagglomerated active inorganic nanoparticles for advanced lubrication. *Tribol. Trans.* **51** 673–678.

[412] Ginzburg B.M., Shibaev L.A., Kireenko O.F., Shepelevskii A.A., Baidakova M.V. and Sitnikova A.A. 2002. Antiwear effect of fullerene c-60 additives to lubricating oils. *Russ. J. Appl. Chem.* **75** 1330–1335.

[413] Yan Q., Feng A.C., Zhang H.J., Yin Y.W. and Yuan J.Y. 2013. Redox-switchable supramolecular polymers for responsive self-healing nanofibers in water. *Polym. Chem.* **4** 1216–1220.

[414] Liu G., Li X., Qin B., Xing D., Guo Y. and Fan R. 2004. Investigation of the mending effect and mechanism of copper nano-particles on a tribologically stressed surface. *Tribol. Lett.* **17** 961–966.

[415] Xiao H.P. and Liu S.H. 2017. 2d nanomaterials as lubricant additive: A review. *Mater. Des.* **135** 319–332.

[416] Sanes J., Aviles M.D., Saurin N., Espinosa T., Carrion F.J. and Bermudez M.D. 2017. Synergy between graphene and ionic liquid lubricant additives. *Tribol. Int.* **116** 371–382.

[417] Lee J.H., Kim S.H., Cho D.H., Kim S.C., Baek S.G., Lee J.G., Kang J.M., Choi J.B., Seok C.S., Kim M.K., Koo J.C. and Lim B.S. 2012. Tribological properties of chemical vapor deposited graphene coating layer. *Korean J. Metals Mater.* **50** 206–211.

[418] Kim K.S., Lee H.J., Lee C., Lee S.K., Jang H., Ahn J.H. and Kim J.H. 2011. Chemical vapor deposition-grown graphene: The thinnest solid lubricant. *ACS Nano* **5** 5107–5114.

[419] Watanabe S., Noshiro J. and Miyake S. 2004. Tribological characteristics of WS2/MoS2 solid lubricating multilayer films. *Surf. Coat. Technol.* **183** 347–351.

[420] Song H., Ji L., Li H.X., Wang J.Q., Liu X.H., Zhou H.D. and Chen J.M. 2017. Self-forming oriented layer slip and macroscale super-low friction of graphene. *Appl. Phys. Lett.* **110** 073101.

[421] Yang Z.G., Zhang J.L., Kintner-Meyer M.C.W., Lu X.C., Choi D.W., Lemmon J.P. and Liu J. 2011. Electrochemical energy storage for green grid. *Chem. Rev.* **111** 3577–3613.

[422] Pomerantseva E. and Gogotsi Y. 2017. Two-dimensional heterostructures for energy storage. *Nat. Energy* **2** 17089.

[423] Chen K.S., Balla I., Luu N.S. and Hersam M.C. 2017. Emerging opportunities for two-dimensional materials in lithium-ion batteries. *ACS Energy Lett.* **2** 2026–2034.

[424] Zhang C.F., Park S.N., Seral-Ascaso A., Barwich S., McEyoy N., Boland C.S., Coleman J.N., Gogotsi Y. and Nicolosi V. 2019. High capacity silicon anodes enabled by MXene viscous aqueous ink. *Nat. Commun.* **10** 849.
[425] Gullapalli H., Kalaga K., Vinod S., Rodrigues M.T.F., George A. and Ajayan P.M. 2017. 2d material integrated macroporous electrodes for Li-ion batteries. *RSC Adv.* **7** 32737–32742.
[426] Zhang Y., Wu Z.F., Gao P.F., Zhang S.L. and Wen Y.H. 2016. Could borophene be used as a promising anode material for high-performance lithium ion battery? *ACS Appl. Mater. Interfaces* **8** 22175–22181.
[427] Li W.F., Yang Y.M., Zhang G. and Zhang Y.W. 2015. Ultrafast and directional diffusion of lithium in phosphorene for high-performance lithium-ion battery. *Nano Lett.* **15** 1691–1697.
[428] Sun D.D., Wang M.S., Li Z.Y., Fan G.X., Fan L.Z. and Zhou A.G. 2014. Two-dimensional Ti3C2 as anode material for Li-ion batteries. *Electrochem. Commun.* **47** 80–83.
[429] Jing Y., Zhou Z., Cabrera C.R. and Chen Z.F. 2013. Metallic VS2 monolayer: A promising 2d anode material for lithium ion batteries. *J. Phys. Chem. C* **117** 25409–25413.
[430] Zhang P.P., Wang F.X., Yu M.H., Zhuang X.D. and Feng X.L. 2018. Two-dimensional materials for miniaturized energy storage devices: From individual devices to smart integrated systems. *Chem. Soc. Rev.* **47** 7426–7451.
[431] Xia H.C., Xu Q. and Zhang J.A. 2018. Recent progress on two-dimensional nanoflake ensembles for energy storage applications. *Nano-Micro Lett.* **10** 66.
[432] Wang B., Luo B., Li X.L. and Zhi L.J. 2012. The dimensionality of Sn anodes in Li-ion batteries. *Mater. Today* **15** 544–552.
[433] Larcher D. and Tarascon J.M. 2015. Towards greener and more sustainable batteries for electrical energy storage. *Nat. Chem.* **7** 19–29.
[434] Goodenough J.B. 2015. Energy storage materials: A perspective. *Energy Stor. Mater.* **1** 158–161.
[435] Bruce P.G., Freunberger S.A., Hardwick L.J. and Tarascon J.M. 2012. Li-O-2 and Li-S batteries with high energy storage. *Nat. Mater.* **11** 19–29.
[436] Ji X.L. and Nazar L.F. 2010. Advances in Li-S batteries. *J. Mater. Chem.* **20** 9821–9826.
[437] Liang J., Sun Z.H., Li F. and Cheng H.M. 2016. Carbon materials for Li-S batteries: Functional evolution and performance improvement. *Energy Stor. Mater.* **2** 76–106.
[438] Fang X. and Peng H.S. 2015. A revolution in electrodes: Recent progress in rechargeable lithium-sulfur batteries. *Small* **11** 1488–1511.
[439] Chung H.C., Chang C.P., Lin C.Y. and Lin M.F. 2016. Electronic and optical properties of graphene nanoribbons in external fields. *Phys. Chem. Chem. Phys.* **18** 7573.
[440] Chung H.C., Lin Y.T., Lin S.Y., Ho C.H., Chang C.P. and Lin M.F. 2016. Magnetoelectronic and optical properties of nonuniform graphene nanoribbons. *Carbon* **109** 883.
[441] Zhu J.X., Yang D., Yin Z.Y., Yan Q.Y. and Zhang H. 2014. Graphene and graphene-based materials for energy storage applications. *Small* **10** 3480–3498.

[442] Chung H.C., Lee M.H., Chang C.P. and Lin M.F. 2011. Exploration of edge-dependent optical selection rules for graphene nanoribbons. *Opt. Express* **19** 23350–23363.

[443] Naguib M., Kurtoglu M., Presser V., Lu J., Niu J.J., Heon M., Hultman L., Gogotsi Y. and Barsoum M.W. 2011. Two-dimensional nanocrystals produced by exfoliation of Ti3AlC2. *Adv. Mater.* **23** 4248–4253.

[444] Dong C.W., Gao W., Jin B. and Jiang Q. 2018. Advances in cathode materials for high-performance lithium-sulfur batteries. *iScience* **6** 151–198.

[445] Liang X., Garsuch A. and Nazar L.F. 2015. Sulfur cathodes based on conductive mxene nanosheets for high-performance lithium-sulfur batteries. *Angew. Chem. Int. Ed.* **54** 3907–3911.

[446] Zhang Q.F., Wang Y.P., Seh Z.W., Fu Z.H., Zhang R.F. and Cui Y. 2015. Understanding the anchoring effect of two-dimensional layered materials for lithium-sulfur batteries. *Nano Lett.* **15** 3780–3786.

[447] Sun Z.H., Zhang J.Q., Yin L.C., Hu G.J., Fang R.P., Cheng H.M. and Li F. 2017. Conductive porous vanadium nitride/graphene composite as chemical anchor of polysulfides for lithium-sulfur batteries. *Nat. Commun.* **8** 14627.

[448] Liu T., Vivek J.P., Zhao E.W., Lei J., Garcia-Araez N. and Grey C.P. 2020. Current challenges and routes forward for nonaqueous lithium-air batteries. *Chem. Rev.* **120** 6558–6625.

[449] Christensen J., Albertus P., Sanchez-Carrera R.S., Lohmann T., Kozinsky B., Liedtke R., Ahmed J. and Kojic A. 2012. A critical review of Li/air batteries. *J. Electrochem. Soc.* **159** R1–R30.

[450] Jung H.G., Jeong Y.S., Park J.B., Sun Y.K., Scrosati B. and Lee Y.J. 2013. Ruthenium-based electrocatalysts supported on reduced graphene oxide for lithium-air batteries. *ACS Nano* **7** 3532–3539.

[451] Liu T., Leskes M., Yu W.J., Moore A.J., Zhou L.N., Bayley P.M., Kim G. and Grey C.P. 2015. Cycling Li-O-2 batteries via LiOH formation and decomposition. *Science* **350** 530–533.

[452] Luo W., Zhou L.H., Fu K., Yang Z., Wan J.Y., Manno M., Yao Y.G., Zhu H.L., Yang B. and Hu L.B. 2015. A thermally conductive separator for stable Li metal anodes. *Nano Lett.* **15** 6149–6154.

[453] Peng Z.Q., Freunberger S.A., Chen Y.H. and Bruce P.G. 2012. A reversible and higher-rate Li-O-2 battery. *Science* **337** 563–566.

[454] Whittingham M.S. 2004. Lithium batteries and cathode materials. *Chem. Rev.* **104** 4271–4301.

[455] Tarascon J.M. and Armand M. 2001. Issues and challenges facing rechargeable lithium batteries. *Nature* **414** 359–367.

[456] Shin W.K., Kannan A.G. and Kim D.W. 2015. Effective suppression of dendritic lithium growth using an ultrathin coating of nitrogen and sulfur codoped graphene nanosheets on polymer separator for lithium metal batteries. *ACS Appl. Mater. Interfaces* **7** 23700–23707.

[457] Foroozan T., Soto F.A., Yurkiv V., Sharifi-Asl S., Deivanayagam R., Huang Z.N., Rojaee R., Mashayek F., Balbuena P.B. and Shahbazian-Yassar R. 2018. Synergistic effect of graphene oxide for impeding the dendritic plating of Li. *Adv. Funct. Mater.* **28** 1705917.

[458] Feng X.N., Ren D.S., He X.M. and Ouyang M.G. 2020. Mitigating thermal runaway of lithium-ion batteries. *Joule* **4** 743–770.
[459] Feng X.N., Ouyang M.G., Liu X., Lu L.G., Xia Y. and He X.M. 2018. Thermal runaway mechanism of lithium ion battery for electric vehicles: A review. *Energy Stor. Mater.* **10** 246–267.
[460] Chung H.C. 2020. Engineering integrations, potential applications, and outlooks of li-ion batteries. *engrXiv*.
[461] Bandhauer T.M., Garimella S. and Fuller T.F. 2011. A critical review of thermal issues in lithium-ion batteries. *J. Electrochem. Soc.* **158** R1–R25.
[462] Balakrishnan P.G., Ramesh R. and Kumar T.P. 2006. Safety mechanisms in lithium-ion batteries. *J. Power Sources* **155** 401–414.
[463] Wang Q.S., Ping P., Zhao X.J., Chu G.Q., Sun J.H. and Chen C.H. 2012. Thermal runaway caused fire and explosion of lithium ion battery. *J. Power Sources* **208** 210–224.
[464] Spotnitz R. and Franklin J. 2003. Abuse behavior of high-power, lithium-ion cells. *J. Power Sources* **113** 81–100.
[465] Mortazavi B., Yang H.L., Mohebbi F., Cuniberti G. and Rabczuk T. 2017. Graphene or h-BN paraffin composite structures for the thermal management of Li-ion batteries: A multiscale investigation. *Appl. Energy* **202** 323–334.
[466] Liu K., Liu Y.Y., Lin D.C., Pei A. and Cui Y. 2018. Materials for lithium-ion battery safety. *Sci. Adv.* **4** eaas9820.
[467] Lim C.H., Jung Y.H., Yeom S.J., Lee H.W. and Kim D.K. 2017. Encapsulation of lithium vanadium phosphate in reduced graphene oxide for a lithium-ion battery cathode with stable elevated temperature performance. *Electrochim. Acta* **253** 208–217.
[468] Zhang F., Nemeth K., Bareno J., Dogan F., Bloom I.D. and Shaw L.L. 2016. Experimental and theoretical investigations of functionalized boron nitride as electrode materials for Li-ion batteries. *RSC Adv.* **6** 27901–27914.
[469] Zhang Q.Q., Liu K., Ding F. and Liu X.J. 2017. Recent advances in solid polymer electrolytes for lithium batteries. *Nano Res.* **10** 4139–4174.
[470] Yuan M.Y., Erdman J., Tang C.Y. and Ardebili H. 2014. High performance solid polymer electrolyte with graphene oxide nanosheets. *RSC Adv.* **4** 59637–59642.
[471] IEA. 2020. Global EV outlook 2020 Report.
[472] Harper G., Sommerville R., Kendrick E., Driscoll L., Slater P., Stolkin R., Walton A., Christensen P., Heidrich O., Lambert S., Abbott A., Ryder K.S., Gaines L. and Anderson P. 2019. Recycling lithium-ion batteries from electric vehicles. *Nature* **575** 75–86.
[473] Dehghani-Sanij A.R., Tharumalingam E., Dusseault M.B. and Fraser R. 2019. Study of energy storage systems and environmental challenges of batteries. *Renew. Sustain. Energy Rev.* **104** 192–208.
[474] Ahmadi L., Young S.B., Fowler M., Fraser R.A. and Achachlouei M.A. 2017. A cascaded life cycle: Reuse of electric vehicle lithium-ion battery packs in energy storage systems. *Int. J. Life Cycle Assess.* **22** 111–124.

[475] Qiao D.H., Wang G.S., Gao T.M., Wen B.J. and Dai T. 2021. Potential impact of the end-of-life batteries recycling of electric vehicles on lithium demand in China: 2010-2050. *Sci. Total Environ.* **764** 142835.

[476] Fan E.S., Li L., Wang Z.P., Lin J., Huang Y.X., Yao Y., Chen R.J. and Wu F. 2020. Sustainable recycling technology for Li-ion batteries and beyond: Challenges and future prospects. *Chem. Rev.* **120** 7020–7063.

[477] Chung H.C. and Cheng Y.C. 2019. Action planning and situation analysis of repurposing battery recovery and application in China. *J. Taiwan Energy* **6** 425.

[478] Hua Y., Zhou S.D., Huang Y., Liu X.H., Ling H.P., Zhou X.N., Zhang C. and Yang S.C. 2020. Sustainable value chain of retired lithium-ion batteries for electric vehicles. *J. Power Sources* **478** 228753.

[479] Olivetti E.A., Ceder G., Gaustad G.G. and Fu X.K. 2017. Lithium-ion battery supply chain considerations: Analysis of potential bottlenecks in critical metals. *Joule* **1** 229–243.

[480] Chung H.C. and Cheng Y.C. 2020. Summary of safety standards for repurposing batteries. *Monthly J. Taipower's Eng.* **860** 35.

[481] Winslow K.M., Laux S.J. and Townsend T.G. 2018. A review on the growing concern and potential management strategies of waste lithium-ion batteries. *Resour. Conserv. Recycl.* **129** 263–277.

[482] Chung H.C. 2021. Charge and discharge profiles of repurposed LiFePO4 batteries based on the UL 1974 standard. *Sci. Data* **8** 165.

[483] White C., Thompson B. and Swan L.G. 2020. Repurposed electric vehicle battery performance in second-life electricity grid frequency regulation service. *J. Energy Stor.* **28** 101278.

[484] Bobba S., Mathieux F., Ardente F., Blengini G.A., Cusenza M.A., Podias A. and Pfrang A. 2018. Life cycle assessment of repurposed electric vehicle batteries: An adapted method based on modelling energy flows. *J. Energy Stor.* **19** 213–225.

[485] Assuncao A., Moura P.S. and de Almeida A.T. 2016. Technical and economic assessment of the secondary use of repurposed electric vehicle batteries in the residential sector to support solar energy. *Appl. Energy* **181** 120–131.

[486] Huang B., Pan Z.F., Su X.Y. and An L. 2018. Recycling of lithium-ion batteries: Recent advances and perspectives. *J. Power Sources* **399** 274–286.

[487] Swain B. 2017. Recovery and recycling of lithium: A review. *Sep. Purif. Technol.* **172** 388–403.

[488] Xu J.Q., Thomas H.R., Francis R.W., Lum K.R., Wang J.W. and Liang B. 2008. A review of processes and technologies for the recycling of lithium-ion secondary batteries. *J. Power Sources* **177** 512–527.

[489] Shi L. and Zhao T.S. 2017. Recent advances in inorganic 2d materials and their applications in lithium and sodium batteries. *J. Mater. Chem. A* **5** 3735–3758.

[490] Wang X.W., Sun Y.H. and Liu K. 2019. Chemical and structural stability of 2d layered materials. *2D Mater.* **6** 042001.

[491] Mannix A.J., Kiraly B., Hersam M.C. and Guisinger N.P. 2017. Synthesis and chemistry of elemental 2d materials. *Nat. Rev. Chem.* **1** 0014.

[492] Ma L., Wang J.L. and Ding F. 2013. Recent progress and challenges in graphene nanoribbon synthesis. *ChemPhysChem* **14** 47–54.
[493] Zhu Y., Peng L.L., Fang Z.W., Yan C.S., Zhang X. and Yu G.H. 2018. Structural engineering of 2d nanomaterials for energy storage and catalysis. *Adv. Mater.* **30** 1706347.
[494] Chen D.Y., Chen W.X., Ma L., Ji G., Chang K. and Lee J.Y. 2014. Graphene-like layered metal dichalcogenide/graphene composites: Synthesis and applications in energy storage and conversion. *Mater. Today* **17** 184–193.
[495] Margine E., Lambert H. and Giustino F. 2016. Electron-phonon interaction and pairing mechanism in superconducting Ca-intercalated bilayer graphene. *Sci. Rep.* **6** 1–6.
[496] Cohen M.L. and Louie S.G. 2016. *Fundamentals of Condensed Matter Physics* (Cambridge University Press).
[497] Dresselhaus G., Dresselhaus M.S. and Saito R. 1998. *Physical Properties of Carbon Nanotubes* (World Scientific).
[498] Togo A. and Tanaka I. 2015. First principles phonon calculations in materials science. *Scr. Mater.* **108** 1–5.
[499] Kittel C. 1976. *Introduction to Solid State Physics.*
[500] Al-Jishi R. and Dresselhaus G. 1982. Lattice-dynamical model for alkali-metal-graphite intercalation compounds. *Phys. Rev. B* **26** 4523.
[501] Miyamoto Y., Cohen M.L. and Louie S.G. 1995. Ab initio calculation of phonon spectra for graphite, BN, and BC 2 N sheets. *Phys. Rev. B* **52** 14971.
[502] Giustino F., Cohen M.L. and Louie S.G. 2007. Electron-phonon interaction using Wannier functions. *Phys. Rev. B* **76** 165108.
[503] Giustino F. 2017. Electron-phonon interactions from first principles. *Rev. Mod. Phys.* **89** 015003.
[504] Hauser W. and Feshbach H. 1952. The inelastic scattering of neutrons. *Phys. Rev.* **87** 366.
[505] Wakabayashi N., Sinha S. and Spedding F. 1971. Phonon spectrum of scandium metal by inelastic scattering of neutrons. *Phys. Rev. B.* 4 2398.
[506] Maultzsch J., Reich S., Thomsen C., Requardt H. and Ordejón P. 2004. Phonon dispersion in graphite. *Phys. Rev. Lett.* **92** 075501.
[507] Nika D.L. and Balandin A.A. 2017. Phonons and thermal transport in graphene and graphene-based materials. *Rep. Prog. Phys.* **80** 036502.
[508] Parlinski K., Li Z. and Kawazoe Y. 1997. First-principles determination of the soft mode in cubic ZrO 2. *Phys. Rev. Lett.* **78** 4063.

# Chapter 13

# PROBLEMS

Lu-Yao Wang, Hsien-Ching Chung, Ching-Hong Ho, Chiun-Yan Lin,
Chih-Wei Chiu, and Ming-Fa Lin

Problem 1: The free-carrier screening phenomenon for 3D [1], 2D [2], layered [3] and 1D-nanotube electron gases [4] in the absence of crystal structures occurs through bi-particle electron–electron Coulomb interactions. The self-consistent field random-phase approximation (RPA) [5] is available for fully understanding the linear relations between the effective Coulomb field/potential and the induced charge density fluctuations in the momentum- and energy-transfer spaces. The longitudinal dielectric function, being well defined as the ratio of field/potential/charge density before and after Coulomb perturbations [6], under RPA gives the analytic formulas. (I) Evaluate their imaginary-part dielectric functions [7] and then the real-part ones through the Kramers–Kroning relations [8]. (II) Thoroughly examine the electron–hole excitation ranges [9] and the Landau dampings associated with the inelastic Coulomb scatterings due to the Fermi-momentum states [10]. (III) Plot the (momentum, frequency)-related Coulomb excitation spectra using the whole intensities of energy loss functions [11]. The close relations between the single- and many-particle Coulomb excitations are discussed in detail, e.g., the critical momenta for the emergence and absence of coherent conduction electron oscillations (plasmon modes in Ref. [12]). (IV) The dimensionality-enriched charge screenings cover the total vacant regions [13], the optical or acoustic plasmons at the long-wavelength limit [14] and the coexistence of single-particle and collective excitations [12]. (V) The sensitive dependences of EGS densities can greatly diversify the featured dynamic/static quasiparticle screenings [1–4]. (VI) Check the specific sum rules related to the bare screening response function of $Im[\epsilon(q,\omega)]$ [11]. Clearly illustrate their physical pictures [15, 16].

Problem 2: It is very interesting to fully understand the charge screening behaviors due to free or valence carriers, as clearly illustrated in the standard textbooks on electric and magnetic fields [17]. Classic electrodynamics can only deal with the dielectric behaviors of semiconductors/insulators using roughly macroscopic viewpoints [17]. The induced multi-pole electric poles, which are created by linear perturbations [17], are proposed to properly define the dielectric constant, but not the (momentum, frequency)-related dielectric functions. It is impossible to adopt this strategy to directly observe the rather quick screening of free conduction electrons in 3D/2D/1D metals since its predictions cannot effectively terminate the long-range Coulomb-field interactions. That is, the classical theory cannot change the sensitive radius dependence [18]. After including the inelastic scatterings of the Fermi-momentum states, the dimensionality-enriched dielectric functions will show unusual screening phenomena due to a charged impurity. For example, dilute magnesium adatoms in bcc sodium metal is capable of establishing a donor-type Coulomb field, (a) experience very efficient charge screenings associated with all conduction electrons [18], (b) display a quickly reduced effective field/potential (the Yukawa potential after charge screenings) [19], and (c) the opposite charges between the perturbed and screened ones (the singular static dielectric function at the long-wavelength limit) [20]. A 3D EGS is very suitable for describing the featured (a), (b) and (c), for which the effective screening length of sodium metal is about 1 Å [19]. (d) Most importantly, the main features of the Fermi-momentum states ($|\mathbf{k_F}|$) [18] have been included in the bare and screened response functions and thus generate the singular scattering behaviors driven by them at the momentum transfers of $|\mathbf{q}| = 2k_F$. (e) This further leads to the well-known Friedel charge oscillations at long distances. The induced charge is conserved by the unusual fluctuations of electron and hole clouds. However, the long-range charge density waves are absent under the Thomas–Fermi model without a discontinuous distribution of Fermi liquid at zero temperature [18]. It is worth conducting systematic investigations on the (f) similarities and (g) differences among 3D, 2D, layered and 1D electron gases in terms of (h) distance- and (i) density-dependent total Coulomb potentials and screening charge distributions [18].

Problem 3: (a) Express the total charge using the carrier density distribution in the momentum space. (b) Charge screenings of metals (Problem 2), which are evaluated through RPA [21], show an infinite static dielectric function under the long-wavelength limit. (c) Such singular behaviors arise in all dimensions [22]. (d) This result is responsible for the conservation of screening charges, as clearly illustrated by both RPA and the Thomas–Fermi model [18].

Problem 4: There are certain important differences between a 2D electron gas and n-type doped graphene in the charged screening behaviors. RPA is very suitable for describing their momentum and frequency dependences, in which the analytic formulas could be obtained through delicate calculations [23]. The latter possesses free conduction and valence $\pi$ electrons, in which the $\pi$-electronic band-structure effects are covered through the tight-binding model of the single-$2p_z$-orbital Hamiltonian [24]. (a) Evaluate the static dielectric functions of two metallic systems and explore their significant similarities and differences. (b) Propose concise screening pictures. (c) Discuss the featured charge screening behaviors in terms of the short, middle and long distances using the effective Coulomb potentials and the induced charge density distributions [25]. (d) The sensitive dependences on Fermi levels are also included in the detailed discussions. (e) The impurity-field scatterings are available for fully comprehending the residual transport properties in any condensed matter system. Their direct combinations with the well-known Boltzmann equation [26], as done for alkali-adatom graphite intercalation compounds, can present the electrical resistivity at zero temperature [27]. (f) Similar studies could be generalized to 1D nanotube electron gas and carbon nanotubes [28].

Problem 5: Consider the magnetic quantization phenomena in 3D, 2D and 1D-nanotube and 1D-nanoribbon electron gases [29–31]. When a uniform magnetic field is perpendicular to the $x$–$y$ plane, it might induce quantized electronic states by solving the critical Hamiltonians only with the kinetic energies according to the translational and rotational symmetries. The vector potentials under suitable gauges play critical roles from perturbation viewpoints. By the exact diagonalization of the magnetic Hamiltonians [32], the magneto-electronic energy spectra and wave functions are revealed in the wavevector [33] and magnetic field strength dependences [34]. (a) Thorough discussions are conducted on the emergence of dispersionless Landau levels in four kinds of EGSs (LLs in Ref. [10]) through the rich non-crossing/crossing/anticrossing behaviors (b) The critical magnetic field strengths, which create the unusual LLs, are quite different among the dimension- and symmetry-based systems. (c) Plot and illustrate the spatial probability distributions, in which the magneto-electronic oscillatory modes are obviously revealed within the standard magnetic length [10]. (d) The well-behaved oscillation zero points are sufficient for characterizing a good quantum number. $n$ will be very useful in accounting for the magneto-selection rules during (e) the Coulomb excitations [35], (f) magneto-optical absorption spectra [36] and (g) the Hall quantum transports [37].

Problem 6: By setting a magnetic field in any direction, all magnetic properties of 1D-nanotube EGSs/carbon nanotubes [38] are greatly diversified by the drastic changes in vector potential in a 1D cylindrical surface [39]. The significant couplings between an external magnetic field and the crystal symmetry are responsible for the well-known Aharovnov–Bohm effects and the magnetically quantized LL states [40]. More interestingly, there exist the strong competition/cooperation phenomenon between these two pristine behaviors during the full variation of field strength and direction. (a) The induced magnetization, defined as the negative first derivative of ground state energy versus magnetic field [41], can be first evaluated under a parallel case. (b) The angular momenta, which are created by the periodic boundary condition, are rather suitable for describing each discrete energy subband. They are drastically changed by the magnetic flux through a nanoscale nanotube, in which their periodic relations can lead to the AB effects, as verified by optical [42] and transport [43] examinations. (c) Explore the carrier-density dependences for the paramagnetic [44] and diamagnetic magnetisms [45]. Furthermore, the singular magnetic responses should be identified from the magnetic phase transitions [46]. (d) Similar calculations could be generalized for any magnetic field direction. The magnetic configurations are expected to become more rich and complicated in a hollow cylinder [39]. (e) It would be very difficult to observe the magnetic quantization phenomena in the perpendicular cases, mainly owing to the vanishing net magnetic flux [40]. (f) This is directly reflected in the other essential magnetic properties, e.g., the unique LL-related magneto-plasmon modes with/without significant inter-LL dampings [47], the existence/absence of magneto-optical selection rules [33] and the quantum Hall transports [37]. (g) Investigate certain important differences between 1D carbon nanotubes and nanotube electron gases in terms of the above-mentioned quasiparticle behaviors [48].

Problem 7: Consider the Coulomb decay rates of the excited conduction electrons and holes in 3D [49], 2D [50], layered [51] and 1D-nanotube electron gases [52]. After Coulomb- or/and electromagnetic-field perturbations, the excited quasiparticle states are generated during the dynamic excitation processes. Their lifetimes and energy spectra are the focus of study for condensed matter systems, which are examined through high-resolution angle-resolved photoemission spectroscopy [53]. Two kinds of approaches, the self-energy method [54] and the screened Fermi golden rule [55], are available for fully understanding the inelastic electron–electron Coulomb scatterings. (a) Evaluate the critical conditions of the Coulomb decays above and below the Fermi level at zero temperature under the conservation of momentum and energy transfers and the Pauli exclusion principle. (b) Clearly illustrate the dimension- and symmetry-enriched de-excitation mechanisms. (c) Qualitatively determine three types of sensitive dependences near/at the Fermi-momentum states, including wavevectors [56], energies [57] and temperatures [58], e.g., $|E - E_F|^2$, $|E - E_F|^2 ln|(E - E_F)|$ and $|E - E_F|$ in 3D, 2D and 1D electron gases, respectively [56, 59, 60]. (d) Thoroughly discuss the single-particle de-excitation channels and the absence of coherent charge oscillations. (e) Express the close relations between the theoretical predictions [61] and the experimental observations [62], and propose the developed viewpoints about the quasiparticle lifetimes [63].

Problem 8: Consider the quasiparticle lifetimes and energy spectra in $n$-type monolayer graphene and graphite intercalation compounds (e.g., the stage-1 alkali graphite intercalation compounds in Fig. 12.5) [64]. The many-body effects might be quite different from those of 2D [65] and layered [66] electron gases in terms of wavevector and carrier-density dependences. As for 2D graphene, (a) separate the screened exchange energy into three parts: (I) an exchange energy, (II) the real line part and (III) the imaginary residual part. (b) Its Coulomb decay rate is inversely proportional to the imaginary-part residual self-energy. This analytic formula could be utilized to describe the inelastic Coulomb scatterings for the excited conduction electrons, conduction holes and valence holes, in which the first and second/third quasiparticles are above/below the Dirac point. (c) Propose certain important differences among the three kinds of quasiparticles through the available de-excitation channels. (d) Present the main features of the Coulomb decay rates, including the anisotropic, non-monotonous and unusual dependences along the high-symmetry paths. (e) Clearly illustrate the unique quasiparticle decay behaviors using the rich (momentum, frequency)-phase diagrams. Roughly, a superlattice model is suitable for the low-energy essential properties of the stage-1 graphite alkali compounds, e.g., the Coulomb excitations and decays [67]. The weak van der Waals interactions are ignored in the band-structure calculations, but the opposite is true for the interlayer Coulomb interactions. (f) Both intralayer and interlayer electron–electron interactions play critical roles in creating the plasmon band structures. (g) Compare the theoretical predictions with experimental observations [68] on the energy loss spectra, where the screened response function must be well defined using the total inelastic scattering probability in the presence of all charge screenings [69]. (h) Evaluate and explore the lifetimes of all excited quasiparticles and their energy spectra. (i) Show the dimensionality-enriched quasiparticle behaviors. (j) Explain the differences in the band-structure effects between Problems 8 and 7.

Problem 9: Calculate the transmittance and reflectance by following the same methods used in the four kinds of graphene systems (Chapter 6) for (a) silicene [70], (b) germanene [71] and (c) tinene [72]. Select one material of monolayer type, (a), (b), or (c), for calculation.

Problem 10: Describe the possible influences of temperature ($T > 0$) and doping ($E_F \neq 0$) on the calculated results of transmittance and reflectance in graphene systems.

Problem 11: Consider a two-band insulator in a 2D lattice, with the Hamiltonian $H(\mathbf{k}) = \epsilon_0(\mathbf{k})\sigma_0 + \mathbf{d}(\mathbf{k}) \cdot \vec{\sigma}$, on the basis of Pauli matrices. Here, the real $\mathbf{d}(\mathbf{k}) \neq 0$ is presumed over the entire 2D Brillouin zone to avoid the existence of the singularity of nodes. Now, define the unit vector field $\mathbf{d}(\hat{\mathbf{k}}) = \mathbf{d}(\mathbf{k})/|\mathbf{d}(\mathbf{k})| = (\cos\phi_\mathbf{k}\sin\theta_\mathbf{k}, \sin\phi_\mathbf{k}\sin\theta_\mathbf{k}, \cos\theta_\mathbf{k})^T$ in terms of the spherical coordinates.

(1) Derive the eigenspinors. Use the result to derive the relevant expressions of the Berry connection, the Berry flux and the first Chern number.
(2) Recalling knowledge about the solid angle, derive the winding number of $\mathbf{d}(\hat{\mathbf{k}})$ over the unit sphere. Show that the result is identical to the Chern number acquired in (1).

Problem 12: If the two-band model has generically a pair of nodal points at $\mathbf{K}^{\pm}$ as a result of certain symmetries, where $\mathbf{d}(\mathbf{K}^{\pm}) = 0$, one usually makes an effective Hamiltonian $H_{\xi}(\mathbf{q})$ in each valley about the nodal point.

(1) Given by $H_{\xi}(\mathbf{q}) = \xi q_x \sigma_x + q_y \sigma_y$, with $\xi = \pm$ corresponding to $\mathbf{K}^{\pm}$ (recall the case of graphene), separately derive the Berry phases, or winding numbers, along arbitrary loops around $\mathbf{K}^{\pm}$. Show that the total Berry phase equals zero by adding the two separate Berry phases.
(2) Add $\exp ik_z$ as a gauge to both the effective Hamiltonians $H_{\xi}(\mathbf{q})$. Show the presence of nodal lines in the 3D Brillouin zone. Repeat as in (1) for massless 2D Dirac fermions at fixed $k_z$.
(3) As we know that 2D $Z_2$ topological insulators are related to two Chern insulators with respect to two spin orientations, from which the total Chern number equals zero, analogously, show how graphene with Dirac nodal points is recognized as a $Z_2$ Dirac semimetal.

Problem 13: Again, consider the two-band model with a pair of Dirac points at $\mathbf{K}^{\pm}$. Consider the zero-mode LL, where only massless 2D Dirac fermions near the Dirac points are involved. Hence, the effective Hamiltonians are given by $H_{\xi}(\vec{\pi}) = \xi \pi_x \sigma_x + \pi_y \sigma_y$ under the Peierls substitution, $\vec{\pi} = \mathbf{q} + e\mathbf{A}$, with $\mathbf{A}$ being a vector potential. Now, the Hamiltonian $H(\vec{\pi}) = \pi_x \sigma_x \tau_z + \pi_y \sigma_y \tau_0$ in the representation $(c_{A\mathbf{K}^+} c_{B\mathbf{K}^+} c_{A\mathbf{K}^-} c_{B\mathbf{K}^-})^T$ is useful, where $A$ and $B$ are the basis of the lattice Hamiltonian, e.g., the sublattices of graphene.

(1) Exhaust all possible chiral operators for $H(\vec{\pi})$ to show if any existing chiral operators cause inter-valley coupling. Does the Hamiltonian $H(\vec{\pi})$ still respect chiral symmetry, leaving the Dirac fermions with different valley flavors $\xi$ decoupled?
(2) Diagonalize $H(\vec{\pi})$. Describe the obtained zero-mode LL.
(3) Try your best to interpret how the zero-mode LL is protected.

Problems 14: For the lattice tight-binding model, as shown in Fig. 13.1, calculate the energy $E$ for the infinite one-dimensional lattice. The effective Hamiltonian is given by

$$H = -t \sum_{n} (|n+1\rangle + |n-1\rangle)\langle n|, \tag{1}$$

and the eigenstate of the system is a Bloch state given by

$$\Psi = N \sum_{m} e^{ikma} |m\rangle. \tag{2}$$

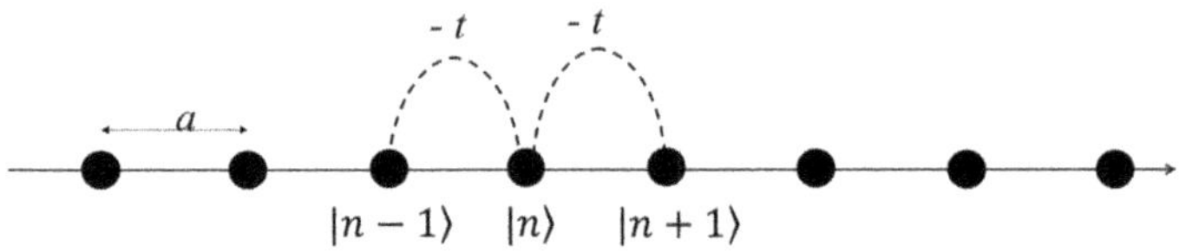


Fig. 13.1 The one-dimensional lattice model has the lattice constant $a$ and "| . . ." denotes the state at the $n$th site.

Here, $N$ is a normalized constant and $k$ is the wavevector parallel to the lattice direction. Assuming that the constant hopping energy is $-t$ for the nearest-neighbor hopping. (Answer: $E(k) = -2t\cos(ka)$.) (Hint: Apply the Schrödinger equation $H|\Psi\rangle = E|\Psi\rangle$ and the orthogonality given by $\langle m|n\rangle = \delta_{m,n}$.)

Problems 15: From Problem 14, obtain the normalized constant given by $N = \sqrt{\frac{1}{N_x}}$, where $N_x$ is the total number of the lattice sites.

Problems 16: To realize the low-energy spin-polarized magneto-electronic properties of buckled monolayer GaAs, the generalized tight-binding model is suggested. Before we go into the model, we should take a look at the geometric structure of the buckled monolayer GaAs. Two types of descriptions are given as follows; it is suggested that the readers draw the geometric structure by hand or using software (e.g., AutoCAD, 3dsMAX, SolidWorks, ProE) for a better understanding of the structure. (3D drawing applications with the advantage of rotating the 3D model are suitable for this work.) Brief description: Monolayer GaAs has a buckled honeycomb lattice with each atom being passivated by an F atom [73]. Detailed description: Both $sp^3$-hybridized Ga and As atoms bond to four atoms (three for As or Ga; one for F) and the Ga–As bond length is about 2.521 Å. The primitive unit vectors $\mathbf{a}_1$ and $\mathbf{a}_2$ are along the two sides of rhombus, with a lattice constant of $a = 4.226$ Å. The altitude of the buckled structure measured from the distance between the Ga and As planes is $l_z = 0.633$ Å. The buckling angle $\theta$ between the Ga–As bond and the $z$-axis is about 104.54°. The F atoms are right above (below) the Ga (As) atoms with the Ga–F and As–F distances of 1.776 and 1.781 Å, respectively [74].

The generalized tight-binding model is developed to study the essential properties of buckled monolayer GaAs, where many critical factors, such as the buckled structure, multi-orbital hybridization, spin–orbit interaction (SOI), electric field and magnetic field, can be involved in the calculation simultaneously. It provides a theoretical framework to study the competition among various critical factors and affords systematic studies of a variety of materials, ranging from multi-dimensional materials to hybrid systems. In buckled monolayer GaAs, the Hamiltonian is built from the tight-binding functions of $4s$, $4p_x$ and $4p_y$ orbitals [73]. It is a good practice for readers to build the Hamiltonian and write down the Hamiltonian matrix elements, which is beneficial for understanding and implementing the generalized tight-binding model. The following order of practice is suggested: (1) Hamiltonian without external fields and SOI: A $6\times6$ Hamiltonian matrix is obtained. (2) Hamiltonian with external electric field and without SOI: On-site energy is considered. A $6\times6$ Hamiltonian matrix is obtained. (3) Hamiltonian with external magnetic field and without SOI: Peierls substitution is applied. An enlarged Hamiltonian matrix is obtained. (4) Hamiltonian without external fields and with SOI: A $12\times12$ Hamiltonian matrix is obtained. (5) Hamiltonian with external electric field and SOI: On-site energy is considered. A $12\times12$ Hamiltonian matrix is obtained. (6) Hamiltonian with external magnetic field and SOI: Peierls substitution is applied. An enlarged Hamiltonian matrix is obtained.

Problems 17: The boundary plays an important role in the essential properties of materials. For example, graphene nanoribbons exhibit width- and edge-dependent electronic properties, e.g., the low-energy band structure and the corresponding energy gap and optical transitions in the armchair, zigzag and chiral directions. The energy gap exhibits a tendency to enlarge as the width increases [75]. Armchair and zigzag graphene nanoribbons are predicted to have entirely different optical selection rules [76]. In the system of buckled monolayer GaAs, edge-dependent features are also predicted in the buckled monolayer GaAs nanoribbons. It is a good practice for readers to draw the geometric structure to realize the boundary and periodic conditions in the system and then establish the Hamiltonian by following the suggested steps in the previous problem.

## References

[1] Lin M.F. and Shung K.W.K. 1992. Screening of charged impurities in graphite intercalation compounds. *Phys. Rev. B* **46** 12656–12663.

[2] Terekhov I.S., Milstein A.I., Kotov V.N. and Sushkov O.P. 2008. Screening of coulomb impurities in graphene. *Phys. Rev. Lett.* **100** 076803.

[3] Mohammadi Y. 2015. Charge screening and carrier transport in AA-stacked bilayer graphene: Tuning via a perpendicular electric field. *Solid State Commun.* **202** 14–18.
[4] Lin M.F. and Shung K.W.K. 1993. Elementary excitations in cylindrical tubules. *Phys. Rev. B* **47** 6617: Chen S., Hseih W. and Lin M.F. 2005. Charge screening of single-walled carbon nanotubes in a uniform transverse electric field. *Phys. Rev. B* **72** 193412.
[5] Ehrenreich H. and Cohen M.H. 1959. Self-consistent field approach to the many-electron problem. *Phys. Rev.* **115** 4.
[6] Lin M.F. and Shung K.W.-K. 1993. Elementary excitations in cylindrical tubules. *Phys. Rev. B* **47** 6617.
[7] Lin M.F. and Shung K.W.-K. 1996. Self-energy of electrons in graphite intercalation compounds. *Phys. Rev. B* **53** 1109.
[8] Grundmann M. 2010. *The Physics of Semiconductors* (Springer).
[9] Huang C.S., Lin M.F. and Chuu D.S. 1997. Collective excitations in graphite layers. *Solid. State. Communs.* **103** 603–606.
[10] Lin C.Y., Ho C.H., Wu J.Y., Do T.N., Shih P.H., Lin S.Y. and Lin M.F. 2019. *Diverse Quantization Phenomena in Emergent Layered Materials* (CRC Press).
[11] Lin M.F. 2000. Optical spectra of single-wall carbon nanotube bundles. *Phys. Rev. B* **62** 13153.
[12] Lin M.F., Chuu D.S., Huang C.S., Lin Y.K. and Shung K.W.-K. 1996. Collective excitations in a single-layer carbon nanotube. *Phys. Rev. B* **53** 15493.
[13] Ma J., Alfe D., Michaelides A. and Wang E. 2009. Stone-Wales defects in graphene and other planar $sp^2$-bonded materials. *Phys. Rev. B* **80** 033407.
[14] Shung K.W.-K. 1986. Dielectric function and plasmon structure of stage-1 intercalated graphite. *Phys. Rev. B* **34** 979.
[15] Shyu F.L. and Lin M.F. 2002. Electronic and optical properties of narrow-gap carbon nanotubes. *J. Phys. Soc. Jpn.* **71** 1820–1823.
[16] Lu C.L., Chang C.P., Huang Y.C., Chen R.B. and Lin M.F. 2006. Influence of an electric field on the optical properties of few-layer graphene with AB stacking. *Phys. Rev. B* **73** 144427.
[17] Jackson J.D. 1975. *Classical Electrodynamics* (Wiley).
[18] Mahan G.D. 1990. *Many-Particle Physics* (Plenum).
[19] Kittel C. 1996. *Introduction to Solid State Physics*, p. 402 (New York: John Wiley & Sons).
[20] Zeppenfeld K. 1969. Wavelength dependence and spatial dispersion of the dielectric constant in graphite by electron spectroscopy. *Opt. Com.* **1** 119.
[21] Ren X., Rinke P., Joas C. and Scheffler M. 2012. Random-phase approximation and its applications in computational chemistry and materials science. *J. Mater. Sci.* **47** 7447–7471.
[22] Lindhard J. 1954. On the properties of a gas of charged particles. *Danske Matematisk-fysiske Meddeleiser.* **28** 8.
[23] Schulze H.-J., Schuck P. and Van Giai N. 2000. Two-dimensional electron gas in the random-phase approximation with exchange and self-energy corrections. *Phys. Rev. B* **61** 8026.

[24] Reich S., Maultzsch J. and Thomsen C. 2002. Tight-binding description of graphene. *Phys. Rev. B* **66** 035412.

[25] Ho J.H., Chang C.P. and Lin M.F. 2006. Electronic excitations of the multilayered graphite. *Phys. Lett. A* **352** 446.

[26] Hai G.Q., Studart N., Peeters F.M., Koenraad P.M. and Wolter J.H. 1996. Intersubbandcoupling and screening effects on the electron transport in a quasi-two-dimensional s-doped semiconductor system. *J. Appl. Phys.* **80** 5809–5814.

[27] Lin M.F. and Shung K.W.-K. 1992. The residual resistivity of a layered system: Intercalated graphite compounds. *Annual Meeting of the Physical Society of the ROC.*

[28] Fischer J.E., Dai H., Thess A., Lee R., Hanjani N.M., Dehaas D.L. and Smalley R.E. 1997. Metallic resistivity in crystalline ropes of single-wall carbon nanotubes. *Phys. Rev. B* **55** R4921.

[29] Jeon S., Zhou B.B., Gyenis A., Feldman B.E., Kimchi I., Potter A.C., Gibson Q.D., Cava R.J., Vishwanath A. and Yazdani A. 2014. Landau quantization and quasiparticle interference in the three-dimensional dirac semimetal $Cd_3As_2$. *Nat. Mater.* **13** 851–856.

[30] Do T.N., Shih P.H., Gumbs G., Huang D., Chiu C.W. and Lin M.F. 2018. Diverse magnetic quantization in bilayer silicene. *Phys. Rev. B* **97** 125416.

[31] Chung H.C., Chang C.P., Lin C.Y. and Lin M.F. 2016. Electronic and optical properties of graphene nanoribbons in external fields. *Phys. Chem. Chem. Phys.* **18** 7573.

[32] Chen S.C., Wu J.Y., Lin C.Y. and Lin M.F. 2017. *Theory of Magnetoelectric Properties of 2D Systems*. IOP Concise Physics (San Raefel, CA, USA: IOP Publishing; Morgan & Claypool Publishers).

[33] Ou Y.C., Chiu Y.H., Yang P.H. and Lin M.F. 2014. The selection rule of graphene in a composite magnetic field. *Opt. Express* **22** 7473.

[34] Ou Y.C., Sheu J.K., Chiu Y.H., Chen R.B. and Lin M.F. 2011. Influence of modulated fields on the Landau level properties of graphene. *Phys. Rev. B* **83** 195405.

[35] Lin M.F. and Shyu F.L. 2000. Low-frequency electronic excitations in doped carbon nanotubes. *Physica B* **292** 117–126.

[36] Lin M.F. and Shung K.W.-K. 1994. Plasmons and optical properties of carbon nanotubes. *Phys. Rev. B* **50** 17744–17747.

[37] Zuev Y.M., Chang W. and Kim P. 2009. Thermoelectric and magnetothermoelectric transport measurements of graphene. *Phys. Rev. Lett.* **102** 096807.

[38] Tsai C.C., Chen S.C., Shyu F.L., Chang C.P. and Lin M.F. 2005. Magnetization of carbon nanotubes. *Physica E* **30** 86–92.

[39] Lin M.F. and Shung K.W.K. 1995. Magnetoconductance of carbon nanotubes. *Phys. Rev. B* **51** 7592.

[40] Ajiki H. and Ando T. 1994. Aharonov-Bohm effect in carbon nanotubes. *Physica B* **201** 349–352.

[41] Huang Y.K., Chen S.C., Ho Y.H., Lin C.Y. and Lin M.F. 2014. Feature-rich magnetic quantization in sliding bilayer graphenes. *Sci. Rep.* **4** 7509.

[42] Lin Y.P., Lin C.Y., Ho Y.H., Do T.N. and Lin M.F. 2015. Magneto-optical properties of ABC-stacked trilayer graphene. *Phys. Chem. Chem. Phys.* **17** 15921.

[43] Do T.N., Chang C.P., Shih P.H. and Lin M.F. 2017. Stacking-enriched magneto-transport properties of few-layer graphenes. *Phys. Chem. Chem. Phys.* **19** 29525.
[44] Vo T.H., Shekhirev M., Kunkel D.A., Morton M.D., Berglund E., Kong L., Wilson P.M., Dowben P.A., Enders A. and Sinitskii A. 2014. Largescale solution synthesis of narrow graphene nanoribbons. *Nat. Comm.* **5** 3189.
[45] Tsai C.C., Shyu F.L., Chiu C.W., Chang C.P., Chen R.B. and Lin M.F. 2004. Magnetization of armchair carbon tori *Phys. Rev. B* **70** 075411.
[46] Landau L.D. and Lifschitz E.M. 1958. *Course of Theoretical Physics*, vol. 5 (London: Pergamon).
[47] Politanoa A., Radovic I., Borkab D., Miskovic Z.L., Yu H.K., Farias D. *et al.* 2017. Dispersion and damping of the interband $\pi$ plasmon in graphene grown on Cu(111) foils. *Carbon* **114** 70.
[48] Lin C.Y., Yang C.H., Chiu C.W., Chung H.C., Lin S.Y., Lin M.F. 2021. *Many-Particle Interactions in Carbon Nanotubes* (IOP Publishing).
[49] Lin C.Y., Wu J.Y., Chiu C.W. and Lin M.F. 2019. *Coulomb Excitations and Decays in Graphene-Related Systems* (CRC Press).
[50] Ho J.H., Chang C.P., Chen R.B. and Lin M.F. 2006. Electron decay rates in a zero-gap graphite layer. *Phys. Lett. A* **357** 401–406.
[51] Ho J.H., Lu C.L., Hwang C.C., Chang C.P. and Lin M.F. 2006. Coulomb excitations in AA- and AB-stacked bilayer graphites. *Phys. Rev. B* **74** 085406(8).
[52] Chiu C.W., Ho Y.H., Chen S.C., Lee C.H., Lue C.S. and Lin M.F. 2006. Electronic decay rates in semiconducting carbon nanotubes. *Physica E* **34** 658.
[53] Ohta T., Robinson J.T., Feibelman P.J., Bostwick A., Rotenberg E. and Beechem T.E. 2012. Evidence for interlayer coupling and moire periodic potentials in twisted bilayer graphene. *Phys. Rev. Lett.* **109** 186807.
[54] Lin C.Y., Do T.N., Huang Y.K. and Lin M.F. 2017. *Electronic and Optical Properties of Graphene in Magnetic and Electric Fields.* IOP Concise Physics (San Raefel, CA, USA: Morgan Claypool Publishers).
[55] Edvinsson T. 2018. Optical quantum confinement and photocatalytic properties in two-, one- and zero-dimensional nanostructures. *R. Soc. Open Sci.* **5** 180387.
[56] Chiu C.W. and Lin M.F. 2008. Temperature-dependent Coulomb scattering rates of Fermi-momentum states in metallic carbon nanotubes. *Physica E* **40** 2720–2723.
[57] Chiu C.W., Ho Y.H., Chen S.C., Lee C.H., Lue C.S. and Lin M.F. 2006. Electronic decay rates in semiconducting carbon nanotubes. *Physica E* **34** 658–661.
[58] Ho J.H., Chang C.P. and Lin M.F. 2006. Electronic excitations of the multilayered graphite. *Phys. Lett. A* **352** 446.
[59] Chubukov A.V., Galitski V.M. and Yakovenko V.M. 2005. Quantum critical behavior near a density-wave instability in an isotropic Fermi liquid. *Phys. Rev. Lett.* **94** 046404.
[60] Chiu C.W., Chung Y.L., Yang C.H., Liu C.T. and Lin C.Y. 2020. Coulomb decay rates in monolayer dopedgraphene. *RSC Adv.* **10** 2337–2346.

[61] Wang F., Dukovic G., Brus L.E. and Heinz T.F. 2004. Time-resolved fluorescence of carbon nanotubes and its implication for radiative lifetimes. *Phys. Rev. Lett.* **92** 177401.
[62] Tang X.-P., Kleinhammes A., Shimoda H., Fleming L., Bennoune K.Y., Sinha S., Bower C., Zhou O. and Wu Y. 2000. Electronic structures of singlewalled carbon nanotubes determined by NMR. *Science* **288** 492–494.
[63] Moos G., Gahl C., Fasel R., Wolf M. and Hertel T. 2001. Anisotropy of quasiparticle lifetimes and the role of disorder in graphite from ultrafast time-resolved photoemission spectroscopy. *Phys. Rev. Lett.* **87** 267402.
[64] Fischer J.E., Bloch J.M., Shieh C.C., Preil M.E. and Jelley K. 1985. Reflectivity spectra and dielectric function of stage-1 donor intercalation compounds of graphite. *Phys. Rev. B* **31** 4773.
[65] Mak K.F., Shan J. and Heinz T.F. 2011. Seeing many-body effects in single- and fewlayer graphene: Observation of two-dimensional saddle-point excitons. *Phys. Rev. Lett.* **106** 046401.
[66] Prezzi D., Varsano D., Ruini A., Marini A. and Molinari E. 2008. Optical properties of graphene nanoribbons: The role of many-body effects. *Phys. Rev. B* **77** 041404.
[67] Hagen A., Steiner M., Raschke M.B., Lienau C., Hertel T., Qian H., Meixner A.J. and Hartschuh A. 2005. Exponential decay lifetimes of excitons in individual single-walled carbon nanotubes. *Phys. Rev. Lett.* **95** 197401.
[68] Bostwick A., Ohta T., Seyller T., Horn K. and Rotenberg E. 2007. Quasiparticle dynamics in graphene. *Nat. Phys.* **3** 36.
[69] Visscher P.B. and Falicov L.M. 1971. Dielectric screening in a layered electron gas. *Phys. Rev. B* **3** 2541.
[70] Wu J.Y., Chen S.C. and Lin M.F. 2014. Temperature-dependent Coulomb excitations in silicene. *New J. Phys.* **16** 125002.
[71] Shih P.H., Chiu Y.H., Wu J.Y., Shyu F.L. and Lin M.F. 2017. Coulomb excitations of monolayer germanene. *Sci. Rep.* **7** 40600.
[72] Chen R.B., Chen S.C., Chiu C.W. and Lin M.F. 2017. Optical properties of monolayer tinene in electric fields. *Sci. Rep.* **7** 1849.
[73] Chung H.C., Chiu C.W. and Lin M.F. 2019. Spin-polarized magneto-electronic properties in buckled monolayer gaas. *Sci. Rep.* **9** 2332.
[74] Zhao M.W., Chen X., Li L.Y. and Zhang X.M. 2015. Driving a gaas film to a large-gap topological insulator by tensile strain. *Sci. Rep.* **5** 8441.
[75] Chung H.C., Chang C.P., Lin C.Y. and Lin M.F. 2016. Electronic and optical properties of graphene nanoribbons in external fields. *Phys. Chem. Chem. Phys.* **18** 7573.
[76] Chung H.C., Lee M.H., Chang C.P. and Lin M.F. 2011. Exploration of edge-dependent optical selection rules for graphene nanoribbons. *Opt. Express* **19** 23350–23363.

# Index

www.ingramcontent.com/pod-product-compliance
Ingram Content Group UK Ltd.
Pitfield, Milton Keynes, MK11 3LW, UK
UKHW021835100926
8556IPUK00004B/345